The New Art of War

Many of war's lethal failures are attributable to ignorance caused by a dearth of contemporary, accessible theory to inform warfighting, strategy, and policy. To remedy this problem, Colonel Geoffrey F. Weiss offers an ambitious new survey of war's nature, character, and future in the tradition of Sun Tzu and Clausewitz. He begins by melding philosophical and military concepts to reveal war's origins and to analyze war theory's foundational ideas. Then, leveraging science, philosophy, and the wisdom of war's master theorists, Colonel Weiss presents a genuinely original framework and lexicon that characterizes and clarifies the relationships between humanity, politics, strategy, and combat; explains how and why war changes form; offers a methodology for forecasting future war; and ponders the permanence of war as a human activity. *The New Art of War* is an indispensable guide for understanding human conflict that will change how we think and communicate about war.

GEOFFREY F. WEISS is an active duty colonel with over twenty-seven years of service in the United States Air Force. He is a graduate of the US Marine Corps School of Advanced Warfighting and a distinguished graduate of the National War College. A former wing commander with multiple tours in the Middle East, Colonel Weiss is presently serving in the Pentagon on the Joint Staff where he advises the Chairman of the Joint Chiefs of Staff on global operations and crisis response from the National Military Command Center.

The New Art of War

The Origins, Theory, and Future of Conflict

Colonel Geoffrey F. Weiss

CAMBRIDGE
UNIVERSITY PRESS

University Printing House, Cambridge CB2 8BS, United Kingdom

One Liberty Plaza, 20th Floor, New York, NY 10006, USA

477 Williamstown Road, Port Melbourne, VIC 3207, Australia

314–321, 3rd Floor, Plot 3, Splendor Forum, Jasola District Centre,
New Delhi – 110025, India

103 Penang Road, #05–06/07, Visioncrest Commercial, Singapore 238467

Cambridge University Press is part of the University of Cambridge.

It furthers the University's mission by disseminating knowledge in the pursuit of
education, learning, and research at the highest international levels of excellence.

www.cambridge.org
Information on this title: www.cambridge.org/9781108837644
DOI: 10.1017/9781108946902

First published 2021

A catalogue record for this publication is available from the British Library.

ISBN 978-1-108-83764-4 Hardback

For my parents, Rick (1939–2019) and Marylou, who have lived life as a harmony of tensions, sometimes at war, mostly at peace, often on my mind, and always in my heart.

&

For those who risk their lives for others.

If you wish to see the truth, then hold no opinions for or against anything . . . as long as you remain in one extreme or the other you will never know Oneness.[1]

Buddhist Proverb

The truth is the Whole.[2]

Georg W. F. Hegel

Contents

Figures

Preface

The journey of a thousand miles begins with one step.[1]

Taoist Proverb

For many, war is an abstraction, a curiosity. For others, it is part of a daily struggle for existence. For still others, it is a livelihood or even an aspiration. But what do we really *know* about war, and why does it matter? In the late 1960s, my father, who due to health reasons never experienced war firsthand, named me after his friend and mentor, Geoffrey Reginald Gilchrist Mure, Warden of Merton College, Oxford (1947–63). Frederick, my father's first name, is also my middle name. Ironically, "Geoffrey" originates from the Old German word for "peace," and "Frederick" means "peaceful ruler," yet I have devoted most of my professional life and this book to the subject of war.[2] Perhaps no one understands fire or disease better than the fireman or physician, professionals dedicated to battling these menaces. The same is often true of warriors and war. And though some warriors are enamored with war, I am no more a fan of it than a doctor is a lover of disease or death. Still, how can we hope to prevent wars or prevail in wars thrust upon us without understanding war? I love peace enough to fight for it, which is why I devoted the majority of my life to military service and ultimately why I wrote this book. *The New Art of War* has been many years in the making, and its creation has been a rewarding though unlikely journey. Despite having settled on a military career (somewhat by accident), I never imagined I would write a book on war; after all, so much has been written already. Yet, the more I have studied history, war theory, and strategy the more I perceived a pressing need for a fresh approach. The result is *The New Art of War.*

The Problem

As I stated, I had never intended to write a book on war. However, there are troubling deficiencies within the current body of war theory and consequently in our approach to strategy and warfighting. War is dangerous, unpredictable, and deadly. It has claimed hundreds of millions of lives, decimated landscapes, and left countless nations, civilizations, and empires in its wake.[3] War is *not* a phenomenon anyone should engage in without knowing its nature.

Clear as Mud

In *A History of Military Thought* (2001), historian and war expert Azar Gat writes, "New and significant intellectual constructions usually emerge at times of fundamental change or paradigmatic shifts, when prevailing ways of interpreting and coping with reality no longer seem adequate."[4] We face that reality now. The complex evolution of war, theory, and strategy, from the classical era to the present, has deposited layers of terms, concepts, and principles stacked like geological strata. As a result, getting at what we really need to know amounts to a laborious dig that, frankly, few have the patience or intellectual discipline for. Instead, most simply pick a few shiny items off the top or attempt to delve deeper into material they do not fully understand. Unfortunately, this state of affairs has caused us to accept some theoretical shortfalls and blinded us to others. Solving these problems and accelerating our understanding of war into the present age demands a holistic reassessment of war theory and strategy. This alone will enable us to resolve the following deficiencies: (1) the inaccessibility of existing theory; (2) the incoherent language and explanation of war's forms; and (3) the lack of contemporary theory development.

Regarding the first, though ample theory is available, its presentation complicates comprehension and application. Ancient origins, disparate styles, and myriad interpretations impede the direct application of traditional war theories to military operations and strategy. As historian Hew Strachan observes, "Strategic theory has therefore failed to provide the tools with which to examine the conflicts which are in hand."[5] This has led some warfighters to misunderstand or misapply war theory, which can cause more harm even than ignoring it. As warrior-theorist Sir Basil Henry (B. H.) Liddell Hart writes, "Misinterpretation has been the common fate of most prophets and thinkers in every sphere. Devout but uncomprehending disciples have been more damaging to the original conception than even its prejudiced and purblind opponents."[6] This undermines trust in otherwise sound theory and frustrates analysis by blurring the line between theoretical defects and execution errors.

History's foremost war theorists are Sun Tzu, Thucydides, and Carl von Clausewitz, a triumvirate responsible for what war and strategy expert Colin Gray calls war's "strategic canon."[7] These savants formulated theory based on scholarship, experience, and astute reasoning. Their writings are profoundly perceptive on war and human nature; yet, they are easily misunderstood. Clausewitz is the most difficult. As Liddell Hart opines of Clausewitz's "opacity": "It must be admitted . . . that Clausewitz invited misinterpretation more than most," and "not one reader in a hundred was likely to follow the subtlety of his logic or to preserve a true balance amid such philosophical jugglery."[8] Adds historian and Clausewitz expert Peter Paret:

It is not surprising that the search for Clausewitz's influence ... has been confused and inconclusive ... if we examine the conduct of war since Clausewitz wrote, we will find little evidence that soldiers and governments have made use of his theories ... nothing has proved more elusive to discover than an application of "lessons" learned from *On War*.[9]

Clausewitz's theory is puzzling for two main reasons. First, he died of cholera before finishing *On War*, leaving his wife to collect and publish his notes posthumously. Second, the incongruity of Clausewitz's dialectic style obscures his conclusions. As Gat writes, "Perplexed by the contradictory ideas they have found in *On War* and failing to follow Clausewitz's tortuous intellectual development ... commentators could never really make up their minds what exactly Clausewitz meant to say."[10] By contrast, Thucydides's *History of the Peloponnesian War* is more straightforward and more history than theory, a chronicle of the sprawling conflict he witnessed and participated in. Even so, the book's remarkable detail can distract readers from the timeless insights into human nature found more in orations and dialogue than battle accounts. Finally, Sun Tzu's work, *The Art of War*, is the most readable, yet also the easiest to dismiss as fortune-cookie mysticism due to its aphoristic simplicity. However, just as the brushstrokes of a master artist may seem unremarkable outside the context of the completed portrait, Sun Tzu's wisdom and relevance can be missed by readers who fail to approach his ideas holistically.

The challenge of grasping these theorists is compounded by a second deficiency, the general incoherence of the vast, diverse body of interpretation and commentary they have engendered. "What's wrongheaded" in this explosion of perspectives, writes the Heritage Foundation's Jay Carafano, is the "presumption ... that, just as farmers must plow the fields and rotate crops every year to get better results, so military thinkers must turn over fundamental military concepts on a regular basis."[11] In fact, Carafano adds, "More military thinking doesn't lead inevitably to better thinking. Indeed, lately it seems to have put more trees in front of the forest."[12] This unending tug-of-war between supporters and critics creates confusion that discourages warriors and strategists from embracing war theory. Competing opinions amplified by the relentless pressure to uncover some revolutionary new nugget from the strata of history and war theory have yielded a plethora of hastily defined terms that impede clear thought, especially for the intellectually intensive art of strategy. For instance, according to Strachan, words like "asymmetric" and "hybrid" lack "either definition or clarity" and reflect "the weakness of war's conceptual vocabulary."[13] To borrow from the Clausewitzian lexicon, the current language of war theory creates cognitive friction that fogs our understanding of war's nature. Clausewitz disparaged the theory of his day, calling it "an ostentatious exhibition of ideas" attended by a "retinue of jargon, technicalities, and metaphors" that posed a "serious menace" if accepted uncritically.[14] Is it any different today?

Clausewitz rightly advises us to critique the language of war, and we should do the same for theory. Has war theory answered all the questions we need it to? In fact, there are many war-related questions that remain inadequately addressed. For example, why do we lack a common definition for war that clearly establishes how war differs from other forms of human violence and competition? How are war and human nature related? How do we distinguish between war's nature and character? How are domain (e.g., maritime and air) and small wars theories related to general war theory? Why is war terminology so unsystematic? For instance, though there are no rigorously defined properties of "regularity" or "conventionality," we routinely deem wars "irregular" and "unconventional." Moreover, how are war's forms (e.g., regular, irregular, and hybrid) related, and how and why do they change? Finally, why has war theory not proposed a methodology for forecasting future war nor settled the question of war's permanence in human affairs?

Answering these important questions requires better theory, and this leads to a third problem: writing war theory is a lost art. While there is ample war-related authorship, including theory interpretation and application, where is the original theory *development*? Amazingly, according to war scholars Tarak Barkawi and Shane Brighton, even the central text used in the war studies department at London's King's College provides "almost nothing in the way of either definitions or theories of war."[15] Indeed, "if something is so important that we cannot understand anything in human affairs without it, either there should be a battery of theoretic approaches, as in any other serious field of social and political inquiry, or developing such approaches should be the first order of business."[16] Instead, asserts Gray, "There are no new ideas in strategy. Most generations produce a few scholars, former strategic practitioners, and popular writers who attempt to update the classics for a better fit with contemporary realities. But the results of their labors are invariably disappointing."[17] Why?

Is There a Theorist in the House?

The fact that no modern writers have managed to rise to [Thucydides's and Clausewitz's] level of sophistication and understanding of the place of war in human affairs is indeed a sad commentary on the intellectual depth of the present age.[18]

Williamson Murray

The renowned military strategist Bernard Brodie attributes some of this to qualification. In his foreword to *On War,* Brodie writes, "While genius has scarcity value in every field of human endeavor, in the field of strategic writing it has special rarity. The reason is that soldiers are rarely scholars, and civilians are rarely students of strategy."[19] Rare is the soldier-scholar or professor-strategist with the balance of education, ability, and inclination to become a theorist, and the prospects for a breakthrough in any field are generally proportional to the number of people engaged. Those few who show promise are too often content to be derivative theorists (i.e., creatively repackaging past principles versus deriving new theories).

Ignorance or inexperience with theory composition are also factors contributing to the dearth of theorists. In his theory construction primer, Paul Reynolds says:

The major factor that thwarts the development of a scientific body of knowledge of social and human phenomena [like war theory] is the character of social scientists themselves. Two major deficiencies are lack of clarity in theoretical writings and ignorance about what scientific knowledge should look like and how it is created.[20]

These deficiencies have often led to ambiguous writing that, according to Reynolds, has "all the clarity and precision of an astrology prediction or political speech."[21] A likely cause for this competence gap is the rarity within traditional academic institutions of programs established for the study of war and military history, genres that remain outside the mainstream of academic vogue. Even dating back to the early twentieth century and Hans Delbrück, arguably the father of modern military history, military historians were seen as "misfits," reflecting a belief that "war is an aberration in the historical process," a trivial "bi-product" of world history.[22] By the twenty-first century, the academy had largely supplanted the practically focused examination of military, political, and economic histories in favor of modish social and cultural themes.

Also contributing to the scarcity of war theorists is the domination of liberal-progressive viewpoints on university faculties. Liberal-progressive ideology tends to marginalize the importance of war and military topics, which are viewed as male-dominated, conservative subject areas, and thus, the reprehensible drivel of the oppressor class.[23] Historian John Lynn says of military historians, "We used to be condemned because we were believed to be politically right-wing, morally corrupt, or just plain dumb."[24] While not all military historians are shunned, they are a rare species. A 2008 *US News and World Report* article asserts:

The field that inspired the work of writers from Thucydides to Winston Churchill is, today, only a shell of its former self. The number of high-profile military history experts in the Ivy League can be counted on one hand. Of the more than 150 colleges and universities that offer a Ph.D. in history, only a dozen offer full-fledged military history programs.[25]

Given these circumstances, many aspiring historians pursue their professional goals in other directions. Of those brave souls who stay the course, most specialize in decoding the truth about specific incidents, epochs, or locations rather than attempting to become strategists or theorists.[26] Military history, though vital to theory, is less pragmatic. Its purpose, says Lynn, is "not practical in any immediate sense; rather it is to achieve, or at least strive for, an understanding of the past as a value unto itself."[27] Still, history is an essential data source theorists and strategists cannot afford to ignore. Regrettably, writes Brodie, "our own generation is unique, but sadly so, in producing a school of thinkers who are allegedly experts in military strategy and who are certainly specialists in military studies but who know virtually nothing of military history."[28] Furthermore, since good strategy must consider both history and theory, doubly inept are the "strategists" who know little of either.

Great war theorists are even rarer than military historians or strategists. In addition to a deep knowledge of war, becoming an effective theorist requires audacity. As legendary author and intellectual Isaac Asimov writes:

The history of human thought would make it seem that there is difficulty in thinking of an idea even when all the facts are on the table. Making the cross-connection requires a certain daring. It must, for any cross-connection that does not require daring is performed at once by many and develops not as a "new idea," but as a mere "corollary of an old idea."[29]

A war theorist must combine the attributes of a detective, historian, philosopher, warrior, and strategist to analyze historical data, experience, and past theory and then make "daring" "cross-connections" that reveal new insights. Theoretical development of this complex phenomenon requires a broad, multidisciplinary, holistic perspective. "In war more than in any other subject," writes Clausewitz, "we must begin by looking at the nature of the whole."[30] This contrasts with modern academia that frequently values narrow specialization, making scholarship an exercise in learning more and more about less and less. Indeed, as Barkawi and Brighton point out in their conclusion to *The Changing Character of War* (2011), "War is strangely decentered and fragmented as an object of inquiry ... [and] attention has been fixed on particular *wars* rather than *war* as a general force."[31]

Warrior-historians are often best qualified to be theorists, and they have produced great work advancing the art and science of warfighting. But they also tend to accentuate historical case studies and lessons from recent events rather than probing war's deeper intricacies. "They have too often sought to simplify and build clear but irrelevant models," writes military historian Williamson Murray, "in the face of what will always be a nightmarish world of complexity and difficulty."[32] Of course, as a practical caste, warriors almost always prefer concrete solutions over flights of fancy down theoretical rabbit holes. Thus, most scholarship on warfare favors the physical over the metaphysical (i.e., how do we win today's fight?). "That is one of the chief reasons why military people are so often disappointed with Clausewitz," asserts Brodie, "for they are particularly accustomed in their training to absorbing against a tight schedule of time specific rules for conduct."[33] Liddell Hart adds: "Great advances in medicine and surgery have been due more to the scientific thinker and research worker than to the practitioner. Direct experience is inherently too limited to form an adequate foundation either for theory or for application."[34]

Given these circumstances, it is not uncommon for warriors and strategists to bypass theory, relying instead on the progeny of theory and experience: tactics and doctrine. But warriors who allow their warfighting doctrine to drift too far from its theoretical roots may sacrifice adaptability and foresight. "With such a limited basis," writes Liddell Hart, "the continual changes in military means from war to war carry the danger that our outlook will be narrow and the lessons fallacious."[35] In the absence of scientific theory, early humans used religion to explain otherwise mystifying occurrences. Ignorance of war theory causes a similar effect in military thought. Every few years, some "high priest" of war articulates a pithy slogan around which zealous acolytes build elaborate doctrines and taxonomies. Examples include "low intensity conflict," "effects-based operations," "net-centric" and "generational" warfare, and "infinite" war. While in fashion, these catchphrases attract widespread praise; but, outside a recent example or two, followers often struggle to explain *why*. Eventually, these "cults" fade away and are replaced by new ones as the original supporters depart and the character of war changes. This repeating cycle can cause practitioners to lose touch with theory entirely, a condition that courts disaster. Of the French in World War II, Liddell Hart says, their "vital weakness" lay "not in the quantity nor in quality of equipment, but in their *theory*. Their ideas had advanced less than their opponents ... [and] as has happened so often in history, victory had bred a complacency and fostered an orthodoxy which led to defeat."[36]

Militaries generally promote superior tacticians who, upon achieving higher rank, aspire to become strategists and advisors to political leaders. Tactical experts accustomed to disregarding theory are naturally less comfortable with it as they rise in rank and require a more comprehensive view of their craft. Once on a headquarters staff, where an understanding of war has the broadest effect, the constraints of personality, process, and politics and the relentless "tick-tock" of meetings and deadlines often preclude reflection on war's true nature. Absent a conscious effort to embrace theory, leaders can struggle to become effective strategists.[37] Ultimately, the best way to appreciate theory is to study it, use it, and every so often, take a crack at creating it.

Easier Said Than Done

This kind of holistic thinking is rare precisely because it is so difficult.[38]
General T. Michael Moseley, former US Air Force Chief of Staff

War's complexity is a key factor inhibiting theory development. Fittingly, "war" comes from the Old English "wyrre" meaning "to bring into confusion."[39] War is an intricate, chaotic phenomenon tied to the elusive depths of human nature. It is "not a chess game," writes counterinsurgency expert David Galula, "but a vast social phenomenon with an infinitely greater and ever-expanding number of variables, some of which elude analysis."[40] War, adds Murray, is "an incredibly complex endeavor which challenges men and women to the core of their souls. It is … not only the most physically demanding of all the professions, but also the most demanding intellectually and morally."[41] Furthermore, as maritime theorist Julian Corbett opines, "war being … a complex sum of naval, military, political, financial, and moral factors, its actuality can seldom offer … a clean slate on which strategical problems can be solved by well-turned syllogisms."[42] This complexity may discourage aspiring war theorists by leading them to conclude that war and strategy *cannot* be expressed theoretically.

The final reason for the shortage of new war theory is the belief that nothing new *needs* to be said about war. To paraphrase patent clerk Charles Holland Duell, perhaps when it comes to war theory, "everything that can be invented has been invented."[43] In his 1883 book, *Das Volk im Waffen* (*The Nation in Arms*), Prussian field marshal Colmar von der Goltz writes:

A military writer who after Clausewitz, writes upon war, runs the risk of being likened to the poet who, after Goethe, attempts a Faust, or, after Shakespeare, a Hamlet. Everything of any importance to be said about the nature of war can be found stereotyped in the works left behind by that greatest of military thinkers.[44]

This stance exemplifies what Nobel Laureate physicist Robert Laughlin calls an "antitheory," the idea that there is no absolute truth and "no fundamental thing left to discover … only a swarm of detail that belongs to no one."[45] Indeed, despite the overt problems with war theory, many experts agree with von der Goltz and focus only on interpreting and expounding upon past wisdom, restating or rearranging it to address specific problems or circumstances. Though often beneficial, these efforts have not produced the holistic interpretation of war we really need. Rare is that field of scholarship so clear and uncontentious that it precludes gain from additional inquiry. War is undoubtedly *not* such a field. Moreover, as philosopher Arthur Schopenhauer says, "Every man takes the limits of his own field of vision for the limits of the world."[46] Thus, while past theorists were visionary, they were still products of their respective eras. Now, aided by advances in history, mathematics, and science unknown to the likes of Sun Tzu, Thucydides, and Clausewitz, war theory is *exactly* the sort of subject for which a re-examination is not only warranted but explicitly needed.

So What?

If we are at war and we get the military strategy about right, people will be killed; if we get it wrong, lots of people will be killed.[47]

General T. Michael Moseley

What are the implications if we refuse this challenge? As Sun Tzu warns, "War is a matter of vital importance to the State; the province of life or death; the road to survival or ruin. It is mandatory that it be thoroughly studied."[48] Problems with war theory can lead to disastrous miscalculations in strategy, warfighting, and diplomacy. "An error in my writings may easily be corrected without harming anybody," writes Machiavelli, "but an error in their practice may ruin a whole state."[49] Indeed, "the cost of slovenly thinking at every level of war," adds Murray, "can translate into the deaths of innumerable men and women ... [and] also leads invariably to the massive waste of national treasure."[50] A brief survey of war's history reveals how painfully often political and military leaders have blundered. Even the mighty US war machine has not been immune. In a 2013 *Foreign Affairs* article, the US Deputy Defense Secretary, Ashton Carter, lamented:

At the outset of the wars in Afghanistan and Iraq, the Pentagon made two fatal miscalculations. First, it believed these wars would be over in a matter of months ... second, the Pentagon was prepared for traditional military-versus-military conflicts. As a result, the military was not well positioned to fight an enemy without uniforms, command centers, or traditional organizational structures.[51]

Colin Gray attributes these miscues to deficiencies in theory saying, "looking back at the 2000s with particular reference to the conflicts in Afghanistan and Iraq ... the truth is that strategic theorists, the inventors and refurbishers of concepts, have failed in their duty to explain the nature and character of contemporary war and warfare adequately."[52] War's complexity, which seems to defy logic and foresight, can induce even seasoned leaders to adopt defective stratagems. Without theory's guiding light, plans may be derived exclusively from personal experiences, case studies, or proximate trends, in essence using the past as a template rather than a reference. As Clausewitz warns: "The danger is that this kind of style can easily outlive the situation that gave rise to it; for conditions change imperceptibly. That danger is the very thing a theory should prevent by lucid, rational criticism."[53] Perhaps more insidiously, ignorance of theory can prevent us from spotting and discrediting bad theory. Contemporary war pundits are quick to proclaim brand-new forms of war or to declare that some innovation has altered war's nature. But in fact, hybrid warfare is *not* new; aircraft, nuclear weapons, and the USSR's collapse did *not* end war; and the twentieth century's explosion of insurgencies has *not* relegated traditional warfare to history's dustbin.

The Solution

It is only afterward that a new idea seems reasonable. To begin with, it usually seems unreasonable.[54]

Isaac Asimov

When theory does not serve its purpose, we must reevaluate and, where appropriate, revise it. This is easier said than done; however, the place to start is at the beginning. A theory of anything must reflect the subject's *nature*, which is determined by those constitutive elements and characteristics that establish its unique identity. Research and study help reveal these elements, thus increasing our depth of comprehension. Ideally, we prioritize research into areas that facilitate safety and well-being. Studying war, as we would disease, famine, or weather, improves our likelihood of avoiding war and mitigating its destructive effects.

According to Paul Reynolds, good theory provides the following benefits:
(1) Typology: a method of organizing and categorizing "things"
(2) *Predictions* of future events
(3) *Explanations* of past events
(4) A sense of *understanding* about what causes events
(5) The potential for *control* of events[55]

In effect, proper theoretical treatment creates a common language with explicit definitions and abstract statements to explain the past, predict the future, and provide insight into causality. Given the right conditions, theory may even permit control. This process is apparent within the sciences, where research and discovery have led to amazing innovations like space travel, advanced medicine, and the Internet.

Though war differs from quantum mechanics or microbiology, it is equally if not more worthy of attention. In the sciences, research and experimentation provide data that forms the basis for theory, which may be expressed mathematically or as a related set of definitions, principles, or axioms. Though its complexity precludes mathematical precision, war has enduring truths and governing principles that we can organize and express in theoretical form. War's "data" comes from a variety of sources including current events, logical deduction, exercises, experiments, and especially *history*. "Historical examples," writes Clausewitz, "clarify everything and also provide the best kind of proof in the empirical sciences. This is particularly true of the art of war."[56] Within this sea of information, which lamentably grows deeper every year, are the clues and themes necessary for theory development. But make no mistake, as political scientist Greg Cashman points out in *What Causes War?* (1993), "'The data never speak for themselves.' Without a theory there is no real explanation."[57] War's complexity substantiates an enduring premise: *no two wars are identical*; therefore, no past war can be a perfect model for a current or future war. Lessons from past or present wars provide glimpses into war's nature that only theory can aggregate into a concise, practical form. Theory also helps bridge the gap between the past and future by illuminating possible futures, a function critical for strategy, which is essentially an exercise in arranging present conditions and resources (ways and means) to yield favorable future outcomes (ends). Finally, theory aids resourcing and warfighting decisions. US Naval War College professor Milan Vego writes, "The commander armed with solid theoretical knowledge would have a firmer grasp of the sudden change of a situation and could act with greater certainty and quickness to obtain an advantage over the opponent."[58] Conversely, strategy and tactics that disregard or misapply theory or leverage defective theory face an *increased risk of failure*, especially when opposed by a competent adversary.

The New Art of War

Theories have four stages of acceptance:
1. *This is worthless nonsense.*
2. *This is interesting, but perverse.*
3. *This is true, but quite unimportant.*
4. *I always said so.*[59]

J. B. S. Haldane

The New Art of War is the admittedly ambitious answer to theory's inaccessibility and incoherence as well as the lost art of theory development. It represents a decade-long exploration of war's nature, including its origins, theory, and future, and its scope exceeds most works in this genre. This is not a book of recommendations for a specific country or audience; instead, it seeks only to express war's universal truths in a useful form; to provide, as Strachan says, "a common doctrine, a common way of thinking about war, and a common method of applying those thoughts in practice."[60] The opening chapter's examination of war's human and metaphysical dimensions and formal definition – including war's twenty dialectics and a new trinity centered on *humanity*, *politics*, and *combat* – set the stage for the analysis, development, and application of war theory in subsequent chapters. Chapters 2 and 3 highlight war's enduring patterns and themes by summarizing and evaluating the seminal ideas of history's master war theorists, including small wars and domain theory. Chapter 4 shapes these concepts into a general, *Unified War Theory* encompassing politics, strategy, and combat; updates the language of war; and redefines war's forms (e.g., regular-irregular) as related elements of a continuum driven by iterative assessments of combatant will and fighting capacity. Finally, Chapter 5 offers a methodology for improving predictions regarding war's future character and concludes with an evaluation of war's permanence as a human activity.

I anticipate some resistance to *The New Art of War*. Those content with the status quo may hesitate to accept a new perspective that challenges entrenched ways of thinking. As Liddell Hart opines, "The direct assault of new ideas provokes a stubborn resistance, thus intensifying the difficulty of producing a change of outlook ... [but] true conclusions can only be reached ... by pursuing the truth without regard to where it may lead or what its effect may be – on different interests."[61] But I welcome the debate, for if controversy accompanies new ideas in proportion to their consequence, then I should wish for plenty! Regardless, I expect this work will stimulate thought and prove beneficial to anyone irrespective of their experience. War experts will likely find affirmation of what they have always known or suspected but also some ideas they may not have considered, while neophytes will benefit from a holistic theory of war unavailable elsewhere. Ultimately, my hope for *The New Art of War* echoes that of Clausewitz, who said of *On War*: "It was my ambition to write a book that would not be forgotten after two or three years, and that possibly might be picked up more than once by those who are interested in the subject."[62]

Acknowledgments

As I highlight in Chapter 1, humans are social creatures. Sure, some prefer solitude; however, as the theme for *Cheers* croons, *sometimes you want to go, where everybody knows your name, and they're always glad you came*. Sharing life's challenges, in battle or book writing, enhances well-being and affords a wealth of life-enriching feedback and encouragement. This book would not have been possible without the contributions and support of family, friends, colleagues, and mentors over the years. I am most grateful to God and *fortuna* for life and the privilege of citizenship in a land that provides the freedom to pursue constructive dreams unimpeded by tyranny. I also have to thank my parents, Rick and Marylou, for blessing me with love and an intellectual inheritance that includes optimism, irrepressibility, and a philosophical spirit; and my wife, Karen, and three gifted children, Alison, Scott, and Brian, for their patience and input as I toiled seemingly forever on "the book." In addition to lending their edits and critiques, they inspire me daily. Other significant supporters include Wray Johnson, who opened my eyes to critical thinking, *Bildung*, and the broader phenomenon of war; Rick Glitz, my early editor-in-chief; Bruce Cooper, my algorithm czar; Alisen Boada, citations and sources guru; Generals Drew Poppas and Glen VanHerck, warrior-leaders and valued role models; my agent, Chris Kepner, who saw the whole elephant and not just the snake; and Michael Watson, Emily Sharp, Ruth Boyes, Raghavi Govindane, Ursula Acton, and the peerless Cambridge University Press for turning big ideas into books. Finally, I am forever indebted to the myriad unnamed colleagues and supporters who have shaped my thinking in (I hope) productive ways.

Introduction: Why Study War?

Of all the common enterprises and of all the collective actions that men have undertaken, war is undoubtedly the most imposing in the amount and quality of the effort men put forth, the most devastating and revolutionary in its consequences. Into war ... man puts all that he has: his wealth, his science, his indomitable will, and eventually his very existence.[1]   Robert E. Park

You may not be interested in war, but war is interested in you.[2] Leon Trotsky

According to Peter Paret, we study war "because wars, the institutions that make them possible, and the ideas that guide their conduct form an important part of the human experience."[3] In other words, why would we *not* study war, this amazing, horrible spectacle that has affected every person on Earth? "Its essential character," writes Bernard Brodie, "remains distinct from every other pursuit of man."[4] "War is a phenomenon," adds military historian John Keegan, "which, however repugnant to all that is sensitive and generous in human nature, is nevertheless universal in the life of mankind."[5]

War invites controversy and raises difficult questions. Is it rational or irrational? For what *reasons* would we fight and kill one another with such regularity and fervor? Philosopher David Hume writes, "Reason is, and ought only to be the slave of the passions," implying that war satisfies the dictates of emotion rather than logic.[6] On the other hand, Hegel says, "The cunning of reason ... sets the passions to work for itself," suggesting that wars ultimately rest upon a rational foundation.[7] Perhaps both are right, for war often seems poised on a knife's edge between reason and emotion, order and chaos. War, writes Liddell Hart, "is horrible and ghastly beyond all imagination ... nevertheless, it has an awe-inspiring grandeur of its own."[8] As Liddell Hart's sentiment affirms, war is enigmatic, replete with contradictions. It repels and attracts. War can involve millions of people and span the globe yet remain intensely personal. It can be waged with modern technology or primitive instruments, and it can be fueled by sophisticated reasoning or primal instincts. "In war," adds philosopher Nikolai Berdyaev, "there intersect the limits of the extreme, and the diabolical darkness is interwoven with Divine light ... we both accept and yet reject war."[9] Indeed, war's complexity and violence conceal metaphysical and ethical intricacies that make it resemble

1

a Rorschach test. Everyone who looks at war sees something different, which complicates our ability to reach consensus regarding war's true nature.

War Is Evil

War always brings about the wreck of everything that is good, and the tide of war overflows with everything that is worst; what is more, there is no evil that persists so stubbornly.[10]   Erasmus of Rotterdam (1516)

War has claimed millions of lives, innocent and wicked, soldier and civilian, young and old, rich and poor. Even those who survive are often scarred in mind, soul, and limb. "War means destruction," says British author and political administrator H. Fielding Hall, "first of laws, conventions, institutions, morality, and next the destruction of life and property."[11] All is fair in war, thus nothing about war is fair. For most of human existence, rape and pillage were war's rewards and atrocity and terror its means. As Shakespeare's Henry V warns the people of Harfleur,

> The gates of mercy shall be all shut up,
> And the fleshed soldier, rough and hard of heart,
> In liberty of bloody hand shall range
> With conscience wide as hell,
> Mowing like grass
> Your fresh fair virgins and your flow'ring infants.   (*King Henry V* 3.3.10–14)

War strips away civilization's veneer, spreading an insidious psychic malignance that numbs us to its horrors and goads us into unleashing our foulest demons upon inhuman enemies. Indeed, "Man is 'predisposed to slide into deep, irrational hostility under certain definable conditions,'" writes pioneering sociobiologist Edward Wilson, conditions invariably present during wartime.[12] As Machiavelli writes of early medieval warfare:

All who were vanquished in battle were either put to death or carried in perpetual slavery ... where they spent the remainder of their lives in labor and misery. If a town was taken, it was demolished or its inhabitants were stripped of their goods, dispersed all over the world, and reduced to the ultimate degree of poverty and wretchedness.[13]

The weapons of modern warfare have not assuaged this effect, for, according to foreign policy experts Angelo Codevilla and Paul Seabury, "the destructiveness, the justice or injustice, the relative mercilessness of any war never depended on the weapons available to the warriors. Rather, it depended on what was on the warriors' minds."[14] Human belligerence has produced a bloody hit list that includes Carthage, Cannae, Antietam, Verdun, Nanking, Dresden, Hiroshima, and countless unrecorded battles and massacres. Consider this gruesome World War II statistic: by June of 1944, *90 percent* of Russian men aged eighteen to twenty-one had been killed in battle![15] As William

Tecumseh Sherman famously opined, "It is only those who have neither fired a shot nor heard the shrieks . . . of the wounded who cry aloud for blood, for vengeance, for desolation. War is hell."[16]

War Is Good

One social reason for the existence of war is that peace is sometimes too costly. There are situations when it is better to send men to die on their feet than have everyone live on their knees.[17]   Lawrence Keeley

And yet, there is another side to war. Ironically, it is against the backdrop of war's chaos and carnage that many of humanity's greatest attributes emerge. "War is part of God's world-order," writes Prussian general Helmuth von Moltke (the Elder), for "within it unfold the noblest virtues of men, courage and renunciation, loyalty to duty and readiness for sacrifice."[18] War is destructive, but it has also encouraged the discipline, industry, and creative energies responsible for nonviolent, life-enhancing innovations like mining, railroads, stainless steel, the Internet, weather radar, digital cameras, computers, penicillin, synthetic rubber, jet engines, canned food, blood transfusions, spacecraft, and nuclear energy. Furthermore, war is an engine of history and often a catalyst for needed change. "The belief . . . that war never solves anything," writes Colin Gray, "soon collapses under the weight of strategic history."[19] In *War! What Is It Good For?* (2014), Stanford archaeologist Ian Morris argues that, "although war is the worst imaginable way to create larger more peaceful societies, it is pretty much the only way humans have found."[20] His research indicates that "10 to 20 percent of all the people who lived in Stone Age societies died at the hands of other humans."[21] However, by 8000 BCE, wars had begun to consolidate humanity into larger, more stable and peaceful societies, which reduced violent death rates by an order of magnitude. "If it had been possible for men to sit still in peace without civilization," writes famed sociologist William Graham Sumner, "they never would have achieved civilization."[22] Hence, Keegan's claim that, "all civilisations owe their origins to the warrior."[23]

War is Darwinian, selecting for survival those with the most powerful ideas, the means to implement them, and the intellect and will to adapt and persevere over competitors. When cultures or leaders become stagnant, corrupt, or malignant, war offers a correcting mechanism. War destroys, but like a forest wildfire, it can burn away decay, leaving a better peace behind. In the right hands, war is a force *against* evil, deterring and defeating those who would employ tyranny and violence. "Greater evils would result," writes theologian Francisco Suárez, "if war were never allowed."[24] Indeed, "War is an ugly thing," adds philosopher John Stuart Mill, but not the ugliest of things: the decayed and degraded state of moral and patriotic feeling which thinks nothing worth a war, is worse. A war to protect other human beings against tyrannical injustice; a war to give victory to their own ideas of right and good . . . is often the means of their regeneration.[25]

Of course, judgments regarding what is right, good, or evil vary, especially between opposing sides in a war, and the ultimate arbiter is war itself; for "in war," writes Austrian legal philosopher Hans Kelsen, "not he who is in the 'right' is victorious, but he who is the strongest."[26]

Seeking Truth

Nowadays, anyone reflecting on war and strategy raises a barrier between his intelligence and his humanity.[27]
                                                                    Raymond Aron

It is aptly said that truth is war's first casualty, a fact that calls into question everything we think we know of war. As a Bedouin poet once wrote, "The sword is truer than what is told in books. In its edge is the separation between truth and falsehood."[28] To find the truth, we cannot avert our gaze when war demands attention. We cannot change the subject because war seems frightening or distasteful. In fact, we should do the opposite. After all, cancer, hurricanes, and pathogens are not studied because we delight in them. Unfortunately, we too often abjure this responsibility. "War is an activity," writes Keegan, "that modern Western man prefers to banish to the remotest corner of his consciousness."[29] Perhaps this reluctance stems from fear that war, unlike other calamities, is *not* an aberration, but rather a window into dark recesses of the human psyche we would prefer to disavow. "In the intellectual and popular culture," writes anthropologist Lawrence Keeley, "war has come to be regarded . . . as a peculiar psychosis of Western civilization."[30] But is this "psychosis" a natural consequence of human interaction or an anomaly catalyzed by modernity? Does admitting our relationship to war create a self-fulfilling prophesy of unending conflict? In *The Island of Dr. Moreau* (1896), H. G. Wells says, "The study of Nature makes a man at last as remorseless as Nature."[31] Does studying war make humanity as remorseless as war?

Though some may believe that ignoring war or embracing pre-civilizational paradigms will drive it away, self-imposed ignorance and societal regression are hardly proven methods for solving anything. No, finding practical answers to life's challenges requires the intellectual honesty and courage to face our demons, for the "peril is not in accepting that the innate nature of war lies in the dark hearts of us all," writes historian Victor Davis Hanson, "but rather in denying it."[32] Ultimately, if we are to gain any mastery at all over war, or any other phenomenon, we must continue to probe its mysteries. Liddell Hart once blamed statesmen, "dangerously ignorant of war," for making "war more difficult to avoid, more difficult to conduct successfully, and more difficult to terminate."[33] This implies that understanding war is the key to avoiding it, reducing its intensity and duration, and winning when necessary. Indeed, Mao Tse-tung believed that comprehending war's "peculiar laws" rendered it "predictable" and improved the probability of preventing, controlling, or even eradicating war forever.[34] Although predictability and eradication will likely remain aspirational, studying war will undeniably improve the *odds* of prevention, mitigation, and even control, and these are goals no rational species should reject.

1   The Origins of War

In all the spheres of the cosmos there storms the fiery and raging element and it brings war.[1]
Nikolai Berdyaev

This book's Introduction explains why we should study war. As for *how* to do this, we turn to Alexander the Great's mentor, Aristotle, who, in *Posterior Analytics*, asserts that understanding something's nature requires that we identify its cause or "why." In *Physics* and *Metaphysics*, he says that this "why" can be expressed as four causes: (1) material (substance); (2) formal (appearance); (3) efficient (source); and (4) final (purpose). For example, a statue's material cause is bronze; its formal cause is the shape; its efficient cause is the artisan and her technique; and its final cause is the satisfaction inspired by the work of art. The "why" of a statue, therefore, is be-"cause" the artisan (efficient) molded it from bronze (material) into a shape (formal) that elicits admiration (final).[2] Applying this technique to war yields a *material* cause (the "stuff" of war), war's physical components; *formal* cause (war's form), force arrangement, movements, and interactions; *efficient* cause (generally, humanity; specifically, warriors), the author of war within whom dwells the will to fight; and lastly, a *final* cause (the war's purpose), what war seeks to achieve. We will refer to these causes throughout our examination of war theory and strategy. Additionally, as implied by our use of an Aristotelian concept, we must recognize that war theory is influenced by great philosophical traditions and, in fact, must be based upon timeless truths. Thus, along with war's greatest figures, we will invoke other premier minds including Confucius, Plato, Shakespeare, Milton, Kant, Hegel, Freud, Hobbes, Rousseau, and many more. "The ideas of military theoreticians have never developed in a vacuum," writes Milan Vego, "but rather have been products of a complex interplay of the scientific, philosophical, and social influences of a given era."[3] For example, Sun Tzu's thinking is emblematic of Confucian simplicity and the balance of Taoism. Thucydides's opus reflects classical Greek insights on politics and human nature. Machiavelli melded the pragmatism and discipline of Rome with Renaissance ingenuity and his own brand of proto-realism. The rivalry and contrasts between Antoine-Henri Jomini and Clausewitz exemplify the collision of Enlightenment and German Romanticist philosophies.

Likewise, a tsunami of "-isms" – including industrialism, fascism, positivism, nationalism, Marxism, futurism, progressivism, modernism, and globalism – influenced theorists of the nineteenth and twentieth centuries, causing some to gravitate to extremes (e.g., the "cult of the offensive" and "the bomber always gets through") and others to embrace the past in search of balance (e.g., Corbett, Liddell Hart, and Mao). Awareness of these philosophical themes not only enhances understanding and analysis of existing theory, it also provides an essential foundation for theory development.

In the Beginning . . .

By cosmic rule, as day yields night, so winter summer, war peace, plenty famine.[4] . . . *The cosmos works by harmony of tensions, like the lyre and bow*[5]
Heraclitus

In all chaos there is a cosmos, in all disorder a secret order.[6]   Carl Jung

War, as with all things, fits within the broader fabric of the universe, which, as the pre-Socratic philosopher Heraclitus suggests, "works by harmony of tensions." In other words, everything in the universe is related, often paradoxically, by the give-and-take between ostensible opposites. "The interplay of opposite principles," says Confucius, "constitutes the universe."[7] What we may perceive as irreconcilable dichotomies are actually natural aspects of a greater whole. Examples are abundant: matter-energy, positive-negative, male-female, light-dark, thesis-antithesis, good-evil, love-hate, peace-war, Cain-Abel, Beowulf-Grendel, Athens-Sparta, Holmes-Moriarty, Jekyll-Hyde, and so on. Of this paradigm Clausewitz writes:

Where two ideas form a true logical antithesis, each complementary to the other, then fundamentally each is implied in the other. If the limitations of our mind do not allow us to comprehend both simultaneously, and discover by antithesis the whole of one in the whole of the other, each will nevertheless shed enough light on the other to clarify many of the details.[8]

And Peter Paret adds:

Purpose and means, strategy and tactics, theory and reality, intent and execution, friend and enemy – these are some of the opposites [Clausewitz] defines and compares not only to gain a truer understanding of each member of the pair but also to trace the dynamic links that connect all elements of war into a state of permanent interaction. War and politics, attack and defense, intelligence and courage, are never absolute opposites; rather one flows into the other.[9]

Eastern philosophy (Taoism) names these opposing elements *yin* and *yang*, which are related parts of an existential continuum. According to Taoism, existence is founded upon the interaction and combination of polarities that fluctuate between states of balance and imbalance, driving change. This applies to the physical world and to interactions within and between human groups, like armies and states. As Clausewitz says, "A state of balance tends to keep the existing order intact … once there has been a disturbance and tension has developed, it is certainly possible that the tendency toward equilibrium will shift direction and try to bring about a particular change."[10]

A little over thirteen billion years ago, the universe's greatest discontinuity, the "Big Bang," created time, matter, and energy within the *cosmos* (Greek for "order"). Since then, the cosmos has hosted a continuous harmony of tensions between matter and energy, creation and destruction, and balance and imbalance. Science has informed us, with its equivalent of an infallible decree, that everything in the physical world must obey a few inviolable "laws." These conservation laws state that mass, momentum, electric charge, and energy can neither be created nor destroyed. They are expressed mathematically as equations; thus, anything added or subtracted from one side must be from the other as well. Therefore, *symmetry and balance are hardwired into the universe.* This means that the cosmos's most basic elements, like matter and energy, are simply different *forms* of the same thing. For example, Einstein's famous equation ($E=mc^2$) proves that energy (E) and mass (m), though appearing to be different, are directly related. One cannot exist without the other.

This paradigm darkens somewhat when applied to human relations. There are no equations governing love and hate or war and peace, and even if there were, they would not resolve all of humanity's mysteries any more than the conservation laws answer every question about physics. Nevertheless, the absence of mathematical precision does not invalidate the notion that human interactions reflect underlying relationships and symmetry between constituent elements. Indeed, the "theoretical distinction between balance, tension, and movement" writes Clausewitz, "has a greater practical application than may at first appear."[11]

As stated previously, asymmetry drives *change*, which generally varies in scale and violence depending upon the energy present. Imbalances and tensions in nature disturb the energy equilibrium unleashing forces that trigger violent phenomena like earthquakes or lightning (in fact, Clausewitz compares tensions in war to those "in the atmosphere between positive and negative electric currents"[12]). Likewise, within human affairs, imbalances in culture, ideology, religion, and politics can heighten tensions causing disorder (or "entropy") that mandates a rebalancing toward equilibrium. The reduction of tension and restoration of order and stability can occur peacefully via diplomacy or arbitration; however, when nonviolent options do not exist or are deemed inadequate, this "disorder," writes Colin Gray, becomes "a fuel of conflict and a context that favors violence and warfare."[13] In war, adds Clausewitz, a "tension of forces builds up" causing a "cycle of tension and decision" between opposing sides.[14] Furthermore, regarding insurgencies, Clausewitz observes, "A state of tension will develop while the two elements interact. This tension will either gradually relax, if the insurgency is suppressed in some places and slowly burns itself out in others, or else it will build up to a crisis."[15] War arises from imbalances and disorder in the political status quo, which may result from blatant injustices, overt aggression, or perceptions that change is imperative. Tensions also build up *within* a war, where "every problem, and every principle," says Liddell Hart, "is a duality. Like a coin, it has two faces."[16] This "duality" causes war to change form (e.g., regular or irregular) as causal factors, like will and fighting capacity, wax and wane.

Paret points out that duality (or "polarity"), including action-inaction, attack-defense, and moral-material, is fundamental to Clausewitz's theory. And in his *Theorie de la guerre* (1777), French strategist Joly de Maizeroy writes that war and strategy require reasoning, "which is the province of *dialectics*" (emphasis added).[17] Indeed, a careful examination of war reveals twenty principal polarities or *dialectics* (Figure 1.1): (1) order-chaos, (2) past-future, (3) peace-war, (4) life-death, (5) creation-destruction, (6) friend-enemy, (7) good-evil, (8) science-art, (9) defense-attack, (10) unlimited-limited, (11) regular-irregular, (12) physical-moral, (13) direct-indirect, (14) certainty-uncertainty, (15) simplicity-complexity, (16) reason-emotion, (17) prudence-boldness, (18) control-autonomy, (19) concentration-dispersal, and (20) rest-movement.[18] As the figure implies, each dialectic relates back to a universal dialectic: order-chaos.

As we will see in future chapters, dialectic symmetries and asymmetries can be found within philosophy and war theory. In fact, "symmetries," writes physicist Michio Kaku, "are now known to be the fundamental guiding principle in creating any new theory"; consequently, war theory and strategy often reflect the effort to explain and reconcile dialectic tensions and asymmetries.[19] For example, "War," writes Liddell Hart, "is always a matter of doing evil in the hope that good may come of it"; thus, what is military strategy if not a creative act, a bridge between the destructive "evils" of war and the creative "good" of a better peace, from the past to a new future?[20] Strategy is a discipline that imposes prudence and order on war's boldness and chaos. When theory abandons the dialectic and exclusively embraces extremes, strategies, plans, and actions become unbalanced. For instance, the "either-or" "binary vision" of regular and irregular forms of war, remarks Hew Strachan, causes "unsupportable tension" that "has the effect of pulling armed forces apart, not providing coherence."[21] The cult of the offensive, which pervaded strategy before World War I, leaned exces-sively toward certainty, attack, boldness, concentration, and unlimited war. On the other hand, Clausewitz's late recognition of his theory's overemphasis

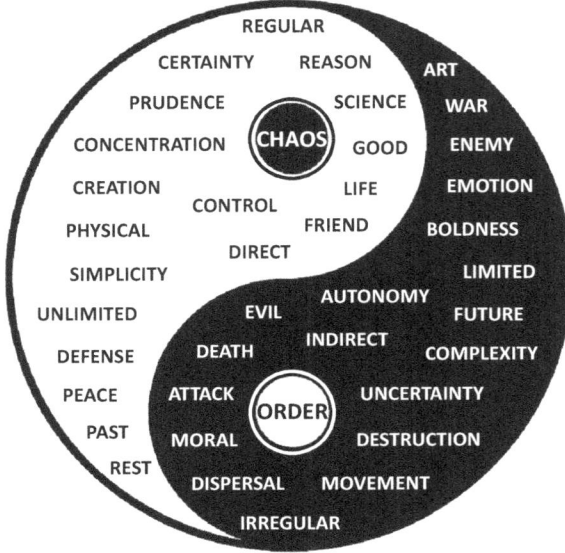

Figure 1.1 The dialectics of war

on attack, concentration, and absolute war led to his "paradoxical trinity" and a more balanced system. These concepts point toward a fundamental truth – that change, driven by the interplay between disparate elements of the greater whole, is part of the cosmos. War, as an expression of this paradigm, is both the result and driver of change. Fully comprehending this reality requires a holistic perspective that includes contextual factors, like human nature, politics, and the environment, and attributes commonly associated with war, like weapons and combat. Only by visualizing these elements and their inherent polarities as related parts of a continuum can we infer new relationships and principles germane to theory. As Clausewitz says, "One may laugh at these reflections and consider them utopian dreams, but one would do so at the expense of philosophic truth. Philosophy teaches us to recognize the relations that essential elements bear to one another."[22] Keeping this in mind, we turn now to war's *most* essential element, its efficient cause: humanity.

WAR AND HUMAN NATURE

> *For though man is indeed an animal, he is an extraordinary animal, differing far more from all other animals than they differ in kind from each other.*[23]
>
> Hugo Grotius (1625)

The human heart is the starting point of all matters pertaining to war.[24]
Marechal de Saxe, *Reveries on the Art of War* (1731)

Justice in our minds is strife. We cannot help but see war makes us as we are.[25]
Heraclitus

As the preceding quotes suggest, war is intimately connected with humanity and a significant, if not defining, aspect of the human experience. Historian Drew Gilpin Faust opines:

As we have sought through the centuries to define ourselves as human beings and as nations through the prisms of history and literature, no small part of that effort has drawn us to the subject of war. We might even say that the humanities began with war and from war, and have remained entwined with it ever since.[26]

"With war being connected to everything else and everything else being connected to war," adds Azar Gat, "explaining war and tracing its development in relation to human development in general almost amount to a theory and history of everything."[27] While our goal is far more modest than a "theory of everything," the place to begin tracing war's origins is with our own. In *The Most Dangerous Animal* (2007), David Livingstone Smith writes, "War can be approached from many angles ... but there is one dimension that underpins them all: the bedrock of human nature. To understand war, we must understand ourselves."[28] Exploring the origins and nature of humanity helps answer a range of pertinent questions: Are humans fundamentally warlike? If so, why? Where does this capacity come from? Is it a natural, beneficial trait or an evolutionary curse? Is war an imperative or a choice? If the latter, why would human groups *choose* to make war, and what motivates individuals to accept the risks of battle?

A Working Definition for War

Clausewitz, for all that he sought to rationalise war's purpose and to systematise its conduct, was interested in war above all as a phenomenon.[29]
Hew Strachan

Before continuing, it is beneficial to establish an initial, shared mental model of war with a working definition. Toward the end of the chapter, we will derive a final definition. For now, we define war as *a uniquely human phenomenon, a rational activity characterized by violent conflict between opposing forces intended to alter or preserve the status quo.* The definition's opening phrase constrains our analysis to *Homo sapiens*, war's efficient cause; thus, it necessarily excludes prehuman eras and nonhuman activities.[30] War is "rational" because it employs "reason" (i.e., it is a choice supported by human faculties). Senseless violence or temporary, violent responses alone are not war; however, using "rational" here is not meant to imply that all wars are desirable, logical, or moral. Next, we call war a "phenomenon" because it is clearly "something that can be observed and studied and that typically is unusual or difficult to understand or explain fully."[31] The definition's latter portion separates war from nonviolent human competition and identifies it as a reciprocal activity with a purpose. In other words, war's "final cause" is to force or resist change. The word "intended" implies that outcomes are unpredictable. To summarize, only humans have the capability to define rational purposes, concoct strategies, and employ violent conflict to serve their ends (i.e., to make war).

On Human Nature

What a piece of work is a man! How noble in reason, how infinite in faculty! In form and moving how express and admirable! In action how like an angel, in apprehension how like a god! The beauty of the world. The paragon of animals.   Shakespeare, *Hamlet* (2.2.295–299)

Inside of me there are two dogs. One is mean and evil and the other is good and they fight each other all the time. When asked which one wins I answer, the one I feed the most.[32]   Sitting Bull

Defining human nature concisely would be a daunting task sure to invite justifiable criticism. For our purposes, we will simply describe human nature as a set of traits characteristic of most *Homo sapiens* and that collectively differentiate our species' aptitudes, attitudes, and behaviors from others. An easy exercise that helps develop a rudimentary sense of human nature is to complete the following sentence: *humans are by nature* ... intelligent, inquisitive, compassionate, ambitious, inventive, social, and so on. Our confidence in the validity of descriptors like these depends on our frame of reference informed by education and experience. That said, should we add *warlike* to the list?

Provided that "warlike" means to fight or make war with regularity, we can state with certainty that humanity *is* warlike. "Fighting," according to behavioral experts E. F. M. Durbin and John Bowlby, "is plainly a common, indeed a universal, form of human behavior."[33] Of course, not every human is pugnacious, but history affirms that enough are inclined to individual or group violence to confidently characterize the species as warlike. Regarding the factors that drive individual attitudes and behaviors, we could wade into the psychological debate surrounding the question of nature versus nurture; but, in reality, *both* have a role, which we can summarize as nature (genes) *loads the gun* and nurture (environment) *pulls the trigger.*[34] Though some human groups and individuals are demonstrably warlike, many are not.[35] This fact is consistent with Heraclitus's philosophy, which suggests that our warlike inclinations must be balanced by *peaceable* tendencies. Indeed, when viewed in aggregate, history proves that humans are at least as peaceable as they are warlike. According to psychologist Mark A. May:

Man's biological nature is neither good nor bad, aggressive nor submissive, warlike nor peaceful, but neutral in these respects. He is capable of developing in either direction depending on what he is compelled to learn by his environment and by his culture. It is a mistake to assume that he can learn war more easily than peace.[36]

Human nature includes attributes for war-making – like the ability to consider the future, plan, organize, and make and use weapons – and balancing traits for peaceful relations, like cooperation, empathy, compromise, and negotiation. Because of this innate dualism, "A *theory of human nature*," asserts anthropologist Leslie E. Sponsel, "is inevitably associated with theories of peace and war."[37] *Homo sapiens*'s straddling of the peace-war, life-death, creation-destruction, and good-evil dialectics is confirmed by Sigmund Freud, who writes: "Human instincts are of only two kinds, those which seek to preserve and unite ... and those which seek to destroy and kill."[38] Ironically, the attribute (rationality) separating us from the rest of nature and enabling us to make war also grants the power to *prevent* war. In fact, while warlike characteristics help human groups compete for resources, the opposing tendencies enable group formation in the first place, a process that, according to ethologist Frans de Waal, requires "a psychological makeup that allows aggression and competition to be switched on and off."[39] The "gregarious and prosocial side of human nature," adds anthropologist Bruce Knauft, is called "reciprocal altruism."[40]

Thus, despite our intrinsic violent propensities, war is not an inevitable consequence of human nature. As anthropologist Bronislaw Malinowski asserts, "war cannot be regarded as a fiat of human destiny in that it could be related to biological needs or immutable psychological drives."[41] In other words, there is no "death gene" that compels us to make war. Furthermore, most humans *prefer* peace. After all, warring is risky, lethal business.[42] Given this, we might ask why humans should be warlike at all. Assuming that life's governing imperatives are to survive and reproduce, then it would appear that war's lethality is fundamentally *unnatural*. However, unless we reject the premise that natural, evolutionary processes have driven human development, this duality, to include our warlike tendencies, *must* be natural. So how did this happen?

The Most Dangerous Animal

I do not see why man should not be just as cruel as nature.[43]   Hitler

Man is the only animal ... that gathers his brethren about him and goes forth in cold blood and calm pulse to exterminate his kind.[44]   Mark Twain

Research indicates that humans, chimpanzees, and bonobo apes descended from a common ancestor about seven million years ago, which explains why

99 percent of our DNA is identical.[45] For apes, monkeys, and other animal species, violent acts are routine, though these are generally instinctual (e.g., for predation, territorialism, or to establish breeding dominance). So, humanity's predilection for violence is not unique in the natural world. Moreover, violent conflict between *Homo sapiens* and now extinct hominid species, such as *Homo erectus*, *Homo denisova*, *Homo floresiensis,* and *Homo neanderthalensis* (Neanderthals), undoubtedly occurred.[46] Our species, powered by superior intellect, social coordination, and adaptability ultimately prevailed. *Homo sapiens*'s martial ferocity, fueled by a large brain and honed against rival brethren-species, supported survival.

But war is an "intra" not "inter" species activity (i.e., *homo homini lupus* – "man preys on man").[47] Research on nonhuman species is somewhat equivocal; however, notable findings reinforce the view that intraspecific killing is "one of the main causes of animal mortality."[48] In other words, animals kill their own kind at scales comparable to humans.[49] Even our closest biological relatives, chimpanzees, organize into proto-political factions and systematically kill other chimps for resources and reproduction.[50] Likewise, alpha male lions kill off their predecessors' offspring to ensure the primacy of their own. Animals instinctively employ violent capacities, otherwise suited for defense or hunting, to attack their own kind. Intraspecies killing that advances superior specimens at the expense of the inferior, a process known as *natural selection* or *survival of the fittest*, serves nature's imperatives, and there is no logical basis for excluding early *Homo sapiens* from this paradigm. As Freud writes, "It is a general principle, then, that conflicts of interest between men are settled by the use of violence. This is true of the whole animal kingdom, from which men have no business to exclude themselves."[51]

Of course, human lethality emanates from intellectual supremacy, not claws or teeth. Natural selection endowed *Homo sapiens* with the capability to neutralize threats and a propensity for the strongest in body *and* mind to eliminate the weaker. Where we differ from the animal kingdom, and where Mark Twain was correct, is in *why* we kill each other. "Our capacity for and use of violence is neither remarkable nor excessive compared with that of many other animal species," writes Keeley, "whereas our sociability and cooperativeness are unique."[52] Intellect combined with aggression and "sociability" grants humans the unique ability to kill for reasons other than biological instinct and to organize and cooperate in larger groups on more complex enterprises, like war.

Hunting and Gathering Answers

The last twenty years have seen a burgeoning of anthropological research on issues of war and peace. One of the most important empirical findings to emerge from this research is that violence is nearly ubiquitous. Not only are there no utopias, but even many well-known cases of 'peaceful' societies turn out to have their share of interpersonal violence, if not warfare.[53]

Thomas Gregor and Clayton A. Robarchek

It should be noted that the amount of violence may be underrepresented in the paleontological record because injury to soft tissues may cause death without leaving any imprint on the skeleton.[54]   Leslie E. Sponsel

Archaeological and anthropological scholarship substantiates the hypothesis that humankind developed the capacity for violent conflict over millennia of Darwinian struggle. Fossil record evidence shows that our story began nearly three million years ago when the genus *Homo* first appeared. Lacking the knowledge or ability for agriculture, 99.5 percent of our distant ancestors' evolutionary development occurred while living the "hunter-gatherer" way of life. Only in the last 10,000 years or so, a speck in the evolutionary hourglass, have we learned to live within the agricultural and animal husbandry pattern. Thus, as Gat asserts, modern humans owe their biological coding to "millions of years in adapting to the selective pressures of hunter-gatherer existence."[55] This means our genetic "wiring" is extremely outdated. Though our minds have enabled us to adapt and make great leaps in fields like science, medicine, art, and even government, our *instincts* are honed for a different context. War, more than any other human activity, taps into primal emotions that have powerful effects on perceptions and choices. Reason may not be a slave to passion, but it is by no means immune to its influence. "Once blood is shed in a national quarrel," opined David Lloyd George, "reason and right are swept aside by the rage of angry men."[56]

Since these violent capacities appeared early in our species' development, they clearly could not have originated in the "recent" invention of organized states, as some have hypothesized. To be sure, the emergence of modern civilization, organization, agriculture, industry, and science profoundly impacted *how* humans fight, perhaps even how often, but these factors are not responsible for our violent nature. In fact, recent scholarship, led by Lawrence Keeley, has "amassed overwhelming evidence" that "has all but demolished the doctrine that pre-state societies were peaceful and, hence, that warfare is a later cultural invention."[57] Gat cites examples from his own research and that of others portraying a hunter-gatherer way of life filled with violence and fighting. And though there is no historical record of these times, cave art depicting "large-scale encounters" and "battle scenes" as well as archaeological finds indicating bones and skulls clearly damaged by human-made weapons are plentiful.[58] Writes economist Paul Collier:

Hunter-gatherer societies are inherently extremely violent, because the technology does not permit anything else: the winning strategy for . . . hunter-gatherers is the preemptive strike . . . through the predawn raid, catching your enemies detached from their weapons . . . so violence is intrinsic to such societies.[59]

Furthermore, evidence for warlike behavior patterns in primitive human societies is not limited only to the archaeological record. Remarkably, several isolated hunter-gatherer cultures, like the primitives of Tasmania and the Aboriginals of Australia, survived into the modern age, thus allowing direct scientific observation. In these pre-civilizational groups, fighting, raiding, and killing were commonplace. In fact, writes Gat, they represent rates of killing "higher than that of industrialized societies" even accounting for our "massively lethal wars."[60] Indeed, "If you were lucky enough to be born in the industrialized twentieth century," writes Ian Morris, "you were on average 10 times less likely to die violently … than if you were born in a Stone Age society."[61]

Collier believes that these violent, warlike origins led to the formation of states and civilization and not the other way around. "Wars, or threats of war, have played a key role in most, if not all, amalgamations of societies," writes Jared Diamond in his Pulitzer Prize winning *Guns, Germs, and Steel* (1997), and "even between mere bands, have been a constant fact of human history."[62] This aligns with Thomas Hobbes's view that, while humans can increase life's advantages collectively, security is "more effectively achieved by Dominion over others" versus cooperation, which is why "the origin of large and lasting societies lay not in mutual human benevolence but in men's mutual fear."[63] Without protection from aggressive rivals, all state institutions and goods are threatened; thus, they ultimately evolved to support and optimize the security function. Nationalism spawned from wars and shared cause further solidified the state construct.[64]

Though the emergence of civilization and nation-states as a security response (coincident with greater population densities and food production) expanded the scope and lethality of war, this was merely an extrapolation of a far-earlier pattern of human violence. According to Malinowski, as the stakes – including "wealth and privilege," "ideological outfit," "and moral experience" – grew higher, "civilized" societies recognized "the pooling of and the reorganization of economic, political, and spiritual resources could lead to things greater than those which had been destroyed through the agency of fighting."[65] Consequently, primitive conflict became modern war as the leaders of growing nation-states increasingly viewed organized violence as an acceptable and even preferred avenue to greater security, wealth, and political power. The net result, as Greg Cashman points out, is that "*social evolution* proceeded in only one direction – toward the creation of ever more powerful and militant societies."[66]

Adapt or Die

Adaptability is the law which governs survival in war as in life, war being but a concentrated form of the human struggle against the environment.[67]

Liddell Hart

Without question, the environment for the earliest humans – a time when our genetic coding was maturing – was exceedingly dangerous. Threats from animals, the environment, and other humans or near-humans abounded, making daily life an existential test. But human development took a fateful turn as our ancestors made a quantum (and not fully understood) intellectual leap around 50,000 BCE. Along with greater powers of imagination, we achieved the ability to transfer knowledge to succeeding generations through example, then through oral and eventually written narratives. This accentuated our advantages but also started an intellectual "arms race" with other humans that continues unabated today. "Man's learning ability," writes Mark May, "is so much greater than that of any other animal that it would be surprising if he had not outstripped them in devising techniques for inflicting punishment on others when it suits his needs."[68] Rather than adapting to the environment, we adapted the environment to us, a power that led to larger, more potent groups, the formation of civilizations and empires, and the ability to wage large-scale wars.

The capacity to harness fire, create tools (and weapons), imagine, and write – four truly momentous adaptations – heralded the ultimate triumph of brains over brawn. We began to perceive *unreal* things and to envision real things not only as they are but also as they *could* be. At present, "*the human being is the only animal that thinks about the future*," writes psychologist Daniel Gilbert, and, "the greatest achievement of the human brain is its ability to imagine objects and episodes that do not exist in the realm of the real, and it is this ability that allows us to think about the future."[69] Fur became clothing, wood became shelter, stone and metal became tools, and tribes became living instruments for changing the future. This transformational gift, fueled by the survival imperative, lent *Homo sapiens* the capability to defend against and proactively eliminate real and perceived threats and to acquire and secure resources and mates.

For Better or Worse

And the Lord God said, "The man has now become like one of us, knowing good and evil. He must not be allowed to reach out his hand and take also from the tree of life and eat, and live forever." So the Lord God banished him from the Garden of Eden to work the ground from which he had been taken.
Genesis 3:22

Interestingly, humanity's maturation has detached us from nature's traditional evolutionary paradigm in two significant ways. First, being a social and compassionate species with power over the environment has enabled the survival and procreation of many who otherwise would perish due to physical or mental deficiency. As civilization evolved, advancements in collective defense, trade specialization, science, and medicine raised life expectancy and quality significantly. In short, humans can short-circuit natural selection in ways that benefit a larger proportion of the species. Unfortunately, we have also developed the ability to bypass natural selection in injurious ways. The biblical account of Adam and Eve is a metaphor for humanity's separation from the rest of the natural world, with free will and knowledge granting us the power to exhibit *beneficial* or *ruinous* behaviors. Individually, we abuse drugs, overeat, inhale poisons, and participate in risky activities; and collectively, we overconsume resources, pollute the environment, and engage in crime, war, and genocide. It is a truth of human nature, writes international relations expert Kenneth Waltz, that "the very instruments that promise security from cold and hunger, a lessening of labor and an increase of leisure, enable some men to enslave or destroy others."[70] It would be arrogant and naïve to think that nature cannot have the last laugh on *Homo sapiens*. Modern humans have unleashed destructive energies far beyond what nature alone could furnish. The nuclear age, biological weapons, and genetic engineering prove that humanity is poised at an extra-evolutionary inflection point where the very traits that once aided survival now threaten it.

Human Organizations and War

War is neither an art nor a science ... rather it is part of man's social existence ... a clash between major interests, which is resolved by bloodshed.[71]
Clausewitz

One tribe ousting another from a hunting ground 20,000 years ago was executing a political act no less than the Allies preventing Germany from conquering Europe in World War II.[72]
Bevin Alexander

War is a collective endeavor in which the "opposing forces" are human groups. Aggregation confers the benefits of specialization, comparative advantage, and burden-sharing for activities like child rearing, hunting, farming, and defense. Moreover, our brainpower endows us with creativity, abstract thinking, and expression that facilitate the bonding and trust-building necessary for more unified, diverse, and extensive organizations. Communication of shared values and beliefs is the glue that binds groups together. "The recognition of a community of interests such as these," writes Freud, "leads to the growth of emotional ties between the members of a united group of people – communal feelings which are the true source of its strength."[73] Moreover, according to Machiavelli, our "military and civic communities ... prevent human beastliness from reducing life to primitive anarchy. Man's primordial and destructive first nature is kept in check by the imposition of a second nature through rational social organization."[74]

Early human needs, communication, and organizations were relatively simple. Tribes maximized well-being, including security from attack, by leveraging each member's diverse abilities. This required a rudimentary political capacity, including awareness and development of individual aptitudes. "Working effectively in groups," writes war and strategy expert Lawrence Freedman, "required understanding the particular characters of other members of the groups, how they were ranked in the hierarchy and with whom they had attachments, and what all this might mean in specific situations."[75] Gender provided the first model for specialization, and beyond gender, then as now, individual skills influenced organizational roles and responsibilities. The consolidation, harmonization, and direction of talents toward productive activities determined a group's overall health and character.

Leadership

The management of a society of specialists requires the imposition of authority and acquiescence to it, and that ethos readily translates into the urge to extend authority over others.[76]
John Keegan

As primitive bands grew into tribes and chiefdoms, specialization, burden-sharing, and leadership were logical developments. Larger organizations composed of autonomous, heterogeneous individuals rarely function efficiently or effectively without a guiding element, the leader. Leadership, whether by one or more persons, directed and prioritized group activities, including combat. The identification and selection of leaders and the transition of power have been and still are among the most contentious activities in human affairs. Leaders may seize power, leading by force of will or just force, or they may be selected to lead by others. Leadership is a natural consequence of human organization, because it is the best mechanism for maintaining cohesion and unity of purpose and action. Good leaders drive and direct organizational activities that can enable a group's productivity to exceed the sum of its inputs. Successful leaders are persuasive and demonstrate competence by resisting personal challenges and effectively providing for a group's needs, including suppressing discord. According to Malinowski, good leaders "rapidly and effectively" eliminate quarrels by exercising their "definite, centralized, and organized authority."[77] While fighting prowess can propel someone to power, remaining there often depends on a wider range of attributes. According to psychologist William McDougall,

Success in combat and survival . . . must have been favored by, and have depended upon, not only the vigor and ferocity of individual fighters, but also . . . the capacity for united action, good comradeship, personal trustworthiness, and the capacity of individuals to subordinate their impulsive tendencies and egoistic promptings to the ends of the group and to the commands of the accepted leader.[78]

Poor leaders may struggle for respect and risk being challenged, and nearly all leaders, good or bad, are reluctant to relinquish power, even in peace. To extend their tenures, leaders may be tempted to use violent force against challengers or to foment war.

Aggression, Power, and Politics

In the first principles of nature there is nothing opposed to war ... for as the end of war is the preservation of our lives and limbs, and the retention or acquisition of things useful to life, it accords well with those first principles.[79]

Hugo Grotius (1625)

Like it or not, humans are predators. Hunting, foraging, and competing with other hominids and wildlife required aggression, seizing resources, killing for food, and fighting anyone or anything that interfered. Acquisition is an offensive act, and we are a highly acquisitive species. Though many use violence only for defensive purposes, others do not. Furthermore, we are awful at self-restraint, particularly with regard to potent capabilities. This is logical because disavowing power in a competitive setting cedes an advantage to rivals, a reality embraced by many leaders who may subsequently become intoxicated by power. However, as Lord Acton warns, *power tends to corrupt, and absolute power corrupts absolutely*, compelling some leaders to prioritize personal gain (perhaps enacted violently) over the greater good.[80] Lord Acton understood that excessive power diminishes the potency of regulatory measures like laws and morals that might otherwise curb violent tendencies. "The practical basis for ordinary moral behavior is the recognition of shared humanity and mutual vulnerability," writes military historian Gwynne Dyer, "which is precisely what is destroyed by the illusion of absolute power."[81] Power is a siren's call to our base nature that can overwhelm the norms of civil orthodoxy, fairness, and altruism. "We can't expect kings to philosophize, or philosophers to become kings," says Immanuel Kant, alluding to Plato, "and it isn't desirable either, because the possession of power inevitably gets in the way of the uncluttered judgment of reason."[82] A word for the apportionment and use of power in human relations and governance is *politics*, and according to Aristotle, man is by nature a political animal.[83] Therefore, leaders often employ aggression to apportion political power to themselves, a practice that may solidify their status but can also lead to the neglect or even persecution of their own people and violence against rivals. With few exceptions, maintaining both *internal* harmony and stability with *external* groups (i.e., preventing war) requires leaders to exercise power effectively and responsibly, if not always ethically.

The Internal Problem

The violent solution of conflicts of interest is not avoided even inside a community.[84]   Sigmund Freud

Men, thinking to better their condition, are always ready to change masters, and in this expectation will take up arms against any ruler.[85]   Machiavelli

Human groupings of any era must balance the tension between self-interest and what Bruce Knauft terms "cooperative affiliation" (i.e., getting along peacefully to promote common interests).[86] Cooperative affiliation is more prevalent within groups than between them, because cooperation is a prerequisite for group formation and sustained cohesion, and the presence of external threats tends to reinforce internal solidarity. However, violence can occur when intragroup tensions become unbalanced. Within any organization, there are many avenues to violent conflict, but perhaps the most probable arises from the relationship between the group and its leadership. These cases have a consistent pattern that begins when an individual or faction questions a leader's competence or suitability. The leader may be cruel, corrupt, or inept, or the disgruntled element may simply be ambitious. This imbalance turns violent when the opposing parties are ready, willing, and able to fight to change or preserve the status quo. The *likelihood* of a coup or rebellion depends on a variety of factors including the organization's culture, antagonist personalities, the level of dissatisfaction, and the probability of success. For example, coups and civil wars are *less likely* in societies that eschew violence, protect their leaders well, or offer nonviolent processes for affecting leadership change. Though the specifics of internal conflicts vary significantly, the basic elements from our working definition of war remain constant: humans fighting each other to change (rebels) or preserve (leader or government) the status quo.

The External Problem

War is a social institution, and it is waged by societies, not only by states.[87]
                                                                        Colin Gray

Human beings have a strong potential to differentiate between "us," members of an ingroup, and "them," members of an outgroup.[88]   Ervin Staub

The logic of violence between groups, the *external* problem, is similar but not identical to that of internal conflict. Disagreements, ill feelings, or conflicts between groups create a similar set of options: accept the status quo or act to change it. However, while the context is analogous to intragroup violence, intergroup relations are exacerbated by their anarchic nature (i.e., the absence of an authoritative arbiter). This explains why internal group relations can be peaceful, while intergroup dynamics are not. Writes McDougall:

The replacement of individual by collective pugnacity is most clearly illustrated by barbarous peoples living in small, strongly organized communities. Within such communities individual combat and even expressions of personal anger may be almost completely suppressed, while the pugnacious instinct finds its vent in perpetual warfare between communities whose relations remain subject to no law.[89]

Within an organization, nation, or society, designated authorities incentivize cooperation, mediate and adjudicate conflicts, and enforce accountability. The same rarely applies between groups, especially when vital interests are at stake. "Among states as among men," writes Waltz, "there is no automatic adjustment of interests. In the absence of a supreme authority, there is then constant possibility that conflicts will be settled by force."[90] When two or more groups quarrel (e.g., over resources, boundaries, or from pure aggression) there are two options. The first is peaceful cooperation to solve the problem, which requires trust and perhaps the creation of intergroup institutions with monitoring and enforcement authorities. The second option, violence, is simpler though more dangerous. With this alternative, one or more groups may opt for war. Furthermore, if three or more groups are in play, alliance-forming can reduce risk and increase the odds of victory.[91]

Besides being simpler than cooperation, wars also tend to beget more wars, because a successful war becomes a model. Denying a proven method for securing resources, power, and prestige contradicts human nature. "Once a strong local group developed a military machine," writes Malinowski, "it would use this in the gradual subjugation of its neighbors and extension of its political control."[92] And Freud adds, "A glance at the history of the human race reveals an endless series of conflicts between one community and another or several others, between larger and smaller units – between cities, provinces, races, nations, empires – which have almost always been settled by force of arms."[93] Once this cycle starts, the peaceful path loses its allure, at least until the belligerents come to realize that war's costs and risks outweigh its benefits.

It Only Takes One

Now Cain said to his brother Abel, "Let's go out to the field." While they were in the field, Cain attacked his brother Abel and killed him.   Genesis 4:8

All men dream, but not equally . . . the dreamers of the day are dangerous men, for they may act on their dreams with open eyes, to make them possible.[94]

T. E. Lawrence

The preceding section illustrates some connections between leadership and war. Internal or civil wars are fought to replace leaders and political systems, while intergroup wars center more on retribution, threat response, or power acquisition. Both instances highlight dissatisfaction with the status quo, which is often a symptom of human diversity. In other words, we have different perceptions of what "right" looks like, even when it comes to war and peace. Indeed, according to anthropologist Thomas Gregor, "The whole enterprise of identifying societies as peaceful or otherwise is an illusion, since what constitutes an aggressive act in one culture may not be perceived that way in another."[95] With human nature, one man's heaven can be another's hell, or as John Milton writes, "The mind is its own place and in itself can make a Heaven of Hell, a Hell of Heaven."[96] Human diversity, a strength when world views (*Weltanschauung*) align, leads to imbalances and divergence when they do not. Not everyone can be a poet or architect, nor will everyone be content with their lot or station in life. What is to stop those with the capability and will to wield violent power to improve their circumstances from doing so? According to Collier, "Some people are more productive than others, and some people are stronger than others. From the resulting four different types of people, ask yourself how one type, the unproductive strong, are going to earn a living?"[97] Adds Keegan, "The violent individual is the principal threat to the norm of cooperativeness within groups, the violent group the principal cause of disruption in wider society."[98]

Indeed, the strong and ambitious learned long ago that they can often take what they want. Consider the Zulu, an early nineteenth-century tribe of southern Africa. The Zulu descended from the Ngunis, a society known for their "gentle, pastoral way of life."[99] The Ngunis were "very civil, polite and talkative . . . kind to strangers . . . notably law abiding."[100] Nguni fighting, while rare, was generally a low-casualty, ritualistic affair. This changed dramatically when Shaka became chief of the Zulus, a small Nguni tribe at the time. Shaka viewed Nguni peacefulness as an opportunity to advance his political aspirations. Through persuasive leadership, Shaka raised and led "an army of savagely disciplined regiments that waged battles of annihilation" against its more peaceful neighbors.[101] When one side fights, writes Clausewitz, "while the other refrains, the first will gain the upper hand. That side will force the other to follow suit."[102] Before long, Shaka's Zulus became the supreme military power in southern Africa. Using war, the charismatic Shaka had transformed the Zulu into an African Sparta.[103] Shaka's story illustrates how a single actor can tip the scales from peace to war. "Much of mankind," writes military historian T. R. Fehrenbach, "abhors competition, and these remain the acted upon, not the actors."[104] Without a balancing power, there is nothing to oppose aggressors.

A Matter of Choice

The very first essential for success is a perpetually constant and regular employment of violence.[105]   Hitler

Non-violence is the greatest force at the disposal of mankind.[106]   Gandhi

War does not just "happen" like an earthquake or plague. It is a choice, but why do we choose to fight wars? What *motivates* us to fight, and what are the deeper psychological needs war satisfies? Transitioning from centuries of atavistic action to rational preference did not happen overnight; thus, the earliest "reasons" were undoubtedly tied to what Gat calls "primary biological ends, directly linked, the one to the organism's existence and the other to its reproduction."[107] To satisfy sustenance, social, and sexual needs, human groups – otherwise internally peaceful and cooperative – risked injury and death in conflicts with other groups. As *Homo sapiens* matured, motivations for war became more complex. Clausewitz writes, "Two different motives make men fight one another: *hostile feelings* and *hostile intentions*."[108] Feelings are emotional and primal, whereas intentions imply rationality. For modern humanity, war's motivations and aims, as Clausewitz implies, are likely the result of emotion *and* reason in variable quantities depending on the circumstances. The *causes* of war (or *casus belli*) and the *motivations* for war are related. Causes are events or circumstances that drive effects. Causes and effects can influence human motivations and decisions, which in turn, become causes in their own right. Human choice prevents any set of causes from making war inevitable, a fact that has bedeviled historians and strategists for centuries. Even invasions do not necessarily precipitate wars as the Nazi *Anschluss* of Austria in 1938 and Russia's *de facto* annexation of Crimea in 2014 prove.

Casus Belli

Where do wars and fightings among you come from? Don't they come from your pleasures that war in your members? You lust, and don't have. You kill, covet, and can't obtain. You fight and make war.   James 4:1–2

The souls of emperors and cobblers are cast in the same mold. The same reason that makes us wrangle with a neighbor creates a war betwixt princes.[109]   Michel De Montaigne, *Essays* (1580)

Casus belli is a tricky concept, not because causes are hard to identify, but because they are hard to identify *in advance*. "History is littered with wars," writes British intellectual Enoch Powell, "which everybody knew would never happen."[110] Even in hindsight, causes are difficult to isolate. For example, was World War I caused by an Archduke's assassination, a paranoid Kaiser, or the war plans of the German General Staff? Was World War II caused by Hitler's megalomania, the draconian Treaty of Versailles, or the feckless diplomacy of France and Britain? After World War I's carnage, many believed peace had come at last, but Hitler disagreed. Then, after World War II, the cognoscenti reversed course, convinced a war between the USA and USSR was inevitable. It never happened. When the USSR broke up, pundits announced a "new world order"; instead, wars broke out all over. This pattern of failing to identify the likely causes of the *next war* is the norm. Though we will never achieve "psycho-historical" premonition as depicted in Asimov's *Foundation* (1951), cultivating greater awareness of motivations for fighting may at least improve our predictive and analytic accuracy.[111]

Classic Causes

When things have arrived at the point where a rational being is convinced that the care for its own preservation is incompatible not only with the well-being of another but with that other's existence, then this being takes up arms and seeks to destroy the other with the same ardor with which it seeks to preserve itself, and for the same reason.[112]   Jean-Jacques Rousseau

Psychologist Lawrence LeShan identifies three classic causes for war. The first is familiar: war is an intrinsic expression of human nature and human relations.[113] As Kant says, "War itself doesn't need any causal explanation, because it seems to be grafted onto human nature."[114] "The events of history are caused by men," adds Fielding Hall, "who in their turn are governed by their minds. The causes of events are therefore in men's minds."[115] LeShan's second cause is economic.[116] This is related to the first cause since economic pursuits and incongruities are often closely related to matters of human nature, power, and politics. Humans are acquisitive, and war is a means for acquiring wealth. As the satirical version of the Golden Rule states, *the person with the gold makes the rules.*[117] LeShan's third cause is structural (i.e., war happens because humans congregate in groups organized and structured for it).[118] This last point is also related to the first, because human groups often exhibit the traits of individuals. Societies in anarchic settings fight for resources and survival against rival societies. In this environment, the security dilemma applies as threats compel martial preparations, which raise the probability for imbalance, fear, and war.[119] "As soon as one state increases ... its troops," writes Montesquieu, "the others suddenly increase theirs, so that nothing is gained thereby but the common ruin."[120]

   In *Man the State and War* (1959), Kenneth Waltz proposes three "images" for locating war's causes: "within man, within the structure of the separate states, [and] within the state system."[121] These correspond to LeShan's first and third causes. Waltz's first image blames war on violent human nature. War happens because it is in our genes, and consequently, in our *psychology.* Regarding individual psychology, according to Durbin and Bowlby, the three simplest motivators for aggression in children include *possessiveness* of things and attention, *intrusion by strangers*, and *frustration.* These factors correlate surprisingly well with reasons *groups* go to war; for example, wars for "things" (like Iraq's Kuwait invasion), to eliminate a hated neighbor (e.g., Arab-Israeli wars), or for relatively trivial or emotional causes (like the War of Jenkins's Ear). In psychological terms, say Durbin and Bowlby, war results from the release of a vast store of transformed aggression, justifiable for any number of reasons.[122]

Not surprisingly, as the prime power wielder and catalyst for action, a leader's psychology is of singular significance. Temperaments prone to ambition, aggression, suspicion, feelings of superiority, and need to control – like dogmatic, authoritarian, domineering, and narcissistic personality types – are not uncommon in those who ascend to positions of high political importance. Such leaders often view and treat other countries as they would individuals. In other words, according to Cashman, "decisions on the use of force at the national level are determined at least in part" by these personality characteristics.[123]

Waltz's second and third images take a more optimistic view of individuals and link war's causes to the nature of internal and external political and social institutions. This view coincides with Mark May's belief that war is caused by social conditioning and environmental factors that "reward pugnaciousness and the use of force more than other forms of adaptive behavior."[124] This social conditioning within state structures, according to Waltz, incentivizes the resolution of external differences by force. Internal political tensions and upheaval entice leaders to conjure *external* threats upon which to transfer animosity. "Peoples involved in civil strife," writes Hegel, "acquire peace at home through making wars abroad."[125] Furthermore, according to Waltz, the anarchic international system of states can precipitate wars. Tensions created through competition and cooperation lead to conflict as states prioritize relative rather than absolute gains. Moreover, leaders can direct states and their institutions, as human constructs created to satisfy shared needs, to benign or violent ends.[126]

Interestingly, nearly every human studies discipline has a "theory" regarding war's causes, and not surprisingly, each favors its own specialty. Economists blame economics, psychologists psychology, political scientists politics, and so on. However, despite mountains of research dedicated to isolating "the true cause of war," no unifying theory has emerged. Myriad proposals – including small-group dynamics; groupthink; rational actor, organizational process, and bureaucratic politics models; incrementalism; stimulus-response, game, and deterrence theory; conflict spirals; security dilemma; power balances, imbalances, polarity, and transitions; governmental and economic systems; and the anarchy of the international system – offer persuasive rationales for how wars may start, but not one of these theories is *sufficient*. Any theory that subordinates human nature, perception, and decision-making to soulless institutions and processes or characterizes human nature as *only* good or *only* evil inevitably falls short of universality. War's true cause is its efficient cause, humanity, in all its variety, irrationality, and inconsistency. Therefore, before deriving a formal definition of war, we must look beneath the grand theories of *casus belli* to the specific motives that drive leaders to start wars and warriors to fight them.

Fear, Honor, and Interest

Tell me not, sweet, I am unkind,
  That from the nunnery
Of thy chaste breast and quiet mind,
  To war and arm I fly.

True, a new mistress now I chase,
  The first foe in the field;
And with a stronger faith embrace
  A sword, a horse, a shield.

Yet this inconstancy is such
  As you too shall adore;
I could not love thee, dear, so much,
  Loved I not honour more.[127]   Richard Lovelace, *Song to Lucasta* (1649)

Despite humanity's penchant for pugnacity, according to Malinowski, " human beings never fight on an extensive scale under the direct influence of an aggressive impulse."[128] Rather, they organize to "assault and kill" for broader, shared purposes like "tribal traditions," religion, patriotism, "cultural values," and "collective hatreds."[129] Thucydides encapsulated these "shared purposes" within three categories: *fear*, *honor*, and *interest*.[130] As we will see shortly, these motives animate political processes within human groups resulting in policies and political aims that rely on war for fulfillment. *Fear*, a visceral state that often clouds higher faculties and judgment and compels violent actions, can motivate an individual or an entire nation. Societies can wage war based on fear – real or perceived – of an uncertain future or an ominous threat. Throughout history, established powers have warred preemptively out of fear that doing nothing would be worse. Examples include Sparta prior to the Peloponnesian War and Kaiser Wilhelm's Germany before World War I. *Honor* is tied to self-awareness and a sense of fairness and justice, and it produces feelings of individual and collective esteem that are not easily disregarded, especially for warriors. "Military honor and the renown of an army and its generals," writes Clausewitz, "are factors that operate invisibly, but they constantly permeate all military activity."[131] As the preceding quote suggests, honor is intangible and perception-based; thus, like a concealed snare, it entangles, leaving no obvious way to escape. Affronts to honor can spur actions to vanquish an oppressor, seek vengeance, or fight beyond practical limits. Thucydides writes:

It is the characteristic of prudent men that they are at rest if they are not wronged; yet characteristic of brave men, once they are wronged, that they leave peace for war and at the right opportunity leave war for reconciliation, and that they are neither carried away by good fortune in war nor victims of injustice because of enjoying the tranquility of peace. For he who holds back for the sake of pleasure will very quickly, if he stays inactive, be deprived of that same blissful state that caused him to hold back, and he who goes too far because of success in war does not realize that he is being carried away by unwarranted confidence.[132]

Here, Thucydides highlights the fine line between appropriate responses to dishonorable aggression (prudence) and brash overreach (boldness). Those who reject war entirely, he warns, may have peace snatched away, and those who live by the sword are likely to eventually die by it. Thucydides's final category, *interest*, is the catch-all, and it seems there are a great many "interests" humans are willing to fight for. Write Durbin and Bowlby, "[people] fight for Protestantism and Mohammedanism, for the emancipation of the world proletariat or for the salvation of the Nordic culture, for nation or for king. Men will die like flies for theories and exterminate each other with every instrument of destruction for abstractions."[133] Fielding Hall believes the pursuit of *freedom*, subject to innumerable interpretations, is humanity's prime interest and the cause of *all* war. According to Hall, everyone wants freedom to live (survival) and to live as they please.[134] Indeed, survival or self-preservation is the principal interest for most societies as well. "Common to the desires of all states," writes Waltz, "is the wish for survival. Even the state that wants to conquer the world wants also, as a minimum, to continue its present existence."[135] Other war-sparking interests include territory, resources, wealth, and commitment to allies. The Japanese attack on Pearl Harbor was propelled by all three Thucydidean motives: fear (of American power), honor (in response to perceived racism and policy insults), and interest (to secure resources for the Greater East Asia Co-Prosperity Sphere).

For Glory, God, and Gold

O war! Thou son of hell,
Whom angry heavens do make their minister,
Throw in the frozen bosoms of our part
Hot coals of vengeance!   Shakespeare, *Henry IV* (5.2.33)

If you prick us, do we not bleed? If you tickle us, do we not laugh? If you
poison us, do we not die? And if you wrong us, shall we not revenge?
                                        Shakespeare, *The Merchant of Venice* (3.1.47)

There are many individual motivational factors that, along with the persuasive messages conveyed by leaders, can convince otherwise ordinary people to fight and kill each other. As discussed earlier, we are a social and communicative species accustomed to grouping for collective support and defense. Subgroups with like talents, abilities, and duties (e.g., guilds, orders, and sects) form within larger groups. For example, in *The Republic*, Plato identifies two categories of people: "producers" and "guardians."[136] Another school of thought partitions humanity into three groups: *sheep* (passive/productive), *wolves* (aggressive/destructive), and *sheep dogs* (aggressive/productive).[137] Regardless of the terminology – caste, rank, or class – nearly every society has some type of warrior sect, but what convinces warriors to brave the dangers of combat?

For war to occur, its reasons must induce collective action. This places a burden on leaders to "make the case" for war. For some, the motivation is simply what Machiavelli deems "the strongest and most operative" of the "various ways of forcing men to fight": giving them "no other alternative but to conquer or to die."[138] Of course, not every warrior is a slave or conscript. Many volunteer to fight for pay, their fellow warriors, ideals, revenge, fear of admonishment, religious reward, love of killing, adventure, or glory. Machiavelli adds that men will often fight for pride – to avoid the shame of cowardice or the perception of inferiority relative to their foe – and "the love of money," which "ordinarily operates as strongly upon men as love of their life."[139] Thucydides lists a variety of motivations in his Sicilian battle narrative:

They advanced, the Syracusans about to fight for their homeland and for each man's individual preservation at the moment and freedom in the future; among their opponents, the Athenians, for making an alien country their own possession and to avoid harming their own country by defeat; the Argives and independent allies, to help them gain what they had come for and, by winning, to see once more the country belonging to them; and the subjects among the allies had their strongest motivation in the immediate survival they could expect only by conquering, and then, as a further consideration, that if they joined in making a further conquest their subjection might be of another and easier sort.[140]

Liddell Hart says,

Man has two supreme loyalties – to country and to family, and with most men the second, being more personal, is the stronger. So long as their families are safe, they will defend their country, believing that by their sacrifice they are safeguarding their families also. But even the bonds of patriotism, discipline, and comradeship are loosened when the family is itself menaced.[141]

Should harm befall the family or other cherished person or thing, revenge can become the prime motive. "The instinct for retaliation and revenge," writes Clausewitz, "is a universal instinct shared by the supreme commander and the youngest drummer boy; the morale of troops is never higher than when it comes to repaying that kind of debt."[142]

Political leaders generally begin wars after communicating a reason that inspires their followers to support the cause, the political aim (which always supports political survival). Politics, the activity through which we apportion power and govern, produces policies, agendas, and objectives, then mobilizes ways and means (including war) to achieve its ends. Liddell Hart writes, "While there are many causes for which a state goes to war, its fundamental object can be epitomized as that of ensuring the continuance of its policy – in face of the determination of the opposing state to pursue a contrary policy."[143] Whether for survival, revenge, territory, or whim, war requires reasons for every participant. These reasons do not have to align, but they do affect the war's outcome. Thucydides writes:

Regarding war and how terrible it is . . . no one is forced into it out of ignorance any more than he is deterred by fear if he believes he will gain an advantage. The situation of the former is that the gains appear greater than the risks, while the latter is willing to face the danger before accepting any momentary deprivation.[144]

Religion

Men and women first construct towering systems of theology and religion, complex analyses of racial character and class structure, or moralities of group life and virility before they kill one another.[145]   Durbin and Bowlby

Mine eyes have seen the glory of the coming of the Lord:
He is trampling out the vintage where the grapes of wrath are stored;
He hath loosed the fatal lightning of his terrible swift sword:
His truth is marching on.[146] Julia Ward Howe, *Battle Hymn of the Republic* (1862)

Religious motivations (and other supernatural belief systems, to include superstition) bear special consideration, because they touch upon fear, honor, and interest and can be extraordinarily powerful individual and collective motivators. Religion looms large in human relations and often in war, for, according to Keegan, "religion was the first immaterial force to elevate the pretext for war to a higher plane."[147] Nevertheless, as religious studies expert Peter Ochs opines, too often "diplomats, international studies scholars and policymakers have failed to recognize that religion is an ongoing, inveterate dimension of human reality."[148]

Humanity has an enduring affinity for religion, which is a natural, emergent aspect of early intellectual and social development. The first religions helped explain our origins, mortality, and natural phenomena. More significantly, religion (like mythology and nationalism) establishes nonkin commonality in larger, heterogeneous groups and a means for *gaining influence* over animals, the environment, and each other. Through experience, our ancestors observed small-scale cause-and-effect and theorized that large-scale occurrences like weather, disease, and even death, must be "caused" by unseen "spirits" (animism) or a higher power (theism).[149] Certainly, they surmised, beings capable of creating and manipulating the world in such wondrous ways must be immensely potent. Accordingly, the person or group most closely affiliated with the mystical or divine wielded tremendous power to unify and control ever-larger human societies. This is likely why theism's rise coincides with the advent of civilization (circa 10,000 BCE). Who would dare challenge clerics or leaders imbued with the unearthly patronage and powers of a god? Individuals or groups, inside or outside the clan, who offended or disobeyed the gods (or their priests) were often subject to mutilation, sacrifice, execution, or war.[150]

Even now, religion influences all aspects of human relations. Sometimes religion and politics overlap binding political (mortal) and religious (immortal) interests together. Indeed, "Those who say religion has nothing to do with politics," says Gandhi, "do not know what religion is."[151] "No religion has ever kept its adherents from fighting each other," adds Sumner, and "religion has been quite as much a stimulus to war as to peace, and religious wars are proverbial for ruthlessness and ferocity."[152] Religion can inspire fear of heavenly or satanic reprisal, arouse feelings of honor and sacrifice, and lead adherents to favor the promise of an eternal reward over their own survival. Moreover, religion aggravates interorganizational relations when dogmatists equate compromise with heresy. Peter Ochs writes,

[Religious warriors] are willing to endure great hardship – including mortal sacrifice – as a demonstration of faith. Religious killers can feel justified or even sanctified in violence and cruelty against those who are considered "other" – including civilians and helpless noncombatants. Religious conflicts resist compromise because the underlying beliefs are . . . not subject to negotiation.[153]

Of course, religion cannot be solely blamed for war, and this analysis is neither intended to be a general condemnation of religion nor an assertion that its elimination would eradicate or reduce war's prevalence. In truth, religion merely reflects the duality of human nature; thus, it often contributes as much or more to advancing peace as it does in provoking conflict.

The Allure of War

Battle is the most magnificent competition in which a human being can indulge. It brings out all that is best; it removes all that is base.[154]

General George Patton

Before concluding this section on war and human nature, we need to look closer at how, as the English classicist Bernard Knox writes in his introduction to Homer's *Iliad*, war's "strange and fatal beauty . . . can call out in men resources of endurance, courage and self-sacrifice that peacetime . . . can rarely command," despite its "monstrous ugliness" and man's "instinctive revulsion from bloodshed."[155] In *The Psychology of War* (2002), LeShan writes, "It is clear that war ... promises to fulfill some need or combination of needs that are at least close to universal."[156] LeShan postulates that war uniquely satisfies four elemental human motivations: (1) displacement of aggression, (2) projection of self-doubts and self-hatred, and the need for (3) meaning or purpose, and (4) greater belonging.[157] "From Oliver Wendell Holmes to Ernest Hemingway," write human studies experts Leon Bramson and George Goethals, "the experience of combat has been proclaimed as the most meaningful epiphany in human existence,"[158] for war offers such a wide variety of extraordinary opportunities including "heroic action, brotherhood, community, dedication, selflessness, order, command, ritual, and aristocracy."[159] Simply put, war more than any other single activity provides an outlet for aggression, insecurity, and thrill-seeking while also satisfying the need for purpose, belonging, and self-actualization.

Leadership's Power

> *Then out spake brave Horatius,*
> *The Captain of the gate:*
> *'To every man upon this earth*
> *Death cometh soon or late.*
> *And how can man die better*
> *Than facing fearful odds,*
> *For the ashes of his fathers,*
> *And the temples of his Gods.*'[160]

Thomas Babington Macaulay, *Lays of Ancient Rome* (1842)

Charismatic leaders instinctively capitalize on the characteristics of war and human nature to influence followers. Their words and example generate a *motivational resonance* in followers, especially in military organizations where shared danger amplifies emotion. "It is necessary that a general should be an orator as well as a soldier," writes Machiavelli, "for if he does not know how to address himself to the whole army, he will sometimes find it no easy task to mold it to his purposes."[161] In fact, with the right words, delivery, and audience, says Machiavelli, a leader "may dispel their fears, inflame their courage, confirm their resolution, point out the snares laid for them, promise them rewards, inform them of danger and of the way to escape it ... and avail himself of all other arts that can either excite or allay the passions and the appetites of mankind."[162] Mao adds, "Political activities depend upon the indoctrination of both military and political leaders ... through them, the idea is transmitted to the troops ... [and] must be an ever-present conviction."[163] And the ancient Chinese Ssu-ma (war minister) writes, "After you have aroused [the people's] ch'i [spirit] ... lead them with your speeches."[164] Motivating warriors to fight is a *push* and a *pull*, a psychological transaction between the individual and the collective, between leader and follower. Trust, like a chain, works in two directions and is forged a link at a time until all the parts become a unified whole. "A cherished cause," writes Jomini, "and a general who inspires confidence by previous success, are powerful means of electrifying an army and conducing it to victory."[165]

Charismatic leaders are especially adept at influencing individuals to subordinate personal interests and standards, including safety or morality, to the greater cause. This results in a "mob mentality" enabled by what cognitive scientist Herbert Simon terms "bounded rationality."[166] In his *La Psychologie des foules* (1895), Gustave le Bon (a positivist who influenced warfare pioneer J. F. C. Fuller) "described the collective behaviour of the crowd as impulsive, driven by primal instincts, irrational, and open to manipulation ... [and] skillful control by a small leading elite."[167] Hitler, Stalin, and Mao exemplified this ability. In the right contexts, like those of late eighteenth-century France and post-Weimar Germany, words can trigger profound changes. Hitler understood this well saying, "All great movements are popular movements. They are the volcanic eruptions of human passions and emotions, stirred into activity by the ruthless Goddess of Distress or by the torch of the spoken word cast into the midst of the people."[168] Thucydides illustrates this effect in his dialogues and orations of the Peloponnesian War. Pericles's funeral oration is a masterwork of mental manipulation that arouses Athenian nationalism and exhorts the masses to fight to preserve the city-state. According to Freedman, Pericles's "success lay in his authority and ability to convince people to follow strategies developed with care and foresight ... [and] to describe a future that could be achieved if his advice was followed."[169] Other noteworthy, motivational orations include:

Robert the Bruce of Scotland to his commanders before the Battle of Bannockburn,

Sirs, we have every reason to be confident of success, for we have right on our side. Our enemies are moved only by desire for dominion, but we are fighting for our lives, our children, our wives and the freedom of our country.[170]

Patrick Henry's plea to arms against Britain,

Is life so dear, or peace so sweet, as to be purchased at the price of chains and slavery? Forbid it, Almighty God! I know not what course others may take; but as for me, Give me Liberty, or give me Death![171]

George Washington on the American Revolution,

The time is now at hand which must probably determine whether Americans are to be freemen or slaves ... The fate of unborn millions will now depend, under God, on the courage and conduct of this army. Our cruel and unrelenting enemy leaves us only the choice of brave resistance, or the most abject submission. We have, therefore, to resolve to conquer or die.[172]

Woodrow Wilson's address to the US Congress in 1917,

> Our motive will not be revenge or the victorious assertion of the physical might of the nation, but only the vindication of right, of human right, of which we are only a single champion ... but the right is more precious than peace, and we shall fight for the things which we have always carried nearest our hearts – for democracy, for the right of those who submit to authority to have a voice in their own governments, for the rights and liberties of small nations, for a universal dominion of right by such a concert of free peoples as shall bring peace and safety to all nations and make the world itself at last free.[173]

And Winston Churchill's defiance in World War II,

> We shall defend our island, whatever the cost may be ... we shall never surrender! ... let us therefore brace ourselves to our duties, and so bear ourselves that, if the British Empire and its Commonwealth last for a thousand years, men will still say "This was their finest hour."[174]

Of course, these are but a few of the better-known speeches. There have undoubtedly been many more, perhaps even more eloquent and persuasive.

Humanity's penchant for deriving psychological benefits from war may seem a defect when viewed through a contemporary moral lens but not when evaluated scientifically. Prehistoric humans required coordinated action and cohesion for survival and this entailed an ability both to motivate and be motivated. Displacement of aggression, projection of hatred, and the need for purpose and belonging helped bind groups together in a kill-or-be-killed world. No other activity could satisfy all these needs like war's crucible of shared danger, high stakes, and glory. Moreover, a propensity for leadership and followership benefits any collective activity, not just war. Responding to a motivational leader when circumstances require coordinated action is no deficiency. Nevertheless, coordinating violent activities requires a stronger leader-follower bond, followers willing to risk their lives, and in some cases, those who even *prefer* a life of violence.

Of Sheepdogs and Wolves

The world is a dangerous place to live, not because of the people who are evil, but because of the people who don't do anything about it.[175]   Albert Einstein

In a world in which the human capacity for violence is ever-present, imagine the consequences if only the unintelligent aggressor, criminal, psychotic, or megalomaniac could summon the will to fight or the charisma and communication to gather others to a violent cause. In this sort of world, civilized people who champion values like freedom, altruism, and peace must accept a special mantle of responsibility. They must be willing to employ the psychology of war and to understand the relationship between war and human nature or risk subjugation by a Shaka or Hitler. Consider the following sentiments from George Washington, "To be prepared for war is one of the most effectual means of preserving peace";[176] Teddy Roosevelt, "Better a thousand times err on the side of over-readiness to fight than to err on the side of tame submission to injury or cold-blooded indifference to the misery of the oppressed";[177] Fehrenbach, "If the free nations want a certain kind of world, they will have to fight for it, with courage, money, diplomacy – and legions";[178] and Ian Morris, "People hardly ever give up their freedom, including their rights to kill and impoverish each other, unless forced to do so, and virtually the only force strong enough to bring this about has been defeat in war or fear that such a defeat is imminent."[179] While the promotion of peace is both noble and wise, for intractable aggressors, we can hardly disagree with Cicero who asks, "For what can be done against force without force?"[180] or Bernard Knox, who warns that Homer's "stern lesson" is "that no civilization, no matter how rich, no matter how refined, can long survive once it loses the power to meet force with equal or superior force."[181] Failure to acknowledge this reality skews the peace-war dialectic and risks subjugation, harm, or death at the hands of wanton aggressors.

DEFINING WAR

What is war? There is no agreed definition today among the multitude offered, and none that works across time and cultures.[182]   Jeremy Black

Having examined war's origins and relationship to human nature using only a general sense of war informed by our working definition, we are now poised to derive a formal definition, which will better summarize war's essential elements; help distinguish between war, nonwar violence, and peace; and establish an organizing concept for theory analysis and development in subsequent chapters.

What Is War?

Defining war is easy, yet doing it well is harder than it seems. The term "war" has many confusing connotations and is frequently applied to any conflict or competition (e.g., "wars" on drugs, terror, and viruses, or corporate, marriage, and sports "wars"). Furthermore, war is often conflated with "warfare." However, as Colin Gray points out, "war and warfare are distinctive concepts, and the distinction matters greatly."[183] In fact, "warfare" pertains strictly to how wars are fought; whereas, "war" applies to the entire phenomenon, including fighting and nonfighting activities like politics, diplomacy, strategy, and logistics.

A "Typical" War

War is commonly defined as organized, large-scale armed conflict between states or factions within a state (civil war). Many wars work something like this: a political dispute triggers animosity between two states. Armed forces mobilize. One country attacks the other. Mutual fighting ensues. One side wins and the other loses, or they sign a conflict-ending treaty or armistice. Peace is restored, at least for a time.[184] Many historical examples follow this general pattern, and many do not. Fighting between peoples within a state is "civil war," which may mimic the pattern of interstate wars, like the US Civil War, or take a more asymmetric form, like an insurgency (e.g., the Afghan *mujahedeen* versus the USSR's puppet regime).

Wars do not have to follow any set pattern or rule, because humans, though adept at creating rules, are not as keen at following them. The American experience in Vietnam is a case in point. Almost everyone agrees that the United States fought a war in Vietnam, but did one side ever formally *declare* war on the other? Was this a civil war pitting South Vietnam against Viet Cong guerrillas or a conventional war between North and South Vietnam, or between the USA and North Vietnam? What role did China, Laos, Thailand, Cambodia, and the Soviet Union play in the war? Which country lost the Vietnam War, the United States or South Vietnam, or both? If the Paris Peace Accords ended hostilities in 1973, why do most histories date the war's end in April of 1975 with the fall of Saigon? The variety and complexity of war, as this case illustrates, complicates the task of creating a satisfactorily inclusive definition.

Existing War Definitions

Surprisingly, many historians and even war theorists (excepting Clausewitz) have abjured the challenge of deriving a formal definition of war. Most generic dictionary or encyclopedia definitions are variations on the same theme: war is armed conflict between states or groups within a state. Some add non-state actors and modifiers like "organized," "long duration," "large-scale," "extreme violence," et cetera. Recall our working definition describes war as a uniquely human phenomenon, a rational activity characterized by violent conflict between opposing forces intended to alter or preserve the status quo. Unfortunately, none of these definitions are sufficient because they are either unacceptably exclusive or ambiguous. For example, why limit war to states or nations when civil wars prove that non-state political groups can wage war? Must there be a threshold of lethality, destruction, duration, scale, or weaponry required for wars, and if so, how could we characterize these objectively and universally? How shall we impartially delimit conditions like "organized," "prolonged," "extreme," "intentional," or "widespread?"

A good definition answers more questions than it raises, includes all varieties of its subject, and has only enough specificity to differentiate its subject from others. Viewed another way, good definitions incorporate the subject's Aristotelian causes: material, formal, efficient, and final. Does our working definition meet this standard? Is war a uniquely human phenomenon? Yes (efficient cause). Is it a rational activity characterized by violent conflict? Yes (formal and material causes implied). Is its purpose to alter or preserve the status quo? Certainly (final cause). This definition excludes war between nonhuman species, and it does not constrain the means for making war, the duration, level of violence, or domain, nor does it limit war-making to states or nations. It diverges from other definitions by specifying that war has a rational purpose, a critical distinction that eliminates random or accidental acts of violence, though it does not remove the possibility of accidents or unintended events *leading* to war. The phrase "opposing forces" is a bit vague, for we have not yet defined "forces."

The Political Dimension

Adult man remains irrationally preoccupied with anxieties and suspicions which center in such questions as who is bigger or better and who can do what to whom.[185]
                                                                        Erik H. Erikson

Our working definition, while a fair start, requires further development relating to the final cause, war's purpose. In *Warless Societies and the Origin of War* (2000), cultural anthropologist Raymond Kelly identifies the calculus of "social substitutability" (i.e., collectively recognized and justified lethal violence against adversaries who may not be personally responsible for reciprocal actions) as a "watershed event in human history."[186] Wars wield this noncriminal, group-sanctioned violence "as *an instrument of the social group* . . for the attainment of group objectives and interests," in other words, for *political aims*.[187]

Politics and governance presume a pact between group members and leaders in which the former accept the latter's legitimacy to govern (i.e., to apportion and wield power) and agree to subordinate individual will to collective aims. In anything but pure democracy, leaders make decisions for the group; thus, choices regarding war and peace may not only reflect collective interests and influences (e.g., cultural, religious, economic, or legal) but also a leader's *personal* motivations (e.g., revenge, glory, or ambition). In either case, if the choice is war, personal or group reasons must be anchored to a cause that a critical mass of the populace, or at least the warrior caste or military, can support. Thus, as we implied earlier, since war is entwined with governance and power apportionment, its broader motives and purposes are subject to *political* processes that are eventually expressed as *policy* and *political aims*. Thus, any valid war definition must include war's *political dimension*, which Clausewitz features as a central pillar of his theory.

Clausewitz's Trinity

Clausewitz, a master of metaphor, employs imaginative and meaningful terms for war, comparing it to a duel, a wrestling match,[188] and a card game.[189] Clausewitz chose these expressions to convey specific ideas about war. These analogues pit two human, thinking sides, each with "a mutual desire or willingness" to "fight," against each other.[190] Furthermore, each side is governed to some degree by rules and influenced by chance, and their outcomes depend on skill and strategy. There is also risk – to life and limb in a duel, to honor and pride in a wrestling match, and to wealth (perhaps) or reputation in a card game. Clausewitz's more specific definition of war is the product of dialectic reasoning composed of a *thesis*: war is "an act of force to compel our enemy to do our will";[191] and *antithesis*: war is "the continuation of policy by other means";[192] leading to a *synthesis*: war is

a paradoxical trinity – composed of primordial violence, hatred, and enmity, which are to be regarded as a blind natural force; of the play of chance and probability within which the creative spirit is free to roam; and of its element of subordination, as an instrument of policy, which makes it subject to reason alone.[193]

This definition requires closer scrutiny. The first and third parts of Clausewitz's "trinitarian" definition echo our prior conclusions (i.e., that war is an extension of humanity's innately violent "primordial" nature *and* capacity for reason and creativity). Clausewitz's use of "blind" is tantamount to "devoid of rationality," which differentiates the first part of the definition from the last. But make no mistake, each individual element of Clausewitz's trinity is a necessary (though insufficient) criterion for war. War, according to Clausewitz, always combines primal *and* rational aspects. There can be no war without violence, chance, *and* reason. Moreover, Clausewitz says war's rationality serves a political purpose to which the fighting is subordinate. Here, Clausewitz identifies the pivotal attribute separating war from other violent conflict.

   Some war scholars interpret Clausewitz's trinity to mean that war "should" be subordinate to policy; however, it is improper to use modal verbs or normative language in a formal definition. Definitions are not maxims. In any case, Clausewitz's point is not that war *cannot* drive or supersede policy. Rather, war results from a human, political decision, expressed as policy (i.e., to use violence for political ends). Clausewitz's inclusion of chance and probability in the second portion of his trinity is actually superfluous since any phenomenon is subject to "the play of chance and probability." Clausewitz undoubtedly knew this; thus, its inclusion merely reflects his intent to highlight uncertainty as a major part of war's nature. As he puts it, "No other human activity is so continuously or universally bound up with chance."[194] With the trinity's second leg, Clausewitz identifies war as a "living" phenomenon rife with uncertainty intensified by creative human minds and the random perturbations of unknown, unseen factors. While he could have omitted this part and still had a commendable definition, Clausewitz wanted his theory to stand apart from those that portrayed war more as a science than an art. He also believed that chance, along with politics, helps explain why war seldom reaches its full destructive potential, a central concern of his theory.

Deriving War

War is fundamentally about humans fighting for opposing political ends. "War," says Clausewitz, "is not the action of a living force upon a lifeless mass ... but always the collision of two living forces";[195] and "Essentially war is fighting, for fighting is the only effective principle in the manifold activities generally designated as war."[196] To ensure that our definition strikes the appropriate balance between specificity and universality, we must disregard elements common to war and nonwar violence including chance, creativity, and probability; and we must remove variable character-of-war details like who fights, what they fight with, how they fight, where they fight, how long they fight, or how many are fighting. Though wars may be fought by large, ordered groups with advanced weapons across vast distances for relatively long periods, none of these qualifications are indispensable; therefore, including factors like force or theater size, weaponry, or duration is unacceptably exclusive. Moreover, while definitions based on the number of casualties or combatants may facilitate database inquiries, they are too arbitrary for theory's purposes. In fact, theoretically, a war only requires one person per side. Thus, a precise and inclusive war definition must omit terms like "group," "state," or "forces" and simply declare that war is combative violence between two opposing *sides*, which could theoretically range from a person to a multinational alliance.

It Takes Two

Every discreet war, regardless of scale or number of participants, will have exactly two sides, no more, no less. "If two or more states combine against another," writes Clausewitz, "the result is still politically speaking a *single war*."[197] The key word here is "combine." Two or more states working closely together against a common enemy become as one; however, two or more states acting independently, even if allied, essentially fight separate wars. Clausewitz uses the following example to illustrate this point: "Should Prussia be attacked by France and Russia simultaneously the effect on the conduct of operations would be as if there were *two separate wars*" (emphasis added).[198] Participants that enter an existing war (e.g., the USA in World War I) do not alter this principle, rather, their contributions create a *net* effect on the balance between opposing sides. There is no such thing as a "three-way" war. According to Thucydides, even the Peloponnesian War, with its complex web of combatants, was essentially a two-sided affair between Athens and Sparta. Moreover, Thucydides labels simultaneous combat with two opponents "double war," which implies a separate war for each adversary. However, if the enemies become allies, the result is a single war.[199] For example, despite the large "number of nationalities converging" on Sicily in response to Athens's expedition, writes Thucydides, the battle consisted of "*two sides*, against Sicily or for it, coming either to join in taking over the country or in preserving it, these were all the nationalities who fought at Syracuse" (emphasis added).[200]

To further clarify this principle of two sides, consider the following example in the hypothetical realm of "Macbethia," which consists of five states: Anger, Bass, Cello, Dulcet, and Echo. The scenario is as follows:

(1) Anger is neutral and observing a war between the Bass-Cello and Dulcet-Echo alliances.
(2) Bass and Cello share borders and have a mutual defense agreement.
(3) Echo started the war to expand its regional influence by acquiring portions of Cello's territory and resources.
(4) Dulcet is economically dependent on Echo and joined the war because it literally could not afford an Echo defeat.

During the war, Echo sinks one of Anger's commercial ships having errone-ously thought it was delivering supplies to Bass. In response, Anger sinks an Echo warship but defers from entering the broader conflict. Nevertheless, a series of escalatory military reactions leads to open war between Echo and Anger. Figure 1.2 graphically represents this dynamic.

There are now two wars: the "War of Sound" (Bass-Cello versus Dulcet-Echo) and the new "War of Fury" (Anger versus Echo). Echo is now waging two separate wars. Scenarios like this normally consolidate into a single war. With its own war aims in jeopardy, Echo's logical option is to gain Dulcet's help against Anger; and with its economic future at even greater risk, Dulcet's best interests are met by doing so. Now facing Echo *and* Dulcet, Anger has three options: sue for peace, fight alone, or join the Bass-Cello alliance. If it chooses to fight, Anger's logical choice is to join the alliance (Figure 1.3). In addition to illustrating the two-sides tenet, this example reinforces two strategic truths: it is unwise to fight more than one war at a time and to fight alone if allies are available.

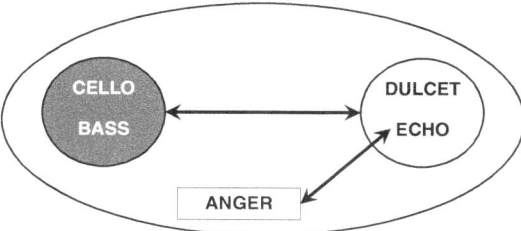

Figure 1.2 The wars of sound and fury

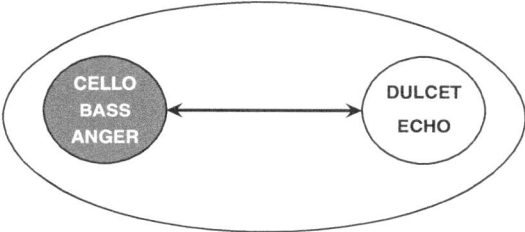

Figure 1.3 The war of sound and fury

War's Why

Generally speaking, one may say that war is politics … in its original, noninstitutional, and nonrational form – a form in which the belligerent states or parties seek by force of arms (a) to extend the territorial limits of their sovereignty and (b) to establish and impose upon the nations and peoples with whom they are in conflict a political and economic order which is in the interest of the dominant party, race, or nation.[201]   Robert E. Park

The final piece of our definition is Aristotle's final cause, war's why or purpose (i.e., a war is fought to achieve *a desired political outcome*). When opposing sides fight for political ends, regardless of motivations or other criteria, they are engaged in war. This means that war is more than just fighting. In fact, the principles of tactical combat are largely independent of purpose, which is why battle-centric strategies may not win wars. Good strategists and generals understand the relationship between war and politics and resist the temptation to equate combat acumen with victory. Achieving political objectives often takes more than military means, and successful campaigning often requires more than tactical proficiency. Consequently, superior soldiers may not make great generals, and competent generals are not always great strategists.

War Defined and the True Trinity

From all this will emerge the concept of "war as an armed contest between two independent political units, by means of organized military force, in the pursuit of a tribal or national policy."[202]   Bronislaw Malinowski

We are now ready to complete our definition, which resembles Malinowski's concept quoted here. War is uniquely human, which implies both primal emotion and rationality; it is characterized by combative violence (fighting) between two opposing sides; and its purpose is to achieve a desired (not just any) political outcome. Therefore, *war is a uniquely human phenomenon characterized by combative violence between two opposing sides for the purpose of achieving a desired political outcome.* In addition to satisfying all our previously stated requirements, this definition introduces war's "True Trinity": *Humanity – Politics – Combat.* These elements contain all our Aristotelian causes: Humanity (efficient), Politics (final), Combat (formal and material), and must always be considered integral to war. Finally, like Clausewitz's trinity, the True Trinity's three "layers" roughly correspond to the People (Humanity), the Government (Politics), and the Military (Combat), see Figure 1.4; but in reality, the human "layer" supports and suffuses the entire Trinity. Any general war theory must articulate in abstract form the indelible linkages and dynamics of these elements.

Our definition begins with war's *sine qua non*, the human element and all its implications, including rationality, creativity, and emotion. "Combative violence" (simplified to "combat" in the True Trinity) is aggressive action intended to harm people or things. Though a small matter, we use "combative violence"

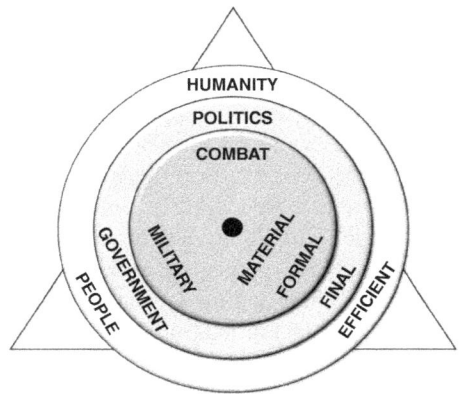

Figure 1.4 War's True Trinity

in the definition because "combat" is often associated with formal, armed forces. In war, any person willing and able to fight for a political purpose, not just military forces, can engage in "combative violence." A "side" is scale-neutral and inclusive of the full range of possible participants, including states and non-state actors. Moreover, as a side effect, this language helps debunk dubious political theories, like Marxism, which propose to permanently end war simply by dissolving states. Finally, as discussed earlier, all discreet wars have *two* opposing sides.

This definition is consistent with Clausewitz's trinity but omits chance, probability, and creativity since they are not exclusive to war. Clausewitz includes the phrase regarding the subordination of war as an instrument of policy to help answer one of his central questions: why war does not always proceed to the maximum use of force (more on this in Chapter 2). However, we do not share Clausewitz's concern here. Defining war as an activity with a political aim implies – without having to specify its subordination to policy – that war's form is influenced by the relationship between the policies it serves and the violence employed to achieve its ends, *and* that war is an *official* rather than a personal or private act.

Testing the Definition

A good definition confirms what is true about war and clarifies areas of uncertainty. The first part of the definition establishes war as an exclusively human activity distinct from other forms of nonhuman violence. The subsequent phrases position war at the nexus of fighting and politics, an activity where opposing sides, distinguishable by their internal cooperation to advance political objectives, vie for supremacy. This distinction excludes nonviolent competition as well as fighting for apolitical purposes. For instance, crime-related violence is not war *unless* the criminals have a political objective *and* their opposition reciprocates. As Gray writes, "Violence without political context can be many things (crime, banditry, sport), cultural expression even, but it cannot be war."[203]

War on Terror?

Explaining Terrorism (2011) author Martha Crenshaw defines terrorism as an activity that "involves the use or threat of physical harm in order to achieve a disproportionately large psychological effect. It is demonstrative or propagandistic violence without significant military value, directed against symbolic rather than utilitarian targets."[204] Though "terrorism" is essentially a tactic, not a political objective or philosophy, it becomes a part of war when at least one side uses it to generate psychological effects that advance a political agenda and another side opposes it with combative violence. Thus, a "war on terrorism" may be a bona fide war, though the title is a misnomer. The war cannot be on "terror" or "terrorism," because a war must be between opposing sides. It is more properly a war against politically motivated "terrorists" or "violent extremists," like al-Qaeda. Consider an insurgent group that finances criminal activity and terrorism to destabilize and overthrow a government. The conflict between law enforcement and criminals is not war *unless* the government fights back to advance its political objective (maintaining power). When Timothy McVeigh bombed a US federal building in Oklahoma City, he believed he was fighting a war against his government; in fact, he was not, because the government viewed him as a criminal, not a political threat or enemy combatant.

War on Drugs?

There cannot be a "war on drugs" any more than there can be a "war on terror." However, if we accept "drugs" as a substitute for "drug criminals," the appearance of war emerges because of military involvement, episodes of combative violence, and the political objective of reducing drug-related mayhem. But a war on drug criminals is generally not a true war for two reasons. First, governments rarely treat drug criminals as enemy combatants, and they usually place restrictions upon the military, using law enforcement personnel to make arrests versus military power to kill and destroy. Second, drug cartels usually pursue profit not political revolution. For instance, since America's "war on drugs" does not feature opposing sides using combative violence to advance political objectives, it is not a real war. Nevertheless, similar struggles can become wars when governments employ combative violence against groups that use criminality and drug profits to advance aggressive political agendas, like the Revolutionary Armed Forces of Colombia (FARC) or the Afghani Taliban.

Cold War?

Even with the aid of our definition, this one is complex. Many consider the twentieth century's Cold War – pitting the USA against the USSR – a real war. It was certainly a competition between opposing sides bent on political change. To make the call here, we need to invoke the definition's "characterized by combative violence" clause. If you count proxy wars in places like Korea, Vietnam, Nicaragua, and Afghanistan, then the Cold War satisfies the definition. But in reality, these are wars in their own right. The Cold War held few instances of direct combative violence between the Soviets and Americans. Here the definition's precise wording helps. Was the Cold War *characterized* by combative violence? It was not, hence the name "Cold" War. Therefore, the Cold War was *not* a real war. "The USA and the Soviet Union," opines historian and Clausewitz expert Raymond Aron, "[did not wage] a war in the Clausewitzian sense";[205] indeed, "However many forces may intrude which themselves are not part of fighting," adds Clausewitz, "it is inherent in the very concept of war that everything that occurs *must originally derive from combat.*"[206]

War and Peace

Your breath first kindled the dead coal of wars
And brought in matter that should feed this fire;
 And now 'tis far too huge to be blown out
With that same weak wind which enkindled it.

Shakespeare, *King John* (5.2.113)

War does not begin when some people kill others; instead, it starts at the point where they themselves risk being killed in return ... where inflicting mortal injury becomes reciprocal.[207]

Martin van Creveld

In addition to helping us identify true wars, our definition aids analysis of the boundary and transitions between war and peace, an area Galula says can be "very gradual and confusing."[208] True peace is not simply the absence of war, because violence and discord occur beyond war's bounds. Therefore, it is helpful to characterize four "zones" related to the peace-war dialectic. Zone one is *true peace*, a condition of human harmony and reciprocal altruism; zone two is *competition* or nonviolent contention; zone three is *conflict*, which includes all forms of violence short of war (e.g., apolitical or unilateral); and zone four is *war*. In other words, true peace ends where competition begins; competition becomes conflict with the introduction of combative violence; and *war* occurs when violent conflict pursues a political purpose.

All wars start with an act of combative violence for a political purpose and, as van Creveld implies, a response from an opposing side. From a strictly theoretical perspective, our definition renders declarations of war, which are merely political statements, irrelevant. The war between the USA and Japan, for example, began before Congress answered President Roosevelt's appeal for a formal declaration. It started at Pearl Harbor with the first acts of reciprocal combat. Likewise, the real war between the USA and Hitler's Germany did not begin until combat began in Africa, long after Congress approved Roosevelt's request.

The Seeds of War

There is deceit and cunning and from these wars arise.[209]

Confucius

War occurs between political entities that perceive nonviolent means of achieving political aims to be inadequate, and it starts when two sides employ combative violence against each other. Politics is often complex, always adversarial, and frequently sparks fierce debates, both domestically and internationally. Flipping Clausewitz's dictum around, politics often resembles war by other means! This competitive political arena produces factions, coalitions, and alliances with divergent objectives. Ultimately, specific political circumstances will narrow or widen the gap between peaceful processes and war. When these processes break down, either within or between political entities, the potential for war increases. As history proves, the ubiquity of inept governments and the lack of authoritative international regulatory institutions have driven unending transitions between war, conflict, competition, and peace.

Regarding war and politics, specific political goals are less important than the willingness of political leaders to risk war to achieve them. After all, policies are subject to the ruler's perceptions and beliefs, which may be based on facts and logic or whim and delusion. "Men act from immediate and interested motives," writes Sumner, which may "have nothing at all to do with the consequences of their action."[210] In fact, as a species we are often bewitched by overly optimistic views of the future, and we become convinced that our designs, including war, are destined to bring them about. Human judgment, writes Thucydides, is often "based more upon blind wishing than upon any sound prediction; for it is a habit of mankind to entrust to careless hope what they long for, and to use sovereign reason to thrust aside what they do not desire."[211] Many wars have begun from these and other characteristic misperceptions.

The War Is Over

Peace is not just the absence of war, but a virtue which comes from strength of mind.[212]
                                                                    Baruch Spinoza (1677)

Finishing off a war is even more painful and problematic than starting one . . . because the cessation of hostilities is fraught with the possibility – even the likelihood – of betrayal.[213]
                                                                    Angelo Codevilla

Denoting a war's end is more difficult than defining its start, because combat is never truly continuous. There are pauses, sometimes lengthy, between violent clashes. Until modern times, battles were fought only during a "campaigning season," when the environment was conducive to soldiers, animals, and equipment. Depending on the locale, a campaigning season could be short. Logistical, seasonal, and even pathogenic challenges can interrupt the fighting leading to prolonged wars that span decades. For instance, the Peloponnesian War lasted nearly thirty years (interrupted by a plague), and the Hundred Years' War (concurrent with the bubonic plague) actually went on for *116* years (though it is sometimes subdivided into three shorter, truce-delineated conflicts). While tempting, it is inaccurate to say that war simply ends when the fighting stops, nor can we claim that one war ends and a new one begins when political aims change. If only all wars ended with a treaty or formal surrender!

Yet, as Clausewitz stresses, "Even the ultimate outcome of war is not always to be regarded as final. The defeated state often considers the outcome merely as a transitory evil, for which a remedy may still be found in political conditions at some later date."[214] Moreover, observes Clausewitz, even after a country's armed forces are destroyed and its lands occupied, the war "cannot be considered to have ended so long as the enemy's *will* has not been broken ... not every war leads to a final decision and settlement."[215] "A peace-treaty may bring to an end [a] particular war," adds Kant, "but it doesn't end the state of war, i.e. the state in which new 'reasons' for hostilities can always be found."[216] Examples include England's first civil war (1642–5) and Napoleon's occupation of Spain (1808–13). Of the first example, Liddell Hart says, "When the war ultimately flickered out in 1646, [it] left the Royalist embers still so numerous and so glowing that, with the aid of discord among the victors, the flames burst out afresh two years later in a greater blaze than before."[217] In the second case, the resistance (guerrillas) kept fighting until the occupiers had fled. Liddell Hart's "fire" metaphor is especially apt. As Sun Tzu affirms, "War is like unto fire; those who will not put aside weapons are themselves consumed by them."[218] Like fire, war can start accidentally or on purpose. Once begun, it spreads, often out of control, and becomes far more difficult to extinguish than it was to spark. Even though the flames of combat may subside, the political "heat" that ignited the fire usually lingers. Given the right tinder, the flames of war can reignite causing the hot cinders to rise into the wind and start new blazes elsewhere.

Victory?

Everything depends on the vanquished foe, how well the victor understands it, and how competently the victor deals with it.[219]   Angelo Codevilla

A war is definitively ended only when peace is ratified by the sovereigns on both sides.[220]   Samuel von Pufendorf (1673)

Clausewitz calls victory "that golden prize ... the fruit that quenches the thirst of ambition."[221] But as the protagonist in Brandon Sanderson's *Oathbringer* laments of war, "there are no unequivocal wins. Just victories that leave fewer of your friends dead than others."[222] Although the philosophical debate over whether anyone really wins a war goes on, in terms of strategic logic, peace can be maintained and wars ended when *all* the relevant "political actors" resolve to achieve their ends in ways that political scientist John Vasquez says are "more efficient, less costly, and more legitimate" than combat.[223] If only one side forms this conclusion, then it must either convince the other to agree, fight on, or surrender.

In the simplest sense, wars end with a victory and a defeat, or (as the previous example implies) when *both* sides agree to stop fighting. In other words, war and peace are equally reciprocal states. Victory cannot be created by one side, asserts Sun Tzu, because sustained resistance is always an option for an opponent who retains fighting capacity.[224] According to Clausewitz, victory in battle has three elements: the enemy's greater loss of (1) material strength and (2) morale; and the enemy's (3) admission of (1) and (2) by giving up its intentions.[225] Though strength and morale also apply to the broader war, the final element (giving up intentions) is most important provided it is followed by actions confirming acceptance of defeat. As Clausewitz makes clear, at the tactical level, "abandonment of the fight remains the only authentic proof of victory."[226] Extrapolated to the political level, this means that only the opposition's unequivocal abandonment of the political goal is "authentic proof of victory." Thus, a war's end is always to be found *beyond the military instrument* and *the fighting*. This might seem paradoxical, but it makes perfect sense vis-à-vis war's relationship to politics and the distinction between war and warfare. As Liddell Hart writes of the victors after World War I:

They did not look beyond the immediate strategic aim of "winning the war," and were content to assume that military victory would assure peace – an assumption contrary to the general experience of history ... pure military strategy needs to be guided by the longer and wider view from the higher plane of "grand strategy."[227]

And defense expert Fred Charles Iklé adds, "The political struggle within each country affects everything that matters in ending a war. It intrudes into the formulation of the war aims, it colors and even distorts military estimates, and it inhibits negotiation with the enemy."[228] Indeed, ending a war is almost *always* more complicated than beginning one. Perhaps the Sun King, France's Louis XIV, is history's poster child for this truth. Throughout his unusually long reign (1643–1715), writes John Lynn, "none of his four wars went as expected," and he made numerous "miscalculations ... bumbling into wars he did not anticipate," staying in wars too long, and ending most out of "sheer exhaustion" rather than the fulfillment of his original political objectives.[229] One can almost hear the minstrel's lyric:

> *Oh, what a noxious brew we pour*
> *When we set off to begin a war.*
> *But fouler yet may the ending be,*
> *For those who survive long enough to see!*

But wars do end, and we can characterize when this occurs by reversing our definition, which yields the following: war ends *when opposing sides cease fighting for their political objectives.* Consider the American Civil War. History records the war's end on April 9, 1865, with Robert E. Lee's surrender at Appomattox Court House; however, there were still active units fighting for the political cause after that date. Though there was no formal government surrender or treaty marking its end, Lee's capitulation crushed the South's will to continue, effectively settling the political question and the war. Fighting after the political settlement has the appearance of war but is no more war than the autonomic spasms of a corpse are indications of life. The Battle of New Orleans (from the War of 1812) is a case in point. It took place in January of 1815, *weeks after* the Treaty of Ghent formally ended the conflict.[230]

In the final analysis, ending wars will continue to be difficult and hard even to recognize, for the line between war, conflict, and competition can be quite thin. Moreover, after a war concludes, as Freedman says, "strategy will not stop. Victory in a climactic event such as a battle, an insurrection, an election, a sporting final, or a business acquisition will mean a move to a more satisfactory state but not the end of struggle."[231] These are realities all leaders and strategists should consider soberly before starting a war.

First Look: Are Wars Inevitable?

While we know that wars end, have we cause to believe that they *all* may end someday? There is certainly more to humanity than war, and in the right circumstances, we can live for long periods in peace. However, history proves that even generations of peace have not prevented the *possibility* of war. Does this mean wars are inevitable? Though we will tackle this question in greater depth in Chapter 5, we can pursue a preliminary answer now by using our definition to analyze the feasibility of the opposite condition (i.e., a world *without* the possibility of war). For now, we will hold human nature (and existence) constant; and human alterations (i.e., remaking *Homo sapiens*, like Asimov's robot, with a nonviolence prime directive) are prohibited.[232] Our definition – *war is a uniquely human phenomenon characterized by combative violence between two opposing sides for the purpose of achieving a desired political outcome* – essentially says that war is reciprocal violence for political reasons, which implies two potential remedies: (1) remove the *reasons* for war, thus preventing the *formation* of opposing sides or (2) remove the *means* for turning reasons into combative violence. The first option is *utopian* (abolishes motivation) and the second is *totalitarian* (denies means).

The utopian option creates a society where everyone is satisfied and therefore peaceful. Here, war is unknown because there are no *reasons* to fight. The people are either content with the status quo or with the efficacy of nonviolent means for altering it. Think Huxley's *Brave New World* (1932) minus the dysfunction. But this solution is clearly idealistic and impractical given humanity's size and diversity. Even the word "utopia" [from Thomas More's *Utopia* (1516) about a fictional, idyllic island society] means "no place" in the original Greek.[233] To paraphrase a well-known adage, *you can please all people some of the time, or some people all of the time, but not all people all of the time*. There can be no reasonable expectation of satisfying all the needs and aspirations of humanity for all of time, and we are notoriously inept at governing. Consequently, for other than small, isolated populations and limited time periods, the utopian option appears unrealistic.[234]

Option two, a totalitarian state [evocative of Hobbes's *Leviathan* (1651) or Orwell's *1984* (1949)], prevents war by physical and psychological restraint, repressing human nature and maintaining order through coercion and violence. In this environment, writes Galula, "political opposition is not tolerated and ... the population is kept under a system of terror and mutual suspicion," within which organized resistance "has no chance to develop."[235] Totalitarianism is more plausible than utopianism because it is simply easier to restrict means than to satisfy wants and needs.[236] Whereas *restraint* can be imposed with a finite variety of methods and means, *gratifying* every person presents an insurmountable challenge. For this and other reasons (e.g., the unsuitability of uneducated lower classes to effectively contribute to political decision making), autocratic and totalitarian regimes have been more common historically than utopias. Examples include many early civilizations and more recently, Nazi Germany, the Soviet Union, and North Korea. As with utopias, pure totalitarianism only succeeds for limited periods and in favorable circumstances, and it frequently fails with respect to external relations. Ultimately, the practical difficulties of absolute control by a single authority for all of time make the totalitarian option unattainable as well.

Societies that have presided over long epochs of relative peace, like Rome, gravitate toward "waypoints" between the extremes of utopian and totalitarian forms of governance (e.g., democracy and autocracy). These intermediate governmental forms provide enough fulfillment or control to minimize the potential for civil war. Limiting the possibility of external conflict is harder; however, a strong, defensively oriented military guided by a principled political vision and shrewd diplomacy can be effective. Still, indefinitely maintaining the balance needed to prevent all wars appears unrealistic and drift between the enticement of utopianism and the iron boot of totalitarianism inevitable. We will examine these themes further in the final two chapters.

Now, thanks to our enhanced appreciation for war's relationship to nature and humanity, the True Trinity, and our definition for war, we are ready for the next chapter, which addresses war theory's accessibility deficiency by introducing, and in some cases, critiquing war's greatest theorists and theories.

2   The Masters of War Theory and Strategy

Every art has its rules and maxims. One must study them: theory facilitates practice. The lifetime of one man is not enough to enable him to acquire perfect knowledge and experience. Theory helps to supplement it, it provides a youth with premature experience and makes him skillful through the mistakes of others.[1]   Frederick the Great

To paraphrase Newton, we see farther because we stand on the shoulders of giants. The timelessness of war theory is compelling evidence for the validity of war's fundamental maxim, that war's essence, based in human nature, is unchanging. This chapter's principal theorists – Sun Tzu, Thucydides, Machiavelli, Jomini, Clausewitz, Liddell Hart, and Mao – remain relevant by offering more than specifics on raising armies and fighting battles. They were students of history, experienced warriors, and experts on war and human nature. Figure 2.1, depicting the relative contributions of this "magnificent seven" and other theorists covered in this book (measured by text references), illustrates our debt to these exceptional minds. As indicated, the extraordinary Clausewitz is *primus inter pares*.

This chapter's purpose is not to explore *everything* there is to know about each theorist. Other biographical and historical works, like Paret's *Makers of Modern Strategy* (Princeton, 1986), Gat's *A History of Military Thought* (Oxford, 2001), and Beatrice Heuser's *The Evolution of Strategy* (Cambridge, 2010), cover that ground in finer detail. Rather, our goal, as the Preface states, is to render these theorists and their theories *more accessible* to the warrior, strategist, and policymaker. To borrow from Alfred Thayer Mahan, while the historical facts are meaningful, no less essential are the "structure and content and accessibility to readers."[2]

We have selected these particular theorists because, collectively, their philosophies are the most consequential and enduring, and they exemplify the full arc of relevant military thought from the classical era through the twentieth century. Chapter 3 will complement and extend the systems offered here by focusing specifically on "small wars" and domain theory (e.g., maritime and airpower). Canvassing a variety of eras, domains, and perspectives establishes a more complete picture of theory that helps us see the whole forest of war and not just the trees.

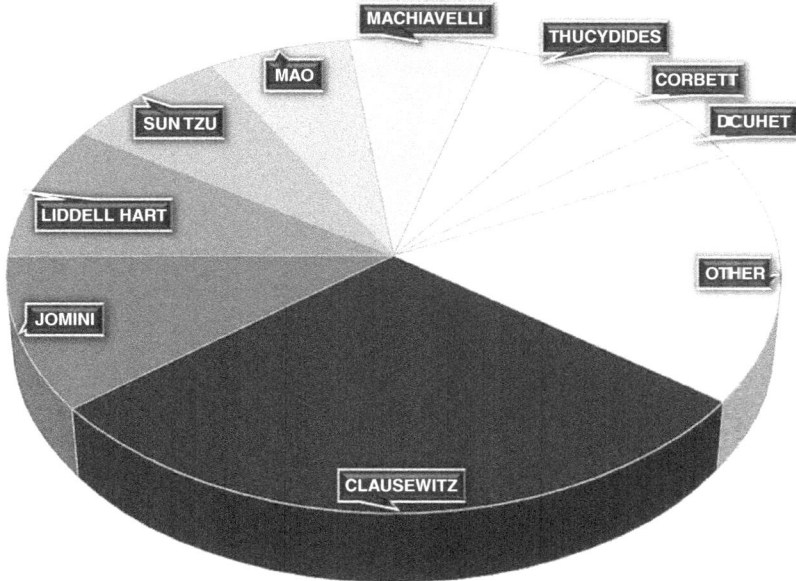

Figure 2.1 Theorist contributions

SUN TZU

Sun Tzu (literally "Master Sun") is synonymous with ancient wisdom, particularly regarding war and strategy; however, his universality has led contemporary authors to adapt and apply his maxims to genres beyond war, including sports and business. The common factor in these ventures is the human dimension, something Sun Tzu understood better than most and why his ideas are likely to remain relevant indefinitely.

On Sun Tzu

Who led a mass of thirty thousand and no one under Heaven opposed him?
Sun-tzu.[3]   We Liao-Tzu

The Chinese leader (Sun Wu) and events that inspired Master Sun's *The Art of War* likely date to an epoch just prior to the "warring states" period (475–221 BCE), and in truth, it is improbable that the whole work is attributable to any single author or theorist. However, Sun Tzu's biographical particulars are not as important as *The Art of War*'s adages, which indicate an intimate familiarity with war worthy of study and reflection. Master Sun's approach, as with poetry, employs economical phrasing that nevertheless retains surprising nuance and depth. With Sun Tzu, there is far more than first meets the eye.

First Principles

Most of Sun Tzu's philosophy rests upon his two most well-known aphorisms:
(1) Know the enemy and know yourself: in a hundred battles you will never be in peril.[4]
(2) To subdue the enemy without fighting is the acme of skill.[5]
These may appear simplistic or even nonsensical. Fight a hundred battles without peril? Subdue the enemy without fighting? What nonsense! However, when considered within the broader context of war and in relation to his other proverbs, their deeper significance emerges.

Know Your Enemy

What does Sun Tzu mean by "know the enemy and know yourself?" Though he begins with "know your enemy," we cannot infer that the former directive is necessarily more important than the latter. Indeed, both are required to realize the advantage: safety from peril. These benefits – victory in battle and safety from danger – confer remarkable potency. Within his philosophy of war, this is Sun Tzu's ultimate imperative, a starting point from which all other considerations follow, even the decision to make war.

To fully know oneself and one's enemy is a practical impossibility. But Sun Tzu, employing a characteristic technique, establishes an *ideal* for which the strategist and warrior strives though never fully attains. The second part of the maxim, "in a hundred battles you will never be in peril," cleverly hints at his intent. Though the statement seems absurd given war's obvious dangers, as an ideal, it effectively delimits Sun Tzu's concept of war. Victory depends upon a thorough, ongoing familiarization with the strengths, weaknesses, capabilities, deficiencies, motivations, and fears of friendly *and* enemy forces. This task never ends nor is perfection attainable; however, peril decreases and the odds of success increase in proportion to the accuracy and thoroughness of one's knowledge of both sides. Knowing the enemy in the fullest sense is to *become* the enemy, in effect, gaining the ability to see through the enemy's eyes, and thus, to *anticipate* his behavior. World-class chess players use this technique. By studying an opponent's habits and the game's rules, they "play the other side," anticipating likely gambits and devising countermoves.

Of course, war is more complex and dynamic than chess and, according to Sun Tzu, cannot be restrained by rules except the laws of nature and the indeterminate limits of human ingenuity. "In the art of war," says Sun Tzu, "there are no fixed rules."[6] Though just the first four moves in chess have nearly 320 billion possible combinations, in war, the possibilities are virtually limitless: so vast, says Sun Tzu, "None can comprehend them all."[7]

Obviously, information gathering and processing, or *intelligence*, is critical to Sun Tzu's theory; thus, it is no accident that he includes an entire chapter on secret agents. For Sun Tzu, enemy intelligence is essential to the successful conduct of war because it helps reveal the relative capabilities of opposing sides and forms the basis for strategy (i.e., identifying ways and means for achieving victory). This is why most countries develop and field advanced means and methods (e.g., spies, satellites, and surveillance aircraft) to gain insights into current and potential enemies.

Know Yourself

The second part of Sun Tzu's imperative, "know yourself," requires no less attention than the first and is arguably more difficult. Accurate, honest self-perception and thoroughness in self-assessment are *not* hallmarks of human nature. We often see in ourselves and our plans what we wish to see, not what is truly there. Emotion and war's uncertainty distort judgment. Even setting aside our penchant for self-delusion, knowing ourselves is not a trivial task, particularly regarding psychology. Strategists and warriors who fail to surmount these difficulties risk dangerous dialectic imbalances, especially regarding prudence–boldness, regular–irregular, and defense–attack.

Relativity

Sun Tzu's philosophy of war is reciprocal and relativistic (i.e., war is a battle of opposing sides), where each side's capacity for war *relative* to the other affects its conduct and character. Sun Tzu's maxims suggest that assessing the characteristics of opposing forces *relative* to each other is a prerequisite for good generalship. He writes: "Now the elements of the art of war are first, measurement of space; second, estimation of quantities; third, calculations: fourth, comparisons; and fifth, chances of victory."[8] Thus, the assessment of relative capabilities (estimation of quantities, calculations, and comparisons) folds into the overall context (space) to provide a sense *ahead of time* of the odds of success. If those odds appear prohibitive, the weaker side may have to avoid or defer costly battle, at least until conditions become more favorable. The commanding general who knows himself and the enemy is best positioned to formulate and execute an effective strategy. In the hypothetical instance that both sides are identical in every respect and all contextual factors are balanced, there can be *no* effective war strategy. Like sine waves 180 degrees out of phase, the two sides cancel each other out, creating a stalemate. Mutual destruction would be the only possible outcome if they opt to fight. However, in reality no two warring sides can ever be completely equal. There are always differences in training, leadership, resources, allies, equipment, experience, goals, and so forth. Therefore, the general that *best* evaluates relative strength gains a strategic advantage and is more likely to select effective courses of action.

Deception

Much of the remainder of Sun Tzu's theory addresses how to fight once the relative capability assessment is complete. For example, his assertion that *all warfare is based on deception*, while perhaps arguable, relates back to this theme.[9] Implicit to deception is the careful consideration of the enemy, both physically and psychologically. True deception is a human, rational, and counter-rational act with elements of art and science, not unlike war itself. Deception is counter-rational because it turns an enemy's rationality into a liability by introducing circumstances that lead to improper decisions. Deception targets decision makers, directing them toward choices that seem beneficial but are, in fact, detrimental. Because the target is a human mind, deception cannot be fully effective without insight into how the target thinks. Deception poisons half the opponent's recipe for victory in war – know the enemy – thus increasing an adept deceiver's odds for success. "I make my enemy see my strengths as weaknesses," writes Sun Tzu, "and my weaknesses as strengths ... I conceal my tracks so that none can discern them."[10] Deception's versatility – it can be active or passive, offensive or defensive, complex or simple – can frustrate an enemy's tactical and strategic designs.

Attacking Strategy

In war, strategy is Sun Tzu's highest priority target. Ranking the psychological over the physical, he explicitly states that generals should first attack the opposing side's *strategy*, then *alliances*, then the *army*, with *cities* as a last resort.[11] Sun Tzu's emphasis on strategy is another consequence of his prime maxim, "know the enemy, know yourself." The general must think like the enemy and answer the questions every commander must ask, first of himself and then in the role of his foe: What do I know of my forces? What are my objectives, and how will I achieve them? What does the enemy know, and what is the enemy likely to do? "Attacks" on strategy and alliances, therefore, are not necessarily physical. As his second maxim indicates, Sun Tzu places great value on the ability to "win" without fighting, to defeat the enemy *before* battle through superior preparation. Sun Tzu's skillful general deduces the opponent's strategy and exploits weak points or vulnerabilities via maneuver, deception, or even diplomacy. The greatest gain at the lowest cost occurs when one side's strategy negates the opponent's, thus degrading or disabling the enemy's mechanism for victory. When one side observes its plans foiled at every turn and sees its strategy collapse, the will to fight often follows. This happened to France in 1940 when Germany sidestepped the Maginot Line and Guderian's panzers drove to Paris.

Sun Tzu's second priority, attacking alliances, complements his first and is most appropriately directed toward political leadership. Alliances can be decisive in war, and the ability to use nonviolent measures to coerce, co-opt, or influence an enemy's allies to stay out of a fight or switch sides can create or preserve an insurmountable capacity mismatch. For instance, in the Peloponnesian War, Athenian and Spartan coercion and diplomacy swayed numerous regional powers to pick sides, playing a pivotal role in the war's conduct and outcome. In the American Civil War, Lincoln's Emancipation Proclamation crippled the Confederacy's hope of an alliance with European powers, who chose to remain neutral rather than aligning with the proslavery South.

Sun Tzu ranks the opposing army as the third priority, not because he feared battle, but more likely out of respect for the *challenge, unpredictability,* and *peril* of combat. Physical force-on-force fighting is always costly, and circumstances beyond control or prediction, like chance and human error, can nullify plans. Moreover, victory in battle, with all its uncertainty and danger, may still fail to produce the desired political outcome. For instance, in Vietnam, both France and the USA enjoyed military superiority, yet neither gained its ultimate objective.

When battle cannot be avoided, Sun Tzu offers the following guidance: with a strength advantage of 10–1, surround; 5–1, fix the enemy in the front and attack from multiple sides; 2–1, divide the enemy to attain a more favorable, localized strength ratio; equal, engage with the caveat that the better army and general will have the advantage; inferior but having the advantage in moral strength, training, organization, and so forth, attack; otherwise, withdraw and avoid until conditions improve.[12] Sun Tzu equates "strength" to numbers, but he understood there is more to combat than the relative size of the armies, and size advantages do not guarantee victory. Even with a 2–1 advantage, he advises generals to seek greater margins, and for less favorable numerical odds, there are numerous moral and qualitative factors that may prove decisive.

Finally, Sun Tzu considers assaults on cities the riskiest option. This is a surprisingly contemporary perspective on an aspect of war that is often misconstrued. Historically, strategists have placed great value on the capture or destruction of an enemy's population centers, especially political capitals, but this practice has proven less decisive than one might expect. In any age, laying siege to a city is a difficult, time-consuming action that can lead to stagnation and leave an attacker isolated and vulnerable. Moreover, fighting *within* cities is always dangerous. Attacking cities can personalize and protract a war by steeling the inhabitants' will to fight. In short, the value of attacking cities is seldom worth the cost. Sun Tzu knew this long before Napoleon took Moscow, the British burned Washington, DC, the Americans bombed Hanoi, or the Germans occupied Paris. In none of these instances did the attacker go on to win the war.

Force Fluidity

Sun Tzu's philosophy of warfighting rests largely upon the concept of combat flexibility. A rigid force, prepared for only a narrow set of circumstances, can rapidly find itself at a disadvantage in the unpredictability of combat. A wise general prepares for a wide range of conditions and trains forces to transition, as required, to an optimum disposition depending on the situation. This appreciation for the *fluidity* of combat is represented by Sun Tzu's use of a water metaphor:

An army may be likened to water, for just as flowing water avoids the heights and hastens to the lowlands, so an army avoids strength and strikes weakness. And as water shapes its flow in accordance with the ground, so an army manages its victory in accordance with the situation of the enemy. And as water has no constant form, there are in war no constant conditions. Thus, one able to gain the victory by modifying his tactics in accordance with the enemy situation may be said to be divine.[13]

Sun Tzu's "divine" army instantly adopts the most advantageous posture based on perfect knowledge. When a powerful enemy prepares to strike, dispersal minimizes the effect. If the enemy is weak or divided, concentrating and striking is the better tactic.

Direct (*cheng*) and Indirect (*ch'i*)

In his chapter on "energy," Sun Tzu introduces a concept consistent with Taoism, force fluidity, and the direct–indirect dialectic. He says warring forces can be of two kinds: *cheng* – normal (or direct); and *ch'i* – extraordinary (or indirect). Sun Tzu writes:

The force which confronts the enemy is the *normal*; that which goes to his flanks the *extraordinary* ... I make the enemy conceive my *normal* force to be the *extraordinary* and my *extraordinary* to be the *normal*. Moreover, the *normal* may become the *extraordinary* and vice versa ... for they end and recommence; cyclical, as are the movements of the sun and moon. They die away and are reborn; recurrent, as are the passing seasons (emphasis added).[14]

Successful generals use *cheng* and *ch'i* as circumstances warrant. The normal force "fixes or distracts the enemy," while the extraordinary forces "act when and where their blows are not anticipated."[15] If the *ch'i* is anticipated, it must become *cheng* and vice versa. This confuses an enemy while improving effectiveness and reducing risk.

Victory without Fighting

The second of Sun Tzu's principal maxims, *the acme of skill is the ability to win without fighting*, is also the most controversial and misunderstood. This guidance is not an affirmation that war can or ought to be bloodless. Rather, as with the previous maxim, Sun Tzu has defined an absolute intended to ensure generals remain mindful of war's *purpose*. Though Sun Tzu firmly believed that generals and not political leaders should be responsible for fighting wars, this adage illustrates his appreciation for war's role as an instrument of policy, which ultimately seeks to create a better peace. War's *object* is not the fighting. Military victories are most effective when they contribute to a political victory. Devising a strategy to realize the political end, and monitoring progress toward it, requires the end state be *defined at the outset* (though it may evolve with time). Therefore, the acme of skill (military or political) is to define what victory looks like and then achieve it as rapidly as possible with the least expenditure of effort and resources.

To better understand Sun Tzu's reasoning, consider the following: the goal of politics is to employ ways and means to achieve a desired political aim, and the goal of war is to employ the ways and means of combat for the same purpose. Political or martial perfection is the ability to realize the desired aim instantly and without cost. Of course, achieving objectives this efficiently is impossible, but there is a method to Sun Tzu's madness. This maxim articulates an ideal that indicates the supremacy of the political objective. Here Sun Tzu demonstrates awareness of war's creation–destruction dialectic, which implies that the aim of war should *not* be destruction for destruction's sake nor simply to prove one's martial prowess (a popular motive in Sun Tzu's time). Rather, war destroys to create a more favorable peace. A conquered enemy is more a penalty than a prize if left burnt, bereft, and embittered. For a case in point, contrast the treatment and dispositions of Germany and Japan after World Wars I and II.

Another implication of Sun Tzu's "win without fighting" dictum is the significance of what happens before battle, including strategy, preparation, logistical support, and maneuver. For example, deterrence (i.e., convincing a potential enemy of the futility of aggression) has the potential to compel an auspicious political reality without fighting. Sun Tzu says, "Those skilled in war subdue the enemy's army without battle. They capture his cities without assaulting them and overthrow his state without protracted operations ... your aim must be to take All-Under-Heaven intact."[16] The greater the skill and capability, the greater the deterrent effect. If deterrence or intimidation alone cannot achieve the aim, the general who is best prepared to fight and keeps the political object in mind has the highest likelihood of succeeding at the least cost. Sun Tzu challenges his readers to recognize that the ability to achieve desired ends in war depends upon careful selection and execution of the ways and means. Tactics and methods chosen only to destroy or win battles could make political victory harder to achieve, not easier. The French military learned this lesson in Algeria, where their heavy-handed methods against National Liberation Front insurgents, including torture and reprisals, achieved tactical gains but ultimately produced a strategic failure.

Conclusion

Sun Tzu's appreciation for war's fluidity, unpredictability, and reciprocity and the resulting premium placed upon good intelligence, sound strategy, and the use of war to pursue well-defined political objectives continues to resonate. He believed the true art of war could only be found by assessing the mind and matter of each side in a conflict and using that data to inform strategic and tactical decisions before and during the war. Furthermore, Sun Tzu prioritized war's psychological component over the physical, highlighting the benefits of pursuing the political objective rather than fighting for fighting's sake. Finally, Sun Tzu established the importance of assessing relative force capacity – which includes fighting ability as well as force size – and context in formulating strategy and preparing and employing forces in war. Many of Sun Tzu's considerable insights into war's nature, especially pertaining to fluidity, *cheng–ch'i*, and the implications of relative capacity assessments, feature prominently in Chapter 4.

THUCYDIDES

Thucydides and Sun Tzu were not quite contemporaries. Thucydides lived roughly a half-century after Master Sun, but there is no evidence that he had any knowledge whatsoever of Chinese history or warfare. While Sun Tzu informs his audience about the nature of war and how to wage it from a theorist's perspective, Thucydides is more a warrior-historian of the Peloponnesian War, chronicling the words and deeds of major participants to illuminate the upheaval from *more* than just a military perspective. Thucydides is unique among our masters of war theory because his purpose is *not* to evoke timeless maxims for strategy and warfighting. Rather, he focused on portraying war's human dimensions, its calculus and character, and the politics and pitfalls of alliance and leadership. His success in this respect has rarely been equaled.

On Thucydides

As a historian, Thucydides exemplified the enlightenment spirit, describing conflict in unsentimental and calculating terms.[17]   Lawrence Freedman

Thucydides was an Athenian historian and general who lived in the latter portion of the fourth century BCE, and his *History of the Peloponnesian War* is among the earliest historical accounts of war. Only works like Homer's *Iliad* and *Odyssey*, *The Histories* of Herodotus, and the Old Testament are earlier. As with Herodotus, there is disagreement about whether Thucydides was a true historian, with some scholars doubting his objectivity. Ironically, Thucydides believed his work might be too bland, saying, "The absence of romance in my history will, I fear, detract somewhat from its interest."[18] In any case, our purpose is not to parse Thucydides for historical inaccuracies. The Peloponnesian War was real and so is Thucydides's incisive perspective.

Just as Darwin had the Galapagos Islands to accelerate his scientific insights, Thucydides had the Peloponnesian War (431–404 BCE), a titanic clash primarily between Sparta of the *Peloponnesian League* and Athens of the *Delian League*, to stimulate his thoughts on human conflict.[19] Thucydides's interpretation is so valuable because of its portrayal of the human condition within the context of war, from the strategic level to "the psychological and physical terror and confusion of the battlefield."[20] Thucydides's war emerges as a profoundly human phenomenon reflecting our greatest and basest qualities and a violent enterprise inseparable from the personalities and politics of its era. During Thucydides's age, Athens, a prototype democracy with a superior navy, and Sparta, renowned for its warrior ethos and brutally disciplined infantry, were the two dominant powers. Leading up to the war, these city-states enjoyed a relatively collegial relationship, but peace disintegrated as leaders and polities on both sides chose war on the basis of fear, honor, and interest.[21]

Thucydides's Trap

An important concept here is what political scientist Graham Allison calls "the Thucydides Trap," an idea that emerges from Thucydides's explanation for the war's "inevitable" cause: "the growth of the power of Athens, and the alarm which this inspired in Sparta."[22] Sparta's *fear* of Athens's growing power motivated them to attack. Thus, Allison's "trap" occurs when an existing power feels compelled to act before a rising power achieves supremacy.

Essentially, Thucydides's Trap is a set of conditions that lead to political insecurity.

For Athens and Sparta, there were actually *two* traps, which are just as apropos today as they were then. The first is the belief that war is an acceptable, and even noble, instrument for realizing political ends. This "traps" ambitious nations into a cycle of competition, conflict, and war for supremacy in power, security, and glory. The second "trap" is more complex and relates to great powers and alliances. This trap is primed when strong states pledge support to lesser powers. During the Peloponnesian War, both Sparta and Athens built influence by forging alliances with other city-states. Thucydides introduces the principle of "the enemy of my enemy is my friend" by stating that, in alliances, the most important shared interest and basis for trust is a common enemy.[23] Alliances are rarely static, and it is not unusual for nations to align with former or future adversaries at some point. To paraphrase Shakespeare, war and politics make for strange bedfellows (*The Tempest*, 2.2.37).[24] For example, the USA joined with historic enemy, England, to fight both World Wars, then collaborated with its adversaries from World War II – Japan and Germany – to oppose Soviet communism and global terrorism.

The alliance-based aspect of Thucydides's Trap is triggered when weaker allies require help, as occurred in the Peloponnesian War when Corinth (allied with Sparta) and Corcyra (allied with Athens) started to fight. The Corinthians entreated Sparta to avenge a Corcyran injustice. Though Sparta initially declined for fear of provoking Athens, the Corinthians argued, "Peace is the most lasting for those who not only use their power justly but also show a clear determination not to submit to injustice."[25] Sparta changed its stance, writes Thucydides, "not so much because they were persuaded by the arguments of their allies as because they feared further increase in the power of the Athenians"; thus, solidarity with Corinth provided a convenient excuse for violently confronting Athens's ascendency.[26] With Corcyra in direct danger from Sparta, Athens was "trapped" into reciprocating. Thus, the two powers fought, not because of direct animosity, but out of *fear* for each other's strength and to *honor* a commitment to allies.

The Plight of the Weak

Another of Thucydides's lessons pertains to power and morality in international relations. As the power of Athens and Sparta grew, regional polities realigned like iron filings around the poles of a magnet.[27] The "love of profit," writes Thucydides, "caused the weaker to submit to the domination of the strong and the more powerful, with their abundant wealth, to make the smaller cities subject to them."[28] Wealth and power often go hand-in-hand, and the powerful frequently use their advantages to impose their will, even by force. This is exactly what Athens does to a weaker Spartan ally, Melia. "Right, as the world goes, is only in question between equals in power," asserts an Athenian delegation to the Melians, "while the strong do what they can and the weak suffer what they must."[29] In this "Melian Dialogue," Athens asserts that superior power abrogates moral obligation, in effect granting license to do as they please, while the Melians "suffer what they must." Principles of morality and law, absent an enforcement mechanism, cannot reliably govern human affairs, especially in war, for "those who are in a position to use force have no need for legal procedures."[30] Indeed, the Athenians affirm the futility of appealing to justice, "an argument that never yet, when there was any opportunity to gain something by might, deterred anyone who propounded it from taking advantage."[31]

Certainly, humanity can do better than to arbitrate all its affairs by the sword, but the threat is always possible without a balancing power. In fact, though the phrase "balance of power" appears well after Thucydides's time, we can trace its philosophical roots to the Melian Dialogue.[32] The strong do what they can, to include making war, unless deterred or stopped by those as strong or stronger; thus, a balance of power can maintain the peace by placing in doubt the probability of successfully attaining political goals through violence.

Fighting for Peace and Justice

For Thucydides, justice is consistent with upholding honor and pursuing interest, though, as the Melian Dialogue illustrates, power can corrupt justice into a form of "might makes right." To help secure the advocacy of their citizens, Athenian and Spartan leaders often cloaked their aggressive designs in the garments of righteousness, defending war as a distasteful but necessary means of protecting rights and values and punishing the unjust. In Thucydides's view, war should be made, "not with fear of the immediate danger but with longing for the lasting peace that will come of it; for after war, peace is more strongly confirmed, while refusal to leave inactivity and go to war is less free from risk."[33] This is consistent with "right intention" in "just war" theory (i.e., just wars must support the common good, not personal agendas). Through Pericles's oration to Athens, Thucydides introduces another just war principle, the right and obligation to self-defense. He writes: "For going to war is great folly for those whose good fortune gave them a choice; but when it was necessary either to become the subjects of others by yielding or to prevail by taking risks, the one who shuns danger deserves condemnation more than the one confronting it."[34]

The Curse of Success

Regardless of a war's initial rationale, its effects on human psychology can alter its conduct and object. Athens eventually lost sight of the peace it had originally pursued, became overconfident, and acted rashly in a quest for greater glory. This overreach ultimately led to Athens's downfall. Thucydides underscores this human inclination that can make success in war almost as dangerous as adversity, saying:

> It is the rule that those cities that encounter the greatest good fortune with the most surprising suddenness are drawn toward insolence; in general, success is safer for men when in accordance with their calculations rather than contrary to their expectations, and it is easier ... for them to fend off adversity than to hang on to prosperity.[35]

When leaders deal imprudently with success, they invite reversal, a psychological "Thucydides Trap" that has enriched many casino owners! To dodge this pitfall, strategy must balance the prudence-boldness dialectic and account for the possibility of good fortune as well as setbacks.

Momentum

Though he does not use the term "momentum," Thucydides recognizes the human proclivity for allowing past results and trends, positive *and* negative, to influence decisions and behaviors. In war, this leads to positive or negative momentum, which can prevent leaders from making prudent judgments as circumstances change. Momentum may lead us to reach for what we cannot grasp or blind us to what we can. Thucydides writes: "It is impossible, and very foolish for anyone to believe that when human nature is eagerly pressing toward some accomplishment, there is some deterrent to stop it by force of law or by any other threat."[36]

Momentum is largely psychological and amplified in large groups where, as discussed in the previous chapter, individuals subordinate personal morality or inhibitions to the leader or collective. Momentum produces and sustains powerful emotions that cloud reason and affect a war's political aims. Thucydides illustrates the peril of "mob" anger in a speech by the Athenian leader, Diodotus, who implores Athens to moderate its response to a rebellion lest their draconian measures create a violent backlash or drive their allies to switch sides.[37] Diodotus knew that the short-term gains realized by employing violent retribution create long-term problems by hardening the enemy's will and producing new rivals. As Thucydides's Plataians say in the face of Sparta's wrath, "The brief action of destroying our lives brings the laborious one of wiping out the infamy."[38] Thucydides, like Sun Tzu, emphasizes the importance of pursuing the political goal despite the temptation to make fighting and retribution their own ends.

Complexity and Emotion

Despite his lack of exposure to a wider variety of conflicts, Thucydides appreciated war's multifaceted nature; and, though he does not elaborate on war's forms, he does say Athens experienced "every variety of war."[39] Thucydides also recognizes – centuries before Clausewitz and von Moltke (the Elder) – the gulf between plans and combat caused by chance and war's complexity. "Everything is uncertain in war," proclaims Archidamos of Sparta, "and attacks usually come at short notice out of passion."[40] Adds his counterpart, Pericles, "The outcome of situations can follow a course as absurd as the plans of man, which is just why we are accustomed to blame chance for whatever turns out contrary to calculation."[41]

Finally, Thucydides laments:

Many poorly made plans have succeeded by chancing on opponents who have planned worse, and on the other hand a still greater number of excellent plans have turned out wretchedly. For no one forms a plan and carries it out with a uniform degree of confidence, rather, we form our opinions in a state of security and in action fall short in the presence of fear.[42]

Emotion, adds Thucydides, only widens the gap between intention and result, for "war is the last thing that follows a prescribed course but draws on itself for a variety of means to meet each situation, leaving the man who approaches it with equanimity in relative security, while the one who deals with it in a passion gets into more trouble."[43] Favoring the emotion extreme of the reason-emotion dialectic clouds awareness and decisions leading to recklessness, a frequent circumstance in Thucydides's account. A competent, dispassionate leader, informed by sound intelligence, often enjoys appreciable advantages over opponents less equipped to adapt to war's uncertainty. Echoing Sun Tzu, Thucydides advises, "Whoever launches his attack by taking the closest possible look at such mistakes by his opponents in conjunction with his own strengths as well will always be the most successful."[44]

Unity of Command

Even when commanders are skilled, having more than one for a given force is detrimental. During their invasion of Sicily, Athens initially enjoyed success because Syracuse's committee of generals performed poorly. "The number of generals and the fragmentation of authority (since they had fifteen generals) had also done great harm," writes Thucydides, "along with the disorganized disunity of their whole force."[45] But that disunity dissolved after Syracuse simplified its chain of command, and they eventually dealt the Greek invaders a stunning defeat.

The Horror, the Horror

There is no form of behavior too ruthless, too brutal, too cruel for adult men and women to use against each other.[46]   E. F. M. Durbin and John Bowlby

Thucydides adeptly contrasts war's primordial violence with its allure as a noble, glorious activity. Though political leaders entreat their citizens to take up arms in a splendid cause, the actual wars and battles are invariably affairs of brutality and carnage. Indeed, war often pushes its participants through a threshold of violence and emotion that unleashes a torrent of blood-lust, particularly in invasions and civil wars. Thucydides describes this context as follows: "And during the civil wars the cities suffered many cruelties that occur and will always occur as long as men have the same nature, sometimes more terribly and sometimes less, varying in their forms as each change of fortune dictates."[47] And of the Corcyrean civil war: "Every form of death prevailed, and whatever is likely in such situations happened – and still worse. Fathers killed sons, men were dragged from the sanctuaries and killed beside them, and some were even walled up in the sanctuary of Dionysos and died there."[48] Like Kurtz in Conrad's *Heart of Darkness* (1899), otherwise ordinary people immersed within violent settings can lose their moral bearings beneath an avalanche of primal emotion. War subordinates morality, prudence, and rationality to its violent imperatives. Thucydides illustrates these imbalances in the reason-emotion, prudence-boldness dialectics as follows:

Men inverted the usual verbal evaluations of actions. Irrational recklessness was now considered courageous commitment, hesitation while looking to the future was high-styled cowardice, moderation was a cover for lack of manhood, and circumspection meant inaction, while senseless anger now helped to define a true man, and deliberation for security was a specious excuse for dereliction. The man of violent temper was always credible, anyone opposing him was suspect.[49]

Fortune Favors the Bold

Building a force capable of inflicting or deflecting rather than absorbing war's brutality is one way to counteract its horrors. Thucydides acknowledges the value of numbers, training, and experience in creating a successful army. "In most cases," he says, "victory is with those who have greater numbers and better preparation."[50] Having a sizable, well-trained, experienced force is not everything, however. According to Thucydides, the moral qualities of courage and boldness are equally significant. "Without courage no skill can prevail against dangers," writes Thucydides, "for fear drives out memory, and skill without heart is useless ... against their greater experience ... set your great boldness."[51] Courage and determination, says Thucydides, can even *offset* enemy advantages in numbers or weapons, "for most opponents, like these, attack with more trust in might than resolve; but some with greatly inferior resources . . . are bold in opposition because in the steadiness of their determination they have something great."[52]

Expeditionary Wars and Colonies

Courage and audacity often accompany the side with the most at stake, like a nation defending its homeland against invasion. Through the words of an Athenian general, Nikias, Thucydides highlights the difficulty of fighting in a distant land, far from support, without a clear vision of the enemy or the peace it can affect. Nikias entreats Athens to reconsider its plans to invade Sicily, saying of the Sicilian peoples,

and yet if we subdued them, we would subjugate them, whereas these people, even if we conquered them, would be ruled with difficulty, thanks to their being a long way off and a lot. It is senseless to move against men if they cannot be subjugated when conquered, and if after failure there will not be circumstances comparable to those before the attempt.[53]

Nikias adds that military superiority cannot guarantee victory when far from home, and that,

we will still have difficulty in conquering and in surviving alike. We must assume that we are going there to found a city in an alien and hostile land, when it behooves men to control the country from the same day they land or, if they fail, to know that they will find every element hostile.[54]

Furthermore,

Few indeed are the great expeditions . . . that succeed when they have ventured a long way from their own land. For the invaders cannot outnumber the inhabitants and their neighbors, since fear drives every element to unite, and if the lack of supplies in a foreign country is their undoing, they still leave the renown with the people they plotted against even when it is more their own fault that they failed.[55]

But his people ignored Nikias, and the expedition failed miserably, just as he had predicted. Athens's calamity stands as a warning to all invaders; yet, like the words of Troy's accursed oracle, Cassandra, it has gone, despite its truth, consistently unheeded.

Conclusion

Thucydides's extraordinary *History of the Peloponnesian War* is a treasure that offers many perceptive glimpses into the essence of human nature and war. This timeless tale proves how little the hearts and minds of humankind and the nature of war have changed. Though he was not a military theorist *per se*, no theorist, strategist, or leader should hope to master their art without having read and pondered Thucydides's masterwork. As we discussed in the last chapter and Thucydides affirms, wars can be easy to start but are often hard to end, for they gain a psychological momentum, especially for the victors, who might ask themselves, *if I have come this far, why not farther? If I have gained this much, why not more?* Indeed, power, ambition, and expediency often engulf ethics and morality, leading the strong to do as they please at the expense of the weak. Thucydides's ultimate message is cautionary. The strong will do what they will, but how long will they be strong? If the goal of war is not a better peace, if it is permitted to vacillate, and if its ambition extends beyond its means, the result is disaster, perhaps even for the "victor." No one really won the Peloponnesian War, for no Greek city-state would again approach its prewar stature. We would all be wise to reflect on this before sounding our own war drums.

MACHIAVELLI

[Machiavelli] taught Europe the art of war; it had long been practiced, without being known.[56]
Voltaire

A Prince, therefore, should have no care or thought but for war, and for the regulations and training it requires, and should apply himself exclusively to this as his peculiar province; for war is the sole art looked for in one who rules.[57]
Machiavelli

On Machiavelli

Born in 1469, Niccolò di Bernardo dei Machiavelli was a man of prominence in his native Florence and achieved greater renown through the years because of his thoughts on a range of subjects including history, politics, and war. He is best known for two works, *The Prince* (1513) and *The Discourses* (1517), completed over three centuries before Clausewitz's *On War*. Indeed, long before the great Prussian declared war an instrument of policy, Machiavelli had proposed that "politics and war constitute a kind of functional unity, with war serving as an instrument of politics."[58] Machiavelli was heavily influenced by the affairs of his fretful age. The late fifteenth and early sixteenth centuries were a time of upheaval as the era's political heavies – Medici, Cesare Borgia, Popes Alexander VI and Julius II, Charles VIII, and Louis XII – vied for control in Italy and surrounding regions. It was "an age of constant warfare, of alliance and counter-alliance, of assassination and *coup d'état*."[59] During this era, war as a political tool was a fact of life, and effective rulers had to be adept at fighting, governing, and diplomacy. "My profession is to govern my subjects well and protect them," writes Machiavelli, "to this purpose, I study the arts of both peace and of war."[60]

Machiavelli served the Florentine Republic in a variety of capacities, including as a minor government official, diplomat, and militia leader, before the ruling Medici family had him dismissed, tortured, and jailed for conspiring against their return to power. These experiences, along with a keen eye for politics and fascination with Roman history, shaped his ideas on leadership, politics, and war. No state could long endure, he believed, that could not defend itself and advance its interests by force. Effective leaders had to be willing to use *any* methods and means, even treachery or brutality, to support those interests. Indeed, thanks to his "ends justify the means" attitude toward ruling, we have the adjective "Machiavellian," which Freedman explains is "a talent for manipulation and an inclination to deceit in the pursuit of personal gain."[61]

The Realist

Wherefore, matters should be so ordered that when men no longer believe of their own accord, they may be compelled to believe by force.[62]   Machiavelli

In Machiavelli's time, few scrupulous leaders ruled successfully because they simply could not acquire and hold the power needed to offer effective stable governance and to resist threats. Only leaders willing to embrace power politics (*realpolitik*) could do so.[63] This viewpoint solidified Machiavelli's reputation as an early father of *realism*, the idea that pursuing power and prioritizing self-interest drive the international system.

Military Theorist

Now the main foundations of all States, whether new, old, or mixed, are good laws and good arms.[64]   Machiavelli

Machiavelli's contribution to political theory often overshadows his legacy as a military thinker. Though never a true field general, he was one of the early martial theorists, and he considered his 1521 treatise, *The Art of War*, his most important work.[65] Indeed, F. L. Taylor, author of *The Art of War in Italy 1494–1529* (1921), called Machiavelli, "the first secular writer to attempt to allot to the practice of arms its place among the collective activities of mankind, to define its aims, to regard it as a means to an end."[66] Both Napoleon and Clausewitz read and admired Machiavelli, with the latter echoing the Italian's thoughts on unlimited war, war's relation to politics, the military genius, the influence of moral factors, and the value of concentrating maximum energy at a decisive point.[67]

In *The Art of War*, Machiavelli argues that armies must be composed of citizens, not mercenaries, if they are to be both effective and loyal to prince and principality. During his lifetime, most Italian armies incorporated hired mercenaries (*condottieri*) because wealthy citizens refused to bear arms. This practice had mixed results, for the *condottieri* were often poor fighters and treacherous, even to their sponsors.[68] Machiavelli despised the *condottieri*, calling them "useless" and "dangerous"; in fact, "Whenever they are attacked defeat follows; so that in peace you are plundered by them, in war by your enemies."[69] He considered war a matter for states only with armies manned by citizens, conscripted if necessary, to serve as a professional fighting force like his exemplars, the Greeks and Romans.[70] "Every state must naturally be more afraid of two enemies than of one," writes Machiavelli, and "a state employing no troops except those composed of its own subjects has only one enemy to fear."[71]

What Would Caesar Do?

The wise man should always follow the roads that have been trodden by the great ... so that if he cannot reach their perfection, he may at least acquire something of its savour.[72]

Machiavelli

Machiavelli esteemed the ancients' discipline, professionalism, and "all's fair" attitude toward warfare. Unfortunately for Machiavelli, medieval armies generally lacked the will to fight their own battles and were often "little more than *ad hoc* collections of local contingents" with virtually no organization, discipline, or training.[73] For them, sustained fighting and campaigning, commonplace activities for Alexander and Caesar, were all but impossible.

What of Chivalry?

According to Machiavelli, Christianity was partly to blame for the Italian military's deplorable state. Christian morality had pervaded the conduct of war, adding an element of fair play and chivalry, which "may have helped to prevent warfare from becoming the very bloody and total kind of activity that it had been among the ancients."[74] On its face, this might seem an auspicious development, but Machiavelli considered it an insidious distortion of war's true nature. For him, Christian influence made defeat seem acceptable, since conquering armies were expected to show mercy. Military combat became ritualistic and defeat more a disappointment than a death sentence. In these circumstances, the stakes of war seldom compelled citizens to risk everything for their country.[75] This disgusted Machiavelli, for he knew the façade of Christian civility could crumble at any time, leaving the unprepared as lambs to the wolf. Machiavelli says:

At present those terrible apprehensions are in a great measure dissipated and extinguished; for after an army is defeated, those falling into the conqueror's hands are seldom or never put to death, and the terms of their ransom are made so easy that they do not long continue prisoners. So, men now no longer care to submit to the rigor and continual hardships of military discipline to ward off evils which they are but little afraid of.[76]

Moreover, Machiavelli noticed that virtuous commanders often lost to more ruthless adversaries. Thus, for Machiavelli, war was properly "a no holds-barred contest" and "victory . . . the aim to which all other considerations on the battlefield must be subordinated."[77] The only rule of war is to win, by any means necessary, including deception and remorseless violence. Only commanders who understood these truths were fit to lead and to be entrusted with the state's defense.[78]

Fortitude and Fortune

I think it may be the case that Fortune is the mistress of one half of our actions, and yet leaves the control of the other half, or a little less, to ourselves.[79]
Machiavelli

Machiavelli defines two terms, *virtù* and *fortuna*, that illustrate his awareness of chance and psychological factors upon war. *Virtù* is "masculine aggressive conduct . . . exhibited in a dangerous and uncertain situation of tension, stress, and conflict," a "tremendous force of will and inner strength . . . boldness, bravery, resolution, decisiveness."[80] *Virtù* pertains to individuals, groups, or nations, and encompasses purpose, order, direction, unity, zeal, energy, and creativity. A people imbued with *virtù*, like Rome at its zenith, are valuable allies and formidable foes possessing an indomitable will to fight. However, *virtù* is impermanent and applies to both sides. No society, thought Machiavelli, can sustain its *virtù* indefinitely, and when it wanes, disorder reigns until discipline and *virtù* are restored.[81]

Fortuna represents all war's unknowns and their impacts, good and bad, on an army's activities. Once the fighting starts, *fortuna* – fate, destiny, and chance – creates unexpected results, challenges, and opportunities that can invalidate war plans. Machiavelli's ideal leader exhibits *virtù*, instills it in his forces, and is flexible, observant, and poised to recognize and adapt to *fortuna*'s effects. This includes knowing when to attack and seize opportunities. Machiavelli writes: "Men who have any great undertaking in mind, must first make all necessary preparations for it, so that when an opportunity arises, they may be ready to put it in execution according to their design."[82]

Thoughts on War

Machiavelli viewed war as an instrument of policy, a means for accomplishing political objectives, and a path to state security, although he considered these ideas axiomatic (i.e., not requiring systematic derivation). As a traditionalist, Machiavelli believed Italy's future glory could be achieved by embracing the past. As such, he considered a "field war" the "proper" and "most necessary and honorable of all wars," and thought generals should "know how to conduct such a war . . . to give battle to an enemy in a proper and soldier-like manner."[83] To do this, field armies required a proper mixture of artillery and cavalry centered on the army's fundamental component and strongest element, the infantry. His ideal army consists of citizen-soldiers, anchored by the infantry, regularly trained but only mobilized to fight in crises. Interestingly, for officers on active service, Machiavelli recommends frequent rotation to different locations (a practice used today in most countries) not only for expanding their abilities, but also to prevent over-familiarization with a particular cohort and to maintain healthy civil-military relations.[84]

Machiavelli knew that disciplined campaigning requires competent staff officers, therefore, "above all, a general should take care to have men of proven fidelity, wisdom, and long experience in military affairs near his person as a sort of council."[85] This council plays a vital role in assessing force status and readiness for both sides, an activity that determines "which of the armies is superior to the other in number, which is better armed and disciplined, which is the stronger in cavalry, [and] which of his own troops are fittest to undergo hard service and fatigue."[86] In particular, it is "of great importance to know the qualities and disposition of the enemy's general and of those about him."[87] Here, Machiavelli identifies an important target, the enemy general, and aligns with Sun Tzu on the value of intelligence and accurate assessments of enemy and friendly capabilities.

Affecting an enemy's physical and moral strength, says Machiavelli, is an essential tactical goal; and, "Above all things, a general ought to endeavor to divide the enemy's strength by making him suspicious of his counselors and confidants, or by obliging him to employ his forces in different places and detachments at once."[88] As a corollary, he takes a page from Julius Caesar's playbook, calling it *unwise* to corner an adversary who still retains cohesion. Do not "reduce an enemy to utter despair," urges Machiavelli, but "open a way for them to escape," for it is "better to pursue them when they [flee], than to run the risk of not beating them while they [defend] themselves with such obstinacy."[89] Moreover, Machiavelli advises that battles be deferred entirely (if possible) unless forces have superiority in numbers, training, and bravery.

Machiavelli recommends intelligence collection on enemy plans to facilitate anticipation of strategy and actions. Good intelligence and planning keeps an enemy off-balance. Machiavelli says: "We see that nothing is as easy to effect as what the enemy imagines you will never attempt, and we see that men are frequently in the greatest danger when they think themselves most secure."[90] He notes that it is possible "to frustrate . . . your enemy's designs" by doing "of your own volition what he endeavors to force you to do."[91] In other words, generals should seize the initiative by acting at a time or place the enemy considers unlikely. Historical examples include Hitler's 1940 Ardennes gambit and MacArthur's improbable landing at Inchon in Korea a decade later.

Machiavelli favored infantry-centered force concentration as a fundamental battlefield tactic. The Romans, he writes, "constantly posted their legions in the center, rightly judging that the forces in which they reposed the greatest confidence should always be compact and united."[92] To Machiavelli, the large but compact, well-armed, and well-defended legions were practically invincible because they presented foes with an immediate dilemma – fight head-on or split to attack from multiple directions. The first choice risks direct battle into the adversary's strength, while the second risks vulnerability to counterattack and isolation.[93]

The Value of Mobility and Dispersal

In his characterization of infantry and artillery, Machiavelli hints at a broader principle beyond concentration of force. The obvious defense against artillery, he says, is to "keep out of reach of its shot or place yourself behind a wall, a bank, or some fence of that kind"; however, "when an army is drawn up in order to fight, it cannot skulk behind a wall or a bank."[94] Defeating cannon, therefore, requires the attacker to assume risk by abandoning barriers and moving rapidly to neutralize the artillery before it can fire. For Machiavelli, the attacker must give up its "close order" to mitigate this risk, for "when your men are thinly drawn up [artillery] cannot do much damage among them"; thus, "a compact body of regular forces is not at all proper for this service."[95] He advises "*velites*" along the wings and "light cavalry" advance quickly while the cannon reloads.[96] Furthermore, says Machiavelli, *virtù* coupled with this "rush up" tactic maintains the infantry's primacy, and "therefore, the invention of artillery is no reason . . . why we should not imitate the ancients in their military discipline and institutions."[97]

Hindsight reveals the flaws in Machiavelli's advice vis-à-vis advances in artillery accuracy, destructive power, reload times, and tactics (like staggered fires), not to mention airpower. As we now know, similar beliefs (e.g., that *élan* can negate robust defensive firepower) pervaded war commands during World War I with ruinous consequences. However, regarding the principle of dispersion, Machiavelli hits closer to the mark. Reexamining the problem of artillery and his desire to emulate ancient tactics and formations, Machiavelli acknowledges the ineffectiveness of concentrated, frontal assaults saying, "war before long will be reduced to the question of artillery."[98] Taking a page from Scipio Africanus (who dodged Hannibal's Carthaginian elephant charges with force dispersal), he recommends *spacing* troops to create clear lines of fire for friendly soldiers while complicating the enemy's targeting problem, "for it is a general rule always to give way to such things as cannot be opposed – as the ancients used to do when they were attacked by elephants and armed chariots."[99] Machiavelli's statement, "always give way to such things as cannot be opposed," implies a linkage between concentration and capacity that we will develop further in Chapter 4.

Theses

A Prince should read histories, and in these should note the action of great men, observe how they conducted themselves in their wars, and examine the causes of their victories and defeats, so as to avoid the latter and imitate them in the former.[100]
Machiavelli

Machiavelli encapsulated his guidance on war into twenty-seven "theses." The following are the most universal of these [with short commentary in brackets]:

➢ Whatever is of service to the enemy must be prejudicial to you; whatever is prejudicial to him must be of service to you. [Avoid or reduce the enemy's strengths and exploit his vulnerabilities.]
➢ If you observe the enemy and take care to drill and train your army, you will minimize danger and maximize your chances of success. [Intelligence-informed training is essential to combat success.]
➢ You must conceal your plans from the enemy to increase your chances of success. [Hiding intent is the key to surprise, one of war's most powerful advantages.]
➢ Few are brave by nature, but discipline and experience make many so. [A well-disciplined, trained, and experienced force overcomes fear.]
➢ If a general knows his own strength and that of the enemy perfectly, he can hardly miscarry. [Like Sun Tzu's "know the enemy; know yourself" adage.]
➢ It is best to accustom your men to the reality of conflict by small actions and skirmishes than to throw them into a great battle all at once. [Battle acclimatization is best accomplished incrementally to prevent the shock of war from shattering an army's courage.][101]

Conclusion

Though not generally heralded as a military theorist, Machiavelli deserves credit for rediscovering the martial attributes of the ancients and literally bringing military thought out of the dark ages. Indeed, many of his themes were adopted by Clausewitz, including the value of concentration, assessing the enemy's will and disposition, and adept generalship. While Machiavelli's advice on emulating ancient armies is not universally sound, and his realist philosophy clashes with the guidance of Sun Tzu and Thucydides, his appreciation for discipline and professionalism endures along with his descriptions of *virtù* and *fortuna*. Furthermore, Machiavelli's discussion regarding artillery touches upon the concentration-dispersal dialectic, a key concept related to combat dynamics. Ultimately, Machiavelli's insights on war, politics, and human nature should remain required reading for warriors, strategists, and theorists.

JOMINI

> *In essence, Jomini fused two of the great cultural currents of the early nineteenth century: a boundless romantic sensibility and an obsession with the power of science, reduced to formulaic statements and prescriptive injunctions.*[102]
>
>   John Shy

Jomini and Clausewitz

Antoine-Henri, baron de Jomini is frequently contrasted with Clausewitz. As we will see shortly, Jomini's general theory of war, which reflects the Enlightenment's faith in human reason to express the essence of nature as a set of universal principles, is clearer and less philosophical than Clausewitz's, which explains Jomini's popularity with military leaders of his and later generations. Moreover, compared to Clausewitz or Machiavelli, Jomini was less interested in the nexus between war and politics. He viewed the political dimension as a given (i.e., politicians start wars and generals win them).

On the other hand, Clausewitz (reflecting German Romanticism) believed war and politics are inextricably linked and that war *cannot* be reduced to a set of principles. Rather, war is a *complex*, violent, reciprocal activity composed of physical *and* psychological elements, subject to *chance, creativity*, and *emotion* as well as reason, and clouded by *uncertainty* born of battlefield fog and friction. Jomini treated many of these factors as givens working on both sides; thus, he placed more emphasis on his cherished principles as a basis for tactics and strategy. Says Jomini, my "desire has been to facilitate the study of the art of war for careful, inquiring minds, by pointing out directing principles," for which "I have not found a single case where correctly applied, did not lead to success."[103] Jomini's view of war tilts to the science side of the science-art dialectic, which was preferred by airpower advocates, for example, whose doctrines stress the identification and rapid destruction of vital points with concentrated force.

Lessons from the untidy "small" wars of the twentieth and early twenty-first centuries have strained Jomini's credibility as a great war theorist. Compared to Clausewitz's theory, Jomini's seems less applicable to the complex, "wicked" problems presented by terrorists and insurgents. Western experiences in Indochina highlighted the drawbacks of using fixed principles to fight wars and of marginalizing political and cultural factors, theoretical ground Clausewitz addresses more effectively. Like Clausewitz and Machiavelli, Jomini was a product of his era; however, because his principles derive largely from observations of Napoleonic warfare rather than independent premises related to war and human nature, some are now obsolete. The result is a body of theory that, while more organized and readable than Clausewitz's, is less timeless. Nevertheless, Jomini's legacy remains strong, and many of his recommendations and principles continue to resonate in contemporary military doctrines.

On Jomini

Jomini was born in 1779 in Switzerland and from a young age was fascinated with military affairs. These were heady times to be alive in the Western world as revolutions in blood and thought were underway on both sides of the Atlantic, and Jomini wanted a piece of the action. Though his professional career started in business and banking, Jomini aspired to become a military officer. As a financier in Paris at the turn of the century, he kept a close eye on Napoleon's early campaigns, seizing every opportunity to turn his disciplined mind to writing about warfare. By 1798, Jomini had parlayed a growing reputation as a military thinker into a position within the Swiss Army. Just three years later, having surveyed and disparaged most eighteenth-century French Enlightenment military thought (including the celebrity-theorist, Comte de Guibert), Jomini moved back to Paris where he began to write a *Treatise on Grand Tactics* (1803), a tome complementary of Napoleon. To Jomini's delight, the emperor took notice and hired the young banker in the grade of colonel.

In 1805, Jomini joined Marshall Ney's staff and became an official member of Napoleon's *Grande Armée*. Jomini's drive and intellect propelled him rapidly up the ranks, earning him the title "Baron of the Empire" within three years and the rank of brigadier general five years after that. But rapid success inflated Jomini's considerable ego, and his supercilious demeanor (character-ized by John Shy as "slippery and presumptuous") created powerful enemies, including Marshal Berthier who, not long after Jomini's ascension to brigadier general, engineered his ouster from the French Army.[104] Nevertheless, through his authorship of such works as the fifteen-volume *Histoire critique et militaire des campagnes de la Revolution* (Paris, 1820–1824), Jomini had already cemented "an international reputation as the preeminent historian and theorist of modern warfare."[105]

Embittered but undeterred by his dismissal, Jomini leveraged his reputation to obtain a position with the Russian Army in the grade of lieutenant general. Though he never participated in direct action against Napoleon on French soil, Jomini's service to Russia was exemplary. By 1826, he had gained the hallowed rank of general-in-chief and was entrusted with tutoring the Tsar and the Grand Duke on military matters. He also found time to establish the Russian military staff college and pen *Vie Politique et Militaire de Napoleon* (Paris, 1827).

Despite all this success, Jomini's signature accomplishment was the consolidation of his thoughts on war into *Precis de l'art de la guerre* (*Summary of the Art of War*), published in 1838. Unlike Clausewitz, Jomini survived to see his masterwork in print and lived long afterwards (he died in 1869) as a valued consultant and renowned military strategist. Indeed, his reach extended even to the United States, where *Summary of the Art of War* was required reading at West Point. Most prominent American Civil War generals, like Lee, McClellan, and Grant, went to battle with Jomini's principles in mind if not also in their jacket pockets.

A Study in Contrasts

As stated earlier, Jomini provides an interesting counterpoint to Clausewitz. During the Napoleonic Wars, they initially fought on different sides, but both eventually served with Russia. Both were noted thinkers, strategists, and warriors, though Jomini achieved greater rank and stature over the course of a longer life (he lived to age 90). Where Clausewitz's purpose was to explain Napoleonic power within a broader theory of war, Jomini's intent was to distill the era's lessons into irrefutable laws.

Europe was small enough, even in the nineteenth century, for Jomini and Clausewitz to know of each other's work. Though Clausewitz never names Jomini in *On War*, he was well aware of Jomini's earlier works (*Summary of the Art of War* was published seven years after Clausewitz's death), and Jomini was undoubtedly one of the maneuver and principle-obsessed theorists Clausewitz disdained. Jomini returned the sentiment and even critiques Clausewitz by name in his opus, saying the Prussian's "logic is frequently defective."[106]

Both men respected Bonaparte and had cause to dislike him at the same time, Clausewitz for the punishment inflicted upon Prussia and its allies, and Jomini for not rescuing his career after Berthier's betrayal. Of Napoleon, Jomini opines: "He loved war and its chances ... it might be said that he was sent into this world to teach generals and statesmen what they should avoid.

His victories teach what may be accomplished by activity, boldness, and skill; his disasters, what might have been avoided by prudence."[107] Whereas Jomini saw Napoleon's example as a trove for the development of prescriptive, rules-based theory, Clausewitz was suspicious of rules and maxims, which he believed superfluous to the military genius, too often subject to misinterpretation and misuse, and too frequently negated by chance, fog, friction, and enemy action.

In many ways, Jominian thought represented the culmination of Enlightenment doctrines pioneered by the likes of the eighteenth-century Welch general Henry Lloyd (whose ideas included "lines of operations" and a belief that war is based on immutable properties only "superficially" chaotic); the Prussian soldier Heinrich von Bülow (whose unilateral "principle of the base" largely equated strategy with geometry); and the Austrian general Archduke Charles. Jomini viewed principles as essential guidelines for generals who exhibited their genius by knowing how and when to best employ them.[108] He writes: "All my works go to show the eternal influence of principles and to demonstrate that operations to be successful must be applications of principles."[109] And, "Theories cannot teach men with mathematical precision what they should do in every possible case, [but] in the hands of skillful generals commanding brave troops, [rules become] means of almost certain success."[110] For Jomini, fog, friction, and chance are real factors but apply equally to both sides; thus, the key to defeating the enemy is to employ the correct principles more accurately, precisely, and rapidly.

Finally, Clausewitz was unafraid of controversy in his search for war's truths. He "wrote for himself," asserts Raymond Aron, "for those who would be prepared to study him; he felt no need to tread warily."[111] Clausewitz wanted to stir things up, and he did not disguise his contempt for the theorists and strategists of his day, a characteristic that won him few friends, even in the Prussian political and military hierarchy. Jomini, however, wanted to be revered, and he was. But this approach precluded Jomini from expanding his thinking in a way that might have lent his ideas the enduring relevance and stature ultimately enjoyed by Clausewitz.

The Parts of War

For Jomini, war has "six distinct parts": (1) Statesmanship; (2) Strategy, or the art of properly directing masses upon the theater of war, either for defense or for invasion; (3) Grand Tactics (major components of battle); (4) Logistics, or the art of moving armies; (5) Engineering, the creation of attack and defense fortifications; and (6) Minor Tactics (considerations for individual soldiers).[112] Like Clausewitz, Jomini understood that war is more than just fighting battles. He writes: "War in its *ensemble* is not a science, but an art. Strategy particularly, may indeed be regulated by fixed laws resembling those of the positive science, but this is not true of war viewed as a whole."[13] Still, though he mentions the broader phenomenon of war, Jomini is more comfortable with the "science" of *warfare* and thus directs the weight of his analysis toward strategy and battle tactics. Consistent with this emphasis, Jomini's target audience is the rank-and-file soldier. The book's light treatment of politics and its style – orderly, precise, and easily followed, with frequent use of enumerated lists – are readily accessible to the military mind and clearly reflect Jomini's banking and staff experience, where well-ordered ledgers and precise plans and orders are hallmarks of the trade.

How Wars Begin

Simply put, wars happen because governments find reasons to start them. Before starting a war, the statesman, writes Jomini, must ascertain if it is "proper, opportune, or indispensable."[114] "Proper" means the war is justified and can reasonably be expected to generate the desired result; "opportune" means timing and circumstances are favorable; and "indispensable" indicates that there are no alternatives for gaining the desired result. These tests are sensible and applicable to any era, though standards for what is proper, opportune, or indispensable may vary widely.

According to Jomini, governments go to war to:
(1) reclaim or defend rights
(2) protect the state's interests (like commerce, manufactures, or agriculture)
(3) support neighboring states important to the government
(4) fulfill alliance obligations
(5) propagate, crush, or defend political or religious theories
(6) gain territories to increase the power and influence of the state
(7) defend the threatened independence of the state
(8) avenge insulted honor
(9) satisfy a mania for conquest[115]

Although supporting allies can cause wars (e.g., Thucydides's Trap), Jomini values alliances and especially the enhanced benefit of turning a potential enemy into a friend. "There is no enemy, however insignificant," says Jomini, "whom it would not be useful to convert into an ally."[116]

War's Poetry and Metaphysics

Like fingerprints and snowflakes, wars may appear alike, yet all are different. Jomini says the "different kinds of war" are linked to its reasons, which "influence in some degree the nature and extent of the efforts and operations necessary for the proposed end."[117] Once a war has begun, influencing factors include,

> the passions which agitate the masses that are brought into collision, the warlike qualities of these masses, the energy and talent of their commanders, the spirit, more or less martial, of nations and epochs ... everything that can be called the poetry and metaphysics of war.[118]

Jomini lists the following permutations affecting "the nature and conduct of war":
(1) A state may make war against another state
(2) A state may make war against several states in alliance
(3) A state in an alliance may make war against a single state
(4) A state may be either the principal party or an auxiliary
(5) In the latter case the state may join the war at the beginning or after it starts
(6) The theater of war may be on the soil of the enemy, an ally, or its own
(7) If the war is an invasion, it may be on adjacent or distant territory
   (The invasion may be prudent and cautious, or bold and adventurous)
(8) It may be a national war against ourselves or against the enemy
(9) The war may be a civil or a religious war.[119]

As this list implies, the character of war (covered in greater depth in Chapter 4) and its "poetry and metaphysics" are functions of who is fighting, why they fight, and the means and location of fighting, but Jomini holds the means (infantry, cavalry, and artillery) and location (Europe) constant. Fortunately, Jomini does not try to evaluate every circumstance and variable, and though he does expound upon a few specific cases, Jomini's most salient assertion is that the political and moral dimensions of war are *separate* from the fighting. "If the principles of strategy are always the same," writes Jomini, "it is different with the political part of war, which is modified by the tone of communities, by localities, and by the characters of men ... the fact of these modifications has been used to prove that war knows no rules."[120] However, "military science rests upon principles which can never be safely violated in the presence of an active and skillful enemy, while the moral and political part of war presents these variations."[121] For Jomini (echoing Henry Lloyd), political and moral factors muddy the waters of war making it *appear* that rules are irrelevant; however, once these factors are accounted for and war begins, "the great principles of war" apply and "must be observed."[122]

The Darker Side of War

Like Clausewitz, Jomini labors to accommodate wars outside the major state-versus-state model. Recall, from an early age Jomini was attracted to the prestige and honor of officership, command, and the profession of arms. He was never a common soldier nor experienced the sting of defeat (as Clausewitz does at Jena). Thus, he judged civil uprisings and insurgencies, which disregard codes of chivalry, "deplorable" affairs and relegates them to the bottom of his list under the heading of "civil" or "religious" wars.[123] Compared to Napoleon's dazzling aura, these small wars must have seemed no more than flickering candles to Jomini. Were they distinct phenomena governed by unique principles or twisted offshoots of traditional warfare? Jomini appeared to prefer the former interpretation, Clausewitz the latter. "Although wars of opinion, national wars, and civil wars are something confounded," writes Jomini, "they differ enough to require separate notice."[124] According to Jomini, "wars of opinion" seek to advance or eliminate a religious or political ideology. These wars, writes Jomini, "enlist the worst passions, and become vindictive, cruel, and terrible,"[125] and religious wars are the "most deplorable" of all.[126] In these conflicts, dogma "is a powerful ally, for it excites the ardor of the people."[127] Jomini doubted the efficacy of resolving these wars by force of arms and likened the attempt to quenching a volcanic eruption or halting an explosion after the powder has ignited. Says Jomini:

War and aggression are inappropriate measures for arresting an evil which lies wholly in the human passions, excited in a temporary paroxysm, of less duration as it is the more violent. Time is the true remedy for all bad passions and for all anarchical doctrines . . . it is far better to await the explosion and afterward fill up the crater than to try to prevent it and to perish in the attempt.[128]

And, "In a military view these wars are fearful, since the invading force not only is met by the armies of the enemy, but is exposed to the attacks of an exasperated people."[129] While these assessments are largely sound, Jomini's characterization of their short-lived nature is suspect. The history of the twentieth and early twenty-first centuries provides ample evidence that such wars can smolder and reignite seemingly without end.

Jomini calls "national wars" "the most formidable of all" as they apply to wars "waged against a united people, or a great majority of them, filled with noble ardor and determined to sustain their independence."[130] In these wars, an invading army is everywhere threatened, not just with a regular army but by the citizenry themselves, and "the consequences are so terrible that, for the sake of humanity, we ought to hope never to see it."[131] Combining regular military with a hostile populace denies the invading army "any ground but that upon which he encamps; outside the limits of his camp every thing is hostile and multiplies a thousandfold the difficulties he meets at every step."[132] These words, though written over a century earlier, would have made an apt epitaph for the French and American wars in Vietnam. In truth, generals of any era would be wise to heed Jomini's warning that "no army, however disciplined, can contend successfully against such a system applied to a great nation, unless it be strong enough to hold all the essential points of the country, cover its communications, and at the same time furnish an active force sufficient to beat the enemy wherever he may present himself."[133] This dilemma faces any general waging counterinsurgency or a war seeking to overthrow a government supported by its people, and the danger increases with distance from the homeland. Indeed, Jomini implies an inverse relationship, perhaps informed by his experiences in Russia, between the validity of his principles and the distance between an invading army and its support bases. He says: "We are far from thinking that any purely military maxims can insure [sic] the success of remote invasions: in four thousand years only five or six have been successful, and in a hundred instances they have nearly ruined nations and armies."[134]

Touching on some key elements of contemporary counterinsurgency theory, Jomini stresses that success is aided by using a large "mass of troops," calming "the popular passions in every possible way," using time and patience to "exhaust" the enemy, displaying "courtesy, gentleness, and severity united," and dealing "justly."[135] Patience and persistence increase the odds of success. Examples include the British resolution of the Malayan Emergency and the USA's counterinsurgency triumphs in the Philippines during the Philippine War (1899–1902) and Hukbalahap Insurrection (1942–54).[136] Each of these benefitted from sustained commitment, presence, indigenous support, and robust resourcing.

The Good Ole Days

Whereas Thucydides, Machiavelli, and Clausewitz *wanted* readers to recognize war's cruelty and dreadfulness, Jomini preferred to view war in a different light. One does not have to be a utopian, writes Jomini, to

desire that wars of extermination may be banished from the code of nations. As a soldier, preferring loyal and chivalrous warfare to organized assassination ... my prejudices are in favor of the good old times when the French and English guards courteously invited each other to fire first ... preferring them to the frightful epoch when priests, women, and children throughout Spain plotted the murder of isolated soldiers.[137]

After the decades of war Jomini witnessed, it is not surprising that he favored the reduction if not eradication of its most dreadful forms; yet his nostalgia for a more benign brand of warfare indicates, if not *naiveté*, at least a blind spot regarding human nature. As American Civil War general William Tecumseh Sherman once said, attempts "to make war easy and safe will result in humiliation and disaster."[138]

The Perfect Army

Jomini's ideal army is highly motivated, skilled, backed by a supportive populace, and commanded by carefully selected generals free to fight wars without political interference. The following are Jomini's twelve conditions for creating the "perfect" army [implied purpose in brackets]:

(1) A good recruiting system [to gain quality soldiers]
(2) A good organization [to provide a structure for the soldiers]
(3) A well-organized system of national reserves [for contingencies]
(4) A good education and training program [to make learned, skilled warriors]
(5) Strict discipline based on conviction not formality [to maintain order]
(6) A system of rewards [to encourage soldiers to improve and excel]
(7) A well instructed engineering and artillery corps [to maximize army effectiveness]
(8) Superior offensive and defensive weapons [to gain an advantage over the enemy]
(9) A capable general staff [to administer these conditions and support the commander]
(10) A good system of commissaries, hospitals, and administration [to satisfy soldiers' basic needs]
(11) A good system of assigning command and directing war operations [to lead the army in battle]
(12) Exciting and keeping alive the military spirit of the people [to sustain the army and its operations][139]

"None of the above twelve conditions can be neglected without grave inconvenience," cautions Jomini.[140] In fact, the world's most effective militaries incorporate all twelve of these elements.

For Jomini, civilized nations raise, prepare, maintain, and lead quality armies to avoid becoming victims of competing political agendas, whether of conquest or intimidation. While he knew that no state could wage war indefinitely, Jomini also believed that there was no such thing as indefinite peace. He saw war as an inseparable thread woven into the fabric of human nature and international relations. Consequently, Jomini thought nations should hold their militaries and warriors in highest esteem. Writes Jomini: "Misfortune will certainly fall upon the land where the wealth of the tax-gatherer or the greedy gambler in stocks stands, in public estimation, above the uniform of the brave man who sacrifices his life, health, or fortune to the defense of his country."[141]

Leadership and Strategy

For Jomini, politics was a ruling-class activity separate from combat. The military commander must understand the war's political purpose, but the commander's duty is to create and execute a war plan, preferably without interference from the sovereign (a sentiment Jomini shares with Sun Tzu). Jomini thought rulers should be educated in politics and military affairs, and officers should study the "military sciences" and the geography and "material and moral capacity" of other states.[142] Furthermore, campaign plans should reflect understanding of the war's object, the enemy forces (including moral and material means and prospects for alliance), the nature and resources of the country, and both the army and state leadership.[143] These guidelines mirror those of Sun Tzu and Machiavelli by recognizing the importance of relative capacity assessments to successful strategy. Turning good intelligence into accurate appraisals of enemy physical and psychological capacity is essential to warfighting. Indeed, Jomini says:

By adopting as a rule an order of battle well closed and well connected, a general will find himself prepared for any emergency, and little will be left to chance; but it is especially important for him to have a correct estimate of his enemy's character and his usual style of warfare, to enable him to regulate his own actions accordingly.[144]

Evaluating the relative capacities of two armies, Jomini asserts, is more than a mere size comparison. Capacity also depends on "military qualities, their *morale*, and the ability of the commander."[145] Here, Jomini emulates Sun Tzu and foreshadows Clausewitz by stressing the importance of war's moral (psychological) *and* material (physical) components.

War on a Map

For Jomini, the art of war and strategy are found in the movement and employment of armies, from an initial base of operations to various objectives, point-by-point like connect-the-dots. "Strategy," writes Jomini "is the art of making war upon the map."[146] Planning defines objectives and lines of operation to control "decisive points" (e.g., a terrain feature, base, road, river crossing, or town), which then generate advantages and produce great victories. Jomini defines "Grand Tactics" as the art of "posting troops upon the battle-field ... of bringing them into action ... and the art of fighting upon the ground."[147] For Jomini, *strategy* is campaigning (i.e., moving forces across a theater of operations), and *grand tactics* pertain to the actual fighting.[148]

The Ideal General

Jomini valued generals with moral and physical courage as well as decisiveness, but not surprisingly, placed more importance on knowing "the principles at the base of the art of war."[149] "He will be a good general," says Jomini, "in whom are found united the requisite personal characteristics and a thorough knowledge of the principles of the art of war."[150] Comprehensive knowledge of "leading principles" and the ability to select the correct principle for the circumstances form the nucleus of Jomini's assessment of military genius.[151] Even so, Jomini concedes that the decisions of some geniuses reflect correct principles even without the generals' foreknowledge of them. After all, he writes, "I lay no claim to the creation of these principles, for they have always existed, but I claim to have been the first to point them out."[152] Conversely, ordinary generals "without a natural talent for war, and who have not acquired that practical *coup d'oeil*, which is imparted by long experience in the direction of military operations" may struggle to reconcile their plans with the real-time dilemmas presented by a "skillful, active, and enterprising adversary."[153]

Valuing Intelligence

Jomini's ideal general appreciates good intelligence, for

how can any man say what he should do himself, if he is ignorant what his adversary is about? As it is unquestionably of the highest importance to gain this information, so it is a thing of the utmost difficulty, not to say impossibility; and this is one of the chief causes of the great difference between the theory and the practice of war.[154]

Here, Jomini suggests the gulf between theory and practice can be filled in part with good intelligence and also indirectly addresses one of Clausewitz's strengths (i.e., war's uncertainty makes it subject to chance and probability). "A good general," affirms Jomini, "should neglect no means of gaining information," but "perfect reliance should be placed on none of these means," for "it is impossible to obtain exact information"; thus, "a general should never move without arranging several courses of action ... based upon *probable* hypotheses" informed by the principle of war (emphasis added).[155] Indeed, "the real secret of military genius consists in the ability to make these reasonable suppositions in any case."[156] In other words, the skilled general, despite incomplete or inexact information, cultivates options based on war's fundamental principles to increase the probability of success.

The Grand Principle of War

Jomini identifies a single "fundamental principle of war," the "one great principle underlying all the operations of war – a principle which must be followed in all good combinations."[157] He initially expresses it in four parts, but later condenses it as follows:

(1) to obtain by free and rapid movements the advantage of bringing the mass of the troops against fractions of the enemy;

And,

(2) *to strike in the most decisive direction,* – that is to say, in the direction where the consequences of his defeat may be most disastrous to the enemy, while at the same time his success would yield him no great advantages (i.e., *throwing the army's mass against a decisive point*).[158]

At first glance, this maxim seems simplistic; however, it stresses that the purpose of the Jominian army is to control the map by occupying and holding positions of importance, and when encountering the enemy force, by forcing a surrender or retreat. *Awareness* (intelligence), *access* (via bases and lines of communication), and *freedom of action* are prerequisites for concentrating against and striking a decisive point.[159] Thus, for two-dimensional, continental warfare, once awareness and access are gained, there are but three problems of strategic movement or tactical maneuver: "whether to maneuver to the right, to the left, or directly in front."[160]

Decisive Points

Decisive points are central to Jomini's theory, but how are they identified ahead of time and in the heat of battle? Jomini admits: "It is easy to recommend throwing the mass of the forces upon the decisive points, but [the] difficulty lies in recognizing those points."[161] Jomini's decisive points are similar to Clausewitz's centers of gravity (more on this later) and include intermediate objectives, like geographical features or infrastructure, as well as objectives like a capital city or the enemy's army. During battle, a portion of the enemy's army or its lines of communication could be decisive points. Jomini says: "The decisive point of a battle-field is determined . . . by the character of the position, the bearing of different localities upon the strategic object in view, and, finally, by the arrangement of the contending forces."[162]

Additional Jominian decisive points include *strategic,* "all those which are capable of exercising a marked influence either upon the result of the campaign or upon a single enterprise"; *geographic,* "those points the possession of which would give the control of the junction of several valleys and of the center of the chief lines of communication in a country"; and *maneuver,* which "are on the flank of the enemy upon which, if his opponent operates, he can more easily cut him off from his base and supporting forces without being exposed to the same danger."[163]

Jomini identifies two types of *objective* points: (1) *maneuver*, determined by enemy force locations and disposition, and (2) *geographical* (e.g., a fortress, river crossing, capital city, or hilltop).[164] Finally, Jomini defines *political* objective points that support "sometimes quite important" political purposes but are also often "irrational" and "frequently lead to the commission of great errors in strategy."[165]

For Jomini, the selection of decisive points "will generally depend upon the aim of the war and the character which political or other circumstances may give it, and finally, upon the military facilities of the two parties."[166] Though his "decisive points" concept is linked to sound strategic principles like assessing the war aim, contextual factors, and relative capacities of the warring sides, Jomini's maxims leave many variables undefined, which limits their practicality.

Winning battles, says Jomini, requires movement to decisive points, concentrating fighting power, and attacking with utmost speed. However, he admits, "this kind of war is not suitable to all capacities, regions, or circumstances," citing Napoleon's invasion of Russia as an example.[167] Regrettably, Jomini does not fully explain what he means by "this kind of war" or which "capacities, regions, or circumstances" are unsuitable. Was Napoleon's failure in Russia unique or indicative of a form of war for which Jomini's principles become unreliable? Did Napoleon fail in Russia because he made a mistake in following the principles, because the Russians employed the principles better, or because, in this case, the principles did not apply? This example highlights a deficiency in all rule-based theories – the problem of how to treat exceptions. As basic and universal as Jomini's great principle is, there are still circumstances where it appears to fail, and in these cases, we must turn to other theorists.

Offense and Defense

The first point to be decided once war is initiated, writes Jomini, is "whether it shall be offensive or defensive."[168] Like Clausewitz, Jomini favors offense over defense, and at the strategic level, the offensive takes the initiative, usually as an invasion. Offensive advantages include forcing a defender to respond or retreat, driving the fighting and resultant devastation to an enemy's homeland, and boosting morale. Jomini's attacker has the advantage of the initiative, of deciding when and where to strike; whereas, "he who awaits the attack is everywhere anticipated."[169] Invasions also have disadvantages, including the potential for long, vulnerable supply lines and arousing the "ardor and energy" of the adversary army and populace.[170] Offensive formations "must possess the characteristics of *solidity, mobility,* and *momentum,* whilst for the defensive *solidity* is requisite, and also the power of delivering *as much fire as possible.*"[171]

At the tactical level, attackers and defenders can concentrate force for a decisive blow. A defender who strikes first essentially becomes the attacker, reversing roles, at least for a time. As for *where* to strike, Jomini calls it a grave error to attack the center *and* extremities unless the enemy is greatly dispersed. Doing so may overextend the attacker and expose lines of communication. Once the target is known and the time is right, the attacker focuses power to deliver effects before the enemy can seize or regain the initiative. Rapid, concentrated attacks work for both the offense and defense if the goal is a decision; otherwise, either side may attempt to withdraw.

Momentum

Jomini's "momentum" characterizes an attacker's tendency to remain on the move after a successful attack, following up the initial action with an advance and pursuit to gain a decision. Momentum has both a physical and "moral effect." The side with positive momentum (i.e., on the advance), gains a burst of energy and sense of invincibility as the enemy falls back; whereas, the retreating force may despair if defeat or death appears imminent. Defenders generally seek to arrest the attacker's momentum and start their own. Jomini distinguishes between a defensive strategy and defensive tactics, where the former can have an active component that "may accomplish great successes," while the latter is generally passive and "pernicious."[172] Furthermore, Jomini believes that a proper defense exploits opportunities to strike at an attacker's weaknesses, to counterattack when the attacker tires or makes a mistake. The defender's objective is to "neutralize the first forward movement" of the attacker, to protract operations without giving the war away, and to defer "a decisive battle until the time when a portion of the enemy's forces are either exhausted . . . or scattered."[173]

Concentration and Interior Lines

Jomini, like Clausewitz, considers rapid concentration of combat power an essential prerequisite for applying his universal principle, because once the decisive point is identified, it must be hit with all available force while the opportunity remains. A force divided is vulnerable, for it becomes susceptible to isolation and destruction by a larger, more concentrated force. Jomini writes: "The essential conditions for every strategic position are that it should be more compact than the forces opposed, that all fractions of the army should have sure and easy means of concentrating, free from the intervention of the enemy."[174]

Nevertheless, Jomini did observe instances where order and concentration were *not* necessarily superior, especially relating to cavalry. He observes that "regular" cavalry rarely get the better of "irregular" cavalry, "which will avoid all serious encounters, will retreat . . . and return to the combat with the same rapidity, wearing out the strength of its enemy by continual skirmishing."[175] And, irregular cavalry "acquire a habit of moving in an apparently disorderly manner, whilst they are all the time directing their individual efforts toward a common object . . . irregular charges may cause the defeat of the best cavalry in partial skirmishes; but . . . are not to be depended upon in regular battles."[176] This account of irregular cavalry is an exception to Jominian maxims; however, he did not believe these exceptions *invalidated* his principles. If a rule and maneuver works the vast majority of the time, writes Jomini, is the "occasional failure . . . a sufficient reason for entirely denying" its value?[177] "Shall a theory be pronounced absurd because it has only three-fourths of the whole number of chances of success in its favor?"[178] Perhaps not absurd, but likely in need of further refinement.

Central Position

One of Jomini's better-known maxims pertains to the advantages of a "central position." In short, an army in the central position can move coherently and concentrate rapidly to assault vulnerable appendages of an enemy force. This principle only applies when force ratios between opposing sides are roughly equal or, when they are not, for the side with greater strength. A weaker force in the central position is vulnerable to isolation and encirclement (i.e., as a "fractional" force, it becomes a decisive point). Jomini says: "If you are very weak and the enemy very strong, a central position, that may be surrounded on all sides by forces superior at every point, is untenable."[179] Even when not in a central position, a weaker force should only accept battle when it can achieve a localized advantage through surprise or maneuver. Jomini does not address the advantages a smaller or otherwise weaker force (e.g., a guerrilla or insurgent army) gains by *avoiding* a central position. These cases appear to contradict his principle. It is difficult for a large, coherent army in a central position to remain cohesive and still move quickly enough to pursue and engage small, mobile foes determined to prolong a war by avoiding decisive battle.

Lessons in Geometry

Since Jomini equates strategy with war on the map, not surprisingly, he develops an elaborate, geometry-based nomenclature and system for map-based strategy. In geometry, the shortest distance between two points is a line, and Jomini defines no fewer than ten different kinds of "lines," including *double, concentric, deep, long, maneuver, secondary, accidental, provisional,* and *defensive*.[180] No wonder Clausewitz dismissed this sort of rule-based minutia. It is not hard to imagine a poor general so preoccupied with all these lines that he loses sight of the actual battle! Jomini's most significant geometrical terms, *zones* and *lines of operation*, describe an army's area of operations and movements within that area, respectively. A *zone of operations* is a fraction of the overall theater of war, and within a zone there may be one or more lines of operation. Jomini's lines of operation apply only to physical, military actions and movements, whereas, *lines of communication* connect elements of an army or an entire force in a theater to its bases of support, command, and control.[181] *Lines of operation* indicate the direction an army travels on the map, and "*interior lines of operations*," writes Jomini, "are those adopted by one or two armies to oppose several hostile bodies, and having such a direction that the general can concentrate the masses and maneuver with his whole force in a shorter period of time than it would require for the enemy to oppose to them a greater force."[182]

Though *interior lines* and a *central position* are "two very different things" according to Jomini, they are still related. The central position is defined relative to other forces and is more of a tactical consideration that may not imply freedom of movement and maneuver. However, a force in a central position *with interior lines* has the freedom to use those lines to concentrate power at a decisive point without threat of force isolation. "Central positions, salient towards the enemy ... are the most advantageous," writes Jomini, "because they naturally lead to the adoption of interior lines and facilitate the project of taking the enemy in reverse."[183] As mentioned earlier, the benefits afforded a force in a central position with interior lines may be negated by an elusive, irregular force (i.e., that fights in a non-Jominian manner).

For Jomini, the proper selection and employment of lines of operation make it possible to achieve his prime maxim – applying concentrated force to a decisive point. Therefore, the "fundamental idea in a good plan of a campaign" is the choice of these lines based upon the geography and the disposition of the enemy.[184] Jomini recommends lines of operation "have a geographic and strategic direction" to threaten the rear or extremities of the enemy while simultaneously allowing an avenue of retreat either to the rear or sides.[185]

The orientation of lines relative to each other produces varied effects. For example, concentric operations converge on a single point, increasing concentration of power; whereas, eccentric operations diverge and expand outward, decreasing concentration.[186] Concentric operations are effective for disrupting enemy cohesion by piercing its lines, and eccentric operations can be used to pursue and destroy the disintegrated enemy force. Jomini describes this tactic as Napoleon's "favorite maneuver."[187] "This alternate application of extended and concentric movements," says Jomini, "is the true test of a great general."[188] Yet, concentric operations can backfire if the enemy moves faster and concentrates its power first. In these cases, if competently led, the faster, better-organized force will claim the central position and interior lines of operations enabling *it* to isolate and attack its enemy's extremities.

Maneuver and Battle

Of course, the raison d'être of lines, points, and maneuvers is to place the army in the best position to fight, to win a decisive battle, which Jomini considered the ultimate arbiter of the "great questions of national policy and of strategy."[189] Strategy "directs armies to the decisive points and has influence on the battles," but it is the battle itself, "aided by courage, by genius and fortune" that "gains victories."[190] Jomini chastises Clausewitz for his criticism of maneuver-based warfare saying, "Clausewitz commits a grave error in asserting that a battle not characterized by a maneuver to turn the enemy cannot result in a complete victory."[191]

Though he receives little acclaim for his insights into war's psychological dimension, Jomini admits that a battle's results often depend upon "a union of causes that are not always within the scope of the military art," including "the nature of the order of battle," the "wisdom" of the plan, the plan's execution, the performance of subordinate officers, the "cause of the contest," "the proportions and quality of the troops," their enthusiasm, the relative strengths of opposing cavalry and artillery, how they are handled, and more than anything else, "the *morale* of armies, as well as of nations."[192] "No system of tactics can lead to victory when the *morale* of an army is bad."[193]

Jomini identifies three kinds of battles: defensive, offensive, and unexpected (when marching armies collide).[194] "In every battle," asserts Jomini, "one party must be the assailant and the other assailed. Every battle is hence offensive for one party and defensive for the other."[195] He goes on to list two "undoubted truths": (1) the more simple a decisive maneuver is, the more sure of success will it be; and (2) sudden maneuvers seasonably executed during an engagement are more likely to succeed than those determined upon in advance.[196] These maxims stress the value of simplicity over complexity, and rapid, real-time battlefield maneuvers over predetermined plans. In the heat of battle, simple maneuvers are easier to execute with rapidity and precision. Implied in the second "truth" is the importance of battlefield awareness, decisiveness, and command flexibility based on the latest available information. Finally, despite his faith in guiding principles, Jomini rejects the notion of a one-size-fits-all approach to strategy and tactics saying, "My long experience has taught me to believe that nothing is impossible; and I do not belong to the class of men who think that there can be but one type and one system for all armies and all countries."[197]

Napoleon's Legacy

In his attempt to reconcile Napoleon's accomplishments with theory, Jomini draws a distinction between the "old" style of warfare, which he calls a "war of positions," and Napoleon's war of long marches and distant invasions. Foreshadowing the carnage of the world wars to come, Jomini saw the range, power, and devastation of Napoleonic war as something to be feared, akin to "the devastations of the barbarian hordes."[198] With Napoleon, says Jomini, "a great truth has been demonstrated ... that remoteness is not a certain safeguard against invasion."[199] Still, Jomini hoped that nations could coexist, only fighting for interests, small territorial gains, or to "maintain political equilibrium," and thus, would require smaller armies that would fight "a mixed system of war" short of the war of marches and position.[200] He adds: "Until then we must expect to retain this system of marches, which has produced so great results; for the first to renounce it in the presence of an active and capable enemy would probably be a victim to his indiscretion."[201]

Impacts of Technology

One of Jomini's biggest deficiencies, shared with Machiavelli, is his failure to anticipate how war's changing character – brought on by rapid advances in technology, armament, organization, and tactics – influences strategy. Jomini writes: "Strategy alone will remain unaltered, with its principles the same as under the Scipios and Caesars, Frederick and Napoleon, since they are independent of the nature of the arms and the organization of the troops."[202] Here Jomini is correct about that portion of strategy based on war's unchanging nature. However, strategy also makes choices regarding force composition, training, staging, support, and employment, and these character-of-war considerations can change dramatically. Jomini goes on to predict that technological improvements "will not introduce any important change in the manner of taking troops into battle"; and, "Two armies in a battle will not pass the day in firing at each other from a distance: it will always be necessary for one of them to advance to the attack of the other"; and "Victory may ... be expected by the party taking the offensive when the general in command possesses the talent of taking his troops into action in good order and of boldly attacking the enemy."[203] Here, Jomini fails to anticipate how war's character must change to address the realities of long-range, accurate, and exceedingly destructive weapons; hence why John Shy writes: "His actual ideas, particularly when seen through the monstrous prism of twentieth-century warfare, lend themselves readily to parody and ridicule."[204]

Conclusion

In truth, the full expanse of Jominian thought on the art of war is due more admiration than ridicule, and he provides a useful foil to Clausewitz. While Clausewitz eschews principles and certainty, Jomini sees principles as fundamental to the education of a good general and key to gaining advantages over enemies who also must deal with the fog of war. Where Clausewitz waxes philosophical, Jomini is more pragmatic. Both men valued the decisiveness of attack, concentrating power, and assessing physical and psychological capabilities.

Jomini's greatest weakness is his belief in the infallibility of his principles. Though they work quite well for most of the warfare of his day, they are less reliable for the small wars he disdained. There were also exceptions even during his era – Allied maneuvers versus Napoleon at Leipzig, for example – that contradicted his trust in the universal superiority of the central position and interior lines. Jomini's second major fault is his failure to appreciate the impact of war's changing character upon strategy and tactics. He does not show proper deference to more universal possibilities and the potential for advancement in technology and weapons that could render portions of his theory obsolete. Indeed, he, perhaps more than Clausewitz, is complicit in the carnage of the American Civil War and World War I, since the armies in those conflicts seemed determined to validate his admonition that no improvements in firepower could match the effect of bold attacks. Jomini was also remiss in marginalizing war's political dimension. He acknowledges that the art of war encompasses all war's facets yet excludes many of them. Therefore, despite many valid insights, Jomini's work falls short of a timeless and holistic theory.

Nevertheless, Jomini is due credit in several areas. His notions about why nations make war and his outline for how to construct a viable military remain relevant. Though he provides little practical advice for dealing with 'wars of opinion," his description of them and discussion of their complications and challenges are apropos. Moreover, the *idea* that we can derive militarily useful principles from history and experience is useful, provided we understand their limitations. The principles themselves, though not completely universal, are beneficial when applied within appropriate contexts. Jomini's main principle of concentrating force at a decisive point (e.g., against extremities of the enemy forces) is valid, but the devil is in the details and a great deal of work goes into identifying and addressing those points. Finally, Jomini had an excellent grasp of the important factors that contribute to battlefield success and for the need to understand oneself and the enemy as a basis for strategy.

One Jominian concept we ought not dismiss is this: Jomini believed the awfulness and complexity of war should never preclude societies from striving to understand it and from harnessing its power, when necessary, to defend civilization. In the face of war, we can and must do better than throw our hands up in despair or wish it away. While Jomini's great principles fall short in some areas, we could do far worse than follow his example in this regard.

CLAUSEWITZ

Clausewitz stands at the beginning of the nonprescriptive, nonjudgmental study of war as a total phenomenon, and On War *is still the most important work in this tradition.*[205]
                                                                    Peter Paret

His is not simply the greatest but the only truly great book on war.[206]
                                                                    Bernard Brodie

Now, having already referred to him extensively, our focus centers on the great Prussian warrior and theorist, Carl von Clausewitz. It is hard to overstate Clausewitz's importance and impact upon the study and practice of war. His ideas touch on nearly every meaningful aspect of war; thus, no examination of war or war theory, least of all this one, could be taken seriously were it to ignore him. As with the other authors presented, our goal is to build familiarity with the theorist; however, because of Clausewitz's primacy and complexity, we will spend more time analyzing and critiquing *On War*. Understanding Clausewitz's theoretical strengths *and* weaknesses is essential for avoiding missteps in the interpretation and application of his ideas.

On Clausewitz

My entrance into the world occurred at the scene of great events, where the destiny of nations was decided.[207]   Clausewitz

Clausewitz was born in 1780 in Prussia (now Germany), and his adult years spanned the first three decades of the nineteenth century (i.e., during the especially violent and dynamic Napoleonic era).[208] As we pointed out in the last section, this was a period of remarkable change in politics, war, and philosophy that included the Enlightenment and German Romanticism. In particular, the "German Movement," led by luminaries like Kant, Goethe, Fichte, and Hegel, and the rejection of Enlightenment precepts represented by romanticism, idealism, nationalism, and historicism would help shape Clausewitz's general theory of war.[209] In addition to these philosophical stimuli, Clausewitz's thought was influenced by Machiavelli, particularly regarding politics and moral factors, as well as fellow Prussians Georg von Berenhorst (on the power of human spirit, emotion, talent, courage, and moral forces)[210] and Clausewitz's principal mentor, Gerhard von Scharnhorst (on theory's importance and connecting war's elements to the whole).[211]

Clausewitz's martial fascination began with his initiation into the Prussian Army as an eleven-year-old lance-corporal and officer (at thirteen). Like many warriors of his day, he coveted the honors and glory of victorious battle. However, unlike his contemporaries, the often-melancholy Clausewitz approached his craft as a sober intellectual more than a swashbuckling dare-devil. He was always thinking about how to think about war. In 1804, he graduated first in his class from Berlin's Institute for Young Officers and became an adjutant to Prussia's Prince Ferdinand. During this time, he fought and lost to Napoleon in the Battle of Jena-Auerstädt (1806) and was held captive by the French until 1807. For many years afterwards, under Scharnhorst's tutelage, Clausewitz worked to reform the Prussian military, but he abruptly resigned in 1812 to join the Russian armed forces after Prussia was compelled to side with hated France. Several years after Napoleon's defeat at Waterloo and despite a prickly relationship with Prussian politicians, Clausewitz was eventually accepted back into military service in his homeland and conferred the rank of major general as head of the Military Academy at Berlin. It was mainly during this period (1818–30) that he organized his revelations into his signature work, *Vom Kriege* (*On War*). In its initial form – prior to an 1827 epiphany triggered by Clausewitz's belated realization that war rarely rises to the absolute – *On War's* purpose was to do what the disappointing body of existing theory could not – explain and provide an antidote to the problem of Napoleon's European domination. From 1827 until his untimely death, Clausewitz labored to rework his opus into a more comprehensive treatise inclusive of limited wars and the pervasive influence of policy and politics.[212]

Regarding Napoleon's influence, it is hard to imagine that Clausewitz (or Jomini) would have become a great theorist had the Corsican never existed. Though he had cause to hate Napoleon, Clausewitz also respected him immensely, dubbing him "the God of War."[213] All told, Clausewitz refers to Napoleon over *170* times in his masterwork. As noted earlier, Clausewitz's initial purpose with *On War* was to create a theoretical model to explain Napoleon's remarkable dominance and to identify ways of countering it. Indeed, Clausewitz's capstone chapter outlines a strategic plan whereby "France can be brought to her knees and taught a lesson any time she chooses to resume that insolent behavior with which she has burdened Europe for a hundred and fifty years."[214] It was only later, as he perceived the need to adjust his theory to address the predominance of war's limited forms, that Clausewitz reexamined and recast his initial premises.

From Clausewitz's perspective, *On War* filled a theoretical void – exploited with devastating effectiveness by Bonaparte – by presenting a timeless philosophy that explains the primacy of decisive battle amidst war's inherent violence and uncertainty while also framing war as a political tool subject to political constraints. Clausewitz expected his theory would arm statesmen and generals (preferably in Prussia) with the necessary knowledge to prevent the recurrence of Napoleonic supremacy and wholly eclipse the deficient military theory of his time (he described the work of his fellow Prussian, von Bülow, as "laughable" and "the children's military companion").[215] As he pondered the appropriate value, form, and purpose of war theory, Clausewitz concluded that it can be no more than a guide, because war's many variables, including chance and human factors, defy prescriptive treatment. Anyone seeking answers to war's riddles, advises Clausewitz, must view with skepticism theories that equate victory with adherence to their principles. Clausewitz was contemptuous of *ancien régime* (~1400s–1789) theorists who believed the highest expression of the art of war was the intricate movement of forces rather than concentration for decisive battle, for "they are convinced that the balanced, sterile, pointless game is the very zenith of development ... a view so lacking in logic and insight that it must be considered a hopeless confusion of values."[216] *On War* is Clausewitz's warning, to prevent leaders from seeing war as merely a ritual, risky pastime, or dangerous game, notions sharply and permanently erased by Napoleon. "It is quite possible," Clausewitz writes derisively,

that at some time in the future, Bonaparte's campaigns and battles will be considered brutalities, almost blunders, while the old-fashioned dress sword of antiquated and desiccated manners and institutions will be relied upon and praised. If the theorist can point out the dangers of this attitude, he will have provided an essential service to those who care to listen.[217]

Regrettably, Clausewitz never finished *On War*. He died in 1831 of cholera, before he could consolidate his notes and edit them into a complete, refined work, particularly regarding the evolution of his thoughts after 1827. As a result, *On War* is not the finished product Clausewitz intended. Nevertheless, when carefully and critically read, Clausewitz's major themes emerge clearly. He teaches us how to identify and think about war's "permanent elements" and, to a lesser degree, presents a guide for how to plan and fight a war.[218] His emphasis on the philosophical and moral versus the practical and prescriptive is where Clausewitz cleanly breaks ranks with his contemporaries and also loses touch with military leaders interested more in battle tactics. Though he believed some leaders have an innate ability or genius for war, Clausewitz knew the rarity of these savants necessitates sound theory for the rest. Thus, *On War* addresses a variety of topics and includes caveats to prevent future strategists from committing the errors he studied and witnessed during his professional life.

Clausewitz's Style

This endemic misinterpretation of Clausewitz's ideas has been mainly due to the failure to grasp fully the origins and nature of the transformation of his thought and, particularly of the new intellectual forms in which this transformation was expressed.[219]

Azar Gat

To reconcile concepts of absolute and limited war within one body of theory, writes Gat, Clausewitz turned to "the most ambitious intellectual attempt at an all-encompassing and integrative explanation of all the contrasts and contradictions of reality; namely, the German idealistic philosophy which was elevated by Hegel."[220] Thus, in *On War*, Clausewitz employs an "ideal-type" methodology[221] and "dialectic reconciliation"[222] to characterize an abstract, "ideal" form of war that serves as a reference for war's "real" forms.[223] He then defines and discusses each element of his theory, keeping in mind its relation to the reference. Not surprisingly, this esoteric approach, which Gat explains is the expression of his late "transformation" of thought, is a key reason Clausewitz's work remains opaque to so many.[224] As Liddell Hart writes: "His theory of war was expounded in a way too abstract and involved for ordinary soldier-minds. Impressed yet befogged, they grasped at his vivid leading phrases, seeing only their surface meaning, and missing the deeper current of his thought."[225] Our purpose, then, is to strike a balance between these "deeper currents" and what is truly of practical value to warfighters and strategists.

Ideal War versus Real War

On War is divided into eight "books." In the first chapter of Book 1 (the only chapter he considered "finished"), Clausewitz introduces his central thesis.[226] War considered only in the abstract, as a purely military clash, has but one object – disarming and defeating the enemy by violent force. Each side must pursue that end without reservation or delay, lest they grant their opponent an advantage. Thus, each seeks to employ the *greatest possible force* (composed of total physical means and strength of will) in the *shortest possible time*.[227] This rationale assumes the swift use of overpowering violent force is always optimum for disarming and defeating the enemy. Clausewitz calls this the "ideal," "absolute," or "perfect" form of war – a form no "real" war could ever attain but might approximate when the aim is the complete defeat of an enemy state and its resources are fully employed toward that end.

Sun Tzu says the acme of skill is to win without fighting; however, for Clausewitz, war *must* include fighting, which is most naturally directed at neutralizing the other side's warfighting instrument, its armed forces. Clausewitz's acme of skill is, therefore, to disarm the enemy as thoroughly and rapidly as possible. Assuming the speed and efficacy of disarming the enemy is proportional to the force advantage, it is illogical to withhold any force or delay its use. Though hypothetical, Clausewitz believed his ideal war concept held the key to unlocking a unique and dominant set of strategic considerations. No previous theory could make the same claim because no previous theoretician had Napoleon's example as model and motivation. Clausewitz muses: "One might wonder whether there is any truth at all in our concept of the *absolute* character of war were it not for the fact that with our own eyes we have seen warfare achieve this state of *absolute perfection*" (emphasis added); [228] and, "In the campaigns of Bonaparte, warfare attained the *unlimited* degree of energy that we consider to be its elementary law. We see it is possible to reach this degree of energy; and if it is possible, it is necessary" (emphasis added).[229]

Having described war's perfect form, Clausewitz continues his dialectic by analyzing factors that cause real war to differ from its ideal. These include "real world" influences like logistics, time and space considerations, alliances, human frailty, imperfect knowledge (the fog of war), unforeseen occurrences and chance (friction), and decisions based upon the relative strengths of each side (more on "fog" and "friction" shortly). Above all, Clausewitz asserts, war fails to achieve the absolute because it is an *act of policy* and therefore inseparable from politics, which often dictates ends short of the total defeat of an enemy and employs ways and means with resource limitations. "The reason [for war]," writes Clausewitz, "always lies in some political situation," and thus, war must be an inseparable subset of politics, "a true political instrument, a continuation of political intercourse, carried on with other means."[230] More concisely, "*war is merely the continuation of policy by other means.*"[231]

Chance, Danger, and the Moral Dimension

War that falls short of its absolute form in intensity, object, and duration, says Clausewitz, becomes subject to chance and probability, forcing leaders to navigate uncertainty and account for moral and physical factors. Clausewitz is the leading theorist regarding the inclusion of "moral" (meaning *psychological* – as in "demoralize," as opposed to *ethical* – as in "immoral") factors, and he emphasizes that "military activity is never directed against material force alone."[232] Of these moral elements, Clausewitz highlights the value of courage and self-confidence, just as Thucydides did, for war's danger and violence inspires paralyzing fear in those unaccustomed to its risks.

The Birth of Strategy

War's dependence on the interplay of innumerable variables, including the characteristics of combatants and their aims, leads Clausewitz to advise:

The first, the supreme, the most far-reaching act of judgment that the statesman and commander have to make is to establish by that test the kind of war on which they are embarking; neither mistaking it for, nor trying to turn it into, something that is alien to its nature.[233]

Clausewitz's "test" pertains to the examination of the "nature" of each side's "motives" and the "situations which give rise to them."[234] Here Clausewitz echoes Sun Tzu's dictum, *know the enemy, know yourself*. He knew that Europe had failed this test with Napoleon, having fought based on *ancien régime* customs (i.e., limited aims and negotiated peace). But Napoleon's strategy was to field the most powerful force, crush his enemies in battle, and dictate political terms to the vanquished.

The Clausewitzian Trinity

Clausewitz concludes his first and best chapter with the "trinitarian" definition introduced in Chapter 1. To recap, Clausewitz's war is

composed of primordial violence, hatred, and enmity which are to be regarded as a blind natural force; of the play of chance and probability within which the creative spirit is free to roam; and of its element of subordination, as an instrument of policy, which makes it subject to reason alone.[235]

The continuous interplay of emotion and violence, chance and creativity, reason and policy creates a chaotic context for war's participants – the people, the government, and the warriors. Here, Clausewitz sets the tone for the rest of his opus, warning us that failure to consider all elements of the trinity is to misunderstand war and court disaster.

War Is Fighting

Throughout the remainder of *On War*, Clausewitz introduces and then expounds upon the various aspects of his theory, defining the elements of strategy, outlining the fundamentals of battle, examining the characteristics of offense and defense, and finally, providing guidance for war planning. He reminds us that *fighting* sets war apart from other acts of politics; and, while the natural object of fighting is the destruction of the enemy's fighting capacity, it can also include objects like seizing territory or surviving until conditions change. Once the fighting begins, Clausewitz says, "The ultimate object is the preservation of one's own state and the defeat of the enemy's."[236] Since, according to Clausewitz, armed forces depend on physical and moral elements, fighting to disarm the enemy (targeting capacity) is only part of the calculus. War must ultimately break the enemy's *will to fight*.

Properties of War

Clausewitz delineates war into "levels" beginning with the highest, which sets the war's political aim and then the supporting levels: "strategy" (coordinating theater battles or engagements to support a campaign); and "tactics" (fighting in a single battle or engagement). Clausewitz defines the "elements" of strategy as "moral" (intellectual and psychological), "physical" (size, composition, armament, et cetera), "mathematical" (movements in time and space), "geographical" (terrain or weather), and "statistical" (relating to logistics, maintenance, and support).[237] "War can be of two kinds," says Clausewitz, "in the sense that either the objective is to *overthrow the enemy* – to render him politically helpless or militarily impotent . . . or *merely to occupy some of his frontier-districts.*"[238] In essence, Clausewitz contends that war can be for *unlimited* (conquest) or *limited* aims (territory or concessions), which includes wars "of extermination down to simple armed observation."[239]

Attack and Defense

At every level of war, Clausewitz writes, each side chooses whether to attack or defend. Using absolute war as a reference, Clausewitz asserts that the obvious route to achieve war's central aim is to attack, but his theory does not allow both sides to be in the same state (attack or defense) simultaneously at the same level of war. Therefore, since there are always two sides in war, one must necessarily be on the defense when the other is on the attack and vice versa. Furthermore, Clausewitz proclaims, somewhat notoriously, that the defense *must* be the *stronger* form of war.[240] In a way, the logic from his first chapter (i.e., by its nature, war rewards those who strike first with the most powerful blow) compels Clausewitz to this conclusion. Without it, war would have no pauses since the stronger side would always be attacking; additionally, not every attack is successful, even with a strength advantage. Therefore, Clausewitz must conclude that the defense is a stronger if less decisive form. In an absolute sense, if the defense were *not* inherently stronger, there would be no reason to go on the defensive, because its *negative* aim (to await the enemy's attack) does not serve war's fundamental purpose: to destroy the enemy force.[241]

Despite his claim that the defense is stronger, Clausewitz affirms that the offensive alone forces a decision in minimal time. The offensive compels the weaker side to choose a defensive form of war to leverage that form's intrinsic strength. The defender will only move to the offensive once its strength surpasses the attacker's. An attacker forced to retreat does not accrue the same advantages of the defensive form as a defender who has had time to prepare its defense and await the attack.

Clausewitz declares the engagement or battle, where offense and defense collide, as the heart of war. Only in battle can two determined foes vie to defeat each other, with victory claimed by the side that gains control of the battlefield. Should the defeated foe maintain enough cohesion to retreat intact, Clausewitz advocates for immediate pursuit to exploit the psychological benefits of victory and the physical strength advantage enjoyed by the attacker.

Military Genius and the Environment of War

A central element of Clausewitz's system is the "military genius." Indeed, Clausewitz's theory without this leads to unsupportable conclusions about war relative to history and his own observations. Genius is the only way to account for great generalship given the uncertainty, chaos, and danger woven into the environment of war, which Clausewitz says may appear simple when defining war aims but, in practice, is extremely complex. War's complexity, amplified by danger, chance, a thinking adversary, and a multitude of other variables, prevents theory from becoming more than a guide.

Fog and Friction

And we are here as on a darkling plain
Swept with confused alarms of struggle and flight,
Where ignorant armies clash by night.[242]

Matthew Arnold, from *Dover Beach* (1867)

Before looking further into Clausewitz's concept of genius, we need to discuss "fog" and "friction," two familiar terms he adapts to describe the effect war's complexity has upon the plans and activities of warfighters. Fog and friction emanate from the "climate" of war, which Clausewitz says is composed of four elements: "danger, exertion, uncertainty, and chance."[243] Moreover, fog and friction are interrelated and can intensify each other. For Clausewitz, the fog of war obscures truth and reality. It exists because we are neither omniscient nor always capable of discerning truth or correctly interpreting sensory stimuli. Moreover, technological aids often fail to function with expected accuracy, precision, or reliability. Clausewitz writes: "Three quarters of the factors on which action in war is based are wrapped in a fog of greater or lesser uncertainty,"[244] and "All action takes place ... in a kind of twilight, which, like fog or moonlight, often tends to make things seem grotesque and larger than they really are."[245]

War's physical domain is composed only of what is real, yet the cognitive domain contains *perceptions* (i.e., what the mind *sees* as real but may not be). In war, the gulf between reality and perception can be extreme. War's inherent "suffering and danger" causes a "psychological fog" that exacerbates this gap.[246] Obviously, the inability to see what is happening or to get accurate reports amplifies this condition. Though he understood the need for intelligence, Clausewitz's experiences caused him to discourage blind faith in it, a practice that can lead to dubious or calamitous decisions.

Contributing to reality-perception and planning-execution gaps is Clausewitz's "friction." Here he borrows a term the physical sciences use to describe the energy-sapping resistance that occurs between objects. Clausewitz defines friction as a negative effect caused by real-world, generally unforeseeable perturbations amongst war's innumerable variables. These are the unseen gremlins that lead Clausewitz to opine, "Everything in war is very simple, but the simplest thing is difficult."[247] "Friction," writes Clausewitz, "is the only concept that more or less corresponds to the factors that distinguish real war from war on paper"; and "War has a way of masking the stage with scenery crudely daubed with fearsome apparitions."[248] Examples of friction include individual responses to danger, fear, exertion, and fatigue; the problem of intelligence or "*accurate recognition,*" which "constitutes one of the most serious sources of friction in war, by making things appear entirely different from what one had expected";[249] and the "countless minor incidents" that defy precise measurement or forecasting.[250] Clausewitz names only one remedy for friction: "combat experience."[251]

Genius Transcends

Genius is a talent for producing that for which no definite rule can be given.[252]
Immanuel Kant

Clausewitz's concept of military genius helps explain why the pervasive influence of complexity, fog, and friction do not completely overwhelm the human capacity for competent war-making, a condition that would cause chance alone to become war's prime determinant (a violation of his trinity). But Clausewitz knew from his own observations that great generalship *was* possible. Reconciling this reality within the context of complex, dangerous, and chaotic war, required Clausewitz to develop a transcendental concept – the military genius, "*which rises above all rules.* Pity the soldier who is supposed to crawl among these scraps of rules, which are not good enough for genius, which genius can ignore or laugh at ... what genius does is the best rule."[253] Clausewitz's military genius assimilates all war's variables, including the adversary's most probable actions, through the fog of imperfect intelligence, human weakness, and the dangers of combat.[254] These rare and talented leaders remain calm and "see" the correct way forward with an "inward eye," intuition, or "*coup d'oeil.*"[255] *Coup d'oeil* (what Germans call *Fingerspitzengafuhl*) grants a leader the "ability to see things simply, to identify the whole business of war completely with himself . . . only if the mind works in this comprehensive fashion can it achieve the freedom it needs to dominate events and not be dominated by them."[256] Clausewitz's genius possesses attributes beyond intellect, including courage (to face personal danger and accept responsibility), strength of mind (decisiveness in the face of changing conditions), boldness (to accept greater risk for greater gain), emotional maturity (to prevent passion from clouding judgment), and character – all instilled by experience. Military genius does not require formal erudition in war theory nor is it restricted by rules. In fact, it is not unusual for a military genius to use rules *against* an enemy, breaking from routine practices or doctrine to achieve an advantage, much like a chess master sacrifices a rook or queen to jeopardize the opponent's king. Genius, says Clausewitz, can eliminate the "gap between principles and actual events that cannot always be bridged by a succession of logical deductions."[257]

Finally, Clausewitz associates true military genius with higher echelons of command. The intellectual demands placed upon military leaders grow with rank. Great tacticians do not necessarily make great strategic leaders where, Clausewitz believed, the mental demands are highest, for "with every step up the ladder; and at the top – the position of commander-in-chief – it becomes among the most extreme to which the mind can be subjected."[258]

Purpose and Objectives

For Clausewitz, the art and science of war require comprehension of fog, friction, and chance; moral and physical strength; war's role as an instrument of policy; and most crucially, the relationship between strategy and the political objective. Clausewitz advises: "No one starts a war, or rather, no one in his senses ought to do so – without first being clear in his mind what he intends to achieve by that war and how he intends to conduct it. The former is its political purpose; the latter its operational objective."[259] When the political aim is total defeat of the enemy, war gravitates toward its absolute form, which necessitates the greatest possible force prepared and concentrated for decisive battle at the right time and place. Strategy coordinates the ways and means to do this, with the ultimate purpose of breaking the enemy's will to fight.

Clausewitz generally denounces strategies that seek victory by indirect actions. For example, he acknowledges the occasional value of surprise and deception, but his own experience and concept of absolute war lead him to counsel against relying on such methods to secure great victories. Additionally, Clausewitz believed that intricate plans and maneuvers and limited objectives siphon power from what should be the main effort – decisive battle. In contrast with Sun Tzu, Clausewitz's experiences with unreliable intelligence and communications and with Napoleon's overt dominance led him to distrust clever schemes. After all, "Bonaparte could ruthlessly cut through all his enemies' strategic plans in search of battle," says Clausewitz, "because he seldom doubted the battle's outcome."[260] When his opponents confronted Napoleon with "subtler (and weaker) machinations, their schemes were swept away like cobwebs."[261]

Center of Gravity

During the Napoleonic Era, defeating the French army was unquestionably the key to subduing France. In keeping with his affinity for co-opting scientific terminology, Clausewitz labeled this prime target, upon which the all the fortunes of a battle or war hinge, the "center of gravity" (*Schwerpunkt*), a term still in use although with an evolved meaning. Identifying a Clausewitzian center of gravity (COG) requires examination of the "dominant characteristics of both belligerents," for the COG is "the hub of all power and movement, on which everything depends. That is the point against which all our energies should be directed."[262]

Examples of Clausewitzian COGs include an enemy's main army, when it has a dominant sovereign as general (e.g., Napoleon, Alexander, or Charles XII).[263] For countries "subject to domestic strife," it is the capital; and in smaller countries, the army of their protector or ally. Interestingly, in an alliance, the COG is the "community of interest," and in popular uprisings, "the personalities of the leaders and public opinion."[264] Clausewitz viewed the COG as a weapon and a target, saying, "A center of gravity is always found where the mass is concentrated most densely. It presents the most effective target for a blow ... [and] the heaviest blow is that struck by the center of gravity."[265]

Lines of Communications

Like Jomini, Clausewitz uses the phrase "line of communications" (LOC) to refer to an army's connection to its headquarters, command and control (C2) elements, supply sources, and avenue of retreat.[266] LOCs are the arteries and nervous system of a fighting force, essential to sustained operations and potentially vulnerable, depending on their position, length, relationship to the enemy, and defensibility.

Culminating Point

One who understands the Tao of Warfare will invariably first plan against the defeats which arise from not knowing where to stop.[267]   Wei Liao-Tzu

Another Clausewitzian term used today is the "culminating point of victory"[268] or "culminating point of attack,"[269] which refer to the time and place an attacker (at the strategic and tactical levels, respectively) loses its physical or moral strength advantage and therefore becomes vulnerable to counterattack. When this happens, a force without a net positive change in strength must become defensive unless the enemy fails to recognize and act on its capacity advantage. An attacking force during combat operations may accrue physical or psychological strength via reinforcements from home or allies, resources siphoned from the enemy, from tactical victories, or even good news. On the other hand, a defender can gain strength by falling back on sources of supply, through the desperation of defending home soil, by reinforcement, or by frustrating the attacker's advance or avoiding a decisive clash. Meanwhile, casualties, weather, disease, and combat fatigue wear down combatants on both sides.

Many other factors contribute to the calculus of culmination, and no simple formula can identify it with perfect clarity. Clausewitz warns: "There is . . . no infallible means of telling when that point has come; a great many conditions and circumstances may determine it."[270] Even so, forecasting and identifying the culminating point are essential to strategy. Indeed, Clausewitz advises: "The natural goal of all campaign plans . . . is the turning point at which the attack becomes defense," for to go beyond that point would not only be "useless" but "damaging."[271] The difficulty of recognizing this point through the fog and friction of war is complicated by nonphysical factors like psychological "momentum," which often occurs after a string of victories. Writes Clausewitz:

Once the mind is set on a certain course toward its goal, or once it has turned back toward a refuge, it may easily happen that arguments which would compel one man to stop, and justify another in acting, will not easily be fully appreciated . . . the action continues, and in the sweep of motion one crosses the threshold of equilibrium, the line of culmination, without knowing it . . . [finding] it less difficult to go on than to stop.[272]

Culmination cannot be viewed only from the attacker's perspective. In war, combatant capacities fluctuate, rising or falling with casualties, destruction of materiel, or reinforcements. A sudden change in capacity for either side, like the infiltration of Chinese soldiers at the Yalu in the Korean War, can rapidly trigger culmination.

Winning Strategy

Though he meticulously defines many of war's key components, like COG, fog, friction, and genius, Clausewitz persistently cautions against reducing war and strategy to a mathematics problem.[273] The successful strategist or general must remember that war is always a bloody and dangerous clash of physical and psychological elements, never a game, and its purest form occurs where military and political objects merge and opposing sides marshal everything to disarm each other. In this unalloyed form, the war's end occurs when one side disarms the other. However, "in practice," writes Clausewitz, the end of war by battlefield decision can "be replaced by two other grounds for making peace: the first is the improbability of victory; the second is its unacceptable cost."[274] In other words, wars end short of full disarmament, because conditions *deter* one or both sides from continuing the fight.

When outright victory is impractical, Clausewitz recommends three methods for convincing the enemy to capitulate: (1) invasion to secure territory or render its resources useless to the enemy; (2) operations that "increase the enemy's suffering"; or (3) "wear[ing] down the enemy ... using the duration of the war to bring about a gradual exhaustion of his physical and moral resistance."[275] Past leaders in insurgencies and defensive wars have exploited this model successfully versus a superior foe reliant on extended lines of support (e.g., Vietnam).

Though not all wars end with disarmament, Clausewitz warns us not to lose sight of war's ideal form. Returning to his central thesis, he reemphasizes that war's aim is "to overcome the enemy's will," with the choice of methods dependent on the circumstances, but most properly by neutralizing the adversary's warfighting capacity.[276] "Everything is governed by a supreme law, the decision by force of arms," says Clausewitz, and "of all the possible aims in war, the destruction of the enemy's armed forces always appears as the highest."[277]

People's War

Despite his conventional war bias, Clausewitz was quite familiar with the characteristics of "national wars" or "popular wars," where a people rise to fight an occupying invader. He could hardly ignore this form of war since Spanish guerrillas orchestrated some of the most effective military actions he witnessed against Napoleon. Still, Clausewitz's chapter on people's war stands apart from the rest of his theory, most likely because he lacked examples of similar conflicts, and because guerrilla war seemed a distant relative to the absolute form of war that preoccupied him. In truth, Clausewitz is due more credit here than he generally receives, especially given his earlier writings and lectures on small wars, but relative to the balance of *On War*, this chapter leaves us craving further development. To his credit, Clausewitz admits his discomfiture with this form of war saying, "This discussion has been less objective analysis than a groping for the truth. The reason is that this sort of warfare is not as yet very common."[278]

During his time, Clausewitz accepted "war by means of popular uprisings" as a legitimate part of the broader context of war.[279] Indeed, he saw its growing prevalence "as an outgrowth of the way in which the conventional barriers have been swept away in our lifetime by the elemental violence of war. It is, in fact, a broadening and intensification of the fermentation process known as war."[280] Yet, Clausewitz does not fully explain how this expanding "fermentation process" integrates within the unlimited-limited war dialectic. A people in arms, says Clausewitz, somewhat paradoxically, "gain some superiority" but may also squander resources that "might be put to better use in other kinds of warfare."[281] Clausewitz correctly links Napoleon's dominance with the emergence of guerrilla war; however, his theoretical system stops short of fully defining the dynamics of this process.

Clausewitz was intrigued by how a people in arms, dispersed throughout a country and often without organized, centralized C2, could effectively wage war against a more concentrated, traditional foe. He characterizes this type of warfare as "scattered," requiring time to "consume the basic foundations of the enemy forces," thus, it is something of an antithesis to war's ultimate form.[282] According to Clausewitz, the nature of guerrilla forces "usually denies them the qualities and virtues . . . vital to concerted action by even moderately substantial forces," because their relatively inferior numbers and lack of combat power would most likely lead to defeat in a face-to-face engagement with a stronger enemy.[283] "The more forces are divided," contends Clausewitz, "the less they can be controlled," and "operations become more and more fragmented"; however, "in war, the sum total of individual successes is more decisive than the pattern that connects them."[284] Here, he admits that what would normally be a fault in battle – divided forces and fragmented operations – is less costly for a people in arms, because their actions rely more on cumulative effects than decisive engagements.

Clausewitz identifies five conditions weaker forces require to remain effective: (1) the fighting is deep in the interior of the country (where the smaller force has greater access to support and the invader operates on extended LOCs); (2) they avoid a decisive battle (where the stronger force has the upper hand); (3) they have a large theater of operations (to provide avenues for maneuver and escape); (4) the national character must "be suited" to a guerrilla war (patience and tenacity); and (5) the country itself offers "rough" and "inaccessible" terrain (which favors the smaller, agile force).[285] Insurgents, says Clausewitz, accrue psychological advantages that amplify effectiveness beyond balance-sheet logic; moreover, by operating in small, widely spread elements "not much is lost if a body of insurgents is defeated and dispersed."[286] Their strength lies in resolve and the will to fight by any means, and as Clausewitz stresses, a strong national character and favorable terrain.[287]

Clausewitz hints at the universality of dispersal tactics saying, "Even a professional standing army will develop something of a national army's qualities" as it uses all options to defend its home soil and preserve for itself "greater independence of action."[288] Still, Clausewitz advises against *remaining* scattered. "There must be some concentration at certain points: the fog must thicken and form a dark and menacing cloud out of which a bolt of lightning may strike at any time."[289]

Critiquing Clausewitz

If an early death should terminate my work, what I have written so far would, of course only deserve to be called a shapeless mass of ideas. Being liable to endless misinterpretation, it would be the target of much half-baked criticism.[290]
Clausewitz

Critiquing Clausewitz entails some risk; however, the unfinished state and expanse of his theory requires careful review so we can distinguish between his best ideas and his less persuasive arguments. Given Clausewitz's stated dissatisfaction with his notes, it is likely the uncritical acceptance of every word in *On War* would only have earned his disdain. To be fair, we will not admonish Clausewitz for having left an unfinished work, for being a product of his era, or for failing to foresee aviation, nuclear weapons, and the digital age. We will also refrain from speculating about how Clausewitz *might* have updated *On War* had he lived longer. Instead, this critique tests Clausewitz's premises, assumptions, methods, and conclusions for consistency, logical validity, and accord with the accepted truths of war and human nature.

The Unreality of Perfect War

He began to perceive that it was logically unsound to assume as the foundation of a strategical system that there was one pattern to which all wars ought to conform.[291]
Julian Corbett

Our critique begins with Clausewitz's ideal-type methodology, which is confusing and leads to some convoluted arguments and arguable conclusions. Much of Clausewitz's theory rests upon the abstract concept of an absolute or perfect form of war that occurs when the only aim is military (i.e., to disarm an enemy by force). This is a theoretical absolute in which each side strives to disarm the other in a decisive clash characterized by continuous action and maximum force. Clausewitz defines war "in the abstract" or the "pure concept of war" as "a clash of forces freely operating and obedient to no law but their own."[292] Real war, on the other hand, can approach the perfect form but generally falls short due to mitigating influences.

Sun Tzu also uses absolutes to describe war: *know the enemy and know yourself: in a hundred battles you will never be in peril*; and *to subdue the enemy without fighting is the acme of skill.* Sun Tzu's absolutes, though as unattainable as Clausewitz's, are reminders of the value of intelligence, self-awareness, and mindfulness of war's ultimate object. At their core, they check the human inclination to see war unilaterally or to view war's fighting as the true object. Therefore, with respect to Sun Tzu's theory, the odds of success generally increase with proximity to the ideal. In contrast, Clausewitz's perfect form omits the human element and political objective by design. In fact, he claims these factors detract from war's purity. Thus, pursuing the Clausewitzian ideal is more about adherence to a soulless imperative than breaking an enemy's will or achieving a political objective. Remember, Clausewitz asserts that war in this form is obedient to its own law alone. This is confusing because we are accustomed to viewing an ideal as something to strive for. While success is not impossible for forces emulating Clausewitz's absolute war, we cannot necessarily conclude that it is always preferable.

Clausewitz presumably adopted this ideal-type method out of admiration for the example set by Newton, Kant, and Hegel and for its potential to unify his absolute and limited forms of war within a comprehensive, dialectical system.[293] Liddell Hart writes, "The influence of Kant can be perceived in Clausewitz's dualism of thought. He believed in a perfect (military) world of ideals while recognizing a temporal world in which these could only be imperfectly fulfilled."[294] Religious doctrine provides another possible influence (though there is no hard evidence to confirm it) as the idea of a perfect form unattainable due to human imperfections has distinctly theological overtones. Finally, it is possible that Clausewitz believed this style would lend his work the philosophical gravitas other war theories lacked. The "obscure and elaborate reasoning" found in *On War*, writes Gat, "only added to Clausewitz's image of profundity" and demonstrated "the 'philosophical' manner of expression that was only to be expected of a philosophical masterpiece on war."[295]

Clausewitz probably favored the ideal-type for many of these reasons, but especially because it enabled him to preserve his preferred view of war as an absolute clash of force while still accounting for war's subordination to policy. Clausewitz surmised that Napoleon's dominance resulted from the confluence of two of his cherished ideas: *the most perfect form of war in the hands of a military genius*. Napoleon was Clausewitz's reference point; likewise, ideal war serves as a reference point for examining real war. Writes Clausewitz: "Only if war is looked at in this way does its unity reappear; only then can we see that all wars are things of the *same* nature; and this alone will provide the right criteria for conceiving and judging great designs."[296]

Clausewitz's theory contrasts real war with the absolute by applying "modifications" (i.e., those innumerable factors across both physical and cognitive domains that cause war to depart from its purest logic). Says Clausewitz: "War can be thought of in two different ways – its absolute form or one of the variant forms that it actually takes."[297] This quote proves that understanding Clausewitzian war requires consideration of both sides of the dialectic: actual *and* absolute war. However, this is frustrating because most of us, particularly pragmatic warriors, prefer the real and tangible to the abstract. "The longer one continues the search for such omnipotent abstractions," writes Liddell Hart, "the more do they appear a mirage, neither attainable nor useful – except as an intellectual exercise."[298] Clausewitz's readers cannot help but ask: is he talking about the characteristics of *real* or *perfect* war?

In fact, his use of the ideal-type requires frequent caveats as he labors to explain the divergence between ideal and real war. For example, though a perfect war is fought with maximum force in the shortest period without pause, in reality Clausewitz says that war is "not a case in which two mutually destructive elements collide, but one of tension between two elements, separate for the time being, which discharge energy in discontinuous, minor shocks."[299] He admits, "The history of warfare so often shows us the very opposite of unceasing progress toward the goal" and that "*immobility* and *inactivity*" are the "normal *state* of armies in war, and action is the exception. This might almost make us doubt the accuracy of our argument."[300] Indeed! Theory that contradicts reality loses its legitimacy – as Clausewitz himself says, "a satisfactory theory of war" is "one that . . . will never conflict with reality," and thus will "end the absurd difference between theory and practice that unreasonable theories have so often evoked."[301]

In the second chapter of his final book, Clausewitz admits that his logical model for war creates a dilemma. He concedes that wars would likely occur as "a matter of degree," but insists that as a theorist, he has "the duty to give priority to the absolute form of war and to make that form a general point of reference."[302] Clausewitz does this because he believed absolute war to be the most dangerous and decisive form, and this reflects his continued partiality for absolute war despite his revised stance after 1827. While this approach has merit, it also marginalizes war's other, often more prevalent forms.

Maximum Force or Farce

Looking deeper into Clausewitz's definition of perfect war, we find more problems. The supposition that war's inherent laws compel each side to bring maximum force to bear to defeat the other rests on the assumption that a relative advantage in material strength or the perception of advantage is the key to defeating the enemy. Recall, Clausewitz says that war has an intrinsic logic independent of human will or context that drives each side to employ maximum force. Setting aside, for now, our unease with a concept of war without humans, does this make logical sense?

Balance of Power

Consider the following scenarios using Clausewitz's perfect war (not real war) assumptions and the relative strengths of opposing sides. First, in a hypothetical instance where each side has equal capacity and equal desire to fight, war becomes illogical, and strategy is stillborn (as we determined in the section on Sun Tzu). Both sides would arrive at the point where all is ready, but neither would have a perceivable advantage and therefore no reason to expect the battle would yield anything other than mutual destruction. A clash between forces of equal power but opposite aims results in the annihilation of both sides (like matter and antimatter). This scenario exemplifies warfare on the Western Front during World War I, nuclear deterrence theory, and the real-war logic of "balance of power."

Imbalance of Power

Unbalanced circumstances are unquestionably the most prevalent, and the only realistic condition between two belligerents. Again, following Clausewitz's perfect-war logic, the weaker side gains nothing and loses everything by marshalling and then expending all its force in a climactic battle, for in the end, the stronger force always wins. Thus, war in this case, like the previous one, would not occur unless the weaker side misperceives the imbalance, is indifferent to losing, or gains an advantage outside the parameters of Clausewitz's perfect-war logic. Only when we reintroduce the "impurities" of human perception, fallibility, politics, and chance does war become possible again, including the likelihood of victory for the "weaker" side. Given the impracticality of these scenarios, we are again left to wonder what necessary purpose the ideal-type serves.

The Inhumanity of War

[Clausewitz] saw that he had been working on too narrow a basis – a basis that was purely theoretical in that it ignored the human factor.[303]

Julian Corbett

As we have mentioned, Clausewitz's ideal war omits the human element. Even as a tool of abstract reasoning, there is something unsettling about a theory that leaves out the phenomenon's most important and causal component. War is *not* a natural phenomenon reduceable to a "pure" form. In the physical sciences, we often eliminate complexity-inducing variables to help solve mathematical problems (e.g., assumptions of a *frictionless* surface, a *pure* substance, or an *incompressible* fluid). But war is different. Rather than starting with humanity, Clausewitz adds the human element back into his theory as an *impurity*. In fact, he cites human weaknesses as prime factors in the reduction of war from its absolute form. Among these are fear, indecision, aversion to danger, the "imperfection of human perception and judgment,"[304] and "that naturally timid creature, man."[305] These factors "reduce war to something tame and half-hearted."[306] "No matter how savage the nature of war," writes Clausewitz, "it is fettered by human weaknesses"[307] and "a practice deeply rooted in the frailties and shortcomings of the human race."[308] Some would argue that human weakness is the *genesis* of war. The history of real war proves that maximum force is not "better" in all cases, and this evidence undermines Clausewitz's position that war "left to its own devices" must progress toward an absolute state of violence. In reality, war is *never* left to its own devices.

Clausewitz asserts that absolute war is actually independent of policy, since policy is an artifact of humanity. This "pure" form of war "would of its own independent will usurp the place of policy . . . and rule by the laws of its own nature . . . but in reality war is not like that."[309] Here he implies that policy, a nonmilitary factor, mitigates war's "independent will," which is driven by the purely military concern of disarming and defeating the enemy by force. Therefore, we can presume a level of tension between ideal war's "independent will" (with only military objectives) and real war. The resultant warfare depends upon the resolution of this tension and is a function of the motives of the belligerents and the degree of correlation between military and political aims. When war is "driven further from its natural course," says Clausewitz, the political object will be more at variance with the aim of ideal war and the conflict will seem more political in character than military.[310]

This reasoning is valid provided we understand that by "natural course" Clausewitz means war for the purpose of disarming an enemy versus achieving a political objective. Even so, Clausewitz does not clearly identify how wars of greater political character differ from those dominated by military thinking. From his theory, we can infer the greater the political influence, the less likely war is to be fought in decisive battles using maximum force. However, these issues tend to be relative. When political aims are absolute – and emotions, tensions, and hatreds are high yet capacity is low – decisive battles on smaller scales for modest objectives may still be fought. As capacity grows, the stakes grow, and war may or may not begin to resemble Clausewitz's ideal. It is also possible for the political character of war to remain constant even as its overall character changes as occurred for the Spanish and Viet Minh guerrillas. Clausewitz covers many of these themes in his chapter on "people's war," but by his own admission, this type of war is an outlier. Within the pure theoretical form of war, says Clausewitz, "a war between states of markedly unequal strength would be absurd, and so impossible . . . but wars have in fact been fought between states of *very unequal strength, for actual war is often far removed from the pure concept postulated by theory.*"[311]

The Conservation Properties of War

In a continuation of his adaptation of scientific analogues (like friction and COG) to war, Clausewitz attempts to articulate a conservation law, or "zero-sum" construct, within his theory. This approach implies that some aspects of war can be characterized algebraically, like science's conservation laws (e.g., conservation of mass or energy), an ironic proposition considering Clausewitz's warning that "continual striving after laws analogous to those appropriate to the realm of inanimate matter [is] bound to lead to one mistake after another."[312] This technique leads Clausewitz into some impractical arguments related to concepts like "polarity," which pertain to attack and defense or political aims and military means. For example, regarding attack and defense, Clausewitz asserts, "a major victory can only be obtained by positive measures aimed at a decision, never by simply waiting on events."[313] Just like "ought" and "should," the words "only" and "never" present problems when used to describe war. With the human mind and will at the heart of war, we can rarely know for certain what measures will determine victory or defeat. The Vietnamese communists and the Afghan *mujahedeen*, for example, foiled the twentieth century's superpowers without a Clausewitzian-style triumph. Though Clausewitz offers several useful concepts during this process, the message is garbled as his perspective vacillates between the operational (what he calls "strategic") and tactical levels of war, where, as he explains it, the most essential element of defense is the counterattack, and where any army on the strategic offensive that is not attacking is on defense.

Counterattacking the Defense

What of Clausewitz's assertion that the defense is the stronger form of war? Clausewitz says: "*The defensive form of warfare is intrinsically stronger than the offensive.*"[314] This dictum has sparked debate, but what exactly does it mean and what leads Clausewitz, who favored offensive force, into making it? In fact, consistency with his own zero-sum, ideal-type logic compels this conclusion. Making a similar claim on the basis of historical analysis alone would have made his reasoning easier to follow but not inarguable. By tying it to the zero-sum logic underpinning his concept of perfect war, he removes any basis for contesting this conclusion provided we accept his original premises.

Ironically, Clausewitz considered the *attack* to be the decisive, preferable form of war, for the attack alone serves a positive purpose, best uses resources, and most closely emulates war's ideal form. Indeed, even as Clausewitz says defense must be a component of any attack, he describes this requirement as "a necessary evil," "its original sin, its mortal disease."[315] Yet, Clausewitz had to admit that wars are rarely fought and won in decisive clashes and forces that reach the culminating point of attack move to a defensive form to survive; therefore, either the defense *must* be stronger or the concept of perfect war is flawed. Here Clausewitz's zero-sum logic corners him, and throughout *On War,* he never really seems comfortable with the maxim.

Following his logic from the original premise of perfect war explains why he came to this conclusion and how it eventually leads to some labyrinthine reasoning. Clausewitz's perfect war will always move inexorably to a maximum clash of power without pause. Given this, how does Clausewitz explain the reality of pauses in war? Clausewitz asserts that the stronger side has the positive aim of using its strength to defeat the enemy without delay, while the weaker side has the negative aim of awaiting the enemy's blow and delaying defeat until conditions change. In cases where the relative strengths of the two sides are nearly equal, the only reason for the stronger side to defer battle is if *the very status of being on the defense conferred an advantage to the weaker side*, thus putting the outcome in doubt.

There are some flaws in this rationale. To begin with, what does Clausewitz mean by the "stronger" form? Given two relatively equal sides, one of which is on the attack and the other on defense, how do we determine which is actually stronger? How much "strength" can a stronger form (defense) confer upon a weaker force? Furthermore, since waiting is not fighting, it is not really war. Thus, for the defense to achieve the aim of war, Clausewitz concludes that the *counterattack*, which reinitiates the fighting, must be an essential element of defense. When the defender counterattacks, should it still be considered a defender or an attacker in its own right, and if so, when does it lose the strength advantage of the form?

According to Clausewitz, counterattack is part of defense. Only when the defender is clearly superior to the original attacker and adopts the positive aim of attacking and destroying the enemy force, and likewise, the original attacker adopts the negative aim of awaiting the attack, can we redesignate the two sides. Therefore, the paused attacker, when forced to defend against the counterattack, does not accrue the inherent strength of that form, and thus, according to Clausewitz, the counterattacking defender need not be concerned about the relative balance in strength between the two sides conferring an advantage to the now-defending attacker. Here, Clausewitz makes an exception to the rule that the defense is stronger.

If all this is not confusing enough, Clausewitz makes the curious claim that the *defender* is the actual *instigator* of war, for in awaiting the attacker's action, the defender creates the condition that enables the attacker to attack in the first place. He says the object of attack is possession, not fighting, therefore, "the idea of war originates with the defense, which does have fighting as its immediate object," and, "the side that first introduces the element of war ... is the side that establishes the initial laws of war. That side is the *defense*."[316] In the sense that war is reciprocal and cannot occur (as we highlight in Chapter 1) unless the defender resists the attack, Clausewitz is absolutely correct. But Clausewitz's presentation is easily misunderstood and can lead to the conclusion that the only way to avoid war is either to attack, defeat, and assimilate every threat or offer no resistance at all. Indeed, militaristic nations like Hitler's Germany and Tojo's Japan adopted this dubious logic to justify attacks on France, Russia, and the USA. In truth, most defenders prefer *not to fight*, and their primary objective is to *deter* the attacker. This applies everywhere from the natural world to the battlefield. Porcupines and turtles evolved defensive attributes to discourage, not invite attack. Likewise, China's Great Wall and the Maginot Line were not invitations to battle. Clearly the attacker, who initially presents the threat (thus forcing the defender to actively defend or capitulate) and then turns threat into action, instigates the fighting and thus, the war. Though war does require the defender to fight, it is equally true there would be no need for defense at all without the aggressor. Looked at another way, defense is only definable in *relation* to the possibility of attack – in terms of preparation for it and reaction to it – whereas the concept of attack need not imply a defense.

Regardless of whether we agree with Clausewitz's assertion about the defense being the stronger form of war, it is fair to ask what purpose this maxim serves in the first place. As we have said, Clausewitz was compelled to include it as a necessary condition to support his theory's underlying premises. But even if defense *is* the stronger form of war, what does that mean for the strategist, soldier, or statesman? Indeed, Clausewitz probably wondered the same thing when he looked back at his notes and wrote, "I am still dissatisfied with most of it, and can call Book Six [Defense] only a sketch. I intended to rewrite it entirely and to try and find a solution along other lines."[317]

Ultimately, the factors that contribute to victory and defeat are so numerous, there is little practical benefit in claiming one form is inherently stronger than another. Indeed, even fellow German, von Moltke (the Elder) "consciously disregarded Clausewitz's insistence that defence was intrinsically stronger than attack. No general answer existed to the question of which of them was stronger, he wrote in his 'Instructions' of 1869."[318] Moreover, what is the utility of calling offense and defense *forms* of war at all? It is more correct to define them as *postures*. Defense and offense pertain to the posture and actions of forces involved in combat, whether they are attacking (moving forward, seeking to move toward enemy positions and territory, and initiating combat);

or defending (holding position or retreating, not seeking to advance on enemy positions or territory, and generally responding to enemy action).

In any case, as Clausewitz affirms, the side choosing the defensive usually does so as a means of preserving its warfighting capacity against a superior attacker. As we now know, advances in mobile war and firepower have reduced the benefits of fixed defenses. The advent of modern airpower, sea power, nuclear, and cyber weapons in the hands of determined, well-equipped attackers has negated most fixed, passive defenses and placed a premium on elusiveness, concealment, and deception. Simply because it *generally* behooves the weaker side to choose a defensive posture does not necessarily mean it is the *stronger* posture. Rather, a defensive posture may *slow the rate of resource expenditure* and corresponding loss of combat power relative to the attacker who assumes the burden of action and advance, thus raising the likelihood that the attack may culminate. Of course, much depends on the magnitude of the strength differential, and we cannot discount the individual and collective psychological effects occurring as forces transition from one posture to another. Even with relatively balanced strength, some battles can become routs as an attacker's psychological momentum swells and the defender's declines. Finally, cross-domain attacks, something Clausewitz's theory ignores, can also alter this equation. For example, air and naval bombardments can devastate land forces.

A side may assume a defensive posture when the attacker owns the initiative and a capacity advantage; however, the defender is not prohibited from mounting an immediate counterattack. As for which posture is *better*, again, the abundance of variables make a relative assessment pointless. To apply a qualifier like "all other factors being equal," is simply impractical. Indeed, "the doctrinal theses which put forward the tactical superiority of defence or attack," stresses Raymond Aron, "arise from an unjustified generalization of a moment which is transitory in the art of war."[319] As we will see in Chapter 4, relative strength assessments not only influence force *postures* but also war's *form*.

Blood on His Hands?

Some believe that Clausewitz's theory helped precipitate World War I by legitimizing war as a policy tool, defining ideal war as the use of maximum power in decisive battle, and implying that weaker states, fearing an inevitable clash (as pre-WWI Germany did), "should attack – but not because attack in itself is advantageous ... but because the smaller party's interest is either to settle the quarrel before conditions deteriorate or at least to acquire some advantages so as to keep its efforts going."[320] While Clausewitz's writings undoubtedly influenced the German General Staff, it is unfair to blame him for their error in taking his theoretical viewpoints literally. Nevertheless, we can question the universality of some of Clausewitz's conclusions, especially considering the realities of the nuclear age. Given our critique and his own misgivings regarding this portion of *On War*, it is fair to conclude that Clausewitz's depiction of attack and defense, particularly regarding relative strength, is not his finest work.

Requiem for Ideal War

Ultimately, Clausewitz's concept of ideal war is confusing and perhaps even unnecessary to his purpose. Most of his major ideas, including the trinity, the environment of war, the key terminology and elements of war, and the military genius, remain valid without the ideal-type. Furthermore, he would not have been compelled to continually contrast "real" and "ideal" war or to view war as a zero-sum, pseudo-algebraic activity with its confounding logic of attack and defense. Ironically, the ideal-type does not even explain Napoleon, nor does it help us better understand the broader phenomenon of real war. According to Gat, Clausewitz had overlooked the "subtlety of conception and maneuver" in much of Napoleon's campaigning, and in fact, Napoleon "had never been as direct" in pursuing great battles as the Prussian had imagined.[321]

In truth, all war is real war and all real war is limited war. Real war does have a logic all its own, but it is not a logic that *necessarily* drives its participants to extremes of violence; rather it is a *human* logic, the logic of *yin* and *yang*, thrust and parry, concentration and dispersal, movement and rest, a harmony of tensions. In real war, regardless of the overall aims, as one side's power increases, the other may try to outmatch *or change the rules* by becoming more elusive and devious. This calculus is based on relative capacity assessments, not politics or human weakness. As for Napoleon, he was dominant because he changed the rules of war abruptly, not because he was able to approximate a supreme form of war.[322] He raised a larger army, devised a means to control, direct, and supply that army better than his adversaries, and employed it with ruthlessness and unlimited political purpose his enemies were slow to match. Napoleon, like Genghis Khan and Alexander, upset the harmony of tensions in his favor, at least for a time.

Although Clausewitz was right to pursue a systematic methodology to reconcile war's apparently disparate forms, as we have discovered in the present era, the forms that actually require unification are *regular* and *irregular* – both true forms of war – unlike the absolute (or unlimited), which is an abstraction. But to echo Gat, whether Clausewitz would have eventually abandoned his preoccupation with absolute war and "developed his new ideas further and in new directions," perhaps to include greater examination of limited forms of war, irregular war, and naval warfare, "no one can tell."[323]

Stopping Bonaparte

Assessing Napoleonic power using Clausewitz's theory reveals two possible ways to deal with the emperor: beat him in decisive battle or surrender; however, Clausewitz witnessed *three* antidotes to Napoleonic power in his time, and only one of these relied on the adoption of superior force. The first (which Clausewitz clearly preferred), was what ultimately dealt the *coup de grace* – decisive battles at Leipzig and Waterloo with forces of comparable moral and physical strength and ability. They finally beat Napoleon at his own game.

The Russians exemplified the second antidote, not limited or absolute war, but rather strategic patience and indomitable will. By refusing to surrender and retreating back into its vast strategic depth, Russia shrugged off severe battlefield losses and even the seizure of its capital. Clausewitz cites the "enormous contribution the heart and temper of a nation can make . . . to its politics, war potential, and fighting strength" and the influence of Russia's sheer size on its resilience in the face of Napoleon's numerous tactical victories.[324]

The third antidote was also an outlier, the "people's war" fought by Spanish guerrillas against Napoleon's regular forces. "The stubborn resistance of the Spaniards," says Clausewitz, "showed what can be accomplished by arming a people and by insurrection."[325] This third option does not follow the ideal-type model, because it does not pursue decisive battle or concentrated force on a large scale. It is built less on bullets and bombs (and military genius) than on emotion, guile, and raw tenacity, and it creates advantages at the *microscopic* level of war, on the fringes, in the gaps and seams, and beyond the boundaries of Clausewitzian ideal war. This "irregular" form of war arises when one side has a deficit in fighting capacity but a surfeit of fighting will and thus counteracts the enemy's superiority by *changing or ignoring traditional rules.*

In the end, Clausewitz's concept of ideal war makes sense only within the strict confines of his abstract, dialectical logic. Taking it beyond these limits causes confusion, as we must then accept the notion of an "ideal" form of war that does not also represent a universally superior path to victory.

On Reciprocity, Intelligence, and Cunning

Clausewitz is justifiably lauded as one of the most enemy-conscious theorists (i.e., he emphasizes war's reciprocal nature). Clausewitz's concept of war, writes Paret, "consists in a relationship of two opposing sides, whose perceptions, emotions, and judgment affect each other's actions and reactions."[326] As much as possible, theory, preparations for war, and campaign planning must account for the enemy's anticipated thoughts and actions and the impact of uncertainty, environmental conditions, and human nature. Yet, by not explicitly conceding that these factors also impact the adversary, Clausewitz seems to imply that fog, friction, probability, and chance only influence one's own forces! This is likely overcompensation for his contemporaries' penchant for ignoring these influences completely. To be sure, fog, friction, genius, uncertainty, probability, and moral factors like fear, morale, and confidence apply to war, but they always apply to *both sides* (if not equally).

Clausewitz is also reluctant to laud the effectiveness of intelligence, surprise, cunning, deception, and maneuver. This perspective arises from his experiences and belief that lesser theorists of his day placed *too* much trust in these elements of war. For example, Clausewitz says, "Surprise has lost its usefulness today," because "modern armies are so flexible and mobile."[327] Also, "We are opposed to bombastic theories that hold that the most overwhelming surprise, the fastest movement or the most restless activity cost nothing."[328] Again, Clausewitz did not consider these factors useless, but he was understandably wary of mistakes caused by undue confidence in intelligence, clever plans, and maneuvers.

Losing Concentration

Clausewitz allowed the two institutions – state and regiment – to dominate his thinking so narrowly that he denied himself the room to observe how different war might be in societies where both state and regiment were alien concepts.[329]   John Keegan

Regarding the importance of concentrating maximum power at a decisive point to destroy the enemy, Clausewitz, spurred on by the ideal-type logic, again creates overemphasis. As "war is the impact of opposing forces," says Clausewitz, "it follows that the stronger force not only destroys the weaker, but that its impetus carries the weaker force along with it."[330] Thus, "the simultaneous use of all means intended for a given action appears as an elementary law of war."[331] Though he is right that activities not directly leading to maximum concentration of force detract from an army's total power, it does not necessarily follow that these activities cannot present a legitimate avenue for achieving desirable effects short of a winner-take-all showdown.

While Clausewitz acknowledges that not *all* forces at the tactical level need be used simultaneously, at the strategic (operational) level, greater force size "is more likely to lead to success"; thus, "it naturally follows that we can never use too great a force, and further, that all available force must be used *simultaneously*."[332] Note his use of the word "must." Clausewitz summarizes: "The rule then, that we have tried to develop is this: all forces intended and available for a strategic purpose should be applied simultaneously" and "will be all the more effective the more everything can be concentrated into a single action at a single moment."[333] He admits that strategy cannot ignore the value of sustained and "successive effort," but only because they arise from logistical limitations, in other words, as a necessary evil of real war.

We have belabored the point that these conclusions, based on the ideal-type model, do not always stand up to the test of history. Even Clausewitz recognized this discontinuity, saying of contemporary warfare, "The aim can no longer be achieved by a single tremendous act of war."[334] Again, here is an example of the confusing juxtaposition of theoretical and actual war.

To further illustrate Clausewitz's difficulty, witness the dubious validity of the following statements vis-à-vis insurgencies: "Only great tactical successes can lead to great strategic ones ... *tactical* successes are of *paramount import-ance* in war";[335] and, "We think it is useful to emphasize that all strategic planning rests on tactical success alone";[336] and, "The complete or partial destruction of the enemy must be regarded as the sole object of all engagements";[337] and, "In the most important cases, the destruction of the enemy's forces must be the main objective ... direct annihilation of the enemy's forces must always be the *dominant consideration*."[338] While these statements make sense within the context of Clausewitz's personal experience, he would almost certainly have wanted to amend them had he observed the twentieth century's small wars. For example, in 1975, an American negotiator, Colonel Harry Summers, Jr., said to his North Vietnamese counterpart, Colonel Tu, "You know, you never defeated us on the battlefield"; to which Tu replied, "That may be so, but it is also irrelevant."[339] Liddell Hart adds: "It was an easy step for Clausewitz's less profound disciples to confuse the means with the end, and to reach the conclusion that in war every other consideration should be subordinated to the aim of fighting a decisive battle."[340] In Clausewitz's defense, his assertions pertain to the *fighting dimension*, not the political-strategic aspects of war. While there is so much more to generating a desired political outcome in war than tactical success, war as fighting *is* based on tactical success in battles. Nevertheless, students of war must recognize the many other factors beyond tactical triumphs that influence a war's outcome. Indeed, for global wars or civil wars, the notion of simultaneous use of all force at any level above the tactical is untenable.

As we discussed earlier, Clausewitz's marginalization of the form war takes when one side's fighting capacity is low but the *will to fight* is strong (e.g., insurgency or guerrilla war) is essentially his theory's Achilles Heel. Destruction of capacity through tactical victories is a reasonable and time-honored approach to warfare, yet history – especially since the nineteenth century – proves that this is not always enough. The enemy must ultimately decide to quit the fight, and the complete elimination of a determined enemy's capacity to fight is easier said than done, especially when a foe fights for homeland and survival. Clausewitz says: "A major victory can only be obtained by positive measures aimed at a decision, never by simply waiting on events."[341] But what of protracted wars like those fought in Vietnam or Afghanistan? In these cases, patience and perseverance are often the best and only paths to victory.

On Trinitarian Analysis

Our section on Clausewitz concludes, not with a critique of his theory, but of what others have made of a portion of it, specifically the so-called "trinitarian analysis" that seeks to fashion an analytical model from the remarkable trinity. Clausewitz explicitly cautions against applying his work to establish "an arbitrary relationship" between the trinity's elements, for they "would conflict with reality to such an extent that . . . it would be totally useless."[342] Therefore, beware references to "trinitarian analysis," which tempt us to ignore Clausewitz's warning. This pseudo-vector analysis model purports to aid in the assessment of wars through the evaluation of relationships between the government, people, and military.[343] While Clausewitz says that each "aspect" of the trinity "mainly concerns" the people, the commander and army, and the government, respectively, he does not imply or define specifics that would make this useful as an analytical tool.[344] Indeed, Paret writes: "Even in Clausewitz's somewhat tentative formulations, these affinities – hatred and violence mainly identified with the people; chance and probability with the army and its commander; rational policy with the government – are of questionable validity."[345]

What has tempted some to use it this way is a misinterpretation of the following sentence: "Our task therefore is to develop a theory that maintains a balance between these three tendencies, like an object suspended between three magnets."[346] Here, Clausewitz is referring to theory development, *not* analysis. His standard for valid war theory, especially his own, is that it must necessarily account for each part of the trinity. No element can be ignored, and the interplay is complex and unpredictable, like the movement of a ferrous pendulum suspended between three magnets. That is what he means by "balance," *not* that strategy must somehow strive to "balance" relationships between the people, military, and government.

The test of *war theory* then is to examine it in relation to all three parts of the trinity: does it remain valid accounting for violence, chance, and politics? The notion of evaluating the characteristics and character of the military, the people, and the government when assessing current strategy or historical wars is only *coincidentally* trinitarian and, in fact, not really Clausewitzian at all. Indeed, Colin Gray calls this approach "a serious misreading of *On War.*"[347]

Trinitarian analysis is not rooted in the nature of war that Clausewitz describes, and there is no logical, objective way to ascertain whether the military, the people, or the government are "in balance" for any given conflict. Therefore, assessing wars and strategy by contrasting the relation-ship between the people, the army, and the government can only yield dubious conclusions that generate more questions than answers. What if there is no formal army? What if the people hate the government but still defend their homeland to the death? How does this apply to non-state actors, and what of the relationship between the trinities of opposing sides, or geography, or economic power, or alliances? What about changes over time? Ultimately, any reservations about the futility of "trinitarian analysis" quickly disappear after trying it.

Conclusion

Incomplete, unfinished, sprawling, and sometimes contradictory, and often as frustrating as it is endlessly rewarding, On War . . . *[is] the greatest monument to military thinking yet constructed.*[348]   Donald Stoker

Though Clausewitz's magnificent *Vom Krieg* greatly expands our understand-ing of war, his fixation on Napoleon, use of the often-perplexing ideal-type, omission of the maritime domain, zero-sum concept of attack and defense, scant treatment of irregular war, and untimely demise (which prevented full characterization and integration of limited war concepts) have left room for further development. Still, his deep reservoir of insights and evocative termin-ology continue to inform contemporary thinking on war because of their grounding in war's unchanging nature – its violence, relationship to human nature, complexity, and importance in human affairs. As Donald Stoker writes, "Clausewitz brought the study of war to a new intellectual level, turning it into a genuine discipline, placing it alongside other fields of study such as art, engineering, or philosophy."[349] And Gat adds:

Clausewitz's real intellectual greatness . . . stems from a unique achievement that has never been equaled. He offered a most sophisticated formulation of the theory of war, based on a highly stimulating intellectual paradigm, and brought the conception of military theory into line with the forefront of the general theoretical outlook of his time.[350]

Clausewitz's contributions have established him as the greatest single figure in military thought.

Throughout the remainder of our exploration of war theory and strategy we will continue to rely heavily upon Clausewitz while keeping in mind that, as Michael Howard writes in his introductory essay to *On War*:

Too much should not be read in Clausewitz, nor should more be expected of him than he intended to give. It remains the measure of his genius that, although the age for which he wrote is long since past, he can still provide so many insights relevant to a generation, the nature of whose problems he could not possibly have foreseen.[351]

Indeed, there can certainly be no unified war theory that neglects his commentary on centers of gravity, lines of communication, genius, concentration and dispersion, uncertainty, politics, and the importance of war's psychological and physical dimensions.

LIDDELL HART

It should be the duty of every soldier to reflect on the experiences of the past, in the endeavor to discover improvements, in his particular sphere of action, which are practicable in the immediate future.[352]   Liddell Hart

Chapter 1 describes a universe composed of a harmony of tensions. This chapter has continued that theme by presenting war theorists with contrasting styles and perspectives. "People like Liddell Hart," writes Gat, "were to apply Corbett's revolutionary theses [covered in the next chapter] to produce a comprehensive reformulation of military theory ... a total rejection of both all-out war and the strategic theory which went with it."[353] Indeed, Liddell Hart was an outspoken critic of Clausewitz, and not surprisingly, complementary of Sun Tzu. Where Clausewitz largely downplays the usefulness of maneuvers and deception, Liddell Hart embraces an *indirect approach* that employs these tactics to put enemies off balance. Moreover, Liddell Hart holds Clausewitz partly responsible for the battlefield carnage from Napoleon to World War II saying, "The outcome of [Clausewitz's] teaching applied by unthinking disciples, was to incite generals to seek battle at the *first* opportunity, instead of creating an *advantageous* opportunity. Thereby the art of war was reduced in 1914–1918 to a process of mutual mass-slaughter."[354] According to Liddell Hart, Clausewitz's theory was "used by countless blunderers to excuse, and even to justify, their futile squandering of life in bull-headed assaults."[355] Only a fool, thought Liddell Hart, would drive directly toward and attack a decisive point or center of gravity. Taking a page from Sun Tzu, Liddell Hart advises military strategists to *always* unbalance and confuse the enemy, striking weaknesses and avoiding strength. Though his is not a systematic nor holistic theory, Liddell Hart's contribution to military thought provides a useful, mid-twentieth-century counterpoint to Napoleonic-era theory and rekindles appreciation for Sun Tzu.

Before delving deeper into Liddell Hart's ideas, we must pause for an important caveat. Many of the concepts presented here were not originally Liddell Hart's, rather, they were first offered by his friend and mentor, the "always breathtaking, provocative ... but often obscure, mystifying, or mystical" fascist, J. F. C. Fuller.[356] However, since clarity and accessibility are central aims of this book, we have sidelined the "obscure" and "mystifying" Fuller in favor of Liddell Hart, who, as Gat confirms, communicates Fuller's ideas "to a wider public in a simplified and marketable form."[357] While Liddell Hart's penchant for self-aggrandizement and frankly, plagiarism, tarnish his legacy, he is still due credit as one of the twentieth century's most impactful military communicators, thinkers, and strategists. Indeed, Liddell Hart, writes Gat, is "the man who first translated the new conditions and sensibilities of the modern world into the language of strategic theory."[358]

On Liddell Hart

Sir Basil Henry Liddell Hart (1895–1970) was a Cambridge-educated warrior, historian, and strategist. He fought in World War I in the British infantry and rose to the rank of captain. Liddell Hart was no sideline warrior or ivory-tower general. He fought on the Western Front and participated in the gruesome Battle of the Somme, where most of his battalion was destroyed, and he was wounded multiple times and exposed to poison gas. Disqualified from frontline service due to his injuries, Liddell Hart returned to Britain and trained infantry until his retirement from the Army in 1927, nearly 100 years after Clausewitz's death.

Deeply affected by the war, Liddell Hart (described by Gat as "self-conscious, highly strung, socially awkward ... [and] highly egocentric") devoted his post–World War I years to thinking and writing about war.[359] Immediately following his wartime service, Liddell Hart largely defended the High Command's strategies of attrition. But gradually, under the influence of Fuller, Liddell Hart's views changed, and by the early 1920s, he had come to reject the idea of "total war and the strategy of destruction" exemplified by the Great War.[360]

As Liddell Hart continued to probe the history of warfare to determine why the tactics of World War I had failed so miserably, his own scholarship (and Fuller's 1923 book, *The Reformation of War*) led him to conclude that contemporary military leaders and strategists had strayed from war's true lessons. Instead of embracing Sun Tzu – who advocated deception and surprise to unsettle an enemy prior to combat – they had misunderstood Clausewitz's dialectic and attempted to emulate his "perfect" form of war. Liddell Hart's writings took a quantum leap forward in style and confidence with his first important book, *Paris, or the Future of War* (1925, essentially an adaptation of Fuller's *Reformation*). However, his strategic vision would not reach full flower until the release of *The Decisive Wars of History* (1929), later recast as *Strategy: The Indirect Approach*, a tome that influenced many tank generals of World War II, including Patton, Guderian, and Rommel. Nearly a decade after that war, in 1954, Liddell Hart published the summation of his ideas in *Strategy*, which was updated to include "the results of twenty-five years' further research and reflection, together with an analysis of World War II," and conceptual themes from Julian Corbett and T. E. Lawrence.[361] *Strategy*'s centerpiece is Liddell Hart's *indirect approach*, a principle so fundamental to the nature of war, in his estimation, it rises to the level of "a law of life in all spheres: a truth of philosophy."[362]

History and the Human Dimension

Liddell Hart believed only a broad historical inquiry, leveraging not just "the experience of another, but of many others under manifold conditions," provides the requisite depth and breadth of data from which to draw valid conclusions about war, particularly regarding its human dimension. Though critical of Clausewitz, Liddell Hart shared the Prussian's belief in the predominance of war's moral elements saying, "Although the moral and physical factors are inseparable and indivisible . . . on [moral factors] constantly turns the issue of war and battle."[363] From his historical survey, Liddell Hart discovered that "physical factors are different in almost every war and every military situation," but moral factors are much more constant, "changing only in degree."[364] Even so, war's human dimension makes each individual conflict complex and unpredictable, for "natural hazards are inherently less dangerous and less uncertain than fighting hazards. All conditions are more calculable, all obstacles more surmountable than those of human resistance."[365]

Levels of Strategy

For Liddell Hart, the ultimate aim of war, what policy and strategy should be directed to achieve and sustain, is a better peace. As the Roman general Belisarius says in a letter to his Persian counterparts, "The first blessing is peace, as is agreed by all men who have even a small share of reason. The best general, therefore, is that one which is able to bring about peace from war."[366] Like Clausewitz, Liddell Hart believes strategy serves policy, and he defines "grand strategy" as the use of nonmilitary "weapons" (like diplomacy or economic power) in support of policy.[367] *Grand strategy*, the highest plane of strategy, directs "all the resources of a nation, or band of nations, towards the attainment of the political object of the war – the goal defined by fundamental policy."[368] Moreover, grand strategy marshals both material *and* moral resources, "for to foster the people's willing spirit is often as important as to possess the more concrete forms of power," and it "looks beyond the war to the subsequent peace."[369] He adds: "The aim of grand strategy [is] to discover and pierce the Achilles's heel of the opposing government's power to make war. And strategy, in turn, should seek to penetrate a joint in the harness of the opposing forces."[370] Furthermore, he advises grand strategists to understand the relationship between ends and resources and to carefully consider the cost of trying to achieve those ends by war. "To forego aims which are not 'worth the candle,'" writes Liddell Hart, "is the difference between grand strategy and grandiose stupidity."[371] After the world wars, Liddell Hart recommended England adopt a defensive grand strategy favoring "containment and deterrence, economic coercion, peripheral war by proxy, blockade, and limited war, in that order."[372]

Liddell Hart equates *military strategy* with "generalship" or "the actual direction of military force, as distinct from the policy governing its employment and combining it with other weapons: economic, political, psychological."[373] Whereas military strategy is "the art of distributing and applying military means to fulfill the ends of policy," *tactics* is "an application of strategy on a lower plane" concerned with "the dispositions for and control of direct action" when "the military instrument merges into actual fighting."[374] Liddell Hart's "military weapon is but one of the means that serve the purposes of war: one out of the assortment which grand strategy can employ."[375]

Objects and Aims

Liddell Hart's "military aim" is the "way that forces are directed in the service of policy," which ultimately seeks "a better state of peace"; however, the political purpose is sometimes subsumed by the military aim, which becomes "as an end in itself, instead of as merely a means to the end."[376] Like Clausewitz, Liddell Hart considered war as its own end a dangerous inversion of its proper, subordinate relationship to policy.

In some respects, the divergence between Liddell Hart and Clausewitz is attributable to the disparity in the character of war between their respective eras. Both ultimately seek to disarm the enemy and break the will to fight; however, in Clausewitz's time this was best done by concentrating power, a strategy that failed in World War I. Therefore, Liddell Hart adopts Sun Tzu's ideal (i.e., "The perfection of strategy would be ... to produce a decision without any serious fighting").[377] He equates major battle with "mauling" and says strategy's true purpose is to leverage the physical effects of movement and the psychological effect of surprise to "*diminish the possibility of resistance.*"[378] Moreover, the strategist "*is not so much to seek battle as to seek a strategic situation so advantageous that if it does not of itself produce the decision, its continuation by a battle is sure to achieve this.*"[379] While this seems to contradict Clausewitz, in fact, both men pursued victory, only from different angles. Liddell Hart, mindful of the stalemate of World War I, exalts the advantages of maneuver and cunning to create the conditions for success, whereas Clausewitz's experience with Napoleon leads him to extol the primacy and decisiveness of the battle itself.

Liddell Hart considers battle one of several possible means for gaining the real objective – compelling an enemy leader to quit the fight. He writes: "The true aim of war is the mind of the hostile rulers, not the bodies of their troops; [and] the balance between victory and defeat turns on the mental impressions and only indirectly on physical blows."[380] Certainly, the point of war is to convince the right person or people – those with the power to make war and sustain a war effort – to cease fighting and accede to agreeable political terms.

Size Matters

Liddell Hart highlights the dangers of "limited strength in an unlimited space," something no other theorist articulates in quite the same way.[381] Force size must be commensurate with the area to be controlled; otherwise, the enemy force may remain elusive. French and German misadventures in Russia, as well as wars in Afghanistan, Vietnam, China, and Spain exemplify the hazards aggressors face when they cannot control all the important points of an expansive country. While advanced technology has made the globe more accessible and intimate, and contemporary powers can project power globally in hours or minutes, Liddell Hart's point remains valid, because the ability to *project* power is one thing, but the ability to truly "control" a theater of war or a warfighting domain (e.g., the air or sea) is quite another.[382]

The Indirect Approach

For Liddell Hart, history proves that "in strategy, the longest way round is often the shortest way home," or in other words, "the lesson emerged that a direct approach to one's mental object, or physical objective, along the 'line of natural expectation' for the opponent, tends to produce negative results ... [for,] to move along the line of natural expectation consolidates the opponent's balance and thus increases his resisting power."[383] Echoing Sun Tzu, he says the direct approach (*cheng*) is counter to the purpose of strategy (i.e., to decrease resistance); therefore, the logical course is to do the opposite, to take an *indirect* approach (*ch'i*) along a line of least expectation. Indeed, "to apply one's strength where the opponent is strong weakens oneself disproportionately to the effect attained. To strike with strong effect, one must strike at weakness."[384]

Borrowing a metaphor from Clausewitz, Liddell Hart likens war to a wrestling match, where "the attempt to throw the opponent without loosening his foothold and upsetting his balance results in self-exhaustion."[385] Unless the attacker has "an immense margin of superior strength in some form," victory is frustrated without "dislocation of the enemy's psychological and physical balance."[386] He calls this the "defensive-offensive," a tactic that draws the opposition off balance or entices overreach, then quickly turns the tide with a crippling counterattack.[387] "The best moment for a major counter-offensive, as for a minor counter-attack," writes Liddell Hart, "is usually when the attacking opponent has fully committed his own strength without having gained his objective."[388] The defensive-offensive lures the attacker toward what Clausewitz would call the culminating point of attack. Liddell Hart writes: "The weaker the defending side, the more essential it becomes to adopt mobile defence. For otherwise the stronger side can make space its ally and gain a decisive advantage through outflanking maneuver."[389]

Mind Game

According to Liddell Hart, generals pursuing military objectives must first mentally dislocate the enemy (if not physically as well). When influencing the mental domain, "to *mystify* the enemy [is] not enough; he must be *distracted* – a term which implies combining deception of the enemy's mind with deprivation of his freedom to move for counter-action, and with the distension of his forces."[390] The best target for psychological manipulation is the leader, who has the greatest impact upon the engagement or war. Liddell Hart says: "A decision is produced even more by the mental and moral dislocation of the *command* than by the physical dislocation of its forces" (emphasis added).[391] "A man killed," he adds, "is merely one man less, whereas a man unnerved is a highly infectious carrier of fear, capable of spreading an epidemic of panic."[392] Fear in a general can paralyze an entire force; and fear in a government, like France (and nearly Great Britain) in World War II, can trigger a sudden surrender.[393] Of course, Liddell Hart endorses the indirect approach as the best way to upset the physical and psychological balance of an opposing person, army, or state, for "direct pressure," he writes, "always tends to harden and consolidate the resistance of an opponent – like snow which is squeezed into a snowball, the more compact it becomes, the slower it is to melt."[394] Dislocation sets the stage for battlefield success, but only if generals place the enemy "at a psychological disadvantage" *before* the battle begins.[395]

Surprisingly Effective

Surprise is perhaps the most potent weapon for unbalancing an enemy. "As the object of all surprise is dislocation," says Liddell Hart, "the effect is similar whether the opponent be caught napping by deception or allows himself to be trapped with his eyes wide open."[396] He applauds Germany's World War I command staff for realizing "how rarely the possession of superior force offsets the disadvantage of attacking in the obvious way. Also, that effective surprise can only be attained by a subtle compound of many deceptive elements."[397] Creating surprise requires the cultivation of feasible options to keep the enemy guessing. The "absence of an alternative," asserts Liddell Hart, is "contrary to the very nature of war"; thus, he advises using alternative objectives in war planning to place the enemy "on the horns of a dilemma" (i.e., where *any* enemy action creates an adverse outcome).[398]

The Best Offense Is a Good Defense

Always wary of direct aggression, Liddell Hart advocates for the strategic and tactical use of the defense. He writes: "In a deeper and wider sense than Clausewitz implied ... the defensive is the stronger form of strategy as well as the more economical."[399] For example, a defensive force can "lure" the enemy by initially retreating, then attacking when conditions are favorable, and an offensive force can "trap" the enemy by initiating an unbalancing attack, then assuming a strong defensive position. "Whatever the *form*," writes Liddell Hart,

the *effect* to be sought is the dislocation of the opponent's mind and dispositions – such an effect is the true gauge of an indirect approach ... we can crystallize the lessons into two simple maxims ... no general is justified in launching his troops to a direct attack upon an enemy firmly in position ... [and] instead of seeking to upset the enemy's equilibrium by one's attack, it must be upset before a real attack is, or can be successfully launched.[400]

In other words, never attack an enemy that has not first been morally or physically degraded.

Liddell Hart defines "strategic dislocation" in the physical sphere as "the result of a move which (*a*) upsets the enemy's dispositions and, by compelling a sudden 'change of front', dislocates the distribution and organization of his forces; (*b*) separates his forces; (*c*) endangers his supplies; [and] (*d*) menaces the route or routes by which he could retreat."[401] When employed successfully, physical dislocation triggers "psychological dislocation" in an opposing commander's mind, and this effect is exacerbated by speed (i.e., the feeling of suddenly being placed at a disadvantage), especially from a "*sense of being trapped*."[402] Sudden, unexpected attacks on an enemy's flanks or rear are proven methods for unbalancing a defender who is forced to turn out of a preferred position to counter the assault.[403]

Liddell Hart shares Clausewitz's and Jomini's appreciation for concentrated force at the appropriate time and place but considers the concentration of *all* force "an unrealizable ideal, and dangerous even as a hyperbole."[404] It is better to produce a "maximum *possible* concentration of force at one place, while the minimum force *necessary* is used elsewhere to prepare the success of the concentration."[405] If the object of the strike has not been dislocated physically or morally, then the attack "may strike an object too solid to be shattered."[406]

Liddell Hart ascribes important roles in the indirect approach to air and sea power. Navies, he asserts, more naturally adopt an indirect approach, because they are accustomed to operating "against the seaborne communications, or means of supply, of opposing *countries*"; whereas land-based forces started to rely upon and target lines of communications later in their development.[407] Since maritime control can affect the land domain, navies provide an important advantage to countries when integrated into plans designed to confuse, cut off, and discourage enemies. Airpower also delivers game-changing capabilities via speed, flexibility, and range. Liddell Hart writes: "The development of air forces offered the possibility of striking at the enemy's economic and moral centres without having first to achieve 'the destruction of the enemy's main forces on the battlefield'. Air-power might attain a direct end by indirect means – hopping over opposition instead of overthrowing it."[408] Additionally, air mobility "could achieve such direct strokes by an overhead form of indirect approach," like dropping forces rapidly into positions that threaten an enemy's lines of communications or other "vital organs."[409] These capabilities, along with advanced mechanized warfare (e.g., tanks), "promised new scope for producing such paralysis of armed opposition," in essence nullifying adversaries without having to fight a direct engagement.[410] This pattern was evident in World War II during America's Pacific island-hopping campaign and German operations in Western Europe and Russia until 1942.

Indirect Approach in History

Liddell Hart substantiates his conclusions regarding the indirect approach with numerous historical examples. Indeed, he organizes *Strategy* as a journey through time, from ancient Greece to the twentieth century, recounting battles and campaigns, highlighting successful deceptions, feints, ambushes, and maneuvers, and pointing out where direct advances and overt assaults often failed. History's successful generals, says Liddell Hart, forced their enemies into disadvantageous positions, put them off balance, and then drove for a decision.

The cat-and-mouse campaigns of the Second Punic War (218–201 BCE) between Carthage and Rome provide an exemplary case study. In this war, Hannibal of Carthage demonstrated the superiority of his army and generalship by ambushing and annihilating a Roman army at the Lake of Trasimene. To counter Hannibal's advantages, the Romans under Fabius employed the "Fabian strategy," avoiding direct battle and attacking with "military pin-pricks" designed to "wear down the invaders' endurance."[411] "Hovering in the enemy's neighbourhood," writes Liddell Hart, "cutting off stragglers ... preventing them from gaining any permanent base, Fabius remained an elusive shadow on the horizon."[412] Fabius's "immunity from defeat" stymied

Hannibal, and his "guerrilla type of campaign also revived the spirit of the Roman troops while depressing the Carthaginians who, having ventured so far from home, were the more conscious of the necessity of gaining an early decision."[413]

By the seventeenth and eighteenth centuries, according to Liddell Hart, successful maneuver and the dislocation of enemy forces had become problematic as offensive firepower and tactical innovation lagged technological advances in defenses. Though Oliver Cromwell orchestrated some brilliant campaigns during the Second Civil War (1648–9) in England, in general, warfare of the age was indecisive because,

the development of fortification had outpaced the improvement of weapons and given the defensive a preponderance such as was restored to it in the early twentieth century by the development of the machine-gun; [and] that armies were not yet organized in permanently self-contained fractions, but usually moved and fought as a single piece, a condition which limited their power of distraction – of deceiving the opponent and cramping his freedom of movement.[414]

Rigid, inflexible armies "made difficult the completion of a strategic maneuver. A general could draw the enemy to 'water,' but could not make him drink – could not make him accept battle against his inclination."[415] Absent the ability to dislocate, surprise, or outmaneuver a foe, the only viable options were to withdraw or drive into the meat-grinder created by small arms and cannons.

Napoleon

Given Liddell Hart's censure of Clausewitz and Jomini, we might expect him to be critical of Napoleon also. However, he actually admired Napoleon's generalship, particularly during his early nineteenth-century campaigning. Liddell Hart attributes Napoleon's remarkable success, not to the size and power of the *Grande Armée* alone, but rather to "a combination of favourable conditions and impelling factors" including fluidity of movement, fractioning the army into self-contained parts, living off the country, and a commander, Napoleon, who studied military history, especially that of Pierre-Joseph Bourcet and Jacques Antoine Hippolyte, Comte de Guibert.[416] Though Clausewitz and Jomini largely dismissed these eighteenth-century theorists, Liddell Hart believed each had positively influenced Napoleon's generalship. He credits Bourcet with promoting the tactics of dispersing an enemy "preparatory to the swift reuniting of his own forces" and also the practices of preparing branch plans and threatening "alternative objectives."[417] Likewise, Guibert advocated for the "supreme value of mobility and fluidity of force, and of the potentialities inherent in the new distribution of an army in self-contained divisions."[418]

The American Civil War

In his synopsis of America's Civil War, Liddell Hart makes some intriguing observations regarding the relationship between war and democracy. He contrasts the objectives of Union generals Ulysses S. Grant, who beset Robert E. Lee's Army of Northern Virginia, and William Tecumseh Sherman, who targeted the city of Atlanta. Liddell Hart considered Sherman's objective "better suited to the psychology of a democracy" than Grant's "armed forces objective," for

> the strategist who is the servant of a democratic government has less rein … he has to work with a narrower margin of time and cost than the "absolute" strategist, and is more pressed for quick profits [since his] military effort rests on a popular foundation … [and] it depends on the consent of the "man in the street."[419]

Unlike Lee's army, Atlanta could not maneuver away from a Union attack; thus, it was an objective Sherman could assault in a manner and at a time of his discretion. In addition to the military significance represented by the Confederate forces and war-making industries resident there, Atlanta presented a strategic prize the Northern constituency could readily appreciate.

Liddell Hart's interpretation of democracy's influence on war is not trivial, for during the lifetimes of most renowned war theorists, democracies were not prevalent. Rare is the democracy in which the out-of-power political element supports the incumbent leadership during a protracted war, and no war goes so smoothly that the opposition party despairs of finding something to exploit for political advantage. The net effect of democracy on war is an amplified role for domestic politics with the mood and disposition of the electorate a vital center of gravity. Prior to an overt attack, democracies can be dovish. For example, before Pearl Harbor, "96 percent of Americans opposed joining the war against Hitler," and "even as late as 1941, with France defeated and England alone, a poll showed 79 percent of Americans still opposed involvement in the war."[420] Democracies can also sour on war quickly if things go poorly and citizens are not experiencing the war's effects directly. Regardless of the outcome of battles, if a critical mass of voters becomes convinced of a war's futility, then even a mighty democracy may quit the fight or sue for peace. Thus, the political cycle of a democracy plays a direct role in the strategy and conduct of war. Kudos to Liddell Hart for saying, "Faced with these inevitable handicaps, it is fitting to ask whether military theory should not be more ready to reconcile its ideals with the inconvenient reality that its military effort rests on a popular foundation."[421] Ultimately, Liddell Hart preferred Sherman's generalship over Grant's, for Sherman's tactics, in Liddell Hart's estimation, were less direct. "Sherman … ingeniously planned to avoid" the limitation of a single, obvious objective, writes Liddell Hart, "by placing his opponent repeatedly 'on the horns of a dilemma' ... ready to take the alternative objective if conditions favoured the change."[422] He considered Sherman's rapid march to the sea, his flexibility, and the unpredictability caused by "the physical and moral effect" of his "deceptive direction," antecedents to the German panzer tactics that shocked Europe nearly eighty years later.[423]

Indirect Approach in the Twentieth Century

Liddell Hart's strategic review of the world wars highlights how the indirect approach helped decide the major engagements. A key vignette from World War I concerns T. E. Lawrence's guerrilla war exploits during his Arab Revolt against the Ottoman Turks. Liddell Hart describes guerrilla war as "by its very nature indirect."[424] Lawrence capitalized on the Arabs' greater mobility to minimize casualties relative to the orthodox Turks, who were less mobile, spread thin, and "depended on a long and frail line of communications."[425] Liddell Hart describes Lawrence's strategy as "the antithesis of orthodox doctrine. Whereas normal armies seek to preserve contact, the Arabs sought to avoid it. Whereas normal armies seek to destroy the opposing forces, the Arabs sought purely to destroy material – and to seek it at points where there was no force."[426] Lawrence's battles preceded a series of decisive thrusts by his theater commander's (General Allenby) regular forces, which "deserve to rank among history's masterpieces for their breadth of vision and treatment."[427] Allenby neutralized Turkish communications, immobilizing their forces in the face of the main assault.

For Liddell Hart, the side that best applied Sun Tzu's maxims in World Wars I and II generally had the most success. "It is wise in war not to underrate your opponents," he writes, and "it is equally important to understand his methods, and how his mind works. Such understanding is the necessary foundation of a successful effort to foresee and forestall his moves."[428] Liddell Hart recommends the government's "advisory organs" have an "enemy department" charged with "studying the problems of the war from the enemy's point of view."[429] In fact, many governments integrate these "departments" into their intelligence services and in military units as "aggressors" and "red teams," which think and act like potential enemies.

Interestingly, two of the twentieth century's villains, Lenin and Hitler, were among the best at pursuing victory *before* fighting. In their own words, they promoted the decay of their enemy's will to fight before firing a shot, rendering victory by force either unnecessary or quick. Lenin, writes Liddell Hart, "enunciated the axiom that 'the soundest strategy in war is to postpone operations until the moral disintegration of the enemy renders the delivery of the mortal blow both possible and easy'"; and Hitler said, "Our real wars will in fact all be fought before military operations begin."[430] Ironically, Hitler's adherence to the indirect approach, like Napoleon's, eventually waned. After initial success versus France, Hitler abandoned the indirect approach. Early tactical triumphs in Russia led to a false sense of invincibility that the Russians later exploited with a series of counterattacks. But Hitler was nothing if not a megalomaniac, and, like a gambler at the end of a hot streak, he refused to accept the turn of events. "Driven on by the spur of insatiable appetite," says Liddell Hart, "by the haunting spectre of lost prestige, and by the instinctive feeling that attack was the only way of dealing with problems," Hitler tried to reverse his fortunes.[431] This mindset led him to place all his chips on the table at Stalingrad, where everything "was sacrificed for a too direct concentration, too directly aimed, against that untaken city."[432] For Germany, Stalingrad was an unmitigated disaster that heralded the beginning of the end for Hitler.

Maxims

Liddell Hart condenses his principal recommendations in *Strategy* into eight "truths of existence which seem so universal, and so fundamental, as to be termed axioms."[433] Underlying each is an "essential truth ... that, for success, two major problems must be solved – *dislocation* and *exploitation*."[434] The following guidance is fairly self-explanatory:

(1) *Adjust your end to your means*
(2) *Keep your object always in mind*
(3) *Choose the line (or course) of least expectation*
(4) *Exploit the line of least resistance*
(5) *Take a line of operation which offers alternative objectives*
(6) *Ensure that both plan and dispositions are flexible – adaptable to circumstances*
(7) *Do not throw your weight into a stroke whilst your opponent is on guard*
(8) *Do not renew an attack along the same line (or in the same form) after it has once failed.*[435]

Guerrilla War

Liddell Hart devotes his final chapter to guerrilla war, and unlike our earlier theorists, he had the advantage of having witnessed an explosion of these "small" wars across the globe. "If you wish for peace," writes Liddell Hart, "understand war – particularly the guerrilla and subversive forms of war."[436] As he does with conventional examples of the indirect approach, Liddell Hart catalogues significant instances of guerrilla warfare during and after World War II, including the resistance to German and Japanese occupation as well as Chinese fighting under Mao (discussed in the next section). After World War II, guerrilla campaigns erupted in Southeast Asia, Africa, Cyprus, and even

Cuba. During these conflicts, threats of nuclear retaliation were as senseless "as to talk of using a sledge hammer to ward off a swarm of mosquitoes."[437] Guerrilla forces were dispersed, attacking in "small bites," giving no substance at which to aim a decisive blow, and existing in "limited density – a multiple infiltration by particles so small that they formed an intangible vapour."[438]

Guerrilla war relies on tactics outside the "normal practice of warfare."[439] First, guerrillas must remain active and fluid to prevent larger but less mobile enemies from forcing a head-on engagement. Guerrillas must fight on their own terms, seeking localized superiority through surprise, deception, and ambushes, before melting away. "Ubiquity combined with intangibility," writes Liddell Hart, "is a basic secret of progress in such a campaign."[440] For the guerrilla, "concentration" is valuable only at the tactical scale, where they may "coagulate like globules of quicksilver to overwhelm some weakly guarded objective"; whereas "dispersion" is the "essential condition of survival and success."[441] Mobility and favorable terrain enable guerrillas to evade encirclement, find safe haven, and resupply as they pursue a strategy designed to "produce the enemy's increasing overstretch, physical and moral."[442]

Darkness, both physical and metaphorical, is the guerrilla's ally. The former provides cover while the latter describes the enemy's perceptions and awareness when faced with guerrilla tactics. Liddell Hart labels this "camouflaged war," fought "by the few but dependent on the support of the many."[443] Finally, Liddell Hart warns that guerrilla warfare can attract "bad hats," unsavory elements that reject rules or authority even after the political objectives are gained. He considers this lesson "too lightly disregarded" during the mid-twentieth century, resulting in "both equipment and stimulus to anti-Western movements in Asia and Africa," a "disease" that has "continued to spread."[444]

Future War

Liddell Hart published *Strategy* shortly after the USA detonated the first hydrogen bomb, and he was an early commentator regarding its effects on future war. Like Jomini, Liddell Hart predicted that technological advances, including nuclear weapons, would have little effect on "the basis or practice of strategy."[445] This judgment is correct for the elements of strategy relating directly to war's unchanging nature; however, Liddell Hart's indirect approach requires modification in the area of thermonuclear war. These weapons are so powerful and difficult to evade that traditional battlefield tactics, no matter how imaginative, are ineffective. Thus, the essence of nuclear strategy for nuclear-armed countries is to determine a proper mix of nuclear forces, including delivery systems and warheads, basing, and deployment such that no rational adversary could sanction their use short of a last resort. In other words, as Liddell Hart opines, "It may be assumed that the H-bomb, would not be used against any menace less certainly and immediately fatal than itself."[446] For all other varieties of war short of nuclear exchange, Liddell Hart's indirect approach applies, and he is right to say that nuclear weapons

would not free us from dependence on what are called "conventional weapons," [but are] likely to be an incentive to the development of more unconventional methods in applying them ... above all, such experience has emphatically borne out the forecast that the development of nuclear weapons would tend to nullify their deterrent effect, thereby leading to the increasing use of a guerrilla-type strategy.[447]

Here he reveals an interesting facet of war, something Sun Tzu grasped but Clausewitz and Jomini (with Napoleon as the nineteenth century's "weapon of mass destruction") only touched upon. Though overwhelming power (exemplified by nuclear weapons) reduces "the likelihood of all-out war," writes Liddell Hart, "it *increases* the possibilities of 'limited war' pursued by indirect and widespread local aggression. The aggressor can exploit a choice of techniques, differing in pattern but all designed to make headway while causing hesitancy – about employing counteraction with H-bombs, or A-bombs."[448] In other words, the direct power of nukes (*cheng*) forces warfare into a more indirect form (*ch'i*). Still, lacking a holistic theory, Liddell Hart can do little more than note, "these problems are of very long standing, yet *manifestly far from being understood* – especially ... where everything that can be called 'guerrilla warfare' has become a new military fashion or craze" (emphasis added).[449]

Critique

Several aspects of Liddell Hart's theory warrant criticism. First, his review and analysis of war's history are heavily influenced by his personal experiences and desire to find an alternative to the direct approach. Thus, his "discovery" of the indirect approach (to the extent it is not Corbett's or Fuller's) does not reflect an objective search for truth. Rather, Liddell Hart emphasizes cases that support his thesis while glossing over those that do not. Second, Liddell Hart never systematically defines the indirect approach. He provides examples and characteristics, but much is left to individual interpretation. In several instances, this produces dubious assertions. For example, in describing the Austro- and Franco-Prussian wars of 1866 and 1870, which were successful for the Prussians and generally characterized by direct advances, Liddell Hart says, "They acquired an unintended indirectness."[450] Without a clear definition and with elements of indirectness credited for a variety of maneuvers and tactics, Liddell Hart ends up attributing nearly all victories to the indirect approach and most defeats to its absence.

The third critique is related to the first two. While valuable, Liddell Hart's theory of the indirect approach is not *always* correct. There are times and circumstances where the direct approach works fine, even against a prepared enemy. When one side enjoys a significant advantage over another, swift, direct movement to battle often produces positive results. In any war, there are many factors that lend one side a disproportionate advantage over another, and not all require a side to place its opponent on the horns of a dilemma. Sometimes, the rapid concentration of forces and striking directly at a decisive point delivers a desired outcome. At other times, the value of the objective and the timing (e.g., Sherman's siege and occupation of Atlanta and Grant's relentless pursuit and harassment of Lee) necessitate direct action *prior* to achieving dislocation.

The final critique concerns Liddell Hart's description of the relationship between concentration and dispersion. He writes: "Effective concentration can only be obtained when the opposing forces are dispersed; and, usually, in order to ensure this, one's own forces must be widely distributed. Thus, by an outward paradox, true concentration is the product of dispersion."[451] This conclusion is debatable, for a concentrated force striking a dispersed force may have little positive effect, like a hammer swishing through the air. It is possible Liddell Hart is ineffectively conflating *dislocation* with *dispersion*. As we will see in the next two chapters, dispersion can be effective against concentration. In fact, Liddell Hart confirms this by calling dispersion an essential tactic versus modern weaponry, especially nuclear arms. "Advancing forces," he writes, "should not only be distributed as widely as is compatible with combined action, but be dispersed as much as is compatible with cohesion."[452] This parallels Machiavelli's advice regarding infantry advances against artillery. Liddell Hart is likely highlighting that concentration is more effective when the enemy is *disorganized* or unprepared (i.e., *demoralized*).[453] Even an organized, concentrated foe might lack the will or capacity to effectively resist a head-on stroke, just as a diamond, the hardest substance known, can be cut with concentrated force at a well-chosen point.

Conclusion

As we implied at the beginning of this section, Liddell Hart is the *yang* to Clausewitz's *yin* (i.e., the ideas of each when considered together paint a more complete picture of war). But Liddell Hart's criticism of Clausewitz, his tendency to represent the ideas of others as his own, and his relatively biased perspective on war have led to justifiable ostracism. Nevertheless, by emulating Sun Tzu and deriving maxims from a wide examination of historical examples, Liddell Hart presents valuable thoughts on war that reemphasize the operational and tactical arts, especially with respect to the primacy of the psychological dimension. Additionally, he offers keen analysis regarding guerrilla war, the impact of strength imbalances on strategy and tactics, and the implications of nuclear weapons and even democracy on war. Liddell Hart's theoretical centerpiece, the indirect approach, commits to battle only after placing the enemy off balance, and while there will never be one best way to do this – and, ironically, a rapid, direct attack may at times be optimum – there is no denying the benefit of weakening or confusing an enemy *before* battle as a significant and desirable component of successful strategy.

MAO

Mao Tse-tung armed the people in a way that only the Prussian patriots had dreamt of (and that Clausewitz himself, and others, had sketched in theory); his methods became the essence of revolutionary war, one leading, through the creation of a regular army consisting of irregulars, to victory and annihilation.[454]   Raymond Aron

The final member of our magnificent seven, like the first, is a Chinese theorist. Of course, Mao Tse-tung is known more as "Chairman Mao," the father of Communist China, than as a military theorist. However, Mao is supreme amongst theorists in terms of his personal military and political successes. He took an idea, and through leadership and tenacity, used it to transform a country, which had been an international whipping dog, into a world power. But we should not forget, this legacy came "from the barrel of a gun." Prior to his political ascension, Mao was an accomplished and underrated general and strategist, a disciple of Sun Tzu. According to Samuel Griffith, Mao was "no novice in the art of war. Actual battle experience with both regular and guerrilla troops has qualified him as an expert."[455] From his experiences and studies, Mao formulated powerful principles regarding guerrilla war. And, perhaps better than any other war theorist, he embraced the holistic nature of war, especially regarding the regular-irregular dialectic. Combining this knowledge with his understanding of human nature, tactics, strategy, and politics, Mao achieved audacious military and political ends.

On Mao

Political power comes out of the barrel of a gun.[456]   Mao Tse-tung (1938)

Mao remains a controversial figure. Without question, he was one of the most consequential political icons of the twentieth century, which included quite a few. His vision, decisions, and actions created momentous change in China that rippled across the world. Millions lived and died within the sphere of his military and political influence. For many, Mao is a hero, and for many more he is one of history's greatest villains. But regardless of how he is depicted today by supporters and critics, his ideas on war are worth examining.

Mao was born a peasant in the Hunan province of imperial China in 1893. He quickly matured into a very bright, voracious reader and student of history. He identified with Chinese nationalism from a young age and had a special affinity for politics. Unrest and rebellion characterized China's political climate in the early twentieth century, and Mao quickly found himself wrapped up in it. As a youth, he even served a stint in a pro-republican rebel army, though he saw no fighting. Through his early twenties, which encompassed World War I, Mao remained physically and intellectually restless, read widely, and developed his leadership acumen while taking a keen interest in war. Though he was initially ambivalent to Marxist Communism, by the 1920s Mao began to see the nascent Party as a vehicle for achieving Chinese independence, throwing off foreign influence, and realizing his own political ambitions. Eventually he joined the Communists, and his charisma, writing ability, and leadership propelled him rapidly up the ranks. Before long, Mao was immersed in the cutthroat politics and bloody battles occurring between Chiang Kai-shek's Kuomintang nationalists and the still-fledgling Chinese Communist Party, and by 1927, he had earned enough respect to be named commander-in-chief of the Workers' and Peasants' Red Army of China.

Mao's early military experience can best be described as "the school of hard knocks," and he repeatedly butted heads and clashed arms with nationalist forces and even his own party leadership, who viewed him as a wild card. Though he was not a dominant general in his early years, Mao was a survivor and a quick study. As a commander, he maintained tight control over his forces and successfully molded them into an effective army skilled at both conventional and guerrilla tactics. Nevertheless, by the fall of 1934, Chiang Kai-shek, who had postponed his campaign against Japan to eradicate the Communists, had Mao's forces on the ropes. Mao narrowly escaped a trap near the city of Jiangxi, and unable to reengage without risking outright defeat, he was forced to flee with the remnants of his army on the "Long March" into the roughs of central China. Though Mao survived the trek, fighting, famine, and disease claimed *90 percent* of his original force, leaving him only 10,000 ragtag troops. Nonetheless, Mao's perseverance and inspired generalship kept the communists a step ahead of the Kuomintang, and he was finally elevated to the pinnacle of Chinese Communism. Moreover, during these arduous trials, Mao had amassed a wealth of experience in warfighting and tactics. With just a bit more time to grow his army, Mao figured he could turn the tables on Chiang, and he got just the break he needed from an unlikely source. Despite their mortal animosity, Mao and Chiang were both Chinese, and Japan was a common enemy, a fact Mao shrewdly exploited to broker a temporary alliance with the Kuomintang. This period bought Mao a respite from civil war, time to write about politics, warfighting, and China's future, and the opportunity to capitalize on his influence with the Chinese peasantry to grow his army. By 1938, his indoctrination, recruitment, and training campaign had added over 500,000 troops to the Communist cause. As expected, the Japanese surrender in 1945 ended Mao's alliance with Chiang, and the civil war resumed. Despite assistance from the United States, the nationalists soon became the underdog to the communists' now large, conventional army, and Mao overwhelmed Chiang in direct warfare, eventually forcing the nationalists to escape the mainland to Taiwan. In defeating the Kuomintang and founding the People's Republic of China, Mao Tse-tung had completed one of history's most extraordinary political and military achievements.

There is much more to Mao's story, but our interest is confined to his ideas as a military leader and theorist. During his development, Mao learned extensively through trial and error. The Long March and his exile offered years to reflect and coalesce his thoughts on war. With Sun Tzu as a main influence, Mao rejected the notion that any single principle or tactic could guarantee success. Rather, he adapted his strategy based on what he knew of his own capabilities relative to the enemy's. As these capabilities changed, Mao altered his approach, moving between irregular and regular forms of war as required by the circumstances.

On Guerrilla War (*Yu Chi Chan*)

In 1937, Mao published a short treatise, *On Guerrilla War*, to define principles that would enable him to gain supreme control over an independent Communist China. To achieve that goal, Mao had to win two wars, one against Japan and another against Chiang. *On Guerrilla War* outlined Mao's military methodology for winning those wars and enhanced his stature in the Party as a leader and military strategist.

Mao emphasized *guerrilla* war out of necessity, for based on his experiences, this form offered the best chance of survival in the initial stages of his campaign. Relative to Chiang, Mao's combat capability was marginal. His forces were inferior in size, quality, and weaponry, but under Mao's tutelage, they developed an unbreakable will to fight based on the shared vision of a better future. Guerrilla war, practiced throughout history but given its name during Napoleon's occupation of Spain, allowed Mao to avoid defeat and reduce his opponents' strength while building his own capacity. Mao saw guerrilla warfare as a starting point for every underdog, rebel, and revolutionary, but it was not the end. Since the Kuomintang were native to China, Mao could not simply outlast them. Instead, he had to *transition* from guerrilla to orthodox warfare to gain a decisive victory. For Mao, guerrilla war was not an aberrant form of war, but a proper way of fighting where the will to fight is ample but resources are not.

The Seeds of Revolution

Mao's war theory has revolutionary origins, and revolutions begin in the hearts and minds of people unhappy with the status quo. Revolutionary guerrilla operations, writes Mao, "are the inevitable result of the clash between oppressor and oppressed when the latter reach the limits of their endurance."[457] Those who favor changing the status quo, as society's "have-nots," generally lack the capacity to defeat their oppressors with force. Rather, they must buy time and gather strength while sapping their enemy's.

Guerrilla (*Yin*) versus Orthodox (*Yang*)

Mao's concept of guerrilla war is rooted in the Chinese philosophy of the "unity of opposites," where, "the *Yin* and the *Yang* are elemental and pervasive. Of opposite polarities, they represent female and male, dark and light, cold and heat, recession and aggression. Their reciprocal interaction is endless. In terms of the dialectic, they may be likened to the thesis and antithesis from which the synthesis is derived."[458] This concept parallels Heraclitus's harmony of tensions, Sun Tzu's *cheng-ch'i*, and war's dialectics (especially regular-irregular, concentration-dispersal, and direct-indirect). The guerrilla ironically draws strength from weakness, remains cohesive despite its dispersion, eludes detection while appearing omnipresent, and makes the enemy's rear its own front. Echoing Sun Tzu, Mao's guerrillas seem "to come from the east and [attack] from the west; avoid the solid, attack the hollow," deliver a "lightning blow," and against a "stronger enemy, they withdraw when he advances; harass him when he stops; strike him when he is weary; pursue him when he withdraws."[459] Revolutionary guerrilla war, writes Samuel Griffith,

is endowed with a dynamic quality and a dimension in depth that orthodox wars, whatever their scale, lack [and] is not susceptible to the type of superficial military treatment frequently advocated by antediluvian doctrinaires. [It is] in fact more sophisticated than nuclear war or atomic war or war as it was waged by conventional armies, navies, and air forces. It can be conducted in any terrain, in any climate, in any weather; in swamps, in mountains, in farmed fields. Its basic element is man, and man is more complex than any of his machines.[460]

Clearly Griffith, a retired US Marine general, admired the power of guerrilla war as Mao did, for it represents the triumph of human resiliency and ingenuity over might. Guerrilla war, writes Mao, "is a weapon that a nation inferior in arms and military equipment may employ against a more powerful aggressor nation."[461] But where guerrilla war against an invader can be a necessary and sufficient tactic, in a civil or revolutionary war, guerrilla action may only be an initial part of a multiphase strategy. When opposing forces are indigenous or neighboring (e.g., Mao-Chiang or Spain-France), guerrilla war alone may lack decisiveness. In these cases, guerrilla in concert with orthodox forces can be more effective. "This warfare must be developed to an unprecedented degree," says Mao, "and it must coordinate with the operations of our regular armies."[462] For example, General Giap's Viet Minh *survived* against France in Vietnam using guerrilla tactics but ultimately *triumphed* in a conventional battle at Dien Bien Phu.

Mao's Three Phases

Mao did not think guerrilla operations should "be considered as an independent form of warfare," but as "one step in the total war, one aspect of the revolutionary struggle."[463] Therefore, he devised a three-phase model for revolutionary war: "Phase I (organization, consolidation, and preservation)," "Phase II (progressive expansion)," and "Phase III decision, or destruction of the enemy."[464] As progress is made through each phase, "a significant percentage of the active guerrilla force completes its transformation into an orthodox establishment capable of engaging the enemy in conventional battle."[465] The remaining guerrillas "operate in conjunction with other units of the regular army."[466] Though guerrilla forces in this model must work with regular forces, the character of guerrilla war remains distinct from conventional warfare, which Mao calls "orthodox hostilities" and "war of position" and "movement."[467]

Unlike regular forces, says Mao, guerrillas are neither fixed, passive, nor static, nor is there "such thing as a decisive battle."[468] Guerrillas are dispersed and independent, which requires physically and mentally agile leaders comfortable operating outside the bounds of traditional, centralized C2. Due to their mobility, guerrilla forces have no front or rear like a conventional army. Guerrillas often live off what they can acquire, steal, or scavenge simplifying their own logistics structure even as they exploit the opposition's. Regular forces seek decisive battle, to hold centers of gravity at risk, and to fix and destroy the enemy's regular forces. Guerrillas, on the other hand, "exterminate small forces of the enemy . . . harass and weaken larger forces . . . attack enemy lines of communication . . . establish bases capable of supporting independent operations in the enemy's rear . . . force the enemy to disperse his strength; and . . . coordinate all these activities with those of regular armies."[469]

Mix and Match

It is possible, though not necessarily easy, for a force to employ both regular and guerrilla forms of war simultaneously, though effectiveness depends on training and objectives. "While it is improper to confuse orthodox with guerrilla operations," asserts Mao, "it is equally improper to consider that there is a chasm between the two."[470] Guerrilla forces and regular forces each serve vital functions, and neither should be wholly ignored by strategy. Mao affirms, "We must promote guerrilla warfare as a necessary strategical auxiliary to orthodox operations, we must neither assign it the primary position in our war strategy nor substitute it for mobile and positional warfare as conducted by orthodox forces."[471]

Mao identifies two types of guerrilla warfare: one as a form of war and the other as a tactic. The former is true guerrilla warfare "based on the masses of the people," and it leverages and relies upon their support for efficacy and potency.[472] On the other hand, guerrilla *tactics* (e.g., small groups, mobility, terrorism, and raiding) can occur without the support of the local populace, and when employed indiscriminately may undermine revolutionary political aims by alienating supporters. "The second type of guerrilla warfare directly contradicts the law of historical development," writes Mao, because it uses guerrilla tactics in a manner "contrary to the true interests of the people" and "must be firmly opposed."[473]

Guerrilla Strategy

The hardest thing of all is to find a black cat in a dark room, especially if there is no cat.[474]
                                                                              Confucius

"Guerrilla strategy," writes Mao, "must be based primarily on alertness, mobility, and attack," adjusting according to "the enemy situation, the terrain, the existing lines of communication, the relative strengths, the weather, and the situation of the people."[475] This strategy relies heavily on information superiority (i.e., acquiring, processing, and disseminating accurate data faster than the enemy). Mao's guerrillas dominate the battle for information and intelligence, keeping the enemy in the dark while gaining and maintaining awareness of the enemy's plans and actions. During battle, the guerrilla attacks only the conventional enemy's vulnerable extremities, becoming corporeal to strike, then evaporating. Griffith writes, "The [conventional] enemy stands as on a lighted stage; from the darkness around him thousands of unseen eyes intently study his every move, his every gesture. When he strikes out, he hits the air; his antagonists are insubstantial, as intangible as fleeting shadows in the moonlight."[476] The guerrilla moves when the outcome is certain and then with deception and cunning to erode the enemy's resistance, confidence, and will to fight. "The mind of the enemy and the will of his leaders," writes Griffith, "is a target of far more importance than the bodies of his troops."[477]

Politics and Guerrilla War

Mao was highly attuned to the primacy of the political object in war, which for him was the realization of personal ambition and the rebirth of China as a world power. Great generalship and battlefield success were prerequisites for turning military power and war into political victory. Sounding Clausewitzian, Mao writes, "It is vital that these simple-minded militarists be made to realize the relationship that exists between politics and military affairs. Military action is a method used to attain a political goal. While military affairs and political affairs are not identical, it is impossible to isolate one from the other."[478] Furthermore, "without a political goal, guerrilla warfare must fail, as it must if its political objectives do not coincide with the aspirations of the people and their sympathy, cooperation, and assistance cannot be gained."[479] Additionally, "guerrilla troops should have a precise conception of the political goal of the struggle and the political organization to be used in attaining that goal."[480]

When the Japanese were China's main threat, Mao forged military ties with Chiang, his bitter enemy. This increased China's chances for independence and allowed Mao to spy on Chiang while building strength for the eventual reckoning. During the occupation, Mao's seven strategic objectives were as follows:

(1) Arousing and organizing the people
(2) Achieving internal unification politically
(3) Establishing bases
(4) Equipping forces
(5) Recovering national strength
(6) Destroying the enemy's national strength
(7) Regaining lost territories[481]

These objectives were overtly palatable to Chiang; however, they also served Mao's ends as the simple substitution of "Communist" for "national" in (5) and "Kuomintang" for "enemy" in (6) reveals the flexibility of Mao's strategic vision.

In addition to his strategic objectives, Mao identifies three fundamental political activities for revolutionary guerrilla forces: (1) the "spiritual unification of officers and men within the army"; (2) the "spiritual unification of the army and the people"; and (3) the "destruction of the unity of the enemy."[482] Mao reasoned that no enemy could long resist an army unified internally and externally to its base of popular support. Mao's use of the term "spiritual" lent his cause an air of divine providence and inevitability that resonated with his followers. Through violent physical and relentless psychological and political pressure, Mao wore down and eventually assimilated or destroyed his enemies.

Leading Guerrillas

Like Clausewitz and Jomini, Mao outlined the characteristics of a perfect leader, modeled, of course, after himself. Mao's guerrillas were not an anarchic rabble. They earned popular support and gained cohesion via effective military and political leaders who were, "unyielding in their policies – resolute, loyal, sincere, and robust ... well educated in revolutionary technique, self-confident, able to establish severe discipline, and able to cope with counterpropaganda."[483] Moreover, officers required "great powers of endurance" and had to cultivate connections and rapport with the locals, encourage unity and "complete loyalty" to the cause, remain tactically astute, and educate and train their personnel to uphold standards and exhibit patriotism.[484] According to Mao, the rank-and-file warriors should be volunteers selected for ability and loyalty rather than social condition or position. People of poor discipline and character, including drug addicts or mercenaries, are not suitable for guerrilla warfare, which demands extraordinary courage and fortitude to "bear the hardships of guerrilla campaigning in a protracted war."[485]

Gaining the Decision

For Mao, the most important strategic planning factor is the comparison between "all the elements of our own strength with those of the enemy."[486] These elements include political structure, industrial capacity, the size and skill of the armed forces, access to raw materials, views of the populace toward the war effort, morale, the scope of the theater, and the percentage of total forces that can be spared against a single adversary.[487] Mao's assessment of these factors led him to conclude that China would inevitably defeat the Japanese, because the invaders lacked the resources and political strength to wage protracted war on the mainland. However, while a pure guerrilla campaign might eventually have expelled Japan, addressing Chiang's nationalist forces required a different approach because Mao's ambitions did not include dying of old age in a secluded village. Only a powerful conventional force could deliver Mao a Napoleonic-style victory in decisive battle.

Principles

Though Mao's guerrilla warfare differed from orthodox wars of movement, he considered guerrilla strategy a subset of the strategy of war in general. He calls the "fundamental axiom of combat on which all military action is based ...

'conservation of one's own strength; [and] destruction of enemy strength,'" a formula that works for guerrilla and orthodox warfare.[488] Mao's basic requirements for successful guerrilla war are:

(1) Retention of the initiative; alertness; carefully planned tactical attacks in a war of strategical defense; tactical speed in a war strategically protracted; tactical operations on exterior lines in a war conducted strategically on interior lines
(2) Conduct of operations to complement those of the regular army
(3) The establishment of bases
(4) A clear understanding of the relationship that exists between the attack and the defense
(5) The development of mobile operations
(6) Correct command[489]

Since they generally lack manpower and firepower, guerrillas must use stealth, speed, and mobility to attack and retreat faster than the opponent can respond. They are less likely to win in a single, decisive battle and must remain on exterior lines to gain safe haven beyond the enemy's sphere of control. Dispersion – anathema to Clausewitz – is life for the guerrilla and, according to Mao, fulfills the following needs:

(1) When the enemy is in overextended defense, and sufficient force cannot be concentrated against him, guerrillas must disperse, harass him, and demoralize him.
(2) When encircled by the enemy, guerrillas disperse to withdraw.
(3) When the nature of the ground limits action, guerrillas disperse.
(4) When the availability of supplies limits action, they disperse.
(5) Guerrillas disperse in order to promote mass movements over a wide area.[490]

Concentration

Though the guerrilla's strategic plan is defensive, it is *tactically* offensive, concentrating force at smaller scales to destroy its targets. "Concentration may be desirable," says Mao, "when the enemy is on the defensive and guerrillas wish to destroy isolated detachments in particular localities. The remaining guerrillas are assigned missions of hindering and delaying the enemy, of destroying isolated groups, or of conducting mass propaganda."[491] Over time, "The total effect of many local successes will be to change the relative strengths of the opposing forces."[492] When this occurs, the guerrilla-to-regular ratio decreases until there is a clear force advantage, after which, the strategy changes to pursue a decisive, force-on-force battle. While it is *possible* for the once-dominant side to transition to guerrilla warfare, this is exceedingly difficult in practice absent significant training and transition time and works best when an indigenous conventional force is defeated but the population rises to resist the invader (e.g., Spain's guerrillas and the French *maquisards*). Because he had already gained the Chinese peasantry, Mao knew that Chiang's army could not start and sustain its own guerrilla campaign after losing its capacity advantage. Chiang's only options were to win a conventional war or withdraw.

Conclusion

Without question, Mao's brief discourse on guerrilla war falls far short of Clausewitz's depth and breadth; however, it features something the Prussian's theory only touched on: guerrilla warfare is a natural and integral part of war, effective and even desirable for materially disadvantaged forces. Guerrilla war is not a perfect form of war, but it can be the *best* form under the right conditions. Though Clausewitz does discuss polarity and the connection between opposites, these concepts seem less fitting in the Prussian's hands than they do in Mao's, a true son of the East. Ultimately, Mao was correct that unorthodox forms of war, driven by powerful political ideologies and nationalism and guided by his blueprint, would proliferate causing a fundamental shift in geopolitics. However, he also overestimated the long-term political viability of socialist communism and underestimated the resiliency of the capitalist democracies he and the Soviets loathed. Moreover, communism is by no means the only cause that can inspire success in guerrilla war. Nevertheless, Mao's outline for guerrilla warfare – including his sense of the relationship between war's forms and the strength of opposing sides, his ideas on dispersal and concentration, and his appreciation for guerrilla forces as a "strategic auxiliary" to orthodox – are critical for understanding war in all its dimensions. We will revisit and reinforce many of Mao's conclusions about guerrilla warfare in the next chapter.

CONCLUSION

Since Mao's "objective" approach is as problematic as Clausewitz's "sub-jective" solution, it is best to seek a balanced answer that integrates both.[493]

Michael Handel

While not every pertinent war theorist or war-related idea has been presented in this chapter, it is hardly hyperbole to propose that others must be derivative. To paraphrase a Napoleonic maxim:

Peruse again and again the theories of [Sun Tzu, Thucydides, Machiavelli, Clausewitz, Jomini, Liddell Hart, and Mao]. This is the only means . . . of acquiring the secret of the art of war. Your own genius will be enlightened and improved by this study, and you will learn to reject all maxims foreign to the principles of these great [theorists].[494]

Though each theorist had a unique approach to the problem of articulating a general theory of war, in aggregate, their ideas confirm the following truths:

➢ History informs the practice of war and the development of theory.

➢ War is reciprocal, a violent clash of physical and psychological capacity between forces driven by opposing political objectives.

➢ Victory (i.e., realizing the war's political aim) is achieved by destroying the enemy's will to fight.

➢ War and strategy have multiple levels, from political to tactical, and plans and actions at each level must account for the characteristics of each side and the environment.

➢ War is uncertain, dangerous, complex, and the province of dialectics like defense-attack, direct-indirect, regular-irregular, concentration-dispersal, and reason-emotion.

➢ In addition to understanding a war's purpose, great generals enforce discipline, inspire followers, and are personally courageous, bold, and intuitive.

➢ Successful strategy is flexible, adaptive, and seeks to unbalance opponents via deception, speed, and surprise.

➢ Successful strategists and warriors understand the value and hazards of alliances, intelligence, momentum, culmination, fog and friction, decisive points, lines of communication, and centers of gravity.

➢ Unorthodox warfare (e.g., guerrilla, irregular, or people's war) employs elusiveness, speed, surprise, and time to wear down orthodox adversaries.

The next chapter complements and expands upon these themes by examining two of war's most significant "sub-theories": small wars and domain theory. This added detail will enable us to place war's major ideas and theorists within a "universe" of theory and strategy delimited by war's dialectics and will point the way toward a truly unified war theory.

3   Small Wars and Domain Theory

*The complete truth about the world is not graspable as any single point of
view, but only resides in the totality of several or many distinct views.*[1]

Lee Smolin

Introduction

To this point, our examination has centered on specific war theorists and
general war theory, but to fully understand all dimensions of existing theory
requires that we expand our gaze to include two additional theoretical paradigms: small wars and domain "sub-theory." Consistent with war's definition,
its unchanging truths, as explained by general war theory, emanate from the
confluence and interaction of humanity, politics, and combat. Provided the
general war theory thoroughly articulates these truths, there can be no other
war-related theory not in some way derivative of the general theory. This
includes small wars (e.g., asymmetric warfare) and domain (e.g., maritime
and air) *sub-theories*. Indeed, as Julian Corbett opines, "It is of little use to
approach naval strategy except through the theory of war. Without such theory
we can never really understand its scope or meaning, nor can we hope to grasp
the forces which most profoundly affect its conclusions."[2] And the US Marine
Corps *Small Wars Manual* confirms, "Although small wars present a special
problem requiring particular tactical and technical measures, the immutable
principles of war remain the basis of these operations."[3]

Commentary on the martial ramifications of innovative tactics (e.g., guerrilla warfare) and advanced technologies (e.g., aircraft) is commonplace;
however, when changes in war's character appear extraordinarily impactful,
new "theories" can spring up to explain them. When this occurs, theorists
may draw upon prominent historical and theoretical movements, as Mahan
does in describing sea power as a maritime analogue to climactic land
warfare. In other cases, theorists may forge an original path, like Corbett,
who expands naval theory beyond the Nelsonian-Napoleonic model.[4]
Regardless, the development and study of war's consequential sub-theories

has the twofold benefit of enhancing warfighting acumen and deepening our understanding of war.

SMALL WARS

> *The conduct of small wars is . . . in certain respects an art by itself, diverging widely from what is adapted to the conditions of regular warfare, but not so widely that there are not in all its branches points which permit comparisons to be established.*[5]
>
> C. E. Callwell

Most war theory, until after World War I and the emergence of the more balanced doctrines of Corbett, Fuller, and Liddell Hart, was dominated by the Nelsonian-Napoleonic model (i.e., traditional states conducting relentless offensives in pursuit of decisive battle), which seemed only partially applicable to the "savage wars of peace" waged by imperial powers against rebellious natives.[6] In 1898, to explain this aberrant, odious sort of conflict, the British warrior-theorist Charles E. Callwell wrote *Small Wars: Their Principles and Practice*, from which we get the English "small wars" phrase (derived from the Spanish word *guerrilla*, meaning "little war"). Between the world wars, the US Marine Corps became embroiled in similar conflicts in the Western Hemisphere, and in 1940 published its own *Small Wars Manual* (Fleet Marine Force Reference Publication 12–15). Also, during this period, T. E. Lawrence (*Seven Pillars of Wisdom*, 1922) and Mao (*On Guerilla Warfare*, 1937) published incisive treatises from the insurgent's perspective.

After World War II, Cold War strategic considerations predominated, and the world wars' political detritus ignited such a proliferation of modern small wars that French counterinsurgency expert, Roger Trinquier, termed the entire phenomenon *modern warfare*. These were mainly wars of national liberation, erupting far beneath the threshold of great-power war set by the threat of suicidal nuclear holocaust. By the latter half of the twentieth century, most great powers, including Britain, France, and the USA, had lost touch with small wars theory and had to relearn much of it in the school of hard knocks. The best-written works of this period appeared in 1964 from Trinquier (*Modern Warfare: A French View of Counterinsurgency*) and his countryman, David Galula (*Counterinsurgency Warfare: Theory and Practice*). Still, despite the explosion of thought and writing on small wars, few saw a need or felt empowered to revisit general war theory. Rather, small war "theorizing" continued apace as loosely related addenda to the works of Clausewitz, Corbett, and others.

After the Cold War ended, revising general war theory seemed unnecessary. The promise of precision weapons and networked warfare reinvigorated what military historian Beatrice Heuser terms the "*matériel* school" of strategic thought, which espouses the view that advanced technologies, like airpower, render historical lessons irrelevant; and once again, the hard-learned small wars doctrines were shelved.[7] The late twentieth and early twenty-first centuries' global extremist insurgencies, fully ignited on September 11, 2001, with al-Qaeda's attack on the USA, rekindled interest in small war strategy. The subsequent wars in the Middle East, especially Afghanistan, Iraq, and Syria, spawned a new generation of guidance, including Bard O'Neill's updated *Insurgency & Terrorism: From Revolution to Apocalypse* (second edition, 2005), the US Army and Marine Corps Field Manual (FM 3–24, *Counterinsurgency* (2006), and David Kilcullen's *Counterinsurgency* (2010), among others. In truth, most small wars sub-theory consists of terminology, tactics, and doctrines more accurately labeled "best practices" than theory. This section highlights the most universal facets of small wars doctrine (some of which we have already explored with the magnificent seven, particularly Mao), and along with the forthcoming section on domain theory, provides a baseline for the next chapter, which will update general war theory to be more inclusive of these sub-theories.

On Small Wars

The struggle will assume two aspects: Political – direct action on the population; and military – the struggle against the armed forces of the aggressor.[8]
Roger Trinquier

Callwell defines "small wars" as "campaigns other than those where both opposing sides consist of regular troops," like "expeditions against savages" and "campaigns undertaken to suppress rebellions and guerilla warfare."[9] If war is the most uncertain and hazardous human activity, small wars (what Galula terms "revolutionary warfare") are the most uncertain and hazardous variation of war. "As a regular war never takes exactly the form of any of its predecessors," advises the *Small Wars Manual*, "so, even to a greater degree is each small war somewhat different from anything which has preceded it."[10] Thus, as T. E. Lawrence opines, service in small wars can be "far more intellectual" and "more exhausting" than service in traditional conflicts.[11] Small wars are often more violently intimate. In the decisive phases, they are generally fought with infantry and embody what Trinquier calls "the hard and pitiless realities of war . . . physical suffering and death individually given and received."[12] Indeed, these conflicts, adds Callwell, are a "protracted, thankless, invertebrate" form of war.[13]

Though major powers are regularly surprised and befuddled when they reappear, small wars are, as Heuser points out, "at least as old as any form of war and predate state formation."[14] And to be sure, small wars are *wars* (i.e., combative human violence for political purposes); thus, they (should) fall wholly within the precepts of general war theory. Nevertheless, as alluded to earlier, the ubiquity of small war "theories" highlights general war theory's present insufficiency, which should not be surprising given that general theory evolved almost exclusively from experiences with traditional, land-centric warfare. As such, it largely equates war with combat and marginalizes war's human and political aspects. Paraphrasing Clausewitz, traditional theory describes war as an offensive act of violent force, concentrated and unleashed at a decisive point upon military capacity, to bend or break the enemy's will. This construct normally assumes that if one side has a big, powerful army and the other does not, the weaker side should have the sense to defer. But this is not how the human mind works. Thus, rather than *the strong do what they can and the weak suffer what they must*; the operative strategic principle in small wars is more often *where there's a will, there's a way.*

As most analysts affirm, there is nothing necessarily small about "small" wars. "Small war," writes Callwell, "has ... no particular connection with the scale on which any campaign may be carried out."[15] Moreover, "If wars are different from each other," writes Aron, "it is because the elements involved ... are in different proportions."[16] In fact, what is small about small wars is the role combat plays in the overall outcome. Whereas traditional forms of war are *centered* on combat, in many small wars, combat may be the "smallest" and *least important* factor relative to human and political considerations. This is true because small wars are invariably *asymmetric* (i.e., between sides of markedly different warfighting capacity – "haves" versus "have-nots"). Given its power deficit, the weaker side would be foolish to wage war as a test of combat energies. Thus, rather than describing scale or violence, the word "small" pertains to combat's significance relative to humanity and politics within the True Trinity (Figure 3.1). "Military operations," in small wars,

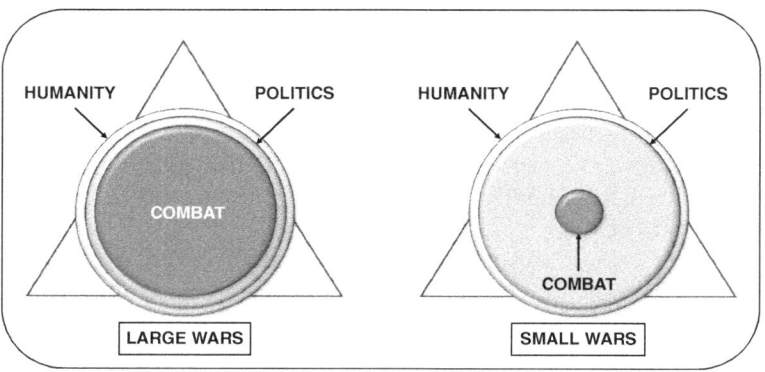

Figure 3.1 – The True Trinity in large and small wars

writes Trinquier, "as combat actions ... are of only limited importance and are never the total conflict."[17] And the *Small Wars Manual* actually defines small wars as "operations ... wherein military force is combined with *diplomatic pressure*" to advance *political interests* vis-à-vis the affairs of "another state" (emphasis added).[18]

To illustrate further, consider the following expansion of Clausewitz's card game metaphor. In small wars, one side is dealt a hand loaded with combat capability. The other side, rather than folding or playing its weaker cards, stacks its hand with valuable human and political cards and plays them instead. If the "stronger" side persists in playing only its combat cards, it is playing the wrong game and must either adapt its strategy, acquiring its own human and political cards, or risk defeat. For instance, Lawrence's "stupid" Turk opponents, thinking "rebellion was absolute like war," dealt with it "on the analogy of war" (i.e., as conventional combat), a misstep Lawrence likens to "eating soup with a knife."[19] On the other hand, a savvy weaker side plays its combat cards selectively, when it can gain surprise or a localized advantage against the opposing military force or more permissive targets, like civilians and public infrastructure.

The most prevalent types of small wars occur between existing governments or invader-occupiers and an indigenous faction (e.g., insurgent) or resistance movement (e.g., partisan). "Have-not" warriors of both types are often called "guerrillas" or "terrorists" for their use of unorthodox or ruthless combat tactics. Small wars caused by invasion are always instigated by the invader; whereas, the domestic faction initiates internal small wars (a variation of civil war), unless started preemptively by the government. Relative to in-power groups, small wars are counterinsurgencies, and for out-of-power factions: insurgencies, rebellions, or revolutions.

In asymmetric warfare, says Galula, the "haves" enjoy "tangible" advantages, like diplomacy; administration and policing; financial, industrial, and agricultural resources; transportation and communications; media access; and military power. Meanwhile, the "have-nots" rely on "intangibles," such as "the ideological power of a cause."[20] In addition to disparate *capacities*, the warring sides may have contrasting *aims* (e.g., unlimited ends, like regime change; or limited goals, like preserving the status quo). This "intangible" usually benefits the "have-nots," who may have less to lose and more to gain than the "haves."

Finally, while politics and policy set a traditional war's strategic aims, leaving the armed forces to focus on military activities and objectives; in small wars, politics often *permeates all war's levels*. The "organization and discipline of guerrilla troops," writes Mao, is *not for fighting*, rather it is for carrying out "the *political activities* that are the life of both the guerrilla armies and revolutionary warfare" (emphasis added).[21] Thus, the fighting forces may be called upon to advance political objectives directly and concurrently with combat. These instances require greater balance in the life-death, creation-destruction dialectics. For example, in a small war, the military may have to fight while also protecting the populace, building a counter-organization, capturing and interrogating spies, providing basic services, building infrastructure, and forging relationships with community leaders. "The soldier," writes Galula, becomes "a propagandist, a social worker, a civil engineer, a schoolteacher, a nurse, a boy scout."[22] Doing all this effectively, says the *Small Wars Manual*, requires "courtesy, friendliness, justice, and firmness" more often than ferocity.[23] Of course, most of these activities are better accomplished by non-military agencies; however, the danger of enemy action prevents this. Since few militaries train for these responsibilities, this is often where the "have-nots" gain the upper hand.

Physical Environment

Geography can weaken the strongest political regime or strengthen the weakest.[24]
David Galula

While the physical environment, including terrain, weather, climate, flora and fauna, topography, urban development, and border geography, is a critical factor in any war, it can play an outsized role in small wars. These wars, says Callwell, "as compared with regular hostilities . . . are in the main campaigns against nature."[25] Neither warriors nor their machines are optimized for every possible environment. Tanks and aircraft, for example, may perform magnificently over the open desert, but poorly in jungle, mountainous, or urban terrain. Likewise, warriors struggle with extremes of temperature and amongst dangerous wildlife and pathogens like malaria or dengue. "The hostilities," writes Callwell, "often take place in unhealthy, and even deadly, climates, in torrid, fever-stricken theatres of war."[26] "Climatic conditions," says the *Small Wars Manual*, "will affect the organization, clothing, equipment, supplies, health, and especially the operations of the intervening forces."[27]

Where two sides are comparable in traditional fighting capabilities, these factors tend to balance. However, in small wars, the weaker side, always seeking advantages, favors terrain that is especially problematic for its enemy. "Rugged terrain," writes O'Neill, "vast mountains, jungles, swamps, forests, and the like – is usually related to successful guerrilla operations."[28] A similar effect occurs in urban conditions, which provide ample hiding areas and non-combatant "human shields," and where adversary mistakes become instant propaganda victories. "The conduct of military operations in a large city," says Trinquier, "in the midst of the populace, without the benefit of the powerful weapons it possesses, is certainly one of the most delicate and complex problems ever to face an army."[29] Border zones are also significant and, as Galula highlights, they are generally "a permanent source of weakness for the counterinsurgent."[30] In short, terrain that conceals while still allowing for freedom of movement or escape (what Clausewitz calls "rough and inaccessible") favors the weaker side.[31] Conversely, the strong benefit from open conditions and settings that restrict a weaker foe's movement or retreat.

Humanity and Politics

The contest was not physical, but moral, and so battles were a mistake.[32]

T. E. Lawrence

Though fighting is part of every war, we cannot dismiss the primacy of war's efficient and final causes – humanity and politics – especially in small wars, which are, in effect, political campaigns with a combat component. The populace "votes" by supporting one side or another, actively or passively, or declines to "vote" and waits to see which side gives up first, then sides with the other. In these conflicts, says Galula, "military and political actions cannot be separated, and military action ... cannot be the main form of action."[33] Since wars of imperial conquest and the pacification of primitive peoples have, thankfully, become historical relics, our concentration narrows to more modern varieties of asymmetric warfare. Again, these conflicts pit the powerful against the weak, where the stronger side is often a traditional state and the opponent is a weaker state or a non- or substate political faction (e.g., Spanish guerrillas or al-Qaeda).

There are few absolutes regarding the human and political "terrain" of war. Therefore, differences in culture, ethnicity, religion, language, demography, and so on, can render lessons from one small war inapplicable to another. In fact, even for the same country, the passage of time alone can alter the strategic calculus. For example, given the French experience in Vietnam, particularly at Dien Bien Phu, the conventional military approach adopted initially by the United States a decade later seemed plausible; however, the Vietnamese communists presented America with a different strategic puzzle. Likewise, the successful counterinsurgency strategy the USA used in the Philippine War of 1899 bore little resemblance to the effective tactics employed to neutralize the Hukbalahap Rebellion there a half-century later.

The ultimate objective in any small war is the political aim, and, according to FM 3–24, this often entails "acceptance of the legitimacy of one side's claim to political power by the people."[34] "Our hostilities must have a clearly defined political goal," writes Mao, for without it, "guerrilla warfare must fail."[35] A well-defined political aim guides strategy and operations, and the entire effort will only be effective when supported by good information. Therefore, prior to and during hostilities, it is essential for both sides to assess the various human and political "cards" in play, including the populations of all participating entities, the relevant factions and governments – friend, enemy, and neutral – the armed forces, and key leaders. This information yields insights into motives and likely actions.

Ultimately, the strategic art of small wars centers on resolving the causes and alleviating the violent symptoms of political illegitimacy. As with medicine, we tend to be better at the latter (treating symptoms), which pertains to combat, than the former (removing causes), which reflects war's human and political facets. In a 2013 study of seventy-one modern insurgencies (*Paths to Victory: Lessons from Modern Insurgencies*), the RAND Corporation found that successful counterinsurgencies supplemented battling the insurgent (symptom) with undermining the insurgent's sources of support (causes).[36] Conversely, insurgents succeed by baiting the counterinsurgent into a combat-centric approach while focusing efforts on intensifying their opponent's illegitimacy. In general, asymmetric wars last a relatively long time (nearly eleven years on average), a fact that often favors the insurgent, unless the counterinsurgent is extremely patient, a difficult requirement for an invader (e.g., Napoleon in Spain) or government supported by a distant ally (e.g., the USA in Vietnam).[37]

The Have-Nots

In small wars, the weak side's primary strategic problem is to find and employ ways and means, other than decisive combat, to compel its opposition to quit. Galula lists two major patterns for insurgents: (1) orthodox, which seeks complete political transformation (e.g., Maoist revolution); and (2) a more radical, spasmodic form that aims to topple the existing political order (e.g., Algerian nationalism). Regarding the former, Trinquier writes, "The guerrilla and terrorism are only one stage of modern warfare, designed to create a situation favorable to the build-up of a regular army for the purpose of eventually confronting an enemy army on the battlefield and defeating him."[38] That favorable situation rests upon a meticulously assembled political foundation constructed to supplant the existing paradigm. The latter pattern employs a temporary organization optimized for severing the populace from its government, primarily via terrorism.[39] The ultimate aim here, according to FM 3–24, is "to win by undermining and outlasting" the adversary's "public support," after which, the insurgents seize and solidify authority.[40] As Galula's patterns imply, small wars gain their causes from political legitimacy crises. This even applies to invasions since most invader-occupiers are considered illegitimate by their victims. Effective small war causes can include expulsion of an occupying power, overthrowing a corrupt or ineffective regime, or championing a favored form of governance. Nationalist (repelling the invader) and religious (repelling the infidel) causes are often the most compelling.

History proves that governing well is hard; thus, no political system is invulnerable to critique, which can lead to unrest and violent challenge by ambitious political factions. Furthermore, when a government, whose main functions are protecting citizens and preserving peace and order, is not performing poorly enough to warrant an uprising, bold rebels can accelerate its demise by threatening the population and, as FM 3–24 says, "sowing chaos and disorder."[41] Since all wars begin with combative violence, we will not dwell on pre-hostility preparation; however, this period is critical as insurgents stock their "hand" with human and political investments – defining and promoting a message, recruiting, building an organization, and planning the active campaign. The longer an insurgent can prevent its opponent from perceiving a threat, the greater the insurgency's chance of success.

Both sides in a small war aim to survive, though this is most in doubt for the weaker side. After combat begins, surviving the war's earliest stages means avoiding defeat and sustaining sources of support, bases, supplies, personnel, and identifying key leaders. Though survival alone may yield victory in the right circumstances, the weaker side must normally increase its military and political strength even as it depletes its opponent's, a process requiring the recruitment and cultivation of human capital (e.g., soldiers, sponsors, and allies). The "have-not's" ability to get stronger depends on the potency of its cause and messaging and the quality of its leadership. An effective cause inspires support, and its attendant messaging should clearly outline the benefits of joining while negatively characterizing the enemy. A charismatic, politically savvy leader, like Mao, can be pivotal. "No matter how favorable the physical milieu or how bad the social, economic, and political conditions," writes O'Neill, small wars will not emerge, let alone prosper, without skilled, "determined opposition leaders."[42] At the tactical level, Mao's perfect guerrilla leaders are "resolute, loyal, sincere, and robust," well-educated, confident, fierce disciplinarians, and incorruptible.[43] Only when it gets cause and leadership right can the weaker side expect to gain the popular and external "sympathy," "cooperation," and "assistance" needed for victory.[44]

The cause, message, and leader are the weaker side's principal "weapons" for engaging a small war's most indispensable target – the population, which must be wrested away from the enemy by persuasion or force. "The *sine qua non* of victory in *modern warfare*," writes Trinquier, "is the unconditional support of a population."[45] Gaining active support, or at least noninterference, from the populace is generally easier for the insurgent, who understands the language, culture, and tendencies better than an alien power or an indigenous government perceived to be inept, uncaring, or oppressive. Winning over the locals means presenting the populace with a choice: join us or face consequences ranging from inconvenience to death. Galula's orthodox pattern generally relies on the first option (persuasion) and the radical variation the second (coercion). According to Trinquier, the weaker side, through "an interlocking system of actions – political, economic, psychological, military," exploits "internal tensions" to exert "a profound influence on the population."[46] Terrorism, defined by O'Neill as "the threat or use of physical coercion, primarily against noncombatants, especially civilians, to create fear in order to achieve various political objectives," can be an effective coercive influence.[47] In fact, Trinquier calls terrorism "the principal weapon of modern warfare," enabling "enemies to fight effectively with few resources and even to defeat a traditional army."[48] Terrorism cows the populace, enrages the "haves" (perhaps eliciting reprisals), and discredits the government's protective ability.

External support is another vital objective for the weak. "Unless governments are utterly incompetent, devoid of political will, and lacking in resources," writes O'Neill, "insurgent organizations must normally obtain outside assistance if they are to succeed."[49] External aid – in the form of funding, supplies, weapons, advice, troops, or basing can be decisive. For powerful entities supporting insurgents, Lawrence offers the following advice: learn the people and their cause ("understanding ... their revolt [and] their present social and political differences were important"[50]); learn the enemy's habits and weaknesses (the Arab rebels sought to "acquire the military knowledge of their masters, and to turn it against them"[51]); be unobtrusive ("your ideal position is when you are present and not noticed"[52]); help them win battles, but do not fight for them; and cultivate active and passive supporters (the Arabs "had a friendly population," active or at least "quietly sympathetic"[53]).

The insurgent *organization* provides the structure for leaders to promote and capitalize on the cause, message, relationship to the people, and external support. The "haves" normally begin with an established organizational structure, while the "have-nots" build their organization even as they execute the war. Even so, insurgents can exploit this nascence to shape the organization's form and function into a political "virus." Leveraging the organization – consisting of soldiers, spies, propagandists, logisticians, and even bureaucrats – to spread its message like a pathogen, the insurgent infects its opponent's "body" politic. By the time the stronger side raises the alarm, the insurgent "disease" has metastasized. Complicating the matter for the powerful, according to Galula, is that, "The actual danger will always appear ... out of proportion to the demands made by an adequate response."[54] Thus, the "haves" must recognize the threat early and act swiftly, forcefully, and adeptly lest they become the new "have-nots."

The Haves

The beating of the hostile armies is not necessarily the main object ... moral effect is often far more important than material success.[55]   C. E Callwell

In most small wars, the strong side's primary strategic problem is to identify and employ ways and means – other than decisive combat – to counter the insurgency's causes (sources of support and political deficiencies) and symptoms (insurgent violence). "The application of purely military measures may not, by itself restore peace and orderly government," says the *Small Wars Manual*, "because the fundamental causes of the condition of unrest may be economic, political, or social."[56] Indeed, "The purpose should always be to restore normal government ... and to establish peace, order, and security."[57] Of course, militaries often lack the psychological and political subtlety required for delegitimizing insurgent movements; however, for the greatest likelihood of success, this is precisely what is needed.[58]

In addition to its inherent complexity, Kilcullen asserts that there are two constants in asymmetric warfare: (1) understanding "in detail what drives the conflict" vis-à-vis the area and population, and (2) prioritizing "respect for local people" over anything else, including "killing the enemy."[59] Therefore, countering an asymmetric foe's human and political elements requires careful analysis. To win against the rebel, writes Trinquier, "We must study his methods, study our own methods and their potential, and draw from this study some general principles that will permit us to detect the guerrilla's weak points and concentrate our main efforts on them."[60] Once the insurgent's cause is understood, sound intelligence yields further insight into the enemy's organization, leadership, sources of support, plans, and activities. "In most cases," says O'Neill, "an effective counterinsurgency program depends on an accurate, substantive, and comprehensive profile of the adversary and the environmental context."[61] This is not trivial, for as Callwell writes, "The nature of the enemy, his strength, his weapons, and his fighting qualities can only be imperfectly gauged."[62] Moreover, in modern small wars, the line between friend and foe can be "a non-physical, often ideological boundary," which complicates discrimination.[63] For example, in the twenty-first century's wars in Afghanistan and Iraq, ambiguities between hostiles and civilians frequently frustrated US counterinsurgency efforts.

To combat these challenges, Trinquier says, "We have to be everywhere informed ... we must have a vast intelligence network ... set up, if possible, before the opening of hostilities."[64] Intelligence, says Trinquier, "should gather information on the physical, economic, and human geography; the current psychological climate, and the disposition of military and police forces."[65] Additionally, O'Neill recommends the stronger side assess popular support, the insurgency's organization and cohesion, and external sponsors. To these we add *leadership* and an area often overlooked, *self*-assessment of military, economic, and political strength, popular support, organization and cohesion, leadership, and prospects for alliances.[66] As mentioned earlier, the stronger side in a small war must reinforce its human and political acumen or risk defeat despite its military advantages. This can be frustrated by overconfidence and, more significantly, because inherent weaknesses in these areas are frequently the opponent's strengths. In Vietnam of the late 1960s, the inept South Vietnamese government struggled to oppose the communist-backed Viet Cong. Though the regime held on with American help, it succumbed rapidly to North Vietnam after the USA's departure.

From its analysis of modern insurgencies, RAND recommends the stronger side prioritize disruption of insurgent support over combat. This includes diminishing the insurgent's motives, and the people's incentive to support them, via reforms, redress, and security. RAND also advises an active campaign to sever the enemy's ties to popular and external support.[67] Winning, says Galula, requires cutting the enemy off from the population, enacting "political, social, economic, and other reforms," demonstrating persistence and resolve, conducting diplomacy from a position of strength, and committing to "a large concentration of efforts, resources, and personnel."[68] Therefore, through "measures that are political, economic, psychological, administrative, and military" and via relentless pressure, adds Trinquier, the guerrilla must be isolated from all support, even favored terrain, for long enough to "yield the desired results."[69]

Since the *specific* human and political actions necessary for defeating an insurgency vary widely from case-to-case, they resist universal characterization; however, we can highlight a few salient best practices. To begin with, the counterinsurgent's own house must be in order. This includes robust resourcing, a strong organization, steadfast commitment, and excellent leadership. "Control of the masses through a tight organization," writes Trinquier, "often through several parallel organizations, is the master weapon of *modern warfare*."[70] Moreover, "small wars demand the highest type of leadership," says the *Small Wars Manual*, "directed by intelligence, resourcefulness, and ingenuity."[71] Turning to the enemy, all "have-nots" rely on a cause, messaging, leadership, organization, and popular support; and many gain external benefactors. Actions and messaging crafted to discredit and delegitimize the adversary are best for undermining cause and message. The objective is to convince all relevant audiences that life will be better if the insurgency fails. To support this, operations must align with the expectations, laws, and values of these audiences. Specifically, Trinquier recommends active policing, "an intensive *propaganda effort*," and social programs that offer "material and moral assistance."[72] Though "propaganda" has a negative connotation and can backfire if used improperly, within the context of modern military information operations, it can be more important than combat victories.

Counter-propaganda, strategic messaging, and information operations in all formats – print, audio, video, social media, and personal engagement – must stress enemy atrocities, terror acts, and other negative impacts to the life and safety of non-combatants, while also offering proof of the prudence, effectiveness, and competence of the government and its allies.

The government must take overt steps to reinforce these perceptions with tangible reforms to include transparency; free elections; accountability and replacement of corrupt officials; robust, fair, and consistent law enforcement and judicial practices; public engagement; and the promotion of competent leaders of character. Additionally, the government must provide adequate services and security for the populace and, when able, attract external allies to increase the appearance of legitimacy. Protecting the public is critical, particularly versus terrorists, because rewards are pointless to people threatened with death for backing the wrong side.

All the preceding measures will *indirectly* affect the enemy's popular support and organization. *Direct* measures should include robust policing, counterintelligence operations, upholding the rule of law (avoiding reprisals), and demonstrating commitment through presence and participation in constructive activities. "The *enemy*," says Trinquier, "consists not of a few armed bands fighting on the ground, but of an organization that feeds him, informs him, and sustains his morale."[73] Thus, Trinquier's "master concept" and key to victory against any insurgency is "through the complete destruction of [the insurgent] organization," not by defeating tactical combat units.[74] To do this effectively, Galula advises the location and recruitment of the population's "favorable minority" (sympathetic to the counterinsurgency) as an antidote to the "insurgent minority" (i.e., fighting fire with fire).[75]

To neutralize the enemy's leadership and organization, the "haves" must identify and target its key personnel. Using sound intelligence, including infiltration, these leaders should be tracked and preferably captured (for information and to avoid the martyr effect) or else eliminated, provided a state of war exists; otherwise, they should be treated as criminals. These tactics are consistent with Galula's three objectives for gaining localized political advantages: (1) assert authority over the population (without treating the people as the enemy), (2) isolate the population, physically if possible, from the enemy, and (3) gather intelligence on the enemy's political cells and eliminate them.[76]

Finally, almost without exception, severing the enemy's connections to external support must be a strategic tenet; and nothing precludes the stronger side from also pursuing external allies. External support can be degraded via diplomacy, economic sanctions, enhanced border security, and armed interdiction. Ultimately, if the "haves" can adopt these measures, they will at least be playing the right game.

Combat

Granted mobility, security ... time, and doctrine ... victory will rest with the insurgents, for the algebraic factors are in the end decisive, and against them perfections of means and spirit struggle quite in vain.[77]   T. E. Lawrence

Though it may be a smaller proportion relative to humanity and politics, combat is still an important aspect of small wars and can be decisive under the right conditions. For the weak, combat diminishes the enemy, influences the population, and may help attract internal and external support. For the strong, combat disrupts the adversary's organization and leadership while imposing manpower and materiel costs. "Killing the enemy," writes Kilcullen, "is, and always will be, a key part of guerrilla warfare."[78]

Because small wars are asymmetric, the combat may bear little resemblance to traditional warfare, and thus, despite being the most common form of war, it is often dubbed "irregular" or "unconventional." Additionally, while total combat deaths in small wars can be far less than those of larger wars, it is not uncommon for a greater *proportion* of casualties to come via atrocities or law-of-war violations. This occurs because the weaker side has less to lose and more to gain by ignoring the rules. Still, how the "have-nots" choose to fight is up to their leadership. If the weak side fights in a traditional manner, they play to the "have's" strengths. As the *Small Wars Manual* advises, "If there is an *organized* hostile force opposing ... the primary objective in small wars, as in a major war, is its early destruction" (emphasis added).[79] If, instead, the "have-nots" concentrate on political objectives, then the stronger side's problem is to find a path to victory against an enemy that refuses battle or fights in an unfamiliar style.

The Have-Nots

The active rebels had the virtues of secrecy and self-control, and the qualities of speed, endurance and independence of arteries of supply.[80] T. E. Lawrence

Whether its goal is a total revolution, the expulsion of an invader, or simply tearing down the existing political order, the weaker side must be selective regarding combat, at least until its warfighting capacity has surpassed its opponent's. Successful insurgents, led with "audacity, initiative, [and] cour-age," must favor the movement extreme of the rest-movement dialectic. With "constant activity and movement,"[81] they use advantages in deception, stealth, speed, mobility, and treachery to avoid detection and to surprise enemy elements that can be engaged successfully at minimum risk (e.g., patrols or convoys).[82] For example, Lawrence's rebels developed "a habit of never engaging the enemy ... never affording a target"[83] and employing "the smallest force in the quickest time at the farthest place."[84] Guerrilla warfare, as Liddell

Hart highlights, "reverses the normal practice of warfare," by favoring quality over quantity, dispersal over concentration, chaos over order, fluidity over rigidity, spontaneity over planning, and economy over mass.[85] "Our strengths," writes Lawrence, "depended upon whim,"[86] and "The smaller the unit, the better the performance."[87] Promoting disorder, adds Galula, diminishes "the strength and the authority of the counterinsurgent ... [it] is cheap to create and very costly to prevent."[88] Furthermore, "The insurgent, having no responsibility, is free to use every trick ... he can lie, cheat, exaggerate."[89] Many of these "tricks," like suicide bombers, "booby-traps," and improvised explosive devices, are exceedingly lethal.

Attacks against larger portions of the enemy force take the form of raids or ambushes orchestrated for greatest effect with less cost and risk. Guerrilla warriors "revel in stratagems and artifice," writes Callwell, "they prowl about waiting for their opportunity to pounce down upon small parties moving without due precaution."[90] Combat from the weaker side's perspective may also include non-military targets (e.g., terror attacks on civilians and public infrastructure). This harassing form of combat produces a political effect by reinforcing the message that the "haves" cannot provide security and are losing the war. Ultimately, guerrillas use combat, often characterized by surprise on the attack and dispersal when defending, to "create a climate of insecurity" that drives out adversaries, expands areas of control, and hangs like a guillotine over the connection between the government and its people, all while preventing the "haves" from inducing unrecoverable losses or landing a fatal blow.[91]

The Haves

In small wars the habits, the customs, and the mode of action on the battlefield of the enemy should be studied in advance ... or the regular troops will assuredly find themselves in difficulties and may meet with grievous misfortune.[92]
                                                                          C. E. Callwell

For the stronger side, a small war's combat component may be more an exercise in restraint and frustration management than tactical prowess. Patience, meticulous intelligence gathering, rigorous analysis, and adaptability frequently yield greater results than advanced weaponry. Additionally, the strong side should try to reduce the weaker side's few advantages by improving its own intelligence, speed, and mobility; and by employing deception, surprise, and leveraging external support. Finally, combat must be synchronized and prioritized within a broader campaign where humanitarian and political actions may be the main effort. Since war-ending battles are unlikely, political lines of effort must continue, with the goal of coaxing or compelling the population to assist with the counterinsurgency or stay out of the way.

With its greater military power, the stronger side's combat objective must be to avoid its antagonist's pin-prick raids, ambushes, and deadly traps, and to compel or lure the weaker side to fight *en masse*. The direct method is search and destroy, but finding guerrillas is difficult, and even when located, forcing them to stand and fight is challenging. As Callwell says, "It is the difficulty of bringing the foe to action which, as a rule, forms the most unpleasant characteristic of these wars."[93] Direct attacks are more effective when achieving surprise, attacking with great speed, or with enough troops to prevent the enemy from escaping. Galula writes, "The destruction of the insurgent forces requires that they be localized and immediately encircled," but often they are "too mobile to be encircled and annihilated easily."[94] If they cannot be contained and destroyed outright, they should at least be expelled.

If guerrillas persistently elude detection, it may be possible to "flush them out" by presenting an enticing target and springing a trap or by threatening a center of gravity, something they feel compelled to defend (e.g., logistics or basing). The latter places the enemy on the horns of a dilemma: defer and lose a valuable resource or fight and risk defeat. Another proven (though resource-intensive) technique includes sealing off borders, dividing the area into sectors, and meticulously clearing each sector of enemy combatants. "The sub-division of the theatre of war into sections," writes Callwell, "may be said to be the first step toward dealing with guerilla warfare effectively."[95] Zones of control help protect the people, ferret out enemy spies and supporters, gather intelligence, and encircle and destroy enemy forces and resources "that would facilitate the existence of the guerrillas ... or which could ... be used by them."[96] Where "natural obstacles are lacking," writes Galula, "consideration must be given to building artificial ones."[97]

Though effective combat depends on information superiority – getting valuable intelligence and delivering supportive messaging while denying the enemy the same – the "haves" must also avoid "paralysis by analysis" (i.e., getting caught flat-footed admiring the problem). "Profound plans," says Callwell, "worked out on paper in advance are very apt to miscarry ... in irregular warfare"; rather, history shows that it "was not elaborate maneuvers, but rapid movement and attack wherever the enemy could be found which paid."[98] Moreover, writes Callwell, "Seizure of the initiative is one of the best assured means of commanding success," for it compels the opponent to react, inducing feelings of "moral inferiority," and demonstrates resolve and potency.[99] Therefore, when the adversary presents a fleeting opportunity, makes a mistake, or the friendly side gains a bit of vital intelligence, it is imperative to act swiftly, decisively, and follow up with vigor.

Effective small wars tactics can contradict traditional principles of war, like mass and concentration. "Small wars," writes Callwell, "often render a separation of force desirable,"[100] but only "when each fraction is strong enough to stand by itself . . . against any force which the enemy will be able to bring against it."[101] Given this reality, the stronger side's ideal structure combines static and mobile forces that can operate independently and are trained, positioned, and equipped to outclass the enemy in any potential engagement, especially in discipline, cohesion, and firepower. This approach was prototyped by Marshal Thomas-Robert Bugeaud, the "founder" of France's nineteenth-century "colonial school" of warfare, who built his forces around the four principles of mobility, morale, leadership, and firepower.[102] The mobile teams, says Galula, seek and destroy the enemy's combat forces, while the static group defends the population and advances human and political lines of effort.[103] To "eliminate . . . the entire enemy organization," Trinquier recommends that armies "operate in light detachments" with "a highly mobile reserve element."[104] If the mobile teams are too large or slow, says Trinquier, they end up "dispersing, rather than destroying" the enemy, like "a pile driver attempting to crush a fly."[105] Likewise, static forces lacking strength and timely intelligence risk disproportionate losses in personnel and equipment and damaged relations with the populace. The detachments charged with killing or expelling the enemy and engaging directly with locals, says Galula, are "the most important . . . in counterinsurgency operations . . . where the war is won or lost."[106] Therefore, per the *Small Wars Manual*, they must demonstrate the utmost initiative, adaptability, leadership, teamwork, and tactical proficiency.[107]

When possible, the "haves" should use a multi-domain approach, as air, sea, space, and cyber power can grant appreciable mobility, logistics, intelligence, and firepower advantages. Unmanned "drone" aircraft are particularly effective due to their fusion of endurance, surveillance, and attack capability; however, they are best employed with ground forces that can hold territory and capture, interrogate, or defeat survivors. Also, direct force cannot be indiscriminate lest it delegitimize the effort. "There is a limit to the amount of license in destruction which is expedient," writes Callwell; success is more likely when employing a "happy combination of clemency with firmness."[108]

Finally, the stronger side must remain committed to its cause until victory is achieved, regardless of the time required. In fact, RAND found that most successful counterinsurgencies require *six or more years*, even after "a substantial balance of good COIN [counterinsurgency] practices is first achieved."[109] "No time limit for the operation should be set," adds Trinquier, instead, operations must continue until the enemy organization and its sustaining apparatus is completely destroyed and "peace has been restored to the entire national territory."[110] Even though things may appear to be going well, "winning over or suppressing the last guerrillas" demonstrates command of the campaign and brings "valuable political benefits both within and without the selected area, on the population, on the insurgents, and on the counterinsurgent's own forces."[111] Ultimately, combat for the stronger side is about prosecuting the offensive as intelligently, aggressively, and lethally as possible *without compromising the human and political lines of effort*.

Conclusion

"Whenever a regular army finds itself engaged upon hostilities against irregular forces," writes Callwell, "the conditions of the campaign become distinct from the conditions of modern regular warfare."[112] This is true insofar as it equates "conditions" of war with "character" of war; however, small wars, like all war, fall within the confines of general war theory. Thus, we must both agree and disagree with Callwell's assertion that "the conditions of small wars are so diversified" and "peculiar" that they must be fought with a "method totally different from the stereotyped system" mandated by "the teachings of great masters of the art of war."[113] We agree that general theory has inadequately addressed small wars, but we disagree that small wars entail a reimagining of war's nature.

Small wars are rife with violence, enmity, and chance. They require assessments of strengths and limitations, timely and accurate intelligence, evaluation and exploitation of the environment, well-defined military and political objectives supported by coherent strategy, identification and pursuit of centers of gravity, dislocation through surprise or speed, concentrated combat power to erode or destroy fighting capacity, security for friendly forces, competent generalship, and breaking the enemy's will to fight. However, relative to traditional wars, the human and political dimensions are often of greater importance than combat. Warriors may have to fight even while accomplishing a myriad of "softer" tasks amid a human and political milieu characterized by unspeakable acts of deceit, provocation, and terror.

Ultimately, a small war is an asymmetric clash between "haves" and "have-nots" ferociously competing for political legitimacy and control, where combat is only one of a broader set of variables that decide the outcome. Like all wars, each small war presents a unique set of circumstances. As Callwell says, "In small wars all manner of opponents are met with," and "in no two campaigns does the enemy fight in the same fashion"; thus, each new case must be thoroughly analyzed as a prerequisite to effective strategy formulation.[114] The stronger side must maintain its potency and composure, honestly address its own political shortcomings, and relentlessly pursue the disintegration of the opponent's organization and fighting forces. Conversely, the weaker side must play the long game, gradually building its political and military strength around a powerful cause, strong leaders, and a resilient, elusive organization – even as it harasses, subverts, and weakens its opponent's military strength and political authority.

Finally, though small war "theory" offers valuable lessons and techniques applicable to asymmetric warfare, it has yet to inspire the fundamental revision of general theory necessary to reveal and define indelible linkages between regular and irregular war, and consequently, to reduce the likelihood that these lessons will be forgotten once again. Establishing this relationship will be our task in the next chapter.

DOMAIN THEORY

The existence of domain sub-theory is partly a consequence of general war theory's failure to satisfactorily incorporate all war's domains. The best theory, much of which is presented in the previous chapter, pertains almost exclusively to land-based warfare. In other words, it is essentially land-centric domain theory! In fact, this has been a criticism of even the best theorists, like Sun Tzu and Clausewitz, who all but ignored naval warfare. Thus, the unique characteristics of war in the sea, air, and now space and cyberspace domains, have engendered sub-theories.

Chapter 1 introduced the True Trinity – humanity, politics, and combat – and proposed that any general war theory must render the linkages and dynamics of these elements in abstract form. Since a general theory cannot vary with domain (or technology), domain theory cannot be a general theory; however, the best domain theory should propose concepts that are at least *universal for that domain*. This is a lofty standard few domain theorists have met. More often, domain theory lacks objectivity (i.e., the theorist is an advocate) or overemphasizes proximate trends in weapons, warfare, and even philosophy. Using general theory as its foundation, sound domain theory must narrow its focus to the domain's *unique impacts* on war's character; explain, when possible, how warfare in that domain *has evolved*; identify *abstract domain-centric principles*; and finally, use those principles to outline the best ways to *exploit the domain* to advance the aims of war. More specifically, domain theory (ideally) leverages history, logic, and general theory to perform three tasks: (1) based upon an assessment of the history, current status, and likely evolution of domain-unique capabilities, deduce abstract principles related to the domain; (2) using the results of (1), determine how to gain a decisive advantage in that domain (this requires the "three a's" of domain theory: *awareness*, *access*, and freedom of *action*); and (3) describe how exploiting that advantage advances the aims of war. Returning for a moment to the "three a's" (introduced in Chapter 2's section on Jomini), *access* and *action* within a domain require *awareness* of the domain including information on the status and disposition of forces within it. *Awareness* comes from multi-domain intelligence, surveillance, and reconnaissance (ISR).[115] No domain can be exploited without *access* (e.g., ports [sea], bases [land and air], launch sites [space], and computers [cyber]). Finally, freedom of *action* (e.g., "command of the sea" or "air superiority") enables domain-optimized forces (e.g., troops, ships, or aircraft) to create effects that advance war aims.

Maritime Theory

He who commands the sea has command of everything.[116]   Themistocles

Compared to land warfare, war in the sea domain is disproportionately influenced by geography, technology, and economics. Many countries lack borders near substantial waters and few can afford a true blue-water navy as battle fleets relative to armies are expensive in any era. These facts, as Heuser asserts, create a hierarchy of sea-going nations, where top-tier states (like Britain, and recently the USA) wield global maritime power; the middle tier settles for regional influence; and the bottom tier manages only the barest local defense, raiding, or piracy.[117] History's dearth of naval powers has contributed to the relative scarcity of maritime war theorists and theory. In fact, there are only two true luminaries: the American, Alfred Thayer Mahan, and Britain's Julian Corbett.[118]

Mahan

Mahan was born in 1840, graduated from the US Naval Academy in 1859, fought in the American Civil War, and eventually totaled forty years of active naval service. In 1884, Mahan was enticed from a Pacific duty station to the new US Navy War College in Rhode Island by Commodore Stephen Luce, the College's founder. Once there, Mahan became the chair for naval history and tactics. Finding himself in an academic setting for which he felt ill prepared, Mahan began an aggressive program of study that stimulated a latent talent for historical analysis and theoretical reasoning. Before long, Luce began to see Mahan as a "master mind" with the potential to do for naval theory what Jomini had done for warfare in general.[119] In fact, Mahan would eventually coin the term "sea power," and from his extensive studies, elaborate on the nature of naval warfare and its relationship to national prominence.[120]

By 1886, only two years after arriving at Newport, Mahan had earned promotion to captain and was elevated to president of the Naval War College. Even more auspiciously, he had begun to write his two signature tomes, *The Influence of Sea Power upon History* (1890) and *The Influence of Sea Power upon the French Revolution and Empire* (1892). These impressive publications, which gained widespread acclaim particularly with the British, vaulted Mahan into the international spotlight. More eloquently than his predecessors, Mahan had identified naval supremacy as "the central link" to national strength, wealth, and stature.[121] This meant that naval strategy must have equal standing with land-based strategy since "maritime strength," like land power, was and is a "profound determining influence . . . upon great issues" and "a great factor in the history of the world."[122] As an unapologetic advocate for sea power, Mahan's intent was to "imbue his hearers with an exalted sense of the mission of their calling" and to "give the service and the country a more definite impression of the necessity to provide a fleet adequate to great undertakings."[123] This mission and message resonated with powerful political icons, including Queen Victoria, Kaiser Wilhelm II, and Teddy Roosevelt. However, though he is called a "theorist," Mahan's work never approaches the systematic form exemplified by the likes of Clausewitz or Jomini. Rather, according to Gat, Mahan was simply the first to collect "the scattered aphorisms previously issued" regarding command of the sea and to portray naval history and sea power "against the background of the wide sweep of historical events, interwoven and interacting with political and economic factors."[124]

Most of Mahan's theoretical insights appear in the introductory portions of *Influence* and some are interspersed with historical analysis in subsequent chapters. Mahan aptly describes war's essence, rooted in human nature, as unchanging, in contrast with war's character, which varies with time and technological advances. "The unresting [*sic*] progress of mankind," writes Mahan, "causes continual change" in weapons and fighting; thus, ferreting out war's foundational principles requires a *comprehensive* view of history, for "precedent is . . . less valuable than a principle. The former may be originally faulty or may cease to apply through change of circumstances; the latter has its root in the essential nature of things."[125]

For Mahan, strategy sets the conditions for successful "engagements," which are tactical affairs; thus, "the dividing line between tactics and strategy" occurs at the "contact" point between engaged forces.[126] In other words, in Jominian (and Nelsonian) fashion, Mahan's strategy ensures that properly prepared naval forces arrive at the appropriate time and place to win sea battles. Naval strategists, according to Mahan, must consider a variety of factors, including the navy's proper function; its true objective, the point or points upon which it should be concentrated; the establishment of depots and lines of communications between them; the value of attacking commerce (*guerre de course*) as a secondary or decisive operation; and the best manner for attacking commerce.[127] More broadly, writes Mahan, "Naval strategy has for its end to found, support, and increase, as well in peace as in war, the sea power of a country."[128]

To Mahan, the sea was "a wide common, over which men may pass in all directions"; however, since the sea itself is rarely a destination, the commonly traversed "lines of travel" or "trade routes" are of greatest significance to navies.[129] Prior to modern air and space surveillance, awareness of the sea domain occurred via intelligence often gathered directly along these routes. Therefore, according to Mahan, a navy's two central functions were (1) to maintain the trade routes or (2) support a nation's "aggressive tendencies ... as a branch of the military establishment" (i.e., for actions against an adversary's homeland or navy).[130] Implicit in the development of naval and national power is the requirement to build and provision a navy, secure *access* via ports and depots, and then to employ the navy to advance the nation's interests, which would naturally include *freedom of action* (i.e., preventing an opposing navy from interfering with those interests or the infrastructure upon which the navy depends). "These combinations of strong points at home and abroad [access]," asserts Mahan, "and the condition of the communications between them [awareness], may be called the strategic features of the general military situation, by which, and by the relative strength of the opposing fleets, the nature of the operations must be determined."[131] And he stresses, "It is not the taking of individual ships or convoys ... that strikes down the money power of a nation, it is the possession of that overbearing power on the sea which drives the enemy's flag from it [freedom of action]. This overbearing power can only be exercised by great navies."[132]

Alluding to France, Mahan warns that pursuing limited objectives over "the control of the sea by the destruction of the enemy's fleets, of his organized naval forces ... is faulty as a rule."[133] Thus, England's rivals failed to match her because they were "habitually defensive" and never sought "to put an end to the empire of the seas by the destruction of the organized force which sustained it."[134] Mahan called the preoccupation with secondary goals "a most dangerous delusion," for "control can be wrung from a powerful navy only by fighting and overcoming it."[135] And this should be done without delay, for once a war is begun, "military wisdom and economy ... dictate bringing matters to an issue as soon as possible upon the broad sea, with the certainty that the power which achieves military preponderance there will win in the end."[136]

In Napoleonic fashion, navies should always pursue the enemy, gain a position of advantage, concentrate effort while dispersing the enemy, and follow up on advantages "with all possible vigor."[137] Only by eliminating its counterpart can a navy gain freedom of action to strangle the opposing state's commerce. In essence, Mahan's theory was the "Nelsonian" heir to the Napoleonic ideals articulated by Jomini and Clausewitz. Not surprisingly, great (England) and aspiring (US) naval powers loved Mahan, but the passage of time, advance of technology, and appearance and maturation of submarines, torpedoes, and aircraft cast doubt on the universality of his ideas, especially his advocacy for battleship-centered fleets.

Critiques of Mahan fall mainly within four areas: causation versus correlation, timelessness, sufficiency, and theoretical validity. Regarding the first point, to be sure, some of history's great powers had great navies, including Athens, Britain, and (eventually) the USA. However, Mahan's histories ultimately lack the rigor required to prove the causality of naval power in the formation of great nations. England was great, but its history, economy, and geography make it something of a special case. Were all powers great because they had great navies, or were other factors responsible? Mahan largely ignores the complementary nature of naval and land power and the effects of shrewd continental diplomacy in Britain's successes. Though Rome had a strong navy at times, its power more often rested upon the discipline of its legions. Khan's Mongols, Tsarist and Soviet Russia, and Napoleonic France were also great powers without great navies. Other, more contemporary examples also call into question the timelessness and necessity of naval power as a prerequisite for national prominence. For example, twenty-first-century China, Brazil, Russia, Japan, Germany, and India are substantial powers without being naval juggernauts. As maritime strategy expert Philip Crowl notes in *Makers of Modern Strategy* (1986), sea power may have been a necessary and important cause "of Britain's triumph over France in the seventeenth and eighteenth centuries. It was not, however, the sufficient cause."[138]

Regarding theoretical validity, Mahan's Napoleonic emphasis on the primacy of concentrated force and great fleet battles too often fails to hold water (pardon the pun) for it to be considered universally valid. As Crowl points out, Mahan's theory "was not systematic"[139] and "mostly disregarded" key naval warfare roles like "power-projection from the sea" and "the utility of naval artillery and of sea-borne infantry."[140] Where the outcome of global competition depends mainly on the sea, as with the administration of imperial colonies in the age of sail, Mahan's ideas are largely sound. But in the modern age of air and cyber-based commerce and with a much different geopolitical context, the influence of sea power slips more toward the margin. Finally, Mahan failed to anticipate the paradigm-changing impacts of new technologies, including steam (and later internal combustion and nuclear) propulsion, submarines, torpedoes, and aircraft; thus, Mahan's stature as a relevant naval theorist has declined precipitously.

Ultimately, Mahan was, despite his fame, less a true theorist than a celebrated enthusiast who had, as Gat describes, "merely applied long-established intellectual and strategic categories to a previously neglected subject."[141] Viewed from the perspective of the three tasks of domain theory, Mahan used a selective view of history, a Jominian theoretical prism, and a dollop of logic to conclude that sea power and command of the seas via a dominant, battleship-heavy fleet were the true keys to winning wars. For a more balanced, objective, and rigorous perspective on maritime theory, we must turn to Corbett.

Corbett

Sir Julian Corbett was a Cambridge-educated lawyer who came to naval theory relatively late in life, and unlike Mahan, he was not a sailor. In fact, he had no personal military experience. However, a stint as a historical biographer and his book *Drake and the Tudor Navy* (1898) elevated his reputation with the Royal Navy resulting in an invitation to lecture young naval officers at the Royal Naval War College in Portsmouth. Corbett took his role as an educator seriously, and thus, like Mahan, he diligently expanded his knowledge on war, naval history, and theory. Corbett was familiar with Mahan and also with two influential English naval veterans and brothers, John and Philip Colomb, both of whom subscribed to the Mahanian "blue water" and "command-of-the-sea" schools of thought. Indeed, Philip's *Naval Warfare* (1891), according to Gat, was even "more impressive" than Mahan's *Influence* books as a work of systematic naval theory.[142] Corbett, however, would come to different conclusions about maritime theory, likely because of his unbiased, outsider's perspective, extensive archival research, and internalization of Clausewitz's ideas on war and policy. Remaining objective throughout, Corbett painted an alternate picture of naval warfare that valued defense, dispersal, deception, and survival more than great fleet engagements. Indeed, he concluded that Britain had been successful, not by "seeking decisive battles at all cost," but by "the flexible adaptation to circumstances, derived from instinct and long experience rather than principle and strategic reflection."[143]

His early works, such as *Fighting Instructions* (1905) and *Signals and Instructions* (1908), reflected Corbett's appreciation for the impact of increasingly powerful armaments, which seemed to call for a withdrawal from the nineteenth century's Nelsonian, confrontational pattern back toward the *ancien régime*'s systems of "control and maneuver."[144] By 1907, informed by his research for *England in the Seven Years War*, Corbett's counter-Mahanian thinking had fully coalesced. For Corbett, control of the sea was necessary but not sufficient to win wars without the accompaniment of effective diplomacy and other military means. For full strategic efficacy, writes Corbett, we must view "the fleet and army as one weapon."[145] Echoing Clausewitz, Corbett proposes that naval actions, guided by political aims and circumstances, may aptly take form as something less than a decisive clash of fleets. While such thoughts were heretical to most naval brass, Corbett's analysis nevertheless earned him a spot as a Naval War College lecturer and strategic advisor to the Admiralty.

Concurrently, Corbett began to summarize his ideas in a series of works culminating in *Some Principles of Maritime Strategy* (1911), described by Gat as "a thoroughgoing and most original revision of the accepted tenets of naval theory and indeed, more implicitly, of strategic theory as a whole."[146] In *Principles*, Corbett addresses the purposes and limitations of theory and, before discussing naval theory, elaborates on general theories of war including those of Clausewitz, Jomini, and two of war's dialectics: defense-attack and unlimited-limited. Corbett describes theory as "a question of education and deliberation, and not of execution at all."[147] Moreover, if discourse on war and strategy, let alone naval warfare, is to become more than "merely verbal contentions which rest on no firm foundation," Corbett asserts that theory must "coordinate our ideas, define the meaning of the words we use, grasp the difference between essential and unessential factors, and fix and expose the fundamental data on which every one is agreed."[148] Only then "[can] we secure the means of arranging the factors in a manageable shape, and of deducing from them ... a practical course of action."[149] Regarding war, he notes, "Certain lines of conduct tend normally to produce certain effects," or in other words, war takes its form based on its object and "its value to one or both belligerents"; however, "a system of operations which suits one form may not be that best suited to another."[150]

Breaking from Mahan and demonstrating a more holistic appreciation for war, Corbett writes, "The paramount concern ... of maritime strategy is to determine the mutual relations of your army and navy in a plan of war," where "maritime strategy" refers to all elements of national strategy relating to the sea domain and "naval strategy" pertains strictly to the military portion.[151] Corbett adds that maritime theory is subordinate to general war theory, and maritime power is a less decisive adjunct to land power since "men live upon land and not upon the sea."[152] Corbett uniquely recognizes that war's forms can vary with domain, even in the same war; for example, while the land component fights an unlimited war, the maritime component may pursue limited objectives. To repudiate the nineteenth century's glorification of decisive clashes between armies or fleets, Corbett leverages Clausewitz's post-1827 modification, which holds that war "is an exertion of violence to secure a political end," and that, in form, "wars will vary according to the nature of the end and intensity of our desire to attain it."[153] In other words, the political object, not some inhuman imperative to fight great battles, determines a war's scope and intensity. The unlimited, "higher form" of war, aims for the complete overthrow of an enemy, and the limited, "lower form," promotes less ambitious ends, like controlling sea commerce.

Corbett further distances himself from Mahan by rejecting the voguish doctrines of the relentless offensive, and, though admiring of Clausewitz, he disavows the Prussian's zero-sum model for attack and defense. In war, writes Corbett, offense and defense "are mutually complementary. All war and every form of it must be both offensive and defensive."[154] Furthermore, a defensive posture is valid and proper "when the offensive would probably lead to its destruction" by a superior enemy.[155] Corbett's concepts better aligned general and maritime theory to historical evidence and affirmed Britain's successes resulted from complementary actions on land and sea in pursuit of limited, not absolute war. Here Corbett takes a "wider view than was open to Clausewitz," recognizing England's isolation as an island state precluded her from bringing the "whole national force ... to bear."[156] Rather, Britain gained its objectives by concentrating force where required, including in homeland defense, and engaging in "direct continental interference [with] a small army acting in conjunction with a dominant fleet."[157] This balanced strategy, coupled with control of the seas, enabled England to uniquely reconcile "the limited method" (i.e., contingents and maritime actions short of fleet-on-fleet battles) "to the unlimited form, as ancillary to the larger operations of ... allies."[158]

Having framed England's strategic circumstances vis-à-vis general war theory, Corbett turns to the maritime realm and how to achieve freedom of action through "command of the sea." For Corbett, command of the sea cannot mean denying the three a's at all times and everywhere to the enemy because the sea is too vast and not subject to occupation, nor can it rest upon pursuing the enemy's battle fleet, a task precluded by a wily adversary. As Corbett points out, "The attempt to seek the enemy with a view to a decisive action was again and again frustrated by his retiring to his own coasts."[159] Rather, "Command of the sea," says Corbett, "means nothing but *the control of maritime communications*, whether for commercial or military purposes" (emphasis added).[160] It is "the special characteristic of naval warfare," writes Corbett, "which always permits action against the ulterior object [communications] when the enemy denies you any chance of acting against his armed force."[161] Attaining Corbett's version of command of the sea grants naval forces the *freedom of action* to oppose an enemy fleet's threats, including attacks on communications and trade (denies *awareness*), invasion, blockade (denies *access*), and safe-guarding expeditions.[162] Controlling sea communications can lead to victory by doing "that which really brings wars to an end – that is, exerting pressure on the citizens and their collective life ... [making] the enemy's country feel the burdens of war with such weight that the desire for peace will prevail."[163] Thus, using history, theory (mainly Clausewitz), and logic, Corbett concludes that naval predominance and winning wars requires awareness of the political object and balanced land and sea operations, where the latter defends and disrupts commerce and the former pursues strategic objectives across the Channel.

Corbett briefly relates the concentration-dispersal dialectic to maritime strategy. Whereas Mahan clearly favors concentration over dispersal, Corbett recognizes an inherent tension between the two, where dispersion enables naval power to "reach" distant ports and trade routes, and concentration through "cohesion" provides the power needed to overwhelm an enemy or defend the homeland. Naval forces, writes Corbett, disperse for flexibility, but should not sacrifice "elastic cohesion, so as to secure rapid condensation ... of the whole at the strategical centre."[164] Echoing Sun Tzu, Corbett advises that "condensation" should avoid a superior force and set "a trap to lure the enemy to destruction. The ideal concentration ... is the appearance of weakness that covers a reality of strength."[165] For a relatively inferior fleet, Corbett admits that full sea control may be impossible; thus, "We have to content ourselves with endeavoring to hold the command in dispute" and to avoid decisive confrontation with the enemy and preserve a "fleet in being."[166] Corbett's "fleet in being" is a seaborne Fabian strategy that avoids "decisive action by strategical or tactical activity, so as to keep our fleet in being till the situation develops in our favour."[167]

In the final analysis, it is hard to fault Corbett beyond the standard complaints regarding his inability to fully anticipate the implications of advanced technologies like the submarine and aircraft carrier. Even here he admits, "Modern developments and changes . . . have indeed so profoundly modified the whole conditions of commerce protection, that there is no part of strategy where the historical deduction is more difficult or more liable to error."[168] Corbett's realization of a broader expanse of possibilities for maritime warfare, contingent upon the political objective and relative fighting capacities of the warring sides, carries Clausewitz's unfinished examination of the absolute war-limited war dialectic another step forward. And ultimately, as Heuser points out, Mahan, Corbett and other early sea power theorists succeeded in identifying three valuable "maritime Strategies": (1) "predominance at sea," which seeks to chase off or defeat "hostile navies"; (2) the disruption and seizure of an enemy's commerce while "avoiding large scale encounters"; and (3) the "fleet in being" that deters and engages only when avoidance is impossible.[169]

Conclusion

Coalescing the best ideas of naval theory with present realities yields the following: Having a navy is a good thing, if you can afford it and can achieve maritime awareness, access, and freedom of action. But naval power is not the *sine qua non* of national prominence. A navy is an instrument of power best used in concert with other instruments, including diplomacy and complementary domain-based military services. For greatest utility, navies ought to be conceived and employed for a variety of roles and missions commensurate with geographic and economic realities, including defending the homeland, ports, and key territories; gaining intelligence; projecting power (ideally in multiple domains); controlling commerce and freedom of navigation; and as a last resort, engaging hostile fleets on the open seas. Moreover, since the whole sea cannot be commanded nor is the vast majority of it of value at any given time, the best aim for naval strategy is to maintain awareness of an enemy's capabilities and disposition, but to refrain from provocative confrontation unless critical military or political aims are threatened. In the nuclear age, replete with airpower, submarines, and fearsome anti-ship weapons, Mahanian great battle has become extinct, yielding to the more sensible approach described by naval historian John Hattendorf: to "establish control of strategic-ally important parts of the sea, and use them for one's own purposes."[170] Thus, a navy's "foremost *raison d'être* [is] no longer major war, but anything short of it."[171]

Airpower Theory

A new weapon arose, an air weapon; a new battlefield opened; the sky; so very present everywhere that a new event took place in the history of war; the principles of war in the air.[172]   Giulio Douhet

"Airpower theory" is something of an oxymoron. Regrettably, it is difficult to name an airpower "theorist" who has offered an objective, systematic, abstract theory suitable for authentication. Perhaps it is no wonder airpower expert Phillip Meilinger opens *The Paths of Heaven* (1997) by saying, "Airpower is not widely understood ... the basic concepts that define and govern airpower remain obscure."[173] In the same work, military historian I. B. Holley, Jr. admits to being "struck by the unsystematic, undisciplined thinking that ... characterized the writings" of early airpower theorists.[174] "Much of what has been written on the subject," adds Holley, "is not ... airpower theory at all but descriptions of varied efforts to implement the then-current conception of such theory ... surely one is surprised by the dearth of really comprehensive thinkers and theorists on airpower."[175] Where is airpower's Corbett? To be fair, land and maritime theorists had centuries of historical data to inform them, while airpower theorists had only the dubious example of World War I and their own imaginations. As such, airpower theory, even more than maritime theory, exemplifies both the triumph and limits of human imagination.

After the Wright brothers' success and as gas-filled airships matured in the early twentieth century, "visionary" warriors and futurists contemplating mili-tary applications for the nascent "air weapon" proliferated. The most notable included Italy's Giulio Douhet (1869–1930), author of *The Command of the Air* (1921, 1925); America's William "Billy" Mitchell (1879–1936), who penned *Winged Defense* (1925); England's Hugh Trenchard (1873–1956), "Father of the Royal Air Force (RAF)"; and Trenchard's disciple, John "Jack" Slessor (1897–1979), author of *Air Power and Armies* (1936). These pioneers were unapologetic zealots – excepting Slessor, who admittedly "tried to abstain from the role of advocate, and to confine himself to a reasoned and impartial examination of the case for air power"; thus, perhaps more even than Mahan, they lacked the objectivity required of great theorists.[176] They *wanted* airpower to succeed. And to gain equivalence with land and sea power and prevent air forces from merely becoming third-dimensional auxiliaries to armies and navies, the air weapon had to be relentlessly characterized as revolutionary.

Given the scarcity of empirical evidence available for making this case, most early airpower theorists employed hyperbole and theater more than hypothesis and theory, a tactic that led to the court-martials of Douhet and Mitchell and to varying degrees of official opprobrium for many of their acolytes.

While there is ample justification for critiquing these men for their audacity, speculative optimism, and embarrassing errors, it is unlikely that airpower would have garnered the attention and resourcing required for full development in the hands of a less brash, persistent, and charismatic cadre. In the words of social theorist Isaiah Berlin, revolutionary thinkers tend "to overstate their central theses. Such exaggeration is neither unusual nor necessarily to be deplored ... nor is it likely that their ideas would have broken through the resistance of received opinion ... if they had not."[177] The air weapon's novelty mandated that these pioneers be defense attorneys more than objective theorists, stacking arguments to sway juries of peers and public opinion. While this practice did not burnish their theoretical bona fides, it ultimately enabled them to prevail over army and navy personalities, ossified bureaucracies, and centuries-old organizational cultures, traditions, and biases.

The Architects of Airpower

Giulio Douhet is the earliest and most important figure in airpower theory, and his larger-than-life persona makes his contemporaries appear as little more than branches off his tree. Notably, "Douhet shifted the dialectic within strategy," says Strachan, "from a discussion between the present and the past, to a discussion between the present and the future."[178] Douhet was an intellectually dynamic "ardent Fascist" fascinated from an early age by science and technology.[179] In 1888, he graduated first in his artillery academy class and was posted to the general staff by 1900. As a captain, Douhet had prophetically "envisaged the mechanization of whole armies," but his imagination soon drifted to aviation.[180] In 1912, he was posted to an aviation battalion and named commander. During these years, he began to realize the "air weapon's" transformational potential, and he envisioned an independent air force that would dispatch bomber armadas to break a nation's will to fight by destroying its cities and demoralizing its civilian population. Douhet's advocacy for airpower, supported by an unshakable faith in the righteousness of his cause and the infallibility of his calculations, led him to overstep in his criticisms of the army; and, in late 1916, he was arrested for insubordination and sentenced to a year-long imprisonment. However, after a series of blunders by the Italian military seemed to confirm Douhet's accusations, he was exonerated and returned to service in a variety of significant capacities culminating with his promotion to general in 1921.[181] Douhet retired shortly thereafter, spending his last decade defending and promoting his theory.

William "Billy" Mitchell was America's Douhet. Described as self-centered, vain, confident, and tireless, Mitchell was well traveled in his early career as an army signals officer. During World War I, he was a pilot, air combat commander, and eventually General John "Black Jack" Pershing's Chief of the Air Service. While flying in combat, Mitchell marveled at the ease with which he could fly over trenches, avoiding the stalemate and carnage below. For Mitchell, the airplane embodied a faster, cheaper, and more humane way to fight a war.[182] In the spring of 1917, hungry for more knowledge about airpower, Mitchell met and was "profoundly affected" by Trenchard, particularly regarding the benefits of a unified air service, aerial warfare's offensive nature, and the war-altering promise of striking targets deep within enemy territory.[183] Mitchell was also "crucially influenced ... by the ideas of Douhet and the Italians."[184] He believed future conflicts would be fought in the air as well as on the surface; thus, airpower would feature prominently. Eventually, like Douhet, he realized the path to independence required an independent mission: strategic bombing. Also, like Douhet, Mitchell's frustration with detractors led him to publish a series of scathing critiques of the US Navy and War Departments that led to his court-martial and early retirement in 1926. Nevertheless, Mitchell is still regarded as "America's foremost airpower prophet," and his powerful legacy included a dedicated band of followers who would fight the air campaigns of World War II and eventually establish and lead the independent US Air Force in 1947.[185]

"Unlike the history of the British Army in which theories were hammered out over the centuries," writes British wing commander H. R. Allen, "the pattern of British air-power was based on the ideas of one man ... the Viscount [Hugh] Trenchard."[186] While Douhet and Mitchell were solidifying their identities as airpower icons in Italy and the USA, Trenchard was single-handedly forging the independent RAF and its airpower doctrines from the crucible of World War I. Prior to his emergence as England's preeminent airpower advocate, Trenchard had served in the British infantry. He saw action and was severely wounded (he lost a lung and was partially paralyzed) in the Boer War. Then, after a freak accident miraculously reversed his paralysis, Trenchard returned to colonial duties in Africa. He eventually learned to fly in 1912, and despite being regarded a poor pilot, he remained with the flying school up to World War I. From that point on, Trenchard rose rapidly, though not serenely, through a succession of high-level jobs and promotions, including appointments as Chief of the Air Staff and service as Commander of the Independent Air Force in 1918. Over the course of his career, Trenchard "wielded a God-like power"[187] over the RAF and influenced many, including Billy Mitchell and Jack Slessor, imparting his three main beliefs: (1) the military necessity of air superiority; (2) the inherently offensive nature of airpower; and (3) its significant physical but even greater psychological effects.[188]

John "Jack" Slessor was one of Trenchard's intellectual heirs, a man Philip Meilinger describes as the most "prescient," "balanced and judicious" of the early theorists.[189] Unlike the others, Slessor was a young man during World War I and a decorated combat pilot. However, Slessor nearly missed out on a military career. His application for army service was initially rejected on medical grounds (he had polio as a youth); then in 1915, leveraging family connections, he secured a spot in the fledgling Royal Flying Corps. It turns out that Slessor's abilities extended beyond his flying skills. Hailed as ' the most intellectually gifted" officer to attend the early RAF Staff College – founded by Trenchard and infused with his "instinctive beliefs" on airpower – Slessor displayed a talent for visualizing, and, unlike his mentor, articulating the theory that made airpower work.[190] Slessor's experience on Trenchard's staff and as an instructor for three years (1931–4) at the British Army Staff College laid the foundation for what would become his seminal work, *Air Power and Armies* (1936), a collection of his lectures described by Meilinger as "perhaps the best treatise on airpower theory written in English before World War II."[191] Regrettably, Slessor admits that he focused *Armies* on how airpower complements surface warfare rather than on broad, objective airpower theory. Had he more thoroughly evaluated airpower's strategic potential, Slessor might have become airpower's Corbett. But a meteoric career (he eventually became Marshal of the RAF and Chief of the Air Staff) and the nuclear age diverted his intellectual efforts in other directions.

History, Theory, and Logic

In the development of air power, one has to look ahead and not backward and figure out what is going to happen, not too much what has happened.[192]
Brigadier General William "Billy" Mitchell

As mentioned previously, domain theory should be informed by history, general war theory, and logic. Due to the lack of relevant historical examples and the revolutionary nature of advanced technologies, airpower theorists largely ignored the past and gravitated toward Heuser's *matériel* school of strategic thought.[193] The early twentieth century's remarkable "engines of a new age," of which the airplane was the darling, drove this *matériel* viewpoint and kindled futurist visions "of boundless potential, moral, and civilization-transforming forces."[194]

As a leading proponent of this perspective and true believer in the revolutionary nature of the airplane, Douhet was particularly dubious of history's value. "The experience of the last war," he writes, can teach us "less than nothing,"[195] because "the character of future wars is going to be entirely different from the character of past wars."[196] World War I proved to Douhet that modern land and sea weaponry had irrevocably granted a combat edge to the defense, a situation airpower alone could reverse. The Great War also convinced Douhet that future war would be total, with "no distinction any longer between soldiers and civilians."[197] Using these premises, Douhet formulated his ideas with themes from past theorists. Like Mahan and sea power, Douhet's air forces were independent of the army, decisive on their own, and would control the Earth's surface with "command of the air," like "command of the sea" controlled ports and commerce.[198] Douhet also borrowed from Jomini: air forces should "always operate in mass . . . [and] effects . . . are greatest when the offensives are concentrated in time and space";[199] and Clausewitz's ideal war: airpower in absolute war aims to "inflict the greatest damage in the shortest possible time."[200]

Mitchell's ideas largely aligned with Douhet's. As a participant in World War I, he was impressed by airpower's ability to fight great battles and rain "bombs down on hostile industrial centers, cities, railroads and ports."[201] Mitchell also echoed Clausewitz (airpower compels an enemy to do our will) and Mahan ("air minded" nations have special geographical and political characteristics; and "command of the air" is gained by defeating the enemy's air "fleets" in great battles in the sky).[202]

For Trenchard, World War I proved the value of aircraft as an aid to expeditionary land forces, an offensive force for projecting power across the Channel, and also a defense against attacks from the continent (such as Britain experienced from Germany's deadly zeppelin raids). However, he eventually favored a bomber-centric force to carry out "a strategic air offensive," the key to "retention of an independent air force."[203] According to H. R. Allen, Trenchard displayed no interest in applying "the principles of war to air-power because . . . [he] . . . never believed that they were relevant to the air."[204] Consequently, rather than repudiating the cult of the offensive, Trenchard merely elevated the philosophy to the skies, believing the offensive (regardless of the human cost) to be the stronger form of war militarily and psychologically.[205]

Slessor best balanced the historical and *matériel* schools, relying "not only on logic – the method employed by most air theorists – but also on history," which he used extensively to support his deductions in *Air Power and Armies*.[206] Drawing from his World War I experiences and research, Slessor envisioned airpower in an independent but complementary capacity that could support land and sea power and also perform strategic missions. Unlike Douhet, Slessor forecast that mechanization would end trench warfare and propel a new era of mobile war, where airplanes would facilitate breakthroughs, harass retreating enemies, and protect friendly withdrawals.[207] Moreover, in contrast to his tech-obsessed peers, Slessor appreciated war's human dimension, saying war "is influenced by factors other than cold military expediency," like "moral, political, social, and economic" circumstances.[208] Intellectual themes and theorists echoed by Slessor include maritime theory: "as in naval warfare the most that can normally be secured under modern conditions is the neutralization of the enemy's battle fleet, so also in the air";[209] Jomini: a key principle is "concentration and employment of the maximum force ... at the decisive time and place,"[210] and air attacks are "in effect always on interior lines";[211] Clausewitz: "perhaps the most important of all – the moral factor";[212] and Fuller: "there are three great fundamental rules ... the principles of *concentration*, of *offensive action*, and of *security*."[213]

Assessing Airpower's Capabilities

Air power is the ability to do something in and through the air, and, as the air covers the whole world, aircraft are able to go anywhere on the planet.[214]

The influence of air power on the ability of one nation to impress its will on another in an armed contest will be decisive.[215]   Billy Mitchell

The domain theorist's first task is to contemplate the unique features of the domain and its weapons. For naval theorists, the vast sea, covering 71 percent of the Earth, connects nations, provides resources, enables commerce, and allows for troop transport and fires against naval vessels, aircraft, and land areas. The atmosphere is simply a larger "sea" of air, a three-dimensional high "ground" with few barriers and from which airpower can place anything on Earth's surface at risk.[216] The melding of naval vessels and airpower – including aircraft, cruise, and ballistic missiles – has blurred the line between air and naval capabilities. Now, naval vessels have expanded *access* to the air domain by acting as mobile airbases with all the advantages and limitations of operating from the maritime domain.

Within the air domain, airpower theorists envisioned swift war machines, detectable only by sight or sound, that could traverse the skies in any direction at high speed (Mitchell: airplanes are "the quickest ... means of communication that the world has ever known"[217]) without fear of effective ground-based interference (Douhet: "nothing man can do on the surface of the earth can interfere with a plane in flight"[218]), and decisively engage any target in the air, on land, or sea with "the most powerful weapons ever devised by man."[219] Air forces, says Slessor, ensure the enemy "will never feel safe anywhere,' and this denial of respite causes "a feeling of permanent insecurity that cannot fail to have a terribly wearing effect on morale."[220]

According to early airpower theorists, the air weapon's intrinsic attributes included *speed*, *flexibility*, *invulnerability*, and *offensive* power, which combine to revolutionize warfare and supplant armies and navies as a *strategic* (Mitchell: "the destinies of all people will be controlled through the air"[221]) and, therefore *decisive* instrument of war (Slessor: "the air is only one, but it is the most decisive one, of a number of factors.")[222] For Slessor, the aircraft represented the third and most important revolution in warfare (after gunpowder and machine guns), and more importantly, "the antidote to modern weapons of surface warfare."[223] Aircraft have "extreme tactical flexibility," adds Slessor, and thus, are "not committed to any one course" or "to any particular task, but can switch ... strength or any proportion of it" from one task to another.[224]

Armed with explosives, incendiaries, and poison gas and employed in total war, Douhet predicted air forces would deter war or cause it to be used only as a last resort; and even then, by raining lethal energies upon helpless populations, aircraft would shorten wars by rapidly breaking the enemy's will to fight.[225] "The menace will be so great," adds Mitchell, "that either a state will hesitate to go to war, or ... will make the contest much sharper, more decisive, and more quickly finished."[226] Though Trenchard agreed with Douhet and Mitchell regarding airpower's ascendancy over land and naval forces and its decisive strategic potential, he also praised its versatility and value, particularly for empire policing, where it performed effectively at a fraction of the cost of gunboats or battalions.[227] Ironically, Mitchell, even more than Trenchard or Slessor, characterized airpower's defensive virtues, claiming that air forces could "hold off any hostile air force" or "any hostile shipping which seeks to cross the oceans."[228] Mitchell even outlined a prototype integrated air defense system composed of "listening and reporting posts," alert fighters, searchlights, and antiaircraft gunnery, all connected by robust communications under a single commander.[229]

According to Mitchell and the rest, only a sovereign military service commanded by an airman could fully exploit the air domain and maximize airpower's versatile, strategic, and decisive attributes. Of course, this idea met with skepticism from army and navy leaders who viewed aircraft more parochially. In response, Douhet, Mitchell, and Trenchard fought for secession from the army and equality on the grounds that air forces would relegate legacy services to secondary status in future war.[230] Though Douhet and Mitchell did not fully renounce "joint" (multi-service) cooperation (Douhet: "the three armed forces constitute an indivisible whole, a single three-pronged instrument of war"[231]), as airpower matured, they predicted bomber-heavy air forces would decide wars faster than armies or navies could. Also, displaying their own brand of competitive insularity, they denigrated the potential of auxiliary airpower, especially regarding naval aviation. Indeed, Mitchell labeled aircraft carriers "expensive" and "useless"[232] and predicted that land-based aircraft would drive ships far out to sea where submarines would finish them off.[233]

In hindsight, while their ideas regarding airpower's offensive evolution were visionary, as Allen asserts, these theorists seemed "to take no cognizance of the reasonably obvious forecast that if the effectiveness of bombers were to increase in the future then so would the effectiveness of the air defenses."[234] Trenchard, says Allen, thought "the bomber deserved to be placed on the plinth as the ultimate weapon" despite "insufficient evidence to test any such assumption."[235] Whether from a lack of imagination or objectivity, these futurists simply did not anticipate incipient detection and engagement capabilities that would invalidate the assumption that "the bomber always gets through." Even Mitchell, who predicted cruise missiles, precision bombs, remote piloted aircraft, jet propulsion, and space travel, did not foresee radar, air-to-air, or surface-to-air missiles.[236]

Gaining Command of the Air

Great contests for control of the air will be the rule in the future. Once supremacy of the air has been established, airplanes can fly over a hostile country at will.[237]
                                                                                        Billy Mitchell

Having characterized the nature of the domain and weaponry, the theorist next identifies ways and means of gaining a decisive advantage over the enemy (i.e., freedom of action) or, in the words of Douhet, "conquering the command of the air." In abstract terms, there are three methods for doing this: (1) intra-domain (e.g., air-to-air), (2) inter-domain (e.g., air-to-surface and vice versa); and (3) via a combination (multi-domain). Freedom of action in the air is often called "air superiority," which Slessor defines as "the capacity to achieve our own object in the air and to stop the enemy achieving his."[238] "To have command of the air," adds Douhet, "means to be in a position to prevent the enemy from flying while retaining the ability to fly oneself,"[239] and to do this, "the enemy must be deprived of the use of all his planes."[240] This was best accomplished, according to Douhet, with heavily armed "battleplanes," which would decimate the opponent's air forces in "air combat, at their bases and airports, or in the production centers . . . wherever they may be found or produced."[241] Likewise, the "root of Trenchard's aerial strategy," according to Allen, was not to pursue air superiority via air battles, but to bypass enemy ground forces, "penetrate the air defences and attack direct the centres of production, transportation and communication from which the enemy war effort is maintained . . . it will be in this manner that air superiority will be obtained."[242]

Mitchell and Slessor preferred intra- or multi-domain approaches, where fighters and bombers won air superiority in air-to-air clashes with an enemy compelled either to rise and fight or passively witness its territory and matériel attacked with impunity.[243] Slessor, mirroring Corbett, considered "air supremacy" (i.e., complete domination of the air domain) "unlikely and unnecessary due to the immensity of space"; thus, air superiority was really a matter of temporary and localized control requiring "constant maintenance."[244] Slessor understood that awareness, access, and freedom of action enabled air forces to wage bombardment and air superiority campaigns in parallel,[245] and he codified this relationship within two principles for gaining air superiority: (1) an active "resolute bombing offensive against the vital centres of the enemy" (to force them to fight in the air or die on the ground); and (2) "by direct action against the enemy air forces in the air . . . [or] upon the ground, their aerodromes, bases, aircraft depots, and technical establishments."[246]

Airpower attacks, advises Douhet, should use the most destructive means available and "keep up violent, uninterrupted action."[247] Furthermore, "No aerial resources should . . . be diverted to secondary purposes, such as auxiliary aviation, local air defense, and antiaircraft defenses,"[248] which he considered "*worthless, superfluous, harmful*."[249] Following this recipe protects friendly territory, equipment, aircraft, surface infrastructure, and personnel by neutralizing the enemy's comparable assets first.[250]

Regarding the composition of an air force, Douhet, Mitchell, and Trenchard preferred bombers (in fact, Douhet's "battleplanes" were "designed for aerial combat and for bombing"[251]) and downplayed the need for fighter escorts.[252] Slessor, on the other hand, valued functional, specialized variants, including fighters, bombers, and attack aircraft.

How Airpower Wins

Airpower theory is not just a matter of defining the various roles and missions of air weapons. Such theory requires the conceptualization of ways to implement it.[253]
                                                                    I. B. Holley, Jr.

The final task for the domain theorist, after describing how to "command" the domain, is to show how this advantage achieves the fundamental object of war that, according to Douhet, "has always been, is now, and always will be, to win – that is, to compel the enemy to bow to one's will."[254] Though some theorists (like Mitchell and Douhet) thought gaining air superiority might actually compel surrender; in reality, paraphrasing Corbett, control of the air is just a prerequisite for influencing events on the ground, where people live. "The air situation," admits Slessor, "has no importance in any form of war except in so far as it affects the situation on the ground."[255] Therefore, "The primary arm of the air force is the bomber," which has the aim of reaching and striking its objective with adequate accuracy and potency to achieve its intent.[256] With the air domain in theory granting access to any point on Earth, the fundamental question for airpower theorists, after gaining air superiority, has always been, "What should I bomb to force the enemy to quit?"

Whereas in the past armies had to fight lengthy, brutal ground engagements to threaten their counterpart's communications or capital city, now airpower could affect multi-domain targets quickly and in depth. Airpower theorists were convinced that, though it could not "hold ground," airpower could break the enemy's will through direct attack on the populace (Douhet) or via a combination of strikes on military forces, supply and support networks, and vital centers of production (Mitchell, Trenchard, and Slessor). Douhet writes, "Having the command of the air, aerial forces should direct their offensives against surface objectives with the intention of crushing the material and moral resistance of the enemy."[257] "Aerial strategy," adds Douhet, considers "military, political, social, and psychological" factors, selects objectives, groups targeting zones, and sequences attacks depending on the overall aim.[258]

Selecting appropriate targets requires detailed analysis of the adversary and awareness of current conditions. It is not enough just to hit targets; they must be *the right targets at the right time*. "The whole art of air warfare," advises Slessor, "is first the capacity to select the correct objective at the time ... that on which attack is likely to be decisive, or to contribute most effectively to an ultimate decision; and then to concentrate against it the maximum possible force."[259] And the air force commander, "must exploit the extreme flexibility ... high tactical mobility, and the supreme offensive quality inherent in air forces ... to threaten his various vital centres as to compel him to be dangerously weak at the point which is *really decisive* at the time."[260] Slessor's "vital centre" is analogous to a center of gravity, "an organ or centre in a man, an army, or a nation, the destruction or even interruption of which will be fatal to ... his continuance of effective operations."[261] Furthermore, selecting and engaging vital centres "presupposes the most exhaustive resources in information and the most meticulous system of day-to-day" military, "political and industrial intelligence."[262] To do this, Slessor recommends consultation with expert technical specialists on areas to be targeted (e.g., logistics officers or food industry authorities).[263] Air action, according to Slessor, should hit "fighting troops" directly (i.e., *close air support*) during active battle periods and (when forces are not engaged) supply, transportation, and production (e.g., *interdiction* targeting war industries, power, fuel, mines, and railroads).[264]

Douhet controversially recommends using "explosive, incendiary, and poison gas" bombs on "industrial and commercial establishments; important buildings, private and public; transportation arteries and centers; and ... [the] civilian population as well."[265] No other theorist condoned gas or incendiary attacks on civilians. In addition to "vital civilian centers," Douhet identifies secondary targets including "railroad junctions and depots, military depots, and other vital objectives," like "naval bases, arsenals, oil stores, battleships at anchor, and mercantile ports."[266] Mitchell not only recommended attacking the opposing army and navy, he also proposed striking economic and military "vital centers" like factories, raw materials, foodstuffs, supplies, transportation nodes, ports, and shipping, which he assumed would "intimidate" and erode the "fragile" will of the population.[267] Finally, he perceived a valuable role for airpower in limited wars to put "down uprisings" and provide rapid transport across wide areas.[268]

For Trenchard, the "airplane was an inherently strategic weapon, unmatched in its ability to shatter the will of an enemy country,"[269] a feat best accomplished by bombing a wide range of targets including industry, roads, rail lines, iron and coal mines, and armament factories.[270] Trenchard's strategy indirectly targets will by strangling industry, like the sea power theorists' *guerre de course*, and undermining the enemy's warfighting capacity.

Slessor believed airpower to be primarily strategic, and that enemy centers of gravity (material and equipment versus population centers) should be attacked "repeatedly, as far from the battlefield as possible," thus "strangling [the enemy] into submission."[271] Perhaps with a Sun Tzu–like eye toward postwar reconciliation, Slessor argues for paralysis through "dislocation and disorganization," *not* destruction, as the proper goal for strategic attack.[272] Finally, as mentioned previously, while Slessor acknowledged the *possibility* of an airpower-centric war, he believed that any war with an active land campaign dictates that "the primary objectives ... those against which action will lead most directly to a decision – will always be the enemy *land forces*, their communications and system of supply."[273]

Seers or Salesmen?

As these airpower theorists surmised, gaining access to the air domain and the emergence of the air weapon were revolutionary developments. But no revolution lasts forever. They should have anticipated that humanity's ability to adapt and innovate would eventually rebalance the scales, causing the air domain to become almost as contested and bloody as the Earth's surface. Additionally, the god-like powers attributed to air forces ignored far too many pertinent variables. Weather, malfunctions, human error, radar, missiles, faulty analysis and intelligence, accuracy and design limits, and resilient enemies – like the sun on Icarus's wings – ultimately brought airpower back to reality and proved the assumption of airplane omnipotence a fantasy. Ironically, when employed in World War II, the ideas of Douhet, Mitchell, and Trenchard did not eliminate carnage on the ground, they just moved it to the skies. As a result, nearly 75,000 Allied airmen were injured, captured, or perished. Additionally, air forces are not always a bargain compared to other forms of military power. They are expensive to develop, build, operate, and sustain, which explains why many contemporary nations rely on a "poor man's air force" via integrated air defense, unmanned aircraft, and surface-to-air missile systems.

Douhet and the others hit closer to the mark with their appreciation for airpower's speed and flexibility. Speed is fundamental to flying (it generates lift), and the airplane's ability to maneuver rapidly in three dimensions and change objectives "on the fly" is unparalleled. A later theorist, Colonel John Boyd (1927–97), would base his ideas regarding action, reaction, and decision speed on his experiences with the swiftness and maneuverability of tactical fighters (we will revisit Boyd in the next chapter). Of slightly less validity are assertions regarding airpower's inherently offensive, strategic, and decisive nature. Prior to the advent of radar and enhanced sensing devices, there was little practical value in launching aircraft for missions other than ground surveillance or attack. However, the air weapon's defensive potential increased in proportion to advances in detection capabilities; and now, airpower is neither inherently offensive nor defensive but retains ample capability for either posture.

Likewise, in the aftermath of World War I, efforts to define a unique role for airpower and secure an independent air service led to overemphasis on a strategic potential that would not be realized for decades. In fact because of the air domain's ubiquity, airpower, after gaining air superiority, retains the exclusive ability to strike at any level of warfare, including fielded forces, supply, transportation, industry, capitals, and leadership. Another modern theorist, Colonel John Warden, author of *The Air Campaign: Planning for Combat* (1988), expressed this in his "five rings" targeting model that highlights how parallel attacks at multiple levels of war (leadership, organic/system essentials/key production, infrastructure, population, and fielded military forces) induces "systemic" paralysis, accelerating a favorable decision.

Yet, as militarily significant as airpower is, historical analysis proves it is not always decisive. There are two main reasons for this. First, aircraft are inherently poor at presence and persistence. They simply cannot remain aloft and militarily potent indefinitely, and they obviously cannot hold ground. Absent a large number of aircraft operating in a very permissive battlespace, enemies confronted only with airpower frequently gain respite. Second, and most importantly, decisions about war and peace occur in the cognitive domain, which depends on a multitude of capricious human factors, including personality, culture, ethnicity, politics, and religion. Thus, it is virtually impossible to know for sure whether the bombardment of any target set will be decisive, and in some cases, bombing may actually stiffen, not break, an enemy's will. "The war aim of the air force," cautions Slessor, "is 'to break down the enemy's resistance ... by attacks on objectives calculated to achieve this end.' The wording is left deliberately indefinite, because in every form of war ... 'the objectives calculated to achieve this end' will vary."[274] Though Warden's paradigm is a valuable reference for planners and strategists, even flawless execution cannot guarantee victory.

Finally, were Douhet and his cohort correct that all air forces must be independent? To borrow a US Air Force Weapons School tenet, "it depends." For nations resourced to develop and employ strategic airpower and to perform a variety of missions exclusive of surface forces, an independent air force is a necessity. However, for less-wealthy nations, auxiliary forms of airpower tethered or even subordinate to surface forces may be more appropriate. Furthermore, when practical, services should retain auxiliary, cross-domain capabilities along domain boundaries (e.g., aviation attached to surface forces). This facilitates access and awareness and covers inter-domain gaps and seams, which explains why the navies and armies of powerful countries retain organic airpower apart from the air force and why air forces (and some navies) employ their own ground combat elements.

Conclusion

As stated earlier, good domain sub-theory aligns with general theory and presents principles that are at least universal for that domain. Excepting Corbett and perhaps Slessor, the domain theorists addressed here mostly fall short of this standard. Good domain theory separates character-of-war variables, like tactics and technology, from nature-of-war constants, like violence and reciprocity. This sets the stage for accomplishing the three tasks outlined at the beginning of this section: (1) assess domain-exclusive capabilities and deduce domain-centric principles; (2) determine the best ways to gain a decisive advantage in that domain using the three a's – *awareness*, *access*, and freedom of *action*; and (3) describe how to exploit that advantage to promote victory.

When considered objectively, the real theory of airpower reflects the nature of war and the uniqueness of the domain and the people and machines that use it to advance war's aims. Airpower provides awareness and influence in all physical domains, moves effortlessly between offensive and defensive roles, and concentrates enormous power with high precision. But airpower is not the panacea Douhet and Mitchell imagined. Though most aircraft are remarkably fast, flexible, and far ranging, they are not invincible. They often rely on distant or vulnerable bases for access, and as technological advances, like stealth or precision-guided munitions, grant an edge, other innovations, like thermal detection or electronic jamming, mitigate that advantage. Thus, great air powers cannot necessarily dominate the world with impunity as Douhet and Mitchell imply. Even relatively weak states can defend themselves with integrated air defense systems, missiles, and unmanned aircraft, which represent an effective, affordable alternative to traditional air forces when power projection from the air is not a military necessity.

As Trenchard and Mitchell deduced early on (but later downplayed), the air weapon's speed and flexibility make it a surprisingly effective asset for operations short of total war, like small wars and humanitarian relief. In these instances, airpower is well suited for logistics support, ISR, no-fly-zone enforcement (air blockade), and direct, tactical action. "Ground support and observation planes of slow speed, high endurance, great firepower ... plus short-takeoff transport planes and helicopters," writes David Galula, "play a vital role in counterinsurgency operations."[275] Finally, airpower can strike almost anything, hence the requirement for exquisite intelligence, rigorous analysis, and prioritization. However, influencing human will is not a science. In general, as with maritime power, the odds of achieving the three a's and securing victory rise when airpower is combined with forces from other domains, and even airpower's staunchest sponsors should acknowledge that having "boots-on-the-ground," putting humans face-to-face, is often necessary for realizing war's ultimate aims.

To conclude, we cannot help but wonder how cyber and space domain theories will evolve provided the example of maritime and airpower theory. The rapid growth of technical capability within these areas is outpacing most efforts to develop sub-theories to characterize them. As these capabilities continue to expand and mature, will any serious theorists emerge? Will they objectively leverage history, general theory, and logic to accomplish the three tasks of domain sub-theory? Will they, like Corbett and Slessor, succeed in balancing the realities of war with the exhilaration of novel technologies, or will they take the more theatrical and less rigorous path exemplified by Mahan and Douhet? Time will tell.

CONCLUSION

> *The essence of a dialectical approach to the study of war may reside less in the pairing of dichotomies and more in their fusion.*[276]
>
> Hew Strachan and Sibylle Scheipers

The Universe of War Theory and Strategy

Embracing the breadth and diversity of war theory, including the preceding chapter's magnificent seven and this chapter's examination of small wars and domain theory, gets us awfully close to seeing the whole "forest" of war. However, though we have now addressed the problem of accessibility regarding existing war theory, we have yet to fully answer all the questions from the Preface. To set the stage for that in the next chapter, we must evaluate war's leading theorists and theories in relation to war's twenty dialectics: order-chaos, past-future, peace-war, life-death, creation-destruction, friend-enemy, good-evil, science-art, defense-attack, unlimited-limited, regular-irregular, physical-moral, direct-indirect, certainty-uncertainty, simplicity-complexity, reason-emotion, prudence-boldness, control-autonomy, concentration-dispersal, and rest-movement.

Examining the major themes and patterns in military thought suggests that most effective theories and strategies reflect a balanced appreciation for *both sides* of these dialectics. This is not meant to imply that choosing middle courses is the key to victory; rather, it means that success is more likely when we have cultivated the ability to *choose the appropriate extreme for a given context*, without losing sight of the whole. For example, our classical theorists, Sun Tzu and Thucydides, are among the most balanced and enduring. Sun Tzu advises us to know friend and enemy and to use war only as a bridge to a better peace. Strategy may be grounded in past lessons but must keep its eye always on the future. He counsels that war be simple for one's own force (with intelligence) and made complex and uncertain for the enemy (with deception). When possessing greater strength, Sun Tzu advises boldness, concentration, and attack, but with a deficit, prudence, dispersal, and defense. For Sun Tzu, war is an art *and* science where *cheng* (direct, regular) and *ch'i* (indirect, irregular) combine to mystify the enemy.

Thucydides probes war's moral and ethical dimensions, showing how the good-evil, friend-enemy, order-chaos, life-death, and peace-war distinctions blur as emotion and boldness often overwhelm reason and prudence. Athens and Sparta were friends, but honor to alliance, fear, and ambition turned Greek "brothers" into enemies. Once the war had begun, neither side could embrace a strategy that would yield a better peace. When prudence (and Nikias) advised Athens to rest, boldness led to the Sicilian debacle, and though Pericles invoked justice to stir the winds of war, what justice did the "good" Athenians mete upon the "evil" Melians?

The medieval period produced only Machiavelli as a relevant theorist, and he rejected the dark ages' chaos and *condottieri* in favor of the ancients' order and discipline. Machiavelli's realist perspective on war and politics endorses "evils" that support the greater good while denouncing the naïveté of Christian morality as a mitigating influence on affairs of war and state. For him, war is complex, chaotic, and bloody and thus requires *virtù* (boldness) in its citizens lest *fortuna* leave them subjugated or worse.

Warfare from Machiavelli's time up through the late eighteenth century exemplified the *ancien régime*'s favoritism for limited aims, intricate stratagems, and maneuver over decisive battle. As the Enlightenment dawned, science and reason in the form of prescriptive, rules-based theories promised to unlock war's secrets. Beneath the unblinking gaze of Minerva's owl, war's chaos, uncertainty, complexity, and irregularity would dissolve, leaving only order, certainty, simplicity, and regularity. But in the nineteenth century, the triumphs of Nelson (at Trafalgar) and Napoleon (in continental Europe) shattered the *ancien régime* paradigm, supplanting it with the supremacy of decisive battle. Jomini and Clausewitz were sons of this epoch, both influenced by Napoleon (unlimited, direct, regular, concentration, boldness, attack, movement), but the former was a child of the Enlightenment (order, reason science, physical, simplicity, certainty) and the latter of German idealism and Romanticism (chaos, emotion, art, moral, complexity, uncertainty). However, Clausewitz eventually achieved greater balance (and relevance) after his epiphany and inclusion of limited aims considerations in *On War*, which as Raymond Aron points out, demonstrated Clausewitz's advocacy for both the "boldness" of war's destructiveness and the "prudence" of its subordination to policy.[277] Moreover, his commentary on the benefits of defense (though tortuous) acknowledged that attack is not war's everything. Still, neither Jomini nor Clausewitz successfully reconciled the regular-irregular dialectic; rather, they continued the era's predilection for unlimited war, the direct approach, and concentration. This reality only intensified in the latter half of the century, leaving theory and strategy perilously unbalanced, as Mahan offered a maritime vision of Jomini's system and the cult of the offensive exalted unlimited, direct, regular, attack, boldness, movement, and concentration above all else. This sharp imbalance portended a precipitous correction.

That reckoning arrived in the early twentieth century. Calamitous experiences in the American Civil and Russo-Japanese Wars, small wars, and World War I sparked a theoretical backlash against the nineteenth century's Nelsonian-Napoleonic model and the cult of the offensive. Corbett, Liddell Hart, Mao, and the small wars theorists retaliated by articulating the value of dispersal, defense, limited aims, and indirect methods. In particular, Corbett helped reconcile Clausewitz's unlimited-limited dialectic, while Mao and Liddell Hart (and Fuller) rediscovered the classical balance of Sun Tzu and Thucydides. Meanwhile, small war theorists struggled to understand and cope with the irregular, uncertain, complex, autonomous, and chaotic (not to mention protracted) warfare earlier theorists had marginalized, realizing paradoxically that success often hinges on life and creation rather than death and destruction.

On the other hand, most airpower "theorists" (excepting Slessor) remained stubbornly oblivious to this theoretical renaissance, which they surmised had little bearing on a development as revolutionary as the aircraft. Despite expressing a preference for all things future over anything past, they put wings on the offspring of the Enlightenment and the cult of the offensive and "carpet-bombed" the dialectics, leaving a lopsided view of war consisting only of order, future, science, reason, unlimited, direct, attack, movement, regular,

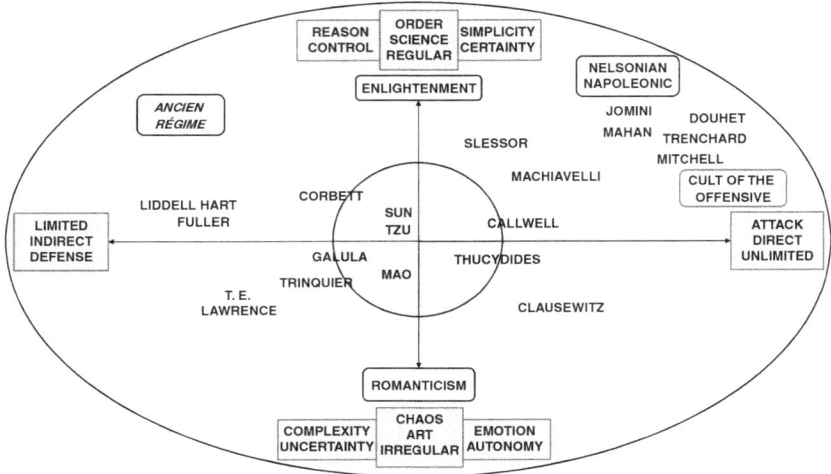

Figure 3.2 The universe of war theory and strategy

physical, certainty, control, simplicity, boldness, concentration, death, and destruction. Along with the positive legacies of magnificent aircraft and independent air forces, there remains a lingering tendency for airpower's practitioners to struggle with war's indirect, irregular, moral, and chaotic aspects and, like Tantalus, to always be reaching for the "fruits" of certainty and simplicity that hover just beyond their grasp.

So where does all this leave us? As we mentioned in the Preface, these conceptual traditions have stacked up like geological strata making them hard to access let alone understand and employ for warfighting and strategy. Figure 3.2 depicts where each theorist and movement appears relative to the others and to war's most representative dialectics in the "universe" of war theory and strategy. The closer to the center, the more balanced the theory. Therefore, our challenge in the next chapter is to articulate a general war theory that hits closer to the "bullseye," or in other words, that *unifies* these theorists and theories, including historical and *matériel* perspectives on strategy as well as small war and domain sub-theory, while maintaining a balanced and inclusive treatment of all twenty dialectics. Among these, the regular-irregular dialectic (and war's formal cause) will attract much of our attention, because a systematic reconciliation of both halves of this dichotomy has eluded even the esteemed magnificent seven.

4   The Unified War Theory

As an issue that dominates our time and as a still imperfectly understood force in our past, war demands much further exploration. That so few scholars and soldiers have taken it up in something of Clausewitz's spirit of objective inquiry, and with his ability to combine reality and theory, is not the least measure of his achievement.[1]   Peter Paret

In a preface to an unpublished manuscript on war theory predating *On War*, Clausewitz says, "Perhaps a greater mind will soon appear to replace these individual nuggets with a single whole, cast of solid metal, free from all impurity."[2] The Prussian master had undoubtedly hoped to accomplish this with his own finished oeuvre. Now, centuries beyond Clausewitz's time, are we any closer to realizing this vision? In fact, with careful analysis of past geniuses – like those of the preceding two chapters – more comprehensive histories, more advanced scientific and mathematical analogues, and more data on contemporary wars, now is the time to develop a *unified* theory of war.

On Theory

This subject [war], like any other that does not surpass man's intellectual capacity, can be elucidated by an inquiring mind, and its internal structure can to some degree be revealed. That alone is enough to turn the concept of theory into reality.[3]   Clausewitz

Before introducing the unified theory, we must clarify what *theory* is and what we can expect of it. In science, a theory is "a well-substantiated explanation of some aspect of the natural world that can incorporate facts, laws, inferences, and tested hypotheses."[4] As noted in the Preface, theory ideally provides a typology, predictions of the future, explanation of the past, enhanced understanding, and the potential for control. Milan Vego describes *military* theory as, "a comprehensive analysis of all the aspects of warfare, its patterns and inner structure, and the mutual relationships of its various components/ elements."[5] War theory may not provide an answer for every situation, but it can establish a baseline for evaluation and comparison. "Every case must be judged on its merits," writes Corbett, "but without a normal to work from we cannot form any real judgment at all; we can only guess," and guessing in war is dangerous.[6] Good theory clarifies a phenomenon in accord with observation, though interestingly, abstract (i.e., independent of time and location) theories cannot be proven true, they can only be proven false.[7] Theory validated to the point of near certainty is a *law* – defined as a descriptive generalization about how some aspect of the natural world behaves under stated circumstances.[8]

Rules of Theory

For any *new* theory to survive scrutiny, it must pass three tests: (1) it must not contradict validated theory or laws without justification; (2) it must still explain what is readily observable or known to be true; and (3) it must offer some new insight into the phenomenon under investigation. As Clausewitz affirms, "A satisfactory theory of war ... will never conflict with reality."[9] An updated war theory must not only acknowledge past theorists; it must also clearly explain points of departure. Fortunately, we can meet these tests having the advantage of significantly more data on war, theory, and history and near-instant access via the Internet. Additionally, we know far more about warfare's evolution, including naval, air, space, and cyber power; nuclear weapons; small wars; digital computing; chaos and game theory; terrorism and non-state actors; globalization; and social media. A general war theory must address these changes while remaining faithful to the True Trinity, and it must express its tenets in thoughtfully defined, unambiguous terms. Furthermore, to fully represent war's enduring truths, all forms of war and sub-theory must be included within the general theory's precepts.

Theory as a Means for Control and Prediction

Sound theory can help us *control* or at least mitigate a phenomenon's negative effects. For example, scientific theory enables us to harness nuclear reactions for safe power generation and to treat and cure disease. However, war's complexity, breadth, and violence complicate theorizing and, consequently, forecasting and control. Since little in war can be known with complete certainty, there are few principles of war we can consider "laws," and there are no theories expansive enough to cover war's every facet. Nevertheless, war presents patterns, hints, and clues we can express theoretically and exploit for the ends of control and prediction. We just have to maintain realistic expectations. As Corbett writes, "[Theory] does not pretend to give the power of conduct in the field; it claims no more than to increase the effective power of conduct."[10]

Jomini

Correct theories, founded upon right principles, sustained by actual events of wars ... will form a true school of instruction for generals. If these means do not produce great men, they will at least produce generals of sufficient skill to take rank next after the natural masters of the art of war.[11]   Jomini

Jomini describes theories as "things so difficult to put in practice, but so easily understood when once exemplified."[12] He centered his theory on principles much easier to state than to apply effectively. Many of history's great leaders studied war and war theory, but not all could turn that knowledge into decisive combat effects. Jomini's statement hints at the gulf between theory and practice and the advantages of hindsight. We are better at evaluating war, like most complex human activities, afterwards, detached from the emotion (and danger) of the actual event. However, with war, we do not always get the chance to live and learn from errors.

Liddell Hart

The predominance of the psychological over the physical, and its greater constancy, point to the conclusion that the foundation of any theory of war should be as broad as possible.[13]   Liddell Hart

On the surface, war appears to be a physical activity, but Liddell Hart recognized that theory cannot omit war's human aspects. Additionally, he knew that war cannot be fully understood by studying its elements in isolation; rather, we uncover its most enduring qualities in the emergent, macroscopic patterns manifest across history. Writes Liddell Hart, "If a specific effect is seen to follow a specific cause in a score or more cases, in different epochs and diverse conditions, there is ground for regarding this cause as an integral part of any theory of war."[14]

Clausewitz

One must never let a cloud of preconceived ideas get in the way ... it is the theorist's most urgent task to dissipate such preconceptions ... the errors intellect creates, intellect can again destroy.[15]   Clausewitz

Clausewitz is supreme in his obsession with theory ("theory" appears over 200 times in *On War*), and he considered the conceptualization of alternative theories an important qualification for the credible *critique* of theory. For Clausewitz, war theory investigates war's basic elements, causes, and means leading to "incontrovertible truth"; and a "working theory is an essential basis for criticism."[16] "Theory," writes Clausewitz,

should cast a steady light on all phenomena so that we can more easily recognize and eliminate the weeds that always spring from ignorance; it should show how one thing is related to another ... if concepts combine ... to form that nucleus of truth we call a principle, if they spontaneously compose a pattern that becomes a rule, it is the task of the theorist to make this clear.[17]

Ironically, Clausewitz says theory's "purpose is to demonstrate what war is in practice, not what its ideal nature ought to be."[18] Yet, much of our Chapter 2 critique centers on his efforts to reconcile the ideal-type with actual war. In considering why real war is *not always* consistent with theory, Clausewitz says a "vast array of factors, forces and conditions" form a "barrier."[19] He adds,

> The reason why the object of war that emerges in theory is sometimes inappropriate to actual conflict is that war can be of two very different kinds ... if war were what pure theory postulates, a war between two states of markedly unequal strength would be absurd, and so impossible ... but wars have in fact been fought between states of *very unequal strength, for actual war is often far removed from the pure concept postulated by theory.*[20]

Though *On War* addresses "less perfect" forms of war, Clausewitz resolves that theory "has the duty to give *priority* to the absolute form of war and to make that form a general point of reference" (emphasis added).[21] Nevertheless, the crux of Clausewitz's work, centered on his paradoxical trinity, is his "real" and most beneficial contribution.

Clausewitz considered theory a remedy for systems that "appear some-times banal, sometimes absurd, sometimes simply adrift in a sea of vague generalization," though one could argue that his own complexity goes too far the opposite direction.[22] Furthermore, although "the primary purpose of any theory is to clarify concepts and ideas that have become, as it were, confused and entangled," Clausewitz was adamant that war's psychological elements not be sacrificed upon the altar of clarity.[23] Theory may become "infinitely more difficult as soon as it touches the realm of moral values,"[24] says Clausewitz, yet any theories that omit these human factors must be con-sidered flawed.[25]

For Clausewitz, theory is guidance, not prescription, and the best leaders, military geniuses, can transcend the need for written theory altogether. Theory is not a precise path, says Clausewitz, but serves as "a frame of reference" leading to "close *acquaintance* with the subject" and "*familiarity* with it" becoming "an active ingredient of talent" that can "light his way, ease his progress, train his judgment, and help him avoid pitfalls."[26] Clausewitz adds, "It is simply not possible to construct a model for the art of war that can serve as a scaffolding on which the commander can rely for support at any time ... *talent and genius operate outside the rules, and theory conflicts with practice.*"[27] And, "What genius does is the best rule, and theory can do no better than show how and why this should be the case."[28] For Clausewitz, war theory neither rises to the level of a natural law nor leads in all cases to a positive doctrine, because war's nature – unpredictable and subject to flawed human perception – precludes such treatment. Clausewitz says, "In the conduct of war, perception cannot be governed by laws ... nor can the theory of war apply the concept of law to action, since no prescriptive formulation universal enough to deserve the name of law can be applied to the constant change and diversity of the phenomena of war."[29]

At war's tactical level, where variables and perceptions are more limited, positive doctrines gain greater traction as combatants employ force according to specific "principles, rules, regulations, [and] methods" as circumstances warrant.[30] But tactics are not always representative of the broader phenomenon of war; thus, says Clausewitz, "All the positive results of theoretical investigation ... will increasingly lack universality and absolute truth the closer they come to being positive doctrine."[31] Ultimately, Clausewitz believed, as we do, that theory combined with experience and education stimulates intuition, even in otherwise ordinary generals. He says,

Theory cannot equip the mind with formulas for solving problems, nor can it mark the narrow path on which the sole solution is supposed to lie by planting a hedge of principles on either side. But it can give the mind insight into the great mass of phenomena and of their relationships, then leave it free to rise into the higher realms of action. There the mind can use its innate talents to capacity, combining them all so as to seize on what is *right* and *true* as though this were a single idea formed by their concentrated pressure – as though it were a response to the immediate challenge rather than a product of thought.[32]

On Theory and Practice

There is a great deal of difference between knowing and understanding. You can know a lot about something and not really understand it.[33]

Charles F. Kettering

Warfighting is a *practical* art; thus, good war theory must have practical application. As Jomini writes, "The study of the principles of strategy can produce no valuable practical results if we do nothing more than keep them in remembrance, never trying to apply them ... to hypothetical wars or to the brilliant operations of great captains."[34] Practice pertains to the *how* and *what*, and theory provides the *why*. Clausewitz says that theory, combined with experience and history, builds "familiarity" with war that, in turn, enables theory to advance "from the objective form of a science to the subjective form of a skill."[35] In other words, theory helps convert knowledge into understanding that informs practice, even where "the nature of the case admits no arbiter but talent."[36] Science is to engineering and medicine what war theory is to warfighting. Rudimentary engineering and medicine existed prior to

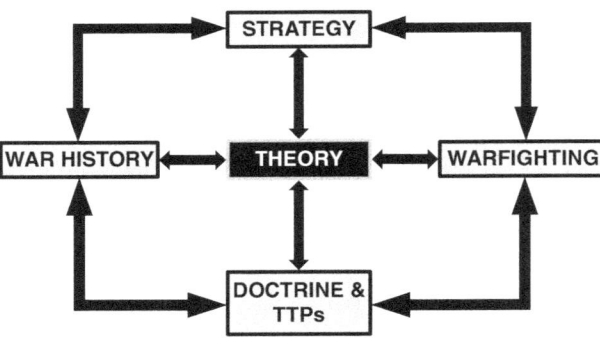

Figure 4.1 Theory, strategy, and warfighting

scientific theory just as warfighting predates war theory. Many successful generals and warriors of antiquity had never heard of Sun Tzu or Clausewitz. However, there have undoubtedly been engineering, medical, and warfighting failures through the ages that better theory might have prevented. Formulating strategy without theory is dangerous. Military doctrine (*general* guidelines for how to use military forces) and tactics, techniques, and procedures or TTPs (*specific* guidance for employing military force) are most effective when grounded in experience *and* validated theory.

Figure 4.1 illustrates the relationship between theory, strategy, warfighting, history, doctrine, and TTPs. Theory is derived from history but can be revised using lessons from strategy, warfighting, doctrine, and TTPs. Furthermore, theory aids the analysis of history even as it informs the other elements. "Theory," writes Clausewitz, "then becomes a guide to anyone who wants to learn about war from books; it will light his way, ease his progress, train his judgment, and help him avoid the pitfalls."[37] Theory has a similar influence on doctrine and strategy, and consequently warfighting. "Knowledge of military theory," asserts Milan Vego, "is essential to understanding and then creatively applying doctrine."[38] Removing theory eliminates the core of these processes and activities.

Ultimately, war theory must produce a framework consistent with experience and history while balancing breadth and specificity. Theory that is too general sacrifices practicality, and theory that is too detailed infringes on doctrine and tactics. Hitting the "sweet spot" for war theory requires careful study of war's character and nature including its four causes, the True Trinity, and a balanced perspective on war's dialectics. Only then will theory finally answer war's stubborn questions and serve the practical needs of analysis, strategy, and warfighting.

The Unified War Theory

Some military innovations involve science and technology ... other equally significant, innovations come in the realm of ideas.[39]

General T. Michael Moseley

Our military today is in a sense operating without a concept of war and is searching desperately for the new "unified field theory" of conflict[40]

General David Barno

The remainder of this chapter introduces the *unified war theory* (UWT) and is intentionally expository. However, fear not! The chapter concludes with a summary of the UWT's main concepts and terms for easy reference. The UWT's "DNA" comes from our Chapter 1 definition: war is a uniquely *human* phenomenon, characterized by *combative violence* between two opposing sides for the purpose of achieving a desired *political* outcome. The emphasized words highlight the True Trinity, which is inclusive of war's four causes. This chapter builds on Chapter 1's analysis of war's human dimension (the Trinity's first layer) by addressing the next two layers, politics and combat, and the activity that links them: strategy.

POLITICS

Politics, Power, and War

I put for a general inclination of all mankind, a perpetual and restless desire of Power after power, that ceaseth only in death.[41]   Thomas Hobbes

A ruler must conquer, because only through power is it possible to obtain security, contentment and happiness.[42]   Chanakya (Kautilya)

Politics is *an activity through which humans gain and apportion power and govern.* Society and politics were born when humans first banded together, elected to share resources and responsibilities, and selected leaders to help maintain order and give direction. "The quintessential political act," write Codevilla and Seabury, is "determining who gets what, when, and how."[43] And Thucydides tells us, "Of men we know, that by a necessary law of their nature they rule wherever they can."[44] Politics is most closely associated with the peace-war, order-chaos, past-future, friend-enemy, control-autonomy, and reason-emotion dialectics. It balances the chaotic emotions of life with the order and reason of laws and establishes governance across a spectrum from control (e.g., totalitarianism) to autonomy (e.g., republicanism). Politics defines friend (citizen) and enemy (threats) and strives to create conditions that transform past and present into a better future. When necessary, politics charts a course from peace to war back to peace.

Politics is an adversarial process that identifies, selects, and empowers leaders; and *policy*, the product of politics and an expression of the leadership's intent, translates human will into governance. Laws, a product of policy, set standards that help leaders advance three universal objectives of governance: order, security, and prosperity. The enactment and enforcement of policies and laws requires *power*. Writes Einstein, "Law and might inevitably go hand in hand, and juridical decisions approach more nearly the ideal justice demanded by the community . . . in so far as the community has effective power to compel respect for its juridical ideal."[45] Political leaders and their constituencies have interests, and they set priorities and define courses of action for achieving them. Power enables those courses of action. Strategy expert Terry Deibel defines power as a "motive force . . . necessary in order to get things done, to accomplish one's purposes, to carry out one's own will despite the resistance that accompanies all . . . endeavors."[46]

Power takes many forms, and individuals, groups, or states can gain and wield power based on their position, wealth, associations, reputation, talents, or violent capacities. State power, writes political scientist John Mearsheimer, "is based on the material capabilities that a state controls . . . [and] latent power is based on a state's wealth and the size of its overall population."[47] Power enables the control of resources, people, and outcomes.[48] Political scientist Joseph Nye defines two types of power: "hard" and "soft."[49] Hard power is overt, direct, and often physical, like military power, which, according to Colin Gray, "is the ultimate form of power."[50] Or as Gwynne Dyer puts it, "Force is the ultimate argument in human affairs, and naked military power is the most effective form of force."[51] Hard power leverages physical or economic advantages to compel political outcomes. Soft power is more subtle, indirect, and psychological. For instances short of warfare, soft power can be more effective than hard power because it promotes *voluntary* support through attraction and persuasion versus coercion.

Once humans gain power, we rarely want to relinquish it. "No ruler," advises Chanakya, trusted minister to Mauryan emperor Chandragupta, "can tolerate a policy that prevents his state from growing in wealth and power. This is deterioration."[52] Leaders use power to get more, turning it to the task of self-preservation and resisting internal and external threats. "The schemes of a prince," writes Rousseau, "form a never-ending upward spiral. He wants to have more power in order to increase his wealth and to increase his wealth in order to have more power."[53]

Influence

Power provides the potential or capacity for a person, group, or state to exert *influence*. While power applies to one side, influence implies a relationship.

A nation, writes famed economist Friedrich List, may leverage its wealth and power – including the diverse accomplishments of its people, its geography, and natural and industrial resources – "to extend its moral, intellectual, commercial, and political influence over less advanced nations and especially over the affairs of the world."[54] Power is the potential to generate an effect, and influence causes the effect. There are five general ways to influence:

(1) Accord: influence by example and shared interests
(2) Persuasion: influence through arguments and logic
(3) Inducement: influence by reward or *quid pro quo*
(4) Coercion: influence by threat or intimidation
(5) Conflict: influence by violent force (includes war)

Influencing things and people are separate matters. Things are governed by physics, which can be precise, and people by psychology, which is not. The gap between intent and effect when attempting to influence people can be very wide.

Ideology and Culture

The important fact that war is cultural does not diminish the logical and historical authority of the argument that war primarily is political.[55]

Colin Gray

The best way to increase power is by influencing other people to act on our behalf to accumulate it. While it is possible to *force* others to support political aims (hard power), it is better to *convince* them (soft power) if we want committed and trustworthy followers. This most effectively happens through an exchange of ideas, which can become an *ideology* that, in turn, forms the basis for political systems. Ideologies take many forms (e.g., cultural and religious beliefs, shared values, or nationalism) that combine to create a worldview or *Weltanschauung*, which forms a basis for effective collaboration in large, diverse societies. Cultural ideologies generally develop over a long period of time and have pervasive effects on a society's *Weltanschauung* reflected in customs, values, institutions, politics, and even warfare. Religion is basically divinely inspired ideology, which makes it especially hard to oppose. Autocratic rulers often solidify their hold on power by linking legitimacy and authority to religion (e.g., theocracy) or by prohibiting religion entirely to prevent challenges by the acolytes of a "higher power" (e.g., god). Some noncultural, secular ideologies (like fascism or communism) can arise relatively abruptly, and since they are human-made, they can (theoretically) be altered or renounced by humans.

In truth, as Gray implies, *Weltanschauung* and culture are almost always linked in some way to power, resource distribution, and governance, and thus, to politics. Political systems and their supporting ideologies generally offer something desirable (e.g., order, freedom, a better life, or security). Religious ideologies often promise divine favor in this life and eternal rewards in the next. Secular ideologies like democracy, republicanism, or communism promise to share political, economic, or social power. It is not uncommon for large, diverse societies to contain a multitude of overlapping ideologies, and the greatest threat to an ideology is a competing ideology. Without a collectively recognized, enforceable process for reconciling ideological friction, clashes can erupt into violence or war. Like a biological organism, a political "organism," or *party*, may employ violence to ensure survival and reproduction (i.e., to spread its ideology). The first and supreme objective of political parties is *survival*.

In most respects, political parties and states act like individuals. Plato says the state is the individual writ large, or more precisely, "We cannot suppose that States are made of 'oak and rock,' and not out of the human natures which are in them . . . States are as the men are; they grow out of human characters."[56] State and party politics mirror the motivations and personalities of their leaders and polities. Political leaders produce policy and specify how and when to use power in pursuit of policy's aims; however, these are seldom simple or precise enterprises. Applying power to create influence, regardless of the target, produces unintended consequences or side effects proportional to the power expended. Powerful weapons (e.g., nuclear, chemical, or biological) are termed "weapons of mass destruction" (WMD) because their influence is so extensive and imprecise. Even the precise application of power does not guarantee a positive target response, and similar targets may react differently to any given application of power. Consider the USA's use of military power to topple Iraq in 2003. Though the invasion's objective was to oust Saddam Hussein, it also provoked an insurgency, stoked sectarian unrest, compelled Libya to abandon its chemical weapons, prompted Iran and North Korea to hasten WMD development, and caused political turmoil in the United States.

Political Perfection

The hypothetical condition of *perfect* politics would eliminate policy, war, and free will except for the political leader who would enjoy omnipotence. As we discussed in the section on Sun Tzu, the perfect politician achieves political objectives instantly and flawlessly because will and policy are indistinguishable. The perfect body politic is like the human body. The brain (leader) does not have to publish a decree or fight a battle to wiggle fingers (the polity). Recall from Chapter 1, a war-free society must either be totalitarian or utopian (extremes of the control-autonomy dialectic), where neither is practically attainable. The same problem viewed from the angle of political perfection – again, a practical impossibility – also eliminates war. The people in such

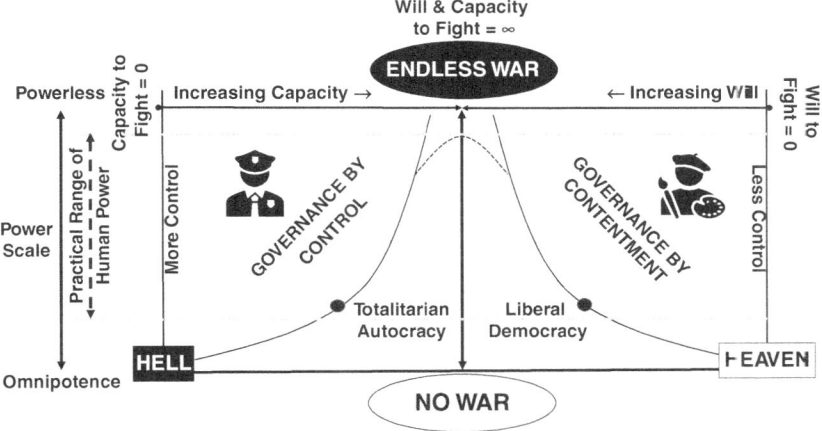

Figure 4.2 War and governance

a world would live either in the equivalent of *heaven* (utopian) – content within the benevolent care of an omnipotent sovereign; or *hell* (totalitarian) – despondent under the capricious thrall of an almighty despot.

Figure 4.2 illustrates the relationship between forms of governance, power, and the people's will and capacity for war. The figure applies to any level of government (i.e., internal-domestic or external-international). Moving from the center to the right, the people's will to fight wanes and capacity to fight becomes irrelevant as governance provides greater contentment. Moving from the center to the left, the people's capacity to fight wanes and the will to fight becomes irrelevant as the government exerts greater control. Power must increase from top to bottom to provide greater contentment or control until, at the bottom and far left ("hell") and right ("heaven"), power becomes infinite and war impossible. The top-middle of the chart represents a hypothetical singularity where governmental power (to control or appease) is zero and will and capacity to fight are infinite; thus here, chaos reigns and war is incessant. The dotted connection bridges the two halves and represents a zone of ineffective governance (e.g., failed states). Since zero power and infinite power are impossible, the range of realistic governmental forms is restricted to the "practical range of human power" depicted by the horizontal dot-dashed lines. Totalitarian autocracy and liberal democracy (includes republics) represent the two prevalent variations of real governance that humanity naturally gravitates toward to reduce the potential for conflict, though all real governments employ varying levels of control *and* contentment. Transitions between forms may be gradual or rapid, and the potential for war depends on the government's power and how responsibly it is wielded. Finally, governments powerful enough to satisfy can also control and may not hesitate to switch to ensure political survival.

As we concluded in Chapter 1 and Figure 4.2 confirms, the "heaven" and "hell" scenarios are impossible in the real world, but so is the infinite war scenario. Humans can neither be fully omnipotent nor totally powerless; therefore, politics requires policy, and war remains a possibility. In fact, violence and its threat are two of the oldest and most ubiquitous forms of power, because they offer a stark choice: comply and live, or resist and suffer or die. "To be proficient in [war]," writes Machiavelli, "is the surest way to acquire power," and the capacity to direct violence on command is power that leads to influence.[57] The greater the power, the greater the potential influence; however, as mentioned previously, no power, unless absolute, can guarantee a satisfactory effect or outcome. The Soviet Union was much more powerful than its Afghan antagonists, yet it could not satisfy its political aims by force. The same was true for England against the American colonies and France versus the Viet Minh.

In the end, war is a tool of politics, and the capacity to wage war is a source of power and influence. Clausewitz and others have proposed that war ought to be subservient to politics, but the line between war and politics is not always distinct, resulting in scenarios where war becomes its own end. Indeed, according to foreign relations expert Edward Mead Earle, for some political systems, the accumulation of power "becomes an end in itself, and all considerations of national economy and individual welfare are subordinated" to the goals of preparing for or waging war.[58] Clausewitz adds, "The only source of war is politics ... but it is apt to be assumed that war suspends that intercourse and replaces it by a wholly different condition, ruled by no law but its own."[59] When this happens, politics and policy serve only to prolong the fighting, the idea of a peaceful future fades, and the warrior and politician become one. This perilous condition can lead to unrestricted escalation, atrocities, and genocide. Yet even this is an impermanent state, for at some point the belligerents will consider an alternative. "Military ferocity is not a fixed quality of any race or culture," writes Lawrence Keeley, "but a temporary condition that usually bears the seeds of its own destruction."[60] Given enough time, the peace-war harmony of tensions will return, though this is small consolation for the dead.

Instruments of Power

States are principally organizations for the accumulation of power in the pursuit of security, and their most significant distinguishing characteristic is the possession of military forces.[61]
                                                                Gwynne Dyer

Power can be hard or soft, physical or nonphysical. Physical power, like military strength, can directly influence the physical world and indirectly influence the psychological. A strong military can win a war, deter one, attract allies, or create favorable perceptions through nonviolent operations like humanitarian relief. Nonphysical power, like wealth or wisdom, can also influence the physical and psychological domains. Leaders use economic power, for example, to create or acquire physical assets, sponsor research and development, or influence others via trade pacts or direct payment.

Instruments of power (IOP) are the means for exerting influence. Writes Friedrich Engels, "Force is no mere act of the will but requires very real preliminary conditions before it can come into operation, that is to say, *instruments*."[62] Though often attributed to states, IOPs apply in different forms to all levels of human interaction. Language and communication skills, intelligence, size, strength, dexterity, and appearance contribute to a person's power and influence. In war, individual combatants use IOPs (e.g., weapons or communication) to influence allies and enemies. Martial IOPs range from personal weapons to an entire nation's armed forces. At the state level, we bin IOPs into four categories: *diplomatic, information, military,* and *economic* (DIME). PMESII-PT (political, military, economic, social, infrastructural, informational, physical, and time) is another common model, and still others add "legal" and "law enforcement." For our purposes, DIME is sufficiently inclusive.

The diplomatic instrument influences others through nonviolent engagement like negotiation, collaboration, and compromise. *Diplomacy* is always on call even in wartime and can preserve or set the conditions for peace and build or reinforce alliances. "Diplomacy," writes von Moltke (the Elder), "avails itself to war to attain its ends, crucially influencing the beginning of war and its end."[63] *Economic* power also influences war and peace. Thucydides writes, "War is a matter not so much of arms as of money";[64] and Corbett adds, "Wars are not decided exclusively by ... force. Finance is scarcely less important."[65] During times of peace, economics influences domestic and international politics, quality of life, and military potential. Economic power funds war readiness, warfighting, and the industry, commerce, and agriculture supporting the war. Harsh economic sanctions can also trigger wars. The *information* IOP is the least tangible. Its effects are hard to predict due to the volume of competing data and the erratic nature of human perception and choices. Information includes public, private, and social media; propaganda; and strategic communications. Private, independent sources, like news agencies, bloggers, and tweets can easily overwhelm "official" communications efforts. Thus, outside hermetic autocracies, government monopolization of the "infosphere" is virtually impossible. Finally, the *military* IOP provides physical power, which may include combative violence, to pursue political outcomes. Like the other

IOPs, the military is normally subordinate to policy. "Policy," writes Clausewitz, "is the guiding intelligence and war only the instrument, not vice versa. No other possibility exists, then, than to subordinate the military point of view to the political . . . it will be perfectly clear . . . that the supreme standpoint for the conduct of war . . . can only be that of policy."[66] This IOP includes all warfighting elements (e.g., armies, navies, and air forces) and multiuse capabilities like logistics, surveillance, engineering, medicine, and law enforcement. Political leaders use the military across a *range of military operations*, from nonwar needs, like peacekeeping and disaster response, to major combat.

Few political leaders explicitly consult the DIME model prior to making decisions; rather, they use IOPs intuitively, mixing and matching means to achieve desired outcomes. IOPs work best in concert. A coherent vision for security and prosperity (e.g., a "national security strategy") facilitates IOP coordination. Ideally, each IOP reinforces the others; however, IOPs can inadvertently clash. Consider World War I's Zimmermann Telegram. German Foreign Minister, Arthur Zimmermann, dispatched this missive to Mexican officials in January of 1917 with an offer of US territory after the war provided Mexico allied with Germany. The USA learned of the deal through British intelligence, and it became a factor in America's decision to enter the war. Germany's bungled use of the diplomatic IOP undermined its political objectives and its other IOPs. Indeed, though the military IOP is central to war, nonmilitary IOPs often play pivotal roles as well.

Political-Military Relations

One of the likeliest places for peace to founder is in the gap between civilian and military assumptions about the predictability of the outcome of war.[67]

Gwynne Dyer

Since war and the IOPs are tools of politics and policy, effective warfighting requires political-military synergy. We need only look to Hitler's Germany and the USA during the Vietnam War for the ramifications of dysfunction here. "In the ideal model of civil-military relations," writes Hew Strachan, "the . . . head of state sets out his or her policy, and armed forces coordinate the means to enable its achievement."[68] This entails a natural give-and-take between politicians and generals as they balance ways and means (and their roles and authorities vis-à-vis the control-autonomy dialectic) to accomplish policy's aims.[69] "The art of wartime leadership is not to destroy the natural tension between political and military requirements," write Codevilla and Seabury, "but to synthesize it."[70] Furthermore, adds Gray, "Policy should direct warfare, but policy usually cannot know what it should ask for until it understands what is [militarily] practicable."[71] If this relationship falters, write Codevilla and Seabury, it "may overwhelm the political leadership, which then . . . stares impotently at the storms of war," or "saddle the generals with political-strategic objectives that they . . . come to view as impossible."[72] Lincoln, Roosevelt, and Churchill exemplified effective political-military balance, and they emerged victorious. Still, there is no formula for ensuring successful political-military relations. They vary, writes historian Gordon Craig, upon factors like "the nature of the political system, the efficiency and prestige of the military establishment, and the character and personality of the political leader."[73] Ultimately, the timbre of the political-military relationship will affect how political and military objectives, like war, are to be realized; and this is the realm of *strategy*.

STRATEGY

> *Without a strategy, facing up to any problem or striving for any objective would be considered negligent.*[74]   Lawrence Freedman

Strategy, like war, is simpler to discuss than to practice effectively, though we are all strategists. As Freedman writes, "There is now no human activity so lowly, banal, or intimate that it can reasonably be deprived of a strategy."[75] Indeed, "The word 'strategy,'" writes Aron, "is no longer applied only to military action and leadership: now, anything may involve strategy."[76] Strategy connects the past and future through coordinated actions in the present. In other words, it takes resources we have and can acquire (things, people, or ideas) and orchestrates them to satisfy an objective. Strategists call the resources "means," the orchestration (how the means are used) "ways," and the objectives (what ways and means achieve) "ends." None of this happens instantly or perfectly; thus, strategists must assess and adjust ways, means, and ends over time. Therefore, we can define strategy as *the art and science of identifying, procuring, implementing, assessing, and adjusting ways and means to accomplish ends*, which can be physical or psychological. IOPs are the means for executing strategies. Strategy is an *art* and a *science* because it incorporates objective (e.g., logic and physics) *and* subjective (e.g., imagination and psychology) attributes. Writes Freedman,

Strategy could be considered a matter of science, in the sense of being systematic, empirically based, and logically developed, covering all those things that could be planned in advance and were subject to calculation. As art, strategy covered actions taken by bold generals who could achieve extraordinary results in unpromising situations.[77]

All good strategies assess *risk*, which is the likelihood of a negative influence or result. Strategists must weigh the risks inherent in strategic execution *and* the risk of overall failure. For example, in the 1990 Gulf War, General Schwarzkopf believed a Marine amphibious assault was too risky; therefore, he used the Marines only as a ruse. But Schwarzkopf *accepted* the risk of executing a vast flanking maneuver to bypass Iraqi defenses. This strategy was wildly successful; however, the coalition's political leaders considered the total destruction of Saddam's army and an advance on Baghdad too risky and ordered a ceasefire after only 100 hours of ground combat. All wars entail risk as unforeseen occurrences, mistakes, and enemy actions elevate the likelihood of negative outcomes. Still, "If we risk nothing," advises Corbett, "we shall seldom perform anything."[78]

Strategy must also assess *cost*, which includes the war's monetary expense as well as the anticipated human (physical *and* psychological) and matériel losses. We are notoriously inept at predicting the costs of war, and no strategy can be deemed successful, regardless of the military result, if the costs outweigh the political benefits. Furthermore, the level of acceptable risk and cost are generally proportional to the stakes; however, this is not a law. Fallible leaders may accept tremendous risk and cost for little gain. Historical examples from World War II include Britain's disastrous assault on Dieppe and Germany's invasion of Soviet Russia.

Types of Strategy

Strategy is the bridge between military power and policy.[79]   Colin Gray

At the highest levels of governance, "the distinction between politics and strategy diminishes," says Winston Churchill. "At the summit true politics and strategy are one."[80] *Political strategy* serves political purposes (i.e., it produces results impacting governance and power apportionment) that are mainly focused internally and directed toward the sustainment of a leader or party. Political strategists in democracies parlay resources, like money and supporters, to influence the electorate, while political strategists in autocracies use IOPs to control the population and eliminate threats. *Grand strategy* is an externally directed form of political strategy that pursues foreign policy objectives related to security, prosperity, and stature relative to other states. Grand strategy is developed by pinnacle leaders like presidents or kings.

Military strategy is the subset of grand strategy centered on the military IOP. As Clausewitz says, "The war itself . . . is governed by political aims and conditions that themselves belong to a larger whole."[81] According to Liddell Hart, military strategy addresses the goal of surmounting an enemy's "adverse will," by force if required, but still must take "its proper place along with the other instruments of grand strategy – which include the more oblique kinds of military action as well as economic pressure, propaganda, and diplomacy."[82] Implicit, as Strachan points out, is a "dialogue between politicians and soldiers" with the aim of harmonizing policy and strategy.[83] Since the military IOP can serve nonwar aims, we can further divide military strategy into war and nonwar varieties. In war, military strategy supports political and grand strategies by unbalancing and then attempting to rebalance the peace-war dialectic, and as we discussed in Chapter 1, this is a daunting task.

Finally, as the preceding discussion implies, strategy is tiered, from the political, to grand, to military, and so on. The "level" of strategy depends on the breadth of its ways and means and the specificity of its ends. In war, the greater the scope and the closer the end state to the war's broadest aims, the higher the level of strategy. The highest level of strategy, ideally supported by all subordinate levels, pursues the overall political objective and draws upon the widest array of ways and means for the achievement of this aim.

The Nature and Character of War

Neither war nor anything else can change in its essentials. If it appears to do so, it is because we are still mistaking accidents for essentials.[84]

Julian Corbett

For war strategy to effectively serve its purpose, leaders and strategists must appreciate the distinction between war's nature and character, the two concepts responsible for war's paradoxical admixture of *continuity* and *change*, respectively. In short, the *nature of war* is everything about war that is unchanging, and its *character* is everything that remains, what Jomini calls war's "poetry and metaphysics."[85] Our analysis of war's four causes, human nature, history, and theory lead us to conclude that war's nature *reflects human nature and interaction within the various contexts of violent political conflict*. Therefore, unless human nature changes, the nature of war must remain constant. Furthermore, war's character, though variable, is still war and cannot escape its lineage. As Clausewitz stresses, the "particular characteristics" of war, "must always be governed by the general conclusions to be drawn from the nature of war itself."[86] In fact, we can visualize war's character as *an image of its nature projected through the continuously changing lens of time and a dynamic universe*. Thus, the character of war can evolve gradually with technology, doctrine, and culture; or suddenly with changing tactics, terrain, weather, and weapons. US military doctrine describes the character of war as its appearance "at any point in time and space; its face, shape, practice, dynamics, intensity, [and] scope," which is composed of three elements: *who* fights, *why* they fight, and *how* they fight.[87] The relationship between human nature and war's nature and character is depicted in Figure 4.3.

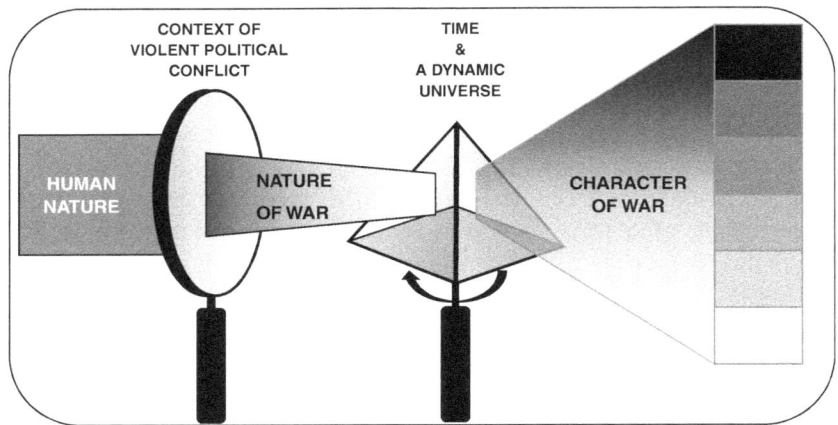

Figure 4.3 The nature and character of war

Nature, Character, and Theory

A distinction between the nature of war and the character of particular wars will prevent the impact of short-term issues from swamping a sense of perspective on long-term continuities.[88]   Hew Strachan

Sound war theory always centers on war's nature, while doctrine and tactics pertain more to its character. Due to the astounding variability of war's character, observing *a* war reveals only a portion of a greater expanse of possibilities. As Clausewitz stresses, we study history not to "assign, in passing, a handful of principles of warfare to each period" but "to show how every age had its own kind of war"; and thus, the futility of attempting to devise a myriad of "theories" to correspond to each war's "peculiar preconceptions."[89] Therefore, we cannot grasp war's nature from a single instance, but only by studying many wars. In this way, says Clausewitz, theorists can find a "more general – indeed, a universal – element" (i.e., war's nature) valid for all "forms" of war.[90] Ultimately, war's nature is like a stage and its character like the play. The lines, scenes, and costumes may change and the actors come and go, but the stage remains, the immovable foundation for the entire drama.

The Four Essential Acts of Strategy

The most important single task for strategy is to understand the nature of the war it is addressing. Its next task may be to manage and direct that war.[91]

                                        Hew Strachan

The first, most essential act of strategy for the political leader contemplating war is to *define the political aim*, the desired ends or peace the war is intended to achieve. No other strategic process should proceed until this happens. Thus, strategy translates policy, which Strachan says, "determines which wars are fought, where they are fought and why they are fought," into action.[92] Addressing the peace-war and past-future dialectics, Saint Augustine writes, "The desired end of war is peace, for everyone seeks peace, even by waging war."[93] For the aggressor, this means a peace characterized by a new, more favorable set of circumstances. For the defender, war's object may simply be survival or restoration of a prior peace, and the political leader must decide whether to fight, seek terms, or surrender. Sound strategy requires that all subsequently defined, subordinate objectives support the overall effort's principal end.

The second essential act of strategy is to *assess whether the IOPs and ways of war*, with consideration of cost, risk of failure or unintended consequences, friend and enemy disposition, and the environment, *yield an acceptable probability of producing the desired ends within a satisfactory period of time*. Strategy, writes Clausewitz, must determine "at the outset of a war its character and scope" based on "the political probabilities" and "not . . . take the first step without considering the last."[94] Moreover, "At the outset, then, we must admit that an imminent war, its possible aims, and the resources it will require, are matters that can only be assessed when every circumstance has been examined in the context of the whole."[95] This is where Clausewitz's "supreme . . . act of judgment" – identifying the "kind of war" – occurs, for it cannot be identified earlier. This step requires extensive study, consultation, and analysis with awareness that changing conditions can invalidate assumptions and conclusions. Creating *effective* strategy hinges on quality intelligence and the strategist's comprehension of war's nature and character. It is also upon this act of strategy that decisions of war and peace hinge. If the cost and risk are too high, war ought to be deferred, or in other words, "When the advantages . . . derived from peace or neutrality and war are equal," advises Chanakya, "a ruler should always prefer peace, for war brings with it both sin and risk."[96]

The third essential act of strategy is to *identify, procure, and implement ways and means to achieve the war's ends*. In other words, once we decide to fight, plans and preparation must transition into specific combat and noncombat actions to defeat the enemy. Furthermore, military strategists should ensure that military strategy and IOPs complement grand strategy and any nonmilitary IOPs supporting the war effort. Here, aggressors gain an advantage having already completed the first three acts of strategy; thus, to avoid a *fait accompli*, defenders must accelerate their strategic development and weigh the pros and cons of resistance versus capitulation.

"All warfare," writes Colin Gray, "is a race between belligerents to correct the consequences of the mistaken beliefs with which they entered combat."[97] Therefore, the fourth essential act of strategy is to *regularly assess and adjust (as needed) the first three acts to account for changing circumstances*. Because the future is unknown, strategy must be dynamic and iterative. As Clausewitz writes, "Most of these matters have to be based on assumptions that may not prove correct, [thus] modifications ... are continuously required."[98] Conditions may necessitate alterations to ways, means, and even ends; however, changes to overall war aims should be made prudently due to their cascading influence.

Tests of Strategy

Good strategy should meet five basic tests:
(1) Suitability: the ways and means are likely to promote the ends
(2) Desirability: the means are commensurate to the ends (reward > risk + cost)
(3) Feasibility: the means are available or acquirable and the ways are workable
(4) Acceptability: the ways and means are morally and legally tolerable
(5) Sustainability: the ways and means will last as long as needed

These tests answer five important questions: Will it work? Can we afford it? Can we resource it? Can we live with it? And will it last? Bottom line: strategy's ways and means must have a realistic and acceptable probability of achieving ends that align with interests and values while accounting for external (allies, enemies, and neutral parties) and internal (domestic populace and institutions) inputs and effects.

Regulating War

There is no doubt a right and wrong in war, but we have not discovered it.[99]
H. Fielding Hall

Wars start when words lose their potency in mediating human relations and laws are supplanted by acts of destructive violence. As arbitration by force, war becomes its own law, suspending issues of right and wrong (regarding the war's causes and how it is fought) until the war ends. As Cicero opines, *inter arma, silent leges* (in war, the laws fall silent).[100] And Kant adds that in war, "Neither party can be condemned as wrong, because that would presuppose a juridical decision. In the absence of such a decision, the question of which side is right is answered by the outcome of the conflict."[101] Once war begins, rules are subject to continuous evaluation against life's highest imperative, survival. If a rule stands in the way of an opportunity to take another's life instead of giving one's own, that rule is usually ignored. Similarly, rules that threaten *political survival* are also likely to be disregarded. This is how our species is "wired." Indeed, according to sociologist Maurice Davie, the majority of observable primitive cultures have recognized "no laws of warfare ... women and children are slain, male prisoners are taken only to be tortured and killed, poisoned weapons are used, treachery is general, and no quarter is given."[102] Civilized humanity, however, has attempted to regulate war by defining "just causes" and circumscribing warfare to prevent excessive suffering. Yet, for the reasons specified, this frequently fails. "The best ordinances in the world," writes Machiavelli, "will be despised and trampled under foot when they are not supported ... by a military power."[103] And Clausewitz adds, "Attached to force are ... self-imposed, imperceptible limitations hardly worth mentioning known as international law and custom, but they scarcely weaken it."[104] In fact, laments Kant, given "the depravity of human nature ... in the unrestrained relations of nations ... it's astonishing that the word 'law' [*recht*] hasn't yet been entirely banished from the politics of war as an academic irrelevance."[105] Specific regulations include Geneva Convention rights, "just war" doctrines, and the law of armed conflict. The Hague Conventions of 1899 presumed to legally restrict wars, but World War I "made a mockery of the Hague movement."[106] After the war, the League of Nations charter encouraged peaceful redress of international disputes, and the 1928 Pact of Paris (also called the "Kellogg-Briand Pact" or "General Treaty for the Renunciation of War") outlawed war between signatories. World War II erupted just a decade later.

In the end, the absence of reliable *enforcement* mechanisms invites warring sides to ignore rules, which typically lose effectiveness in proportion to the war's duration and direness. "All treaties between great states," says Otto von Bismarck, "cease to be binding when they come in conflict with the struggle for existence."[107] With the state's safety at risk, Machiavelli recommends the prince pay no heed to "justice or injustice, to kindness or cruelty," and adopt "that alternative ... which will save the life and preserve the freedom of one's country."[108] War's life-or-death stakes and lack of effective enforcement authorities mean that we can never be *certain* that rules will be observed. Therefore, strategists must recognize that limitations on war are always subject to change or rejection by the opposition. We will analyze this relationship between ethics and war in greater depth at the end of the chapter.

Levels of War

To simplify the complex task of planning, executing, and evaluating large-scale military operations, we define three "levels" of war: "strategic," "operational," and "tactical." Though strategy occurs at all levels, the terms "strategy" and "strategic" apply to the highest levels (i.e., grand strategy and military strategy).[109] Senior military leaders formulate military strategy just as senior politicians make grand strategy. Subordinate leaders plan and execute activities at lower levels to achieve intermediate objectives that contribute to the overall war aim but seldom gain it by themselves. They also lead their forces, monitor conditions, adjust strategy, and advise superiors. Operations and "operational art" pertain to military activities between the strategic and tactical levels of war (i.e., *campaigning*: the planning and execution of multiple engagements in one or more theaters of operation). *Operational art*, what Jomini termed "Grand Tactics," is an intermediate level of strategy that orchestrates ways and means to achieve military objectives via synchronized engagements. As Clausewitz says, operational "strategy" "decides the time when, the place where, and the forces with which the engagement is to be fought."[110] Tactics encompass smaller-scale battles and engagements and are effectively the "mini-strategies" employed to win them. Though they support higher-level goals, tactical aims are narrower in scope (e.g., "take the hill" or "sink the ship"). Tactical leaders are often the least experienced but also the most innovative and dynamic.

Strategy, operations, and tactics are interrelated. The strategic level has the broadest effects, both positive and negative. Hitler's strategic choice to invade Soviet Russia, for instance, doomed Germany, nullifying many tactical successes. However, lower-level victories or failures have strategic implications as well, sometimes counterintuitively. For example, Japan's tactical triumph at Pearl Harbor turned out to be a strategic blunder, whereas North Vietnam's 1968 Tet Offensive was an operational fiasco that turned into a surprising strategic success.

Fractal War

Nature is regulated not only by a microscopic rule base but by powerful and general principles of organization.[111]   Robert Laughlin

Chaos theory's "fractal" geometry concept offers a useful way to visualize war's levels. A *fractal*, derived from the study of chaotic systems, is a pattern within a complex phenomenon that reoccurs because the degree of irregularity remains constant regardless of scale.[112] Chaos theory pioneer, Edward Lorenz, discovered that chaotic systems contain "suggestions of structure amid seemingly random behavior."[113] A similar effect occurs in war, as properties and patterns recur from the smallest element to the largest and from the highest levels to the lowest. Figure 4.4, a fractal construct known as a Sierpinski Triangle, illustrates how the levels of war and strategy "nest."

Interestingly, Clausewitz's war-as-a-duel metaphor captures war's fractal nature: "War is nothing but a duel on a larger scale. Countless duels go to

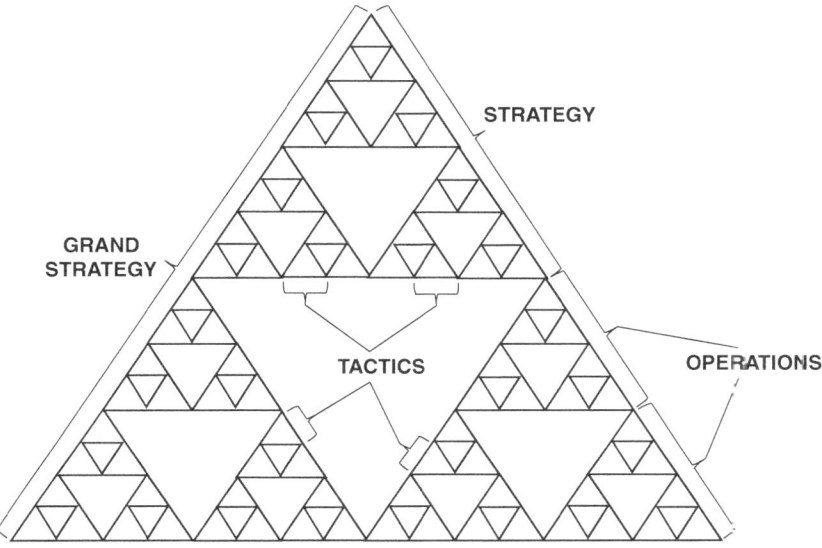

Figure 4.4 Fractal levels of war

make up war, but a picture of it as a whole can be formed by imagining a pair of wrestlers."[114] And, "If fighting consisted of a single act, no further subdivision would be needed. However, it consists of a greater or lesser number of single *acts, each complete in itself,* which ... are called 'engagements' and which form new entities."[115] With these statements, Clausewitz reveals the linkages between tactics and strategy that emerged after armies, which had previously fought as indivisible monoliths, evolved into "a many jointed entity which was pliant and flexible," with units "easily detached and reattached without disturbing the order of battle."[116] Irregular cavalry and insurgents best exemplified this in Clausewitz's time. He hints at their fractal qualities saying, "Above all, the most characteristic feature of insurgency in general will be *constantly repeated in miniature*" (emphasis added).[117] Visualizing war's levels "fractally" helps strategists and warriors create more effective, integrated multi-level campaign plans.

The Cycle of War

The "Cycle of War" (Figure 4.5) illustrates the iterative relationship between strategy, planning, and operations. The cycle's elements – strategy, planning, execution, and assessment – occur in sequence for any specific action, but may run in parallel for simultaneous or phased actions within a campaign. Military

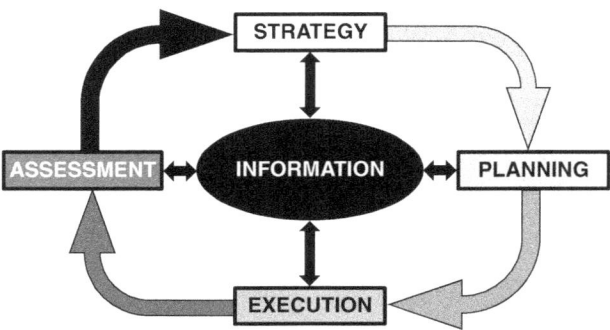

Figure 4.5 The cycle of war

strategy outlines ends, ways, and means that support higher-level strategy and policy. Plans draw from and support strategy by developing and arranging the specific details required to accomplish strategic goals. Plans answer the *who*, *what*, *where*, *when*, *how*, and *then what* questions. Once plans are final, execution (i.e., actions that drive outcomes) commences. The warring sides assess the results and adjust strategy and planning as needed. The cycle repeats until the war is over.

Information – including objectives, commander's intent, and intelligence – feeds each stage of the cycle. The side that collects, processes, exploits, and disseminates the most pertinent and accurate information and more rapidly incorporates it into the cycle has the best chance of gaining an advantage. Finally, since information feeds each part of the cycle, bad information is worse than no information, because it corrupts the entire process.

The Open System of War

The strategist knows that only with a grasp of the contexts of war are we able to explain and understand most of what is happening.[118]   Colin Gray

The "open system of war" (Figure 4.6) consists of the two warring sides and all the variables and factors, internal and external, that influence the war's conduct. The system is "open" because wars are subject to external influence and can affect their surroundings as well. The two warring sides can control some variables and inputs within this system but not all as they strive to influence the other with IOPs. According to the *principle of reciprocity*, factors within the system of war apply to *both* sides, though not necessarily equally, and the enemy "always gets a vote"; thus, strategy cannot be one-sided. Each side employs IOPs to create influence that ideally drives the system to a desired end state. *Forces* are the human and material assets (e.g., weapons, equipment, and

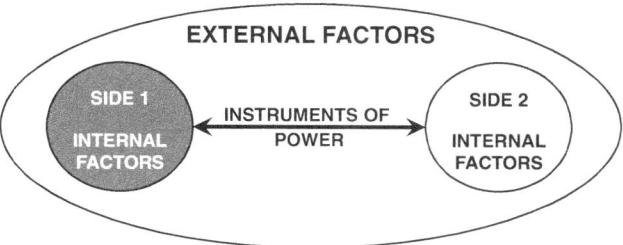

Figure 4.6 The open system of war

people) used to do this. External inputs come from any source not engaged in the war (e.g., neutral parties, foreign media, and the environment). Internal factors are unique to each side and include public and private institutions, media, personalities, and domestic politics. Each side employs IOPs *intended* to influence the other; however, using IOPs invariably creates a "radiation" of influence beyond the intended target. Thus, *radiation* is influence resulting from the imprecision of using IOPs (e.g., unintended consequences). For example, though a well-aimed rifle shot has minimal radiation, the bullet could pass through the target and hit something else (collateral damage), or the target's demise may spark a response (second-order effect). In general, the probability and extent of radiation are proportional to the level of war and the power or complexity of the IOP.

The Tyranny of Proximity

All strategy is geostrategy: geography is fundamental.[119]   Colin Gray

The natural enemy is the one whose territory is the closest.[120]   Chanakya

The system of war's "boundaries" are more figurative than physical. Still, the potential for belligerent interaction increases with physical proximity (e.g., Athens-Sparta, France-Germany, and Iran-Iraq). But in reality, conflict may be caused less by physical proximity than the *political proximity* of contrasting ideologies, goals, and interests. As technology has extended spheres of influence, political proximity has grown, and wars have expanded in scope. During the centuries of colonization following Columbus's voyage, political proximity became globalized, and European countries fought native peoples and each other in distant locales. Furthermore, the arbitrary rearrangement of borders after the world wars (e.g., Iraq and Israel) created conflict-triggering physical *and* political proximity. By the twenty-first century, tech-inspired globalization has further eroded the insulating effect of physical separation, increasing the potential for war between distant antagonists (e.g., the USA and Iraq).

Momentum

A man who is accustomed to act in one particular way, never changes. It thus comes about that a man's fortune changes, for she changes his circumstances but he does not change his ways.[121]   Machiavelli

Momentum in war is *the tendency for trends in actions and results to continue or to become more pronounced over time.* As you might expect, culmination and momentum are related – the former usually precedes a reversal of the latter. For example, momentum in the Pacific theater of World War II shifted from Japan to the USA with Japan's defeat and subsequent culmination at the Battle of Midway in 1942. Momentum happens at all levels of war, and its power increases with the level (i.e., strategic momentum is harder to stop than tactical momentum). War is always costly, and since strategy requires resources, leaders are often reluctant to change course, even when presented with contradictory information (e.g., Napoleon and Hitler in Russia). Changing strategy is tantamount to admitting that past assumptions were wrong and investments wasted. Thus, the longer a strategy remains in effect and the greater the resource investment and political stakes, the harder it is to stop or redirect. Momentum can be good or bad depending on the circumstances. Operation *Overlord*, for instance, granted the Allies strategic momentum in World War II that persisted until Germany's surrender. On the other hand, the strategic and operational momentum created by MacArthur's Inchon landing in Korea led him to overreach, provoking China's momentum-arresting entry into that war.

The Environment of War and Warfighting Domains

It is a powerful general strategic truth that each geographically focused kind of military power recognizes that it needs the others.[122]   Colin Gray

The "environment of war," its physical context, is a subset of the open system and can be subdivided into physical and nonphysical *warfighting domains.* War's character can change dramatically with the environment. Therefore, defining warfighting domains – physical (land, sea, air, and space) and nonphysical (cyberspace and cognitive) – enables strategy, planning, and tactics to account for their unique attributes. Most formal militaries have service branches that specialize in domain-centric warfare (e.g., armies and navies). Because most of what we value is on land and the majority of naval, air, and space forces originate and return there, the land is *primus inter pares* in war. As Billy Mitchell opines, "Of course, everything begins and ends on the ground.

A person cannot permanently live out on the sea nor can a person live up in the air, so that any decision in war is based on what takes place ultimately on the ground."[123]

Pursuing Warfighting Perfection

We have multiple warfighting domains as a result of humanity's unending quest to become more effective warfighters. Earlier we proposed that perfect politics generates *political effects* instantly; thus, we may presume that warfighting perfection produces *military effects* similarly. In the real world, the closest approximation occurs when every *feasible* military action gains its objective with maximum speed and efficiency. In fact, the purpose of all warfighting advances, including technological innovation and *expansion into multiple domains*, is to reduce the gap between real and ideal (i.e., to maximize potency, speed, and efficiency). The quest for perfection has spurred the development of faster, more powerful, long-range weapons. Rocks and clubs led to arrows, bullets, missiles, warships, aircraft, submarines, tanks, rockets, lasers, and even cyber "weapons." These achievements seem revolutionary but are, in fact, *evolutionary*. Airplanes and missiles are essentially powerful, swift, versatile, accurate "rocks" (i.e., human-launched objects intended to affect distant targets). This process will continue to drive weapons innovation in all domains; however, a domain's *nature* (like war's) remains constant. Domain theory that remains "weapons agnostic" and focuses on its *raison d'etre*, the *domain*, retains its relevance. Moreover, as we learned in the last chapter, the three a's – *awareness*, *access*, and freedom of *action* – are essential for unlocking domain-unique, war-affecting advantages.

Nonphysical Domains

Machines don't fight wars. Terrain doesn't fight wars. Humans fight wars. You must get into the minds of humans. That's where the battles are won.[124]

Colonel John Boyd

The quest for warfighting perfection now extends into cyberspace, where a greater proportion of modern fighting capability (and the infrastructure of daily life) is located. Cyberspace is unique in that humans can neither physically exist nor fight within it; however, it has become a strategic thoroughfare for virtual interaction, including commerce, communications, and military C2. Furthermore, since the "brains" that direct advanced machines reside in cyberspace, its exploitation can create powerful psychological and physical effects.

The final and most critical of all warfighting domains is the *cognitive* domain. "An atomic bomb, a cavalry incursion into the enemy's rear, a lone saboteur, an agent whispering in the right ear, or a frightening set of battle cries is useful in warfare," write Codevilla and Seabury, "only insofar as it forms or breaks ideas in the enemy's mind."[125] War is decided in the human mind where perceptions, emotion, and intellect influence strategy, leadership, and the will to fight. Warring sides use the "tangible" warfighting domains to manipulate "hearts and minds" (the reason-emotion dialectic) to promote political and military ends. But this is more art than science, which contributes to war's unpredictability.

Physical Context

Know the ground, know the weather; your victory will then be total.[126]

Sun Tzu

War's physical context, including weather, tides, vegetation and wildlife, climate, and terrain, is a major factor in the character of war. Consider the following examples: a rainstorm in 9 CE in the forests of Germany aided Germania's (under Arminius) annihilation of Quinctilius Varus's 20,000 legionnaires; storms at sea smashed the Spanish Armada and prevented Kublai Khan's Mongols from invading Japan; plagues rocked the Peloponnesian War, the Hundred Years' War, and killed far more Aztecs and Incas than the *conquistadors*; and Russia's winters thwarted invasions by Sweden's Charles XII, Napoleon's *Grande Armée*, and Hitler's *Wehrmacht*. Strategists and tacticians must assess physical factors in all domains and at each level of war.

Uncertainty, Chance, Chaos, and Probability

Nowhere are our calculations more frequently upset than in war.[127]   Livy

Chaos always is either actively present in strategic history, or, at the least, ready in the wings.[128]

Colin Gray

Uncertainty, chance, chaos, and probability have starring roles in any war. "Of all the activities in which mankind engages," affirms Williamson Murray, "it is the conduct of war that envelopes it with the greatest degree of uncertainty, ambiguity, and friction."[129] Clausewitz was so concerned with these factors that he included two – chance and probability – in his trinity. *Uncertainty*, a condition that permeates war (and one extreme of the certainty-uncertainty dialectic), is the inability to know without doubt, a condition that permeates war. Uncertainty takes several forms: what we think we know, but do not (errors); what we know we do not know (known unknowns); and what we are unaware that we do not know (unknown unknowns). In general, certainty increases for events we can directly control or influence, are near in time and space, or are governed by inviolable laws. Many factors, like fog and friction, ambiguity, and *fortuna*, cause uncertainty. Furthermore, uncertainty usually increases with the level of war. "In a tactical situation," writes Clausewitz, "one is able to see at least half the problem with the naked eye, whereas in strategy everything has to be guessed at and presumed. Conviction is therefore weaker."[130]

Chance relates to unexpected and unanticipated incidents, which contribute to uncertainty. Clausewitz's "friction" occurs when chance happenings interrupt plans and operations. As Liddell Hart says, *luck* (another word for chance) "can never be divorced from war, since war is part of life."[131] Favorable chance is *good* luck, and inconvenient or harmful chance is *bad* luck. Clausewitz writes, "From the very start there is an interplay of possibilities, probabilities, good luck and bad that weaves its way throughout the length and breadth of the tapestry."[132] Though we cannot eliminate chance, we can decrease the potential for bad luck by improving awareness, knowledge, and control. Psychologist Richard Wiseman recommends reliance on intuition (e.g., *coup d'oeil*), "creating good luck" via positive expectation, and cultivating a resilient attitude.[133] In war, intelligence, training, provisioning, contingency plans, and technical aids are also beneficial. For example, supercomputing, Monte Carlo methodology, and game theory can improve predictive modeling of complex systems.[134]

Chaos, one extreme of the order-chaos dialectic, is a *lack of order* and implies unpredictability and uncertainty. Complexity (one extreme of the simplicity-complexity dialectic) and nonlinearity are related to chaos. Chaotic, nonlinear systems, like the surface of a choppy sea or a raindrop's path from cloud-to-ground, defy deterministic modeling. War is a complex system of complex systems with the added dimension of human psychology. Indeed, Clausewitz states that "personality" and "personal relations" raise the number of ways of "achieving the goal of policy to infinity."[135] Fortunately, war is not *purely* chaotic (a condition of total disorder that would obviate strategy). Strategy, planning, and training-instilled discipline impose greater balance on war's order-chaos, certainty-uncertainty, and simplicity-complexity dialectics.

Probability is the likelihood of any specific event happening. Clausewitz affirms that war's "own peculiar laws" drive it from the certainty of an ideal model and subject it to the "*laws of probability*."[136] He writes, "In war, all action is aimed at probable rather than at certain success. The degree of certainty that is lacking must in every case be left to fate, chance, or whatever you like to call it."[137] Though the gap between reality and omniscience creates unknowns, our knowledge and experiences help us assess *probabilities*, some with great precision (e.g., a coin toss always has a 50 percent probability of "heads"). We can also calculate probabilities regarding weapons effectiveness given quantifiable parameters. But probabilities are less precise when applied to war. For example, in 1944, Germany thought an Allied landing at Normandy less *probable* than a landing at Pas de Calais, in part due to an active deception campaign. Later, Field Marshall Montgomery's complex Operation *Market Garden* (an airborne assault to seize the Arnhem Bridge over the Rhine) had a low probability of succeeding, and indeed, friction and enemy action disrupted its precise timelines causing failure. As these examples illustrate, risk accompanies wartime decisions regardless of the assessed probability of success, and risk always increases when assessments are inaccurate or when the probability of attainment is low.

Per the principle of reciprocity, all these factors – uncertainty, chance, chaos, and probability – apply to *both sides* in war, though not necessarily equally. "The battlefield," writes Napoleon, "is a scene of constant chaos. The winner will be the one who controls that chaos, both his own and the enemies'."[138] Indeed, as David Chandler writes, "No one was ever more conscious of the influence of chance ... than Napoleon, [and] no one ever left so little to chance."[139] In war, writes Dyer, "He who is less badly informed and less disorganized wins."[140] However, since we cannot eradicate uncertainty, we must avoid the temptation to stand pat until everything is known (i.e., "paralysis by analysis"), a condition that skews the rest-movement and prudence-boldness dialectics. Successful commanders *balance* these extremes, understanding that sometimes "discretion is the better part of valor," and at others, as Clausewitz says, "The utmost daring is the height of wisdom."[141]

Implications for Planning

I remember [Maj Gen] Urquhart asking for questions and nobody raised any. Everybody sat ... looking bored. I wanted to say something about this impossible plan, but I just couldn't ... anyway who would have listened?[142]   Major General Sosabowski (on Operation *Market-Garden*)

Chaos, chance, and uncertainty can rapidly render plans moot. As von Moltke (the Elder) said, "No battle plan survives contact with the enemy."[143] However, this paradoxically makes planning even more important for, as Dwight Eisenhower commented nearly a century later, "Plans are nothing; planning is everything."[144] Although chance and enemy action often throw plans into disarray, good planning improves the probability of success, because the knowledge gained during the process combined with timely intelligence forms the basis for adaptability later on. Adaptability must also apply to the planning process itself. It is not uncommon for leaders or planners to become "wedded to the plan," meaning they refuse to start over regardless of circumstances or conflicting intelligence. In war planning, failure to reassess assumptions or to keep up with changing circumstances can be disastrous. Indeed, Japan in 1941 became so enamored with its plans to *start* a war, its leaders failed to consider how they might actually *win* it. They approved the Pearl Harbor attack despite the reservations of their own Navy Chief of Staff who admitted, "Our empire does not have the means to take the offensive, overcome the enemy, and make them give up their will to fight."[145] Other examples include World War I's Battle of the Somme (over a million casualties) and Sosabowski's Operation *Market-Garden* (1,200 Allies killed and 6,378 missing, wounded, or captured).[146] In each instance, leaders ignored intelligence and subordinate misgivings that should have prompted changes or cancellation.

The Engine of War: Will, Capacity, and Cause

It's easier to kill these Russians than to defeat them.[147]

Frederick the Great (1758)

One of the central concepts in the UWT is the "engine of war," which as the phrase implies, makes war "go." War can only occur if opposing sides have a decision-making capability (strategy and leadership) to specify aims and direct actions, and IOPs (including forces and support) to translate intent into combative violence. When combined, these components constitute warfighting "capacity." The *total combat capacity* is the sum of these elements for a given side and the *relative capacity* is the ratio of one side's capacity to the other's. The *will* to fight, initiated and supported by a *cause*, ignites capacity into action. Clausewitz says the essence of "war is fighting," a trial of *moral* and *physical* forces (war's physical-moral dialectic).[148] Together, *will* and *cause* (moral) and *capacity* (physical and moral) comprise the *engine of war*.

The engine of war has three main components – the "apparatus" (capacity), the "fuel" (will), and the "ignition" (cause). All three must be present for war to occur. We can prevent the engine from starting by removing the ignition; but, once it gets going, only draining the fuel or destroying the apparatus stops the engine. Thus, there are two primary paths to victory in war: (1) eliminate the enemy's fighting capacity (break the apparatus), or (2) destroy the enemy's will to fight (drain the fuel). Clausewitz writes, "If you want to overcome your enemy you must match your effort against his power of resistance, which can be expressed as the product of two inseparable factors, viz. *the total means at his disposal and the strength of his will*."[149]

In other words, "Victory consists not only in the occupation of the battlefield, but in the destruction of the enemy's *physical* and *psychic* forces."[150] Defeating the engine of war is challenging because its fuel (will) and ignition (cause) can power almost any apparatus; thus, even eliminating *all capacity* (a practical impossibility) may only suspend war temporarily if the will to fight remains. As

John Milton warns, "Who overcomes by force, hath overcome but half his foe."[151] And Clausewitz adds,

The animosity and the reciprocal effects of hostile elements, cannot be considered to have ended so long as the enemy's *will* has not been broken: in other words, so long as the enemy government and its allies have not been driven to ask for peace, or the population made to submit.[152]

Therefore, destroying the will is the only *certain* path to victory.

Will

In the long run the sword will always be conquered by the spirit.[153]

Napoleon

Our battle is more full of names than yours,
Our men more perfect in the use of arms,
Our armour all as strong, our cause the best;
Then reason will our hearts should be as good.

Shakespeare, *Henry IV* (4.1.154)

Will is the root of human thought and action. "In the human will," writes Liddell Hart, "lies the source and mainspring of conflict."[154] Will is a mental attribute, the psychological "fuel" that sustains us in any endeavor. It can originate and be supported by reason or emotion and is of paramount importance in war. "Force can always crush force," writes Liddell Hart. "It cannot crush ideas."[155] The engine of war is ignited by an idea, the war's cause or purpose, which, as discussed extensively in Chapter 1, motivates warriors into action. "A good cause," writes Liddell Hart, "is a sword as well as armor."[156] And Clausewitz adds, "The stronger motive increases willpower, and willpower ... is always both an element in and the product of strength."[157] Since all war is political and survival is the political prime directive, the root of any war's cause is *political survival.*

In grand strategy, the cause drives the will to employ IOPs to achieve the political end; and at lower levels, it supports the will to keep fighting, to face danger, and to kill. Will is the chief attribute within the "moral" portions of war, which Clausewitz says "constitute the spirit that permeates war as a whole, and at an early stage they establish a close affinity with the will that moves and leads the whole mass of force, practically merging with it, since the will itself is a moral quantity."[158] In war, he adds, "One might say that the physical seem little more than the wooden hilt, while the moral factors are the precious metal, the real weapon, the finely-honed blade."[159]

Capacity

Get there first with the most.[160]

Nathan Bedford Forrest

Capacity has physical and psychological elements, the tangible means and moral aptitude for warfighting. While we may equate capacity with armies and weapons, even one person with a computer can create military effects. Of fighting capacity, Clausewitz says, "The art of war is the art of using the given means in combat . . . and includes all activities that exist for the sake of war, such as the creation of the fighting forces, their raising, armament, equipment, and training."[161] Additionally, the resources employed in war include "*the fighting forces proper, the country*, with its physical features and population, and its *allies*."[162] In all its forms, capacity *arms* the will. Though the will to fight is paramount, it cannot create combative violence alone. Warfighting requires capacity, which includes forces and intangibles like education, training, and experience. Moreover, while capacity is proportional to force size, a bigger force does not necessarily equate to greater capacity. For example, during the early North Africa campaign of World War II, a small British force drove 500 miles in under two months and routed an Italian army five times its size, capturing 130,000 soldiers and neutralizing "nearly 400 tanks and over 800 guns."[163]

Relative Capacity

After considering the strength of your own forces . . . the factors of the Enemy and Own Strength *are compared in order to arrive at a definite conclusion as to the relative combat efficiency of the opposing forces.*[164]
Small Wars Manual

Assessing relative capacity (sometimes called "net assessment" or "comparative advantage") is a theoretical cornerstone for most of our Chapter 2 theorists.[165] "Everyone tries to assess the spirit and temper of his own troops and of the enemy's," writes Clausewitz.[166] Making *accurate* assessments requires self-awareness, experience, intelligence, intuition, and even luck. Indeed, writes Clausewitz, "If we remember how many factors contribute to an equation of forces, we will understand how difficult it is in some cases to determine which side has the upper hand. Often it is entirely a matter of the imagination."[167]

Evaluating relative capacity is a task for all levels of strategy. Clausewitz advises, "We must first examine our own political aim and that of the enemy. We must gauge the strength and situation of the opposing state. We must gauge the character and abilities of its government and people and . . . our own."[168] Furthermore, "It is crucial that the general decide from the start whether his opponent is both willing and able to outdo him by using stronger, more decisive measures."[169] And finally, at the tactical level, Clausewitz highlights the significance of the battle's tactical pattern, terrain, force composition, and "the *relative strength* of the opposing armies" (emphasis added).[170]

Relative capacity assessments are common in sports, but sporting events are governed by enforceable rules and have competitive categories to preclude mismatches. As we highlighted earlier, rules in war are often disregarded, and when both sides decide to fight, regardless of relative capacity, war becomes a life-or-death sprint to create exploitable mismatches. When capacities are balanced, decisions may be difficult, and factors like the environment, chance, or morale become decisive. In these cases, the advantage often goes to the side that strikes first, fastest, or as Jomini advises, when forces "by a well-combined strategic plan, unite upon and overwhelm successively the fractions of the adversary's forces."[171] Germany's *Blitzkrieg* is a case in point. When differences in relative capacity are stark, as in small wars, the weaker side must find a nontraditional path to victory, and war's human and political aspects can eclipse combat. In these cases, the weaker side may turn to guerrilla tactics and terrorism. As these instances imply, capacity imbalances relate to war's direct-indirect, regular-irregular, and concentration-dispersal dialectics and influence war's forms. We will revisit this theme shortly.

Center of Gravity

As we learned from Clausewitz, a COG is a source of strength (capacity and will). Thus, weakening or eliminating a COG degrades the enemy's engine of war. Strategy defines friendly and enemy COGs to help prioritize and direct offensive and defensive efforts. For our purposes, a COG is *a person or group, location, or thing an adversary* most *relies upon to sustain its will to fight*. Examples include political and military leaders, cities, military units and bases, and industrial, resource, or supply centers. COGs exist at all levels of war, though their significance increases with level, and they may be attacked or influenced directly or indirectly.

Hierarchy of COGs

Since stopping the enemy's will to fight is the only certain way to win a war, the *prime COG* is *the person or group most responsible for motivating or sustaining the war effort*. In a battle, the prime COG is likely the opposing military chain of command and *secondary COGs* the armed forces. At the grand strategic level, the prime COG is often a political leader, a population, or another entity (e.g., legislative body) with the power to decide matters of war and peace. Things and locations are always secondary COGs because they only indirectly influence the will to fight. This explains why neither occupying an enemy capital nor destroying an army guarantees victory.

Lines of Effort

"Lines of effort" (LOE) or "operations" (LOO) are tools of strategy and operational art. LOEs are generally composed of tasks, missions, and objectives associated with specific functional or geographic areas (e.g., "reinforce local governance" or "secure the southern province"). They are sub-efforts of a broader campaign that may incorporate combat and noncombat activities. Campaign planners devise LOEs to apportion capacity and achieve intermediate military objectives that, when aggregated, advance the overall war aims. Defining LOEs aids the coordination and synchronization of large-scale operations where forces pursue multiple COGs in parallel. Of course, COGs and LOEs work both ways (i.e., we must resource and protect our own even as we imperil the enemy's).

The Object of War

In the broadest sense, the war's object is the political aim, which military strategy supports by using combative violence to end the enemy's will to resist. For the aggressor, war ends opposition to the political aim and sets the conditions for its realization. For the defender, war stops the aggressor's gambit. Though strategic will is the prime COG, it is often intangible and its bearers are frequently well defended. Therefore, influencing the strategic COG in war, directly or indirectly, typically requires combative violence against some portion of the enemy's fighting capacity. Thus, fighting, writes Clausewitz, is "the essential military activity, which by its material and psychological effects comprises ... the object of the war."[172]

Improving the odds of victory requires an accurate account of the will and capacity of both sides. This supports strategy and operations, which proceed along a variety of paths, from direct attacks on political leadership (strategic decapitation) and military forces to indirect operations against alliances, infrastructure, and logistics. Sun Tzu recommends attacking *strategy* to compel the enemy to recognize the futility of continued fighting. Where direct disarmament is impossible, Clausewitz agrees saying,

> It is possible to increase the likelihood of success without defeating the enemy's forces. I refer to operations that have direct political repercussions, that are designed in the first place to disrupt the opposing alliance, or to paralyze it, that gain new allies, favorably affect the political scene, etc. ... they can form a much shorter route to the goal than the destruction of the opposing armies.[173]

Nevertheless, adds Clausewitz, victory more often requires targeting "the *armed forces*, the *country*, and the *enemy's will*," for a disarmed opponent has three options: flee, surrender, or die.[174] Besides disarming an opponent, Clausewitz advises commanders to "influence the enemy's expenditure of effort ... to make the war more costly."[175] To do this, he recommends invading and devastating territory to "increase the enemy's suffering" and, where possible, prolonging *"the duration of the war to bring about a gradual exhaustion of his physical and moral resistance."*[176]

Of course, *there is no single best way to win a war.* As Clausewitz says,

In war many roads lead to success ... they range from *the destruction of the enemy's forces, the conquest of his territory, to a temporary occupation or invasion, to projects with an immediate political purpose, and finally to passively awaiting the enemy's attacks.* Any one of these may be used to overcome the enemy's will: the choice depends on circumstances.[177]

Successful strategists must gather accurate information about the enemy, assess each side's engine of war, identify COGs, and outline ways and means to compromise the enemy's will to fight at every level of war. Strategists should also refrain from broadcasting strategic predilections. For example, publicizing a "way of war" (e.g., Liddell Hart's *The British Way in Warfare*, 1932; and Russell Weigley's *The American Way of War*, 1973) is antithetical to sound strategy because it feeds an enemy's strategic calculus. Moreover, "strategic cultures" that base military policies exclusively on "political, economic, social, and cultural modes" create a *fait accompli* approach to strategy that risks exploitation by those whose strategies rest only on the impartial appraisal of relevant factors.[178]

Will-Capacity Model

Figure 4.7, the will-capacity model (WCM), is a graphic representation of the relationship between will and capacity in war. It depicts the primacy of war's

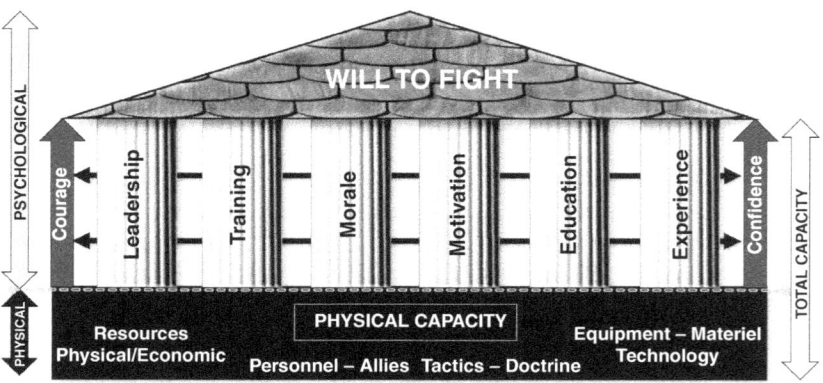

Figure 4.7 The will-capacity model

psychological components and the will to fight and illustrates the linkages between capacity, will, and the major factors associated with each. This model applies to everyone in war, from the soldier to the sovereign. Each combatant, unit, and state has a unique WCM composed of similar elements though influenced by personal and subjective factors.

In war, the will to fight is like a roof supported by pillars that rest upon physical capacity's foundation. The will cannot stand without capacity (a purely abstract condition), and capacity absent will is a foundation without a structure to support, though capacity alone may deter or assure.[179] The will to fight, pillars, and *moral enablers* (courage and confidence) reside within war's psychological realm. Total capacity includes physical and psychological (or moral) factors and attributes. According to Clausewitz, the *moral* factor motivates, inspires, and fuels the will, and "is the most fluid element of all, and therefore spreads most easily to affect everything else."[180] Thus, as the model illustrates and Clausewitz confirms, the "psychological forces exert a decisive influence," though the exact proportions are neither constant nor quantifiable.[181] Furthermore, the pillars and moral enablers are interrelated. Since they operate in the psychological domain, combatants perceive them concurrently, which amplifies their effects, positive or negative, on the will. The powerful influence of moral factors helps explain why underdogs can win and even large, well-armed forces can lose. Physical capacity indirectly supports the will by providing the foundation for each pillar and the moral enablers. Elements of physical capacity include physical and economic resources; personnel and allies; tactics and doctrine; and equipment, materiel, and technology. Within these are unlisted sub-elements like industry, transportation, infrastructure, agriculture, and so on.

The six pillars – *leadership, training, morale, motivation, education,* and *experience* – along with physical capacity, contribute to courage and confidence (like Machiavelli's *virtù*), which fuel the will and reinforce motivation and morale. *Courage* is the ability to overcome fear, a requisite for war-fighting. Clausewitz adds that courage and morale "have always increased [an army's] physical strength," and that superior organization, experience, familiarity with terrain, equipment, mobility, tactics, training, leadership, education, and force size have similar effects on psychology.[182] *Confidence* is a belief in one's abilities, a factor closely tied to physical capacity and to each pillar. "If an army is to win the day," writes Machiavelli, "it is essential to give it confidence so as to make it feel sure that it must win."[183]

Strong, competent *leadership*, balancing boldness with prudence, fortifies the other pillars and the will. As we learned in Chapter 1, leadership is vital to motivation and drives how and how well a military unit fights. Write Codevilla and Seabury, "Anyone who has ever been in the military has noticed that ships or battalions, sometimes whole divisions or armies, reflect the personality and competence of their commanding officer."[184] A military genius, like Alexander or Napoleon, can have a transcendent effect on the will and the effectiveness of the military IOP. As Alexander is reputed to have said, "I am more afraid of an army of sheep led by a lion, than an army of lions led by a sheep." Leadership's four major characteristics are *competence*, *character*, *communication*, and *charisma*. *Competence* encompasses the leader's acumen in the art and science of warfighting and strategy. Followers perform best for leaders who make sound decisions and have a proven record of success. Knowledge, experience, intellect, and attitude all contribute to competence and can arm the best leaders with *coup d'oeil*. *Character* relates to psychological fortitude and how a leader's personal qualities, beliefs, and motives align with followers. Attributes associated with character, like integrity, honor, determination, resilience, and reliability under stress, fortify the leader-follower bond, which is essential in war where conditions may not permit detailed explanations. People will follow a competent leader they trust, who recognizes their contributions and shares risk. Good leadership is impossible without effective *communication*. Leaders cultivate a range of communication skills so followers understand the *who*, *what*, *where*, *when*, *how*, and *why* of their role in war. Communication is fundamental to warfighting and enables command, control, and the exercise of authority. Finally, *charisma* is the intangible "it" factor, a quality of personality that complements and reinforces the intellectual leader-follower connection with an emotional aspect. Charisma accompanies traits like courage, confidence, boldness, and decisiveness. Though all four attributes of leadership are important, the power of charisma, especially combined with superior communication skills, can trump the rest. Military geniuses, like Alexander, Genghis Khan, and Napoleon, embodied the "four C's" of leadership. Writes Paret, "The impact of [Napoleon's] charisma and the belief in his absolute superiority extended to his troops and their officers and generals to his opponents. Wellington thought his presence was the equal of *forty thousand soldiers*" (emphasis added).[185]

Coupling *training* and *education* with strong leadership enables warriors to build the knowledge, skills, discipline, and confidence essential for success in war. "Good discipline," writes Machiavelli, instilled in training, "stimulates courage and ardour, in that it strengthens the hope of victory, which is never wanting so long as discipline remains."[186] Of course, actual combat *experience* is the best teacher because it culls the weak, rewards the strong, and builds confidence, courage, and morale.

Motivation is the most essential and powerful pillar because it reflects the engine of war's "ignition," the *cause*. Roger Trinquier writes, "In each country, within each race, in every social stratum, we must find a reason, an idea ... capable ... of constituting adequate motivation for the assumption of necessary risks."[187] The will, adds Clausewitz, "can only be gauged approximately by the strength of the motive animating it."[188] When all else appears lost, the motivation to survive, to stand with companions, to protect loved ones, or to serve a nation or a god, can sustain the will to fight.

Morale is the most ephemeral but certainly not the least meaningful pillar. Indeed, "It is no exaggeration to claim," writes Gray, "that morale is the single most important contributor to military success."[189] According to General George C. Marshall, "It is not enough to fight. It is the spirit which we bring to the fight that decides the issue. It is morale that wins the victory. It is staying power, the spirit which endures to the end – the will to win."[190] "Military spirit," adds Clausewitz, "is one of the most important moral elements in war," and it arises from success in battle and sharing the trials of combat.[191] Finally, warriors with high morale, says Sun Tzu, "love to fight, their ambitions soaring as high as the azure clouds and their spirits as fierce as hurricanes."[192]

Implications for Strategy

The WCM is a visual reminder that war's origins, conduct, and results are determined *within the human mind*. The cognitive domain's complexity causes tension within war's science-art, reason-emotion, certainty-uncertainty, and simplicity-complexity dialectics. Writes Liddell Hart, "However far our knowledge of the science of war be extended, it will depend on art for its application, [but] this complicates the calculation, because no man can exactly calculate the capacity of human genius and stupidity, nor the incapacity of will."[193] And Clausewitz warns, "The art of war deals with living and with moral forces. Consequently, it cannot attain the absolute, or certainty, it must always leave a margin for uncertainty, in the greatest things as much as the smallest."[194]

Before War

> *It may even be reasonably said that the intensely sharp competitive prepar-*
> *ation for war by the nations is the real war, permanent, unceasing, and that the*
> *battles are only a sort of public verification of the mastery gained during the*
> *"peace" interval.*[195]   William James

War technically begins when the fighting starts; however, strategy, which lays the foundation for success or failure, begins well beforehand. During "peace-time" (including true peace, competition, and conflict short of war), all the essential acts of strategy may be exercised, including the military IOP, only without the combative violence element. Most components of the WCM, including capacity, leadership, education, and training, are built prior to hostil-ities as partners and allies *cooperate* and rivals *compete* for resources and status. To paraphrase Chanakya, most wars are won or lost *before* the fighting starts, and "intrigues" caused by ignorance or poor preparation can weaken the will to fight.[196] Intelligence gathering, planning, and establishing or strengthening alliances are as beneficial while cooperating and competing as they are in conflict. Failure to appreciate this can lead to an exploitable strategic deficit. Consider Russia's annexation of Crimea in 2014, a case that demonstrates command of the peace-war and certainty-uncertainty dialectics. Russia coveted Crimea for its Black Sea ports; however, rather than overt invasion, Vladimir Putin executed a subversion strategy that basically rendered the annexation a *fait accompli*. According to foreign affairs expert Nicholas Fedyk, Russia prevailed using influence versus "destruction; inner decay over annihilation; and culture over weapons and technology."[197] In Russian grand strategy, says Fedyk, "*The main battlespace is the mind,*" and "there is no artificial binary between war and peace, but simply war at all times, in all places, and with all resources."[198] In fact, Russia's actions *before war* enabled them to capture Crimea "silently . . . under the guise of humanitarian intervention," "with little local resistance or bloodshed," or interference from the international community.[199]

Preparing for War: Know Yourself

A Prince, therefore, ought never to allow his attention to be diverted from warlike pursuits, and should occupy himself with them even more in peace than in war. This he can do in two ways, by practice or by study.[200]

Machiavelli

Preparing for war requires the development of physical and moral capacity to support the will to fight, in other words, constructing a robust engine of war. Building capacity and enhancing IOPs depend on a multitude of variables, many of which are not easily altered, like natural resources and population size. However, conscription or incentives for armed service, diplomacy, research and development, military exercises, and military education programs all increase capacity. Likewise, cultivating and promoting competent, well-educated leaders and warriors is the most tangible way to reinforce the other pillars in times of peace.

Know Your Enemy

When I'm getting ready to reason with a man, I spend one third of my time thinking about myself and what I am going to say – and two thirds thinking about him and what he is going to say.[201]   Abraham Lincoln

To stop the enemy's engine of war, we need to know what it looks like, where it is, and how it works, all of which requires intelligence. Maximizing the war effort's probability of success demands accurate intelligence to identify COGs and exploitable vulnerabilities related to the base and pillars supporting the enemy's will. As a picture of the enemy's engine of war takes shape, strategy can begin to assemble ways and means for dismantling it. For instance, options could include phased, sequential, or simultaneous operations, in multiple domains, against various COGs, like enemy leaders, training camps, supply hubs, or lines of communication. Strategists should evaluate these preliminary plans with the tests of strategy as they refine, within resource and time limits, targeting, movement, and engagement decisions.

Targeting Leaders

It is clear that the higher the rank of the persons killed, the more likely they are to be carriers of the purpose that is the legitimate target of hostilities.[202]

Angelo Codevilla

"The responsibility for a martial host of a million lies in one man," writes Sun Tzu. "He is the trigger of its spirit."[203] Obviously, removing that "trigger" can significantly impact the will to fight. Leadership can be targeted in a variety of ways including physically with lethal force or mentally through deception. Human factors analysis and focused intelligence collection can provide details on leaders of interest, including biographical data, personality traits, and habits that might furnish useful insights into enemy actions and strategy. Killing enemy leaders not only neutralizes their valuable experience, it also influences the will of others up and down the chain of command; thus, it has always been a staple of warfighting. For example, in the fourth century BCE, the Theban general Epaminondas created the "oblique order," which directed forces into enemy flank and rear areas expressly for killing opposing leaders.[204] This decapitation tactic led Epaminondas and the Arcadian League to huge victories over the formidable Spartans. For obvious reasons, political and military leaders tend to renounce the practice of assassination; however, as Codevilla and Seabury opine, "Assassination is often the most morally defensible of any lethal act in warfare – as well as the most effective."[205] Imagine the impact upon history if Washington, Napoleon, Lenin, Hitler, or Mao had been killed early in their careers. Finally, when targeting leaders, strategists must consider the "martyr effect," which actually *reinforces* the enemy's will and cause with a powerful vengeance motive. It may be unwise to target especially charismatic leaders who have a strong personal or religious connection to their people and a robust chain of succession.

Targeting Will

It was one thing to recognize the importance of the cognitive domain, but quite another to assume that it was susceptible to straightforward manipulation. Human minds could be capable of remarkable feats of denial, resistance, recovery, and adaption, even under extreme stress.[206]

Lawrence Freedman

When Clausewitz writes, "All war presupposes human weakness, and seeks to exploit it," he is referring mainly to the human psyche.[207] Attacking the enemy's psychology is fundamental to sound strategy; yet, as the Freedman quote implies, this is not easy. Taking the initiative and forcing an enemy to react using persistence, surprise, speed, and deception, is the place to start. Even seasoned warriors experience resolve-eroding doubts when they cannot predict what is coming next. Additionally, writes Clausewitz, inflicting physical losses, seizing enemy territory or lines of communication, attacking relentlessly, or presenting an impenetrable defense induce "the loss of order, courage, confidence, cohesion, and plan," and ultimately morale.[208] Over time, this causes feelings of helplessness, which "[induce] hopelessness," adds Liddell Hart, "and history attests that loss of hope, not loss of lives, is what decides the issue of war."[209]

Targeting Capacity

Attacking physical capacity, due to its tangibility, remains the best understood and most often used method for winning wars; but, as we learned with small wars, it is not always the *best* method. Nevertheless, physical capacity – as the WCM's foundation, war's material cause, and the engine of war's secondary COG – is generally an appropriate target. Degrading or destroying the enemy's ability to degrade or destroy us makes sense, and combat success *is* the best way to increase experience, confidence, morale, and motivation. However, strategists must not forget the impossibility of eliminating *all* physical capacity, and therefore they should ensure that strategy balances the physical-moral dialectic by targeting *total* capacity including the will-supporting pillars.

Efficiency versus Effectiveness

The absence of strong, highly mobile reserves – from the platoon level to that of an empire – must be counted as one of the most fatal errors in warfare.[210]

Angelo Codevilla

Capacity apportionment is critical to strategy implementation, but is it better to emulate Sun Tzu's guidance and husband capacity by striving to win without fighting (efficiency) or Clausewitz's dictum to hold nothing back and apply maximum force in a decisive battle (effectiveness)? Efficiency delivers results without waste, and in business, maximizes profit. In war, efficiency preserves resources for contingencies or follow-on actions. As Clausewitz says, "The belligerent is again driven to adopt a middle course.

He would act on the principle of using no greater force, and setting no greater military aim, than would be sufficient for the achievement of his political purpose."[211] Here Clausewitz echoes Sun Tzu's advice that wars should pursue a stable peace quickly and efficiently. War for war's sake clouds that objective and endangers the political aim. However, war is also far more deadly and unpredictable than commerce. In war, warns Clausewitz, "Imperfect intelligence, the threat of catastrophe, and the number of accidents, are incomparably greater than in any other human endeavor."[212] This makes it impossible to identify the line between efficiency and inefficiency, or where pursuing efficiency jeopardizes effectiveness. The difference between "just enough" and "not enough" in war is life and death. Thus, strategists must *always favor effectiveness over efficiency* because war awards no trophies for second place. "In war," affirms Clausewitz, "too small an effort can result not just in failure but in positive harm."[213] In reality, effectiveness in war *is* efficiency, which explains why efficiency-effectiveness is not one of war's dialectics.

On Total War and Genocide

Death solves all problems – no man, no problem.[214]   Anatoly Rybakov

Now go and smite Amalek, and utterly destroy all that they have, and spare them not; but slay both man and woman, infant and suckling, ox and sheep, camel and ass.   Samuel 15:3

The WCM helps explain two of humanity's seemingly irrational inclinations, total war and genocide, which push warfare to the starkest extremes of the good-evil, friend-enemy, peace-war, life-death, creation-destruction, and unlimited-limited dialectics.[215] According to historians Stig Förster and Jörg Nagler, *total war* occurs when at least one side enlists all its people and resources to annihilate its adversary's engine of war (i.e., to destroy the whole WCM – all capacity and will – with overwhelming violence, like Rome versus Carthage).[216] Total wars, which cannot end in settlements, occur when the peace-war and unlimited-limited dialectics yield to the unlimited-war extreme. The empires of Persia, Rome, and Khan's Mongols practiced total war. In modern times, Napoleon's *levée en masse* and the two world wars come closest.

Though the UN coined "genocide" in the twentieth century (to characterize the Nazi-led Jewish Holocaust), the practice spans human existence.[217] Genocide (which tends to be undocumented because few victims live to tell the tale and even fewer perpetrators offer a truthful account) can be an aim or outcome of total war, or it can occur against defenseless civilians outside the bounds of war. Examples from the twentieth century where, as Freedman points out, "The elimination of whole groups of people of supposedly inferior race or dangerous belief was adopted as a war aim," include the Holocaust, Rwanda, Bosnia, Cambodia, and the Ukraine.[218]

On its face, genocide seems a completely irrational, evil act; yet, the WCM provides an explanation. The aim of war is the political objective, and the purpose of fighting is to eliminate the enemy's capacity and will to resist. But destroying *all* capacity is impractical and, in some cases, can actually *reinforce* the enemy's will. The only certain path to victory is to neutralize *all* the pillars by eliminating the will's source, the *people*. Against enemies "from whom peace can never be hoped for on any terms," writes sixteenth-century theologian Francisco de Vitoria, "the only remedy is to eliminate all of them who are capable of bearing arms against us."[219] Depending upon a war's circumstances and the depth of its animosities, one or both sides might view the other's *existence* as an unacceptable risk requiring the most extreme measures, total war or genocide. These instances shatter the friend-enemy, good-evil, and life-death dialectics, and the opponent is seen as purely evil-enemy, inhuman, with no rights, not even to life. When this occurs, as history confirms, genocide becomes strategy's "way." And we cannot dismiss the chance that genocide or total war might recur, for these are evils that have their own dark logic.

Beyond Fighting

You will not find it difficult to prove that battles, campaigns, and even wars have been won or lost primarily because of logistics.[220]

General Dwight D. Eisenhower

Global strategic history always has been governed in practice by logistics, meaning the science of supply and movement.[221]

Colin Gray

Before transitioning from strategy to combat, we must acknowledge the essential, noncombat elements of war, which comprise a good portion of war's material cause. *Logistics*, the art and science of keeping a force supplied and fit to fight, is an intellectually intensive discipline as vital to war as combat skill. As Omar Bradley once said, "Amateurs study strategy, professionals study logistics."[222] Though we equate "military" with warriors and weapons, we cannot dismiss critical enablers like maintenance, training, supply, transportation, medical care, and even finance. Remember this unbreakable rule: there is no victory in battle without weapons, water, food, and fuel! Marginalizing these functions elevates risk and may negatively impact physical and psychological strength including morale and the will to fight. Failure to consider and integrate logistics into strategy creates a self-imposed, potentially unrecoverable handicap.

COMBAT

> *Combat at every level is an environment that requires officers to make*
> *decisions on inadequate information, in a hurry, and under great stress, and*
> *then inflicts the death penalty on many of those who make the wrong deci-*
> *sions – and on some of those who have decided correctly as well.*[225]
>
> Gwynne Dyer

We are now ready to discuss the True Trinity's "top layer," combat. The purpose of preparation and logistics, as Clausewitz says, "*is simply that he should fight at the right place and the right time.*"[224] Combat occurs because war presupposes that each side will use violence to dismantle its opponent's engine of war and advance its political aims. However, since combative violence can occur for nonpolitical ends, many tactical fighting fundamentals are not necessarily exclusive to war (i.e., many combat principles are valid in other contexts). Distinctions between war and nonwar "combative violence" are found in its character (e.g., scope, weaponry, and constraints). For example, police must uphold the law even when fighting and soldiers may follow "rules of engagement;", whereas terrorists and criminal organizations rarely adhere to any restrictions.

Battle

> *Armies exist ultimately to fight battles – the most complex, fast-moving, and*
> *essentially unpredictable collective enterprises (not to mention the most*
> *dangerous and confusing) that large numbers of human beings engage in.*[225]
>
> Gwynne Dyer

Military strategy employs destructive force to disable or destroy an enemy's engine of war. We call the ensuing violent clash of opposing forces and wills *combat*, and a specific instance of combat is a *battle*. Every discrete battle is tactical, but the aggregation of many battles distributed in time and space signifies higher levels of war. Strategists and commanders naturally want to win every battle; however, prudence dictates consideration of positive *and* negative outcomes. Following a successful battle, a wise commander seeks to capitalize on tactical momentum while being wary of overextension and counterattack. Furthermore, prudent commanders cultivate options for disengagement and withdrawal should battles go poorly.

Attack and Defense

> *One defends when his strength is inadequate; he attacks when it is*
> *abundant.*[226]   Sun Tzu

In Chapter 2, we classified attack and defense as *postures* often dependent upon the relative strengths of the opposing sides (as Sun Tzu implies).

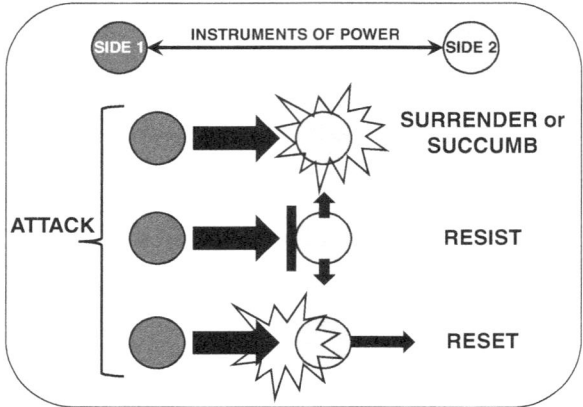

Figure 4.8 Response options to attack

Defense-attack is also one of war's dialectics. Attackers initiate the action with a positive aim, to defeat or dislodge the opposing force or otherwise compel a change in the status or position of the fight. Defenders receive the action with a negative aim, to hold position and oppose the attacker's objective. Through coordinated actions, each side tries to force the other into a position of greater risk and disadvantage. Many principles introduced previously, including surprise, deception, speed, maneuvers, and concentrating power at a decisive point or COG, are operative in combat. Intelligence helps identify the best places and times to strike, and doctrine and TTPs guide activities to maximize the destructive effects of each action.

Battle postures are not mutually exclusive; they can change rapidly and occur simultaneously (i.e., both sides attacking or defending). Attackers employ IOPs to affect the adversary's engine of war. Defenders have four response options when attacked (Figure 4.8): (1) *surrender* (comply willingly); (2) *succumb* (comply unwillingly); (3) *resist* (fight back); or (4) *reset* (withdraw). Only (3) and (4) permit the possibility for a defender to win, which Clausewitz says occurs by defeating the attacker in the present battle or having the attacker succumb "*by his own exertions*" (resist), or winning a future engagement (reset).[227] When the defender counterattacks, the situation reverses, and when both sides attack, postures become fluid and effects accumulate until there is a decision or both sides reset. In the latter case or when both sides adopt defensive postures, the battle ends, at least temporarily.

Deciding which posture to adopt requires consideration of a variety of factors. At the strategic level, the offensive is usually taken by the side that starts the war. However, it is possible to take different postures at different levels of war (e.g., strategically defensive while tactically offensive). Postures often reverse if the attacker's capacity drops below the defender's. As Napoleon says, these changes present operational and leadership challenges, because "the transition from the defensive to the offensive is one of the most delicate operations in war."[228]

Fire for Effect

In battle, regardless of domain, attackers employ IOPs against defenders (or other targets) to advance the immediate aim and the overall war effort. The *engagement sequence* (or "kill chain") is how violent force is applied against an enemy. The links in the chain are *find*, *fix*, *track*, *target*, *engage*, and *assess*. The first four steps use awareness via ISR to identify and locate targets. The essence of combat, the engagement, resides in the gap between targeting and assessment. Consider the following engagement examples at each level of war:

(1) A soldier (attacker) uses a grenade (IOP) to kill (effect) another soldier (defender) and neutralize a gun emplacement (objective).
(2) A cyber warrior (attacker) uses computer code (IOP) to disable (effect) a satellite (defender) to disrupt enemy surveillance (objective).
(3) A general (attacker) uses an army (IOP) to defeat (effect) another army (defender) to gain control of a city (objective).
(4) A state (attacker) uses DIME (IOP) to compel (effect) another state (defender) to withdraw from its territory (objective).

To be effective, attacks at all levels must contribute to the overall objective. For example, a bombing sortie's immediate aim of destroying a command post supports the broader objectives of degrading the enemy's leadership pillar, increasing the enemy's vulnerability to future attacks, and reducing the will to fight.

The OODA Loop

I believe we have uncovered a Dialectic Engine that permits the construction of decision models needed by individuals and societies for determining and monitoring actions in an effort to improve their capacity for independent action.[229]
                                                                    Colonel John Boyd

Clausewitz says there is a "logical hierarchy that governs the world of action."[230] Accordingly, the dynamics of battle require the ability to collect and process information, assess variables, decide what to do and how and when to do it, act on that decision, assess results, and adjust or repeat. This iterative method, a tactical analogue to the cycle of war, continues for both sides until one has met its objective or they mutually withdraw. US Air Force Colonel John Boyd calls this model the "decision cycle" or *Observe-Orient-Decide-Act* (OODA)

Loop (Figure 4.9). Boyd believed the key to success in any human-versus-human competitive context, *but especially in battle*, is to cycle through this algorithm faster and more accurately than the opponent, thus forcing him into a reactive posture.[231] The side that achieves OODA Loop superiority is more likely to gain its objective. In fractal style, the OODA Loop nests within the cycle of war's "execution" element (Figure 4.10).

Command and Control (C2)

C2 – the exercise of authority and direction over military forces – is enabled by a chain of trained personnel and technical systems that links commanders to their forces. C2 leads and manages the cycle of war and OODA Loop, determines the balance between control and autonomy, and its criticality expands with the scope of war. Robust C2 mitigates war's fog and friction granting commanders OODA Loop awareness, responsiveness, and adaptability even during large-scale, complex operations.

Branches and Sequels

When we act rationally, we consider alternative versions of the future, making explicit those expectations about the future that are so often buried in the realm of hunch.[232]

Harold D. Lasswell

Military engagements take time during which conditions inevitably change. Both the wrong action at the right time and the right action at the wrong time raise the likelihood of failure, so prudent commanders must consider timing and attempt to anticipate future conditions. Seizing the initiative with full awareness of our own plans and the environment of war permits rough forecasting of likely enemy responses, which in turn, enables the development of *branches* and *sequels. Sequels* are actions that occur after initial objectives are met (e.g., "after you capture the bridge, move on to the next town"). *Branches* are conditional actions and objectives (e.g., "if the missile radar goes active, strike it; otherwise, bomb the factory"). As mentioned earlier, this planning technique was encouraged by Napoleon's mentor, Pierre-Joseph Bourcet, who "advanced the idea of a 'plan with branches'" to increase flexibility and confuse enemies.[233]

Surprise, Speed, and Deception

Nothing so enhances your forces or detracts from the enemy's as knowing his plans while leading him to think what you will about your own.[234]

Angelo Codevilla

What is of greatest importance in war is extraordinary speed.[235]   Sun Tzu

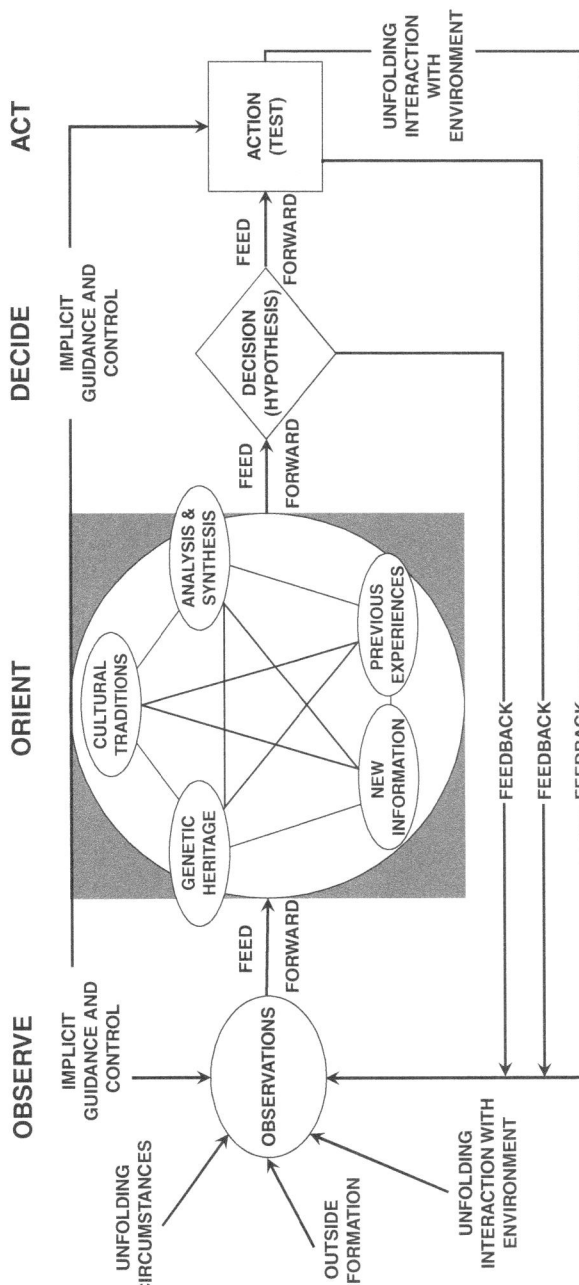

Figure 4.9 Boyd's OODA Loop

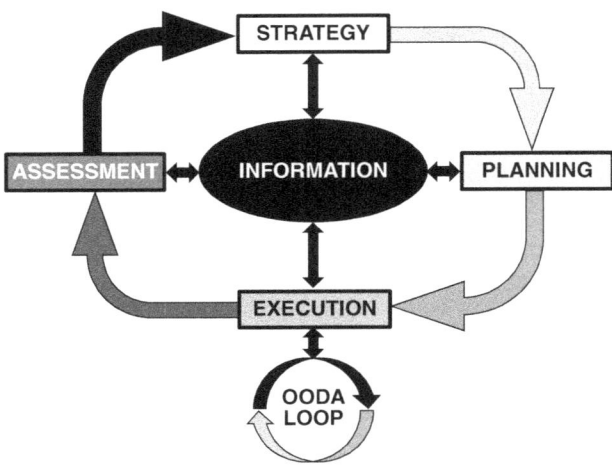

Figure 4.10 The cycle of war and OODA Loop

Information feeds the cycle of war and OODA Loop, so gaining information while denying it to the enemy is central to strategy and tactics. Proven methods for doing this include *operational security* (OPSEC): passive measures to protect information; *counter-intelligence*: actions that disrupt enemy intelligence gathering; *camouflage* and *concealment*: actions that disguise and hide forces and structures; and *deception*: actions that fool an enemy into making poor decisions. These methods along with *speed* contribute to *surprise*: catching an enemy unawares. Clausewitz writes, "Surprise lies at the root of all operations without exception," and, "the two factors that produce surprise are secrecy [OPSEC] and speed."[236] Speed accelerates the OODA Loop, reducing an enemy's ability to keep pace, and deception and surprise corrupt enemy observation and orientation resulting in delayed or defective decisions and actions. When combined, surprise, speed, and deception amplify capacity, at least temporarily, granting an immediate, localized advantage to the side employing them.

Tactical Containment

The idea was to commit a strong force to hold the enemy army in place on his main line by a fierce attack or threat and to send a powerful column around the enemy's flank onto his rear and there establish a strategic barrage or barrier across his line of supply and retreat.[237]

Bevin Alexander on Napoleon's *manoeuvre sur les derrière*

In war, particularly the asymmetric variety, it can be difficult for a strong force to compel a weak opponent to fight. To land a decisive blow, the strong must "contain" the weak. *Tactical containment* prevents an opponent from resetting or avoiding the attack, and thus, sets the conditions for the effective use of violent force. In some cases, like a static defensive posture, the enemy contains itself. Otherwise, the attacker must establish containment. This can be done with surprise or what Callwell terms "a containing force" that holds the enemy with a frontal attack then cuts off its avenues of retreat (e.g., Napoleon's *manoeuvre sur les derrière*).[238] Clausewitz says, "The commander who wishes to *retreat* and is able to do so can hardly be forced into battle by his opponent . . . there are two principal ways of accomplishing this: first, to *surround* the enemy and cut off his retreat . . . and second . . . *by surprise*."[239] Furthermore, "A unit that is attacked in flank and rear will soon reach the stage where its chance of retreat has vanished: in such a situation it comes close to being completely unable to continue the fight, [thus] all tactical arrangements aimed at envelopment are highly effective."[240] While Clausewitz's comments generally apply to the stronger force, weaker forces can also capitalize on speed, stealth, surprise, and mobility to gain tactical containment. Like a mosquito on an elephant, a small, swift attacker creates localized pockets of containment by striking the main force or its isolated elements and departing before the enemy can respond. We will revisit the concept of "containment" in greater depth shortly.

The Forms of War

A nation that does not prepare for all the forms of war should then renounce the use of war in national policy.[241]
                                                                         T. R. Fehrenbach

There are no old wars and no new wars . . . there are only wars which are all variants of the single species – war.[242]
                                                                         Colin Gray

As mentioned in the Preface, one of war theory's biggest deficiencies relates to war's formal cause: what are war's forms and how and why do they change? Theory's incoherence on these questions has led to the surprising failure of otherwise gifted leaders. For example, Hannibal was unprepared for Fabius, Napoleon for the Spanish guerrillas, Chiang for Mao, America for the Viet Cong, the Soviets for the *mujahedeen*, and so on. The remainder of this chapter addresses this deficiency by updating the current language of war, deriving a unifying property to reconcile the regular-irregular dialectic, and explaining the dynamics of and relationships between the engine of war and war's forms.

Chapter 2's most "balanced" theorists, Sun Tzu and Mao, best grasped the dynamics of war's forms. They valued an army's ability to optimize form based on a war's circumstances. Clausewitz and Jomini, in contrast, focused more on Napoleonic warfare and thought other forms less consequential and decisive.

But what *is* a form of war? Clausewitz (and others) suggest that war's forms are related to its political aims (e.g., limited and unlimited). In this paradigm, according to Paret, "each specific conflict [is] shaped and guided by the kind and intensity of its political motives."[243] For example, unlimited aims, like the conquest of a state, lead to an absolute form: the enemy's complete disarmament; whereas limited aims, like territory acquisition, result in more limited forms that employ less extreme means. But this model has three flaws. First, the boundary between limited and unlimited war is imprecise. In fact, all *real* war is limited to some degree; thus, truly unlimited war, at least regarding form as opposed to aim, is notional. Second, this idea seems to violate the principle of reciprocity. If *our* form of war is based on *our* political aim, then where does "know the *enemy*" fit in? Finally, there are wars for which the aims and means do not seem to align with the unlimited-limited form model. For example, revolutions and insurgencies often pursue unlimited political aims with limited means, at least initially. Conversely, the US-led coalition's aims in the 1990 Gulf War were limited, but the massive military clash hardly seemed limited in form. Exceptions like these raise doubts about the suitability of political aims and the unlimited-limited aim dialectic as the prime determinants for war's forms.

The US military's *Joint Publication 1* (JP 1) identifies two forms of war, *traditional* and *irregular*, discernable by their strategic purposes and consequently dissimilar conduct.[244] Although US doctrine describes war as a "unified whole," it neglects to define or develop the relationship between traditional and irregular war, calling it a "useful dichotomy."[245] In fact, war's forms and character are related, which helps explain why the forms of war vary. Furthermore, prior analysis implies a relationship between war's form and the relative capacity of opposing sides (i.e., "regular" forms are associated with higher capacity and "irregular" with lower). However, before developing this connection further, we must first evaluate existing terminology.

The Language of War

If concepts are to be clear and fruitful, things must be called by their right names.[246]
                                                                            Clausewitz

A theory is a set of related ideas, and according to Paul Reynolds, "the only acceptable means" of conveying ideas and fostering comprehension and acceptance "is through the use of a written language."[247] Ambiguous language prevents theory, no matter how profound, from achieving its purpose, especially for strategy and warfighting, which rely upon common terms of reference to build shared perceptions and produce operational synergy. "Clear-cut nomenclature," says Liddell Hart, "is essential to clear thought."[248] Without this, adds Corbett, strategic deliberations devolve into "verbal contentions which rest on no firm foundation, and end either in every one retaining his own opinion. or in a compromise from considerations of mutual respect – a middle course of no actual value."[249] Unfortunately, war's lexicon is littered with ambiguity. Writes Gray, "Literature of just the past fifty years would have us believe that the following adjectives are akin to DNA-affirmed identities for war: total, general, limited, unlimited, nuclear, conventional, regular, irregular; low-intensity, guerrilla, insurgency and counter-insurgency, and now hybrid."[250] Most of these neologisms are coined "on the fly," without any analytical rigor. "Literally every allegedly distinguishing adjective," laments Gray, "did more or less violence to reality," resulting only in greater confusion.[251]

By Any Other Name ...

There is more semantic confusion in this subject area than anywhere else, as terminology has changed so often, even though the phenomenon of asymmetric war ... is perennial.[252]
Beatrice Heuser

The following is a summary of relevant terms related to war's forms:

(1) *Regular (Conventional or Traditional) War*: This is what Clausewitz and Jomini focused on: open warfare between formal state militaries using conventional weapons and tactics.

(2) *Irregular (Unorthodox or Unconventional) War*: Joint Publication 1–02 (JP 1–02) defines "irregular warfare" as "a violent struggle among state and non-state actors for legitimacy and influence over the relevant population(s)."[253] The same publication defines "unconventional warfare" as "activities conducted to enable a resistance movement or insurgency to coerce, disrupt, or overthrow a government or occupying power by operating through or with an underground, auxiliary, and guerrilla force in a denied area."[254] These definitions are too specific and restrictive for a general theory of war. Just as non-state actors can adopt traditional forms of war, traditional military forces can fight in an irregular manner. Moreover, irregular or unconventional forms of war can be fought for any political purpose, not just to overthrow a government or control a population.

(3) *Asymmetric War*: JP 1–02 defines "asymmetric war" as "the application of dissimilar strategies, tactics, capabilities, and methods to circumvent or negate an opponent's strengths while exploiting his weaknesses."[255] This definition hinges entirely on the word "dissimilar," which is unacceptably ambiguous.

(4) *Guerrilla War*: JP 1–02 does not explicitly define "guerrilla war" but does refer to it as a subset of unconventional war and defines a "guerrilla force"

as "a group of irregular, predominantly indigenous personnel organized along military lines to conduct military and paramilitary operations in enemy-held, hostile, or denied territory."[256] This "definition" is too specific for theory and laden with hazily defined terms like irregular, military, paramilitary, enemy-held, hostile, and denied.

(5) *Insurgency*: JP 1–02 defines an "insurgency" as "the organized use of subversion and violence to seize, nullify, or challenge political control of a region. Insurgency can also refer to the group itself."[257] Insurgencies are a variety of civil war that can include external powers and feature irregular (e.g., guerrilla) or regular forms of war.

(6) *Civil War*: These are fought between political factions within a state or nation. The warring sides in a civil war can fight with any form.

(7) *Hybrid War*: This includes both traditional and nontraditional forces, which makes "hybrid war" a meaningless term without better definitions for its constituent forms.

(8) *Unlimited and Limited War*: JP 1–02 does not define these terms. Extrapolating from Clausewitz, unlimited war pursues an unlimited political aim: the total defeat (or disarmament) of the opposing state; whereas a limited war is fought for any purpose short of that. We can infer that the former requires maximum force to annihilate the enemy's ability to resist and the latter a reduced effort to secure more modest aims.

(9) *Nuclear War*: Nuclear war is *war between forces that employ nuclear weapons*. Nuclear war has connotations of global obliteration, but in truth, no one knows how a nuclear war would play out.

The preceding forms of war are not exclusive. For example, while war cannot be simultaneously regular and irregular or limited and unlimited, and a nuclear guerrilla war is highly unlikely, a civil war or insurgency can have regular, irregular, and asymmetric attributes. Ultimately, concludes Gray, we can characterize all war's forms as variations of regular and irregular war.[258] Provided we agree with this and that, as Gray asserts, war "has a single nature" (we do), then there *must* be an underlying property of war linking all its forms.[259] Therefore, the key to "unifying" the regular-irregular dialectic within general war theory is to define this *underlying property* and its drivers.

Holistic Language

Absent a holistic approach, our universe of possible constructs would be little more than a series of disconnected loose ends.[260]

General T. Michael Moseley

Our language must avoid terms that suggest some forms of war are "better" than others. Thus, rather than imagining the totality of war as a rectangular table with preferred forms (e.g., regular) "seated" at the head, we must view war as a *continuum*, like an Arthurian round table where there is no irregularity and all war's forms have equal stature befitting their kinship. In this model, war's forms are analogous to the *states* of a physical substance (e.g., solid, liquid, or gas); thus, all are related forms of the same substance, though the states themselves, like the forms of war, *appear* quite different.[261] For example, water exists as liquid, solid (ice), and gas (steam). Water, ice, and steam are neutral terms and all are equally valid *forms* of H_2O. Furthermore, scientific theory explains that water's forms depend on measurable variables like temperature and pressure. While war is more complex than water, this physical model provides a blueprint for revising how we think and communicate about war and its forms. If war is like water with forms that are analogues of ice and steam, we have only to select war's most inclusive form dialectic (i.e., regular-irregular), recast it using neutral terminology (our underlying property or dependent variable), and define its independent variables (like temperature and pressure) to complete our task. We will begin by contextualizing war's forms within an ancient but familiar philosophical paradigm.

Yin-Yang

There is a right time for everything . . . a time to kill; a time to heal; a time to destroy; a time to rebuild . . . a time for war; a time for peace.

Ecclesiastes 3:1–8

It was the best of times, it was the worst of times . . . it was the season of Light, it was the season of Darkness, it was the spring of hope, it was the winter of despair.[262]
Charles Dickens, *A Tale of Two Cities*

Chapter 1 explains that war has a dialectic nature, and history and war theory confirm the variability of war's character. Sometimes, a direct path, concentration, speed, and ethical restriction facilitate victory. In other cases, an indirect path, dispersion, patience, and deceit are more effective. War theory must forge these dialectics into a coherent whole. War's dialectics and humanity's dual nature reflect the writings of Heraclitus and Taoism, philosophies that characterize the cosmos as a continuum of elements existing in various states of balance and polarity. *Yin* and *yang* are opposing yet complementary energies that constitute the "language of atoms, cells . . . ecosystems . . . galaxies, and life itself" (see Figure 4.11).[263] In his *Yin Yang Primer*, Edward Esko writes, "Reality is a unified field of countless interrelationships, all of which are defined by polarity. Polarities, or complementary opposites, are a common factor unifying all of existence."[264] Heraclitus was also fascinated by dichotomies, which he viewed as a motive force for change and representative of a universal whole. "Many

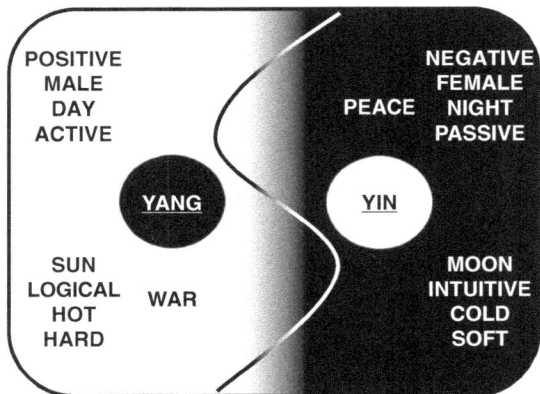

Figure 4.11 Yin-yang

who have learned from Hesiod the countless names of gods and monsters," writes Heraclitus, "never understand that night and day are one."[265] And, "From the strain of binding opposites comes harmony."[266] Greek mythos introduces the concepts of *metis* (cunning or wisdom, e.g., Odysseus) and *bie* (brute force, e.g., Achilles), which are akin to war's prudence-boldness, rest-movement, and reason-emotion dialectics. Greek fighting was most successful with the *metis-bie* dialectic in balance. Much later, Machiavelli adopted a similar metaphor: the lion (strength) and the fox (cunning). When using force, writes Machiavelli, "A Prince should . . . choose both the lion and the fox."[267]

War's Yin-Yang

The unorthodox and orthodox are the mutual changes of yin *and* yang *in Heaven and man.*[268]
Li Ching

Interestingly, the Chinese ideograph for military (武 or *wŭ*) is a composite of two separate ideographs, one that means "spear" and the other "prevent"; thus, *even this Chinese word related to war is a dialectic.*[269] Indeed, "war," writes Clausewitz, "consists of a continuous interaction of opposites."[270] Consider the following statement from Corbett: "We are accustomed, partly from convenience and partly from lack of a scientific habit of thought, to speak of *naval strategy* and *military strategy* as though they were distinct branches of knowledge which had no common ground. It is the theory of war which brings out their intimate relation" (emphasis added).[271] Replacing the italicized words with Sun Tzu's *cheng* and *ch'i* or "regular war" and "irregular war" illustrates how Corbett's statement exemplifies our problem. Since all war's forms are variations of regular or irregular and the regular-irregular dialectic aligns to the prime dialectic (order-chaos), these two forms are war's *yin-yang*. However, these terms are unsuitable, for there is no principle of "regularity" upon which to define and develop the relationship, and the terms have inappropriate connotations. Even US military doctrine admits their inadequacy in the following note from JP 1:

It is recognized that the symmetry between the naming conventions of *traditional* and *irregular* warfare is *not ideal*. Several symmetrical pair sets – regular/irregular, traditional/nontraditional (or untraditional), and conventional/unconventional – were considered and discarded. Generating friction in the first two instances was the fact that most US operations since the 11 September 2001 terrorist attacks have been irregular; *this caused the problem of calling irregular or nontraditional what we do routinely* (emphasis added).[272]

Returning to our water analogy, if regular and irregular are war's "ice" and "steam," then our remaining tasks are to name war's underlying property and identify the variables (e.g., temperature and pressure) that influence it. Fittingly, we can find our answers in fluid dynamics.

Fluidic War

Suppose they were an influence, a thing invulnerable, intangible, without front or back, drifting about like a gas?[273]
                                                                                    T. E. Lawrence

In the latter half of the twentieth century, the scientific and mathematical renaissance – which enabled us to split the atom, visit the moon, and invent the microcomputer – opened new avenues for comprehending the impossible (e.g., chaotic and complex nonlinear systems like weather). Indeed, chaos theory originated from the effort to predict weather in the fluidic atmosphere, and it required a new conceptual framework and language.[274] Like weather, war can be chaotic, complex, and nonlinear. Its forces are dynamic. They move, rest, concentrate, and disperse, flowing from form to form. In other words, they behave *similarly to fluids*.[275]

While fluid dynamics and war are hardly identical, there are many similarities. Large units are composed of smaller units and individuals, like fluids are composed of molecules and atoms. Though combatants can make individual decisions, they are guided by their training, tactics, and doctrine, standards analogous to the physical laws governing fluid dynamics. Thus, neither forces in war nor fluids are *entirely* chaotic.[276] Furthermore, the larger a volume of fluid, the less the individual molecules affect the attributes of the whole. Similarly, the larger a fighting force, the less each individual influences the rest, unless that person is a key leader like a general. Still, there are instances – for fluids and war – where small perturbations can result in disproportionate effects (e.g., the "butterfly effect").[277]

Associating war with fluids, in fighting and theory, is not new. Sun Tzu and many others have described war in fluidic terms. Sun Tzu says, "The control of forces is like water: do not fight rigid battles, water's movement avoids the high and runs downward ... avoids superiority and strikes inferiority ... water conforms to terrain in determining movement and forces conform to the enemy in determining victory."[278] In Homer's *Iliad*, armies clash "wildly as two winter torrents raging down from the mountains, swirling into a valley, hurl their great waters together."[279] "Fluid" often describes rapidly changing circumstances or mobile, adaptable forces. Lawrence's guerrillas "added fluidity to speed" to achieve "maximum disorder."[280] Mao's mobile warfare was known for its "flexibility and fluidity"; and, he credited the "fluidity of guerrillaism [*sic*]" for confounding Chiang Kai-shek.[281] Moreover, writes Mao, guerrillas "move with the fluidity of water and the ease of the blowing wind."[282] These forces, writes Samuel Griffith, "avoid static dispositions; their effort is always to keep the situation as fluid as possible, to strike where and when the enemy least expects them."[283] Furthermore, Galula says, "The insurgent is *fluid* because he has neither responsibility nor concrete assets; the counterinsurgent is *rigid* because he has both" (emphasis added).[284] Finally, irregulars, affirms Liddell Hart, "can operate only in minute particles, though these may momentarily coagulate like globules of quicksilver ... for guerrillas the principle of 'concentration' has to be replaced by that of 'fluidity of force.'"[285]

Fluids are represented mathematically by the nonlinear Navier–Stokes equation.[286] Except for special cases (e.g., incompressibility), this equation is unsolvable.[287] Indeed, in 2012, the scientific community declared the challenge of describing fluidic behavior the biggest unsolved mystery of physics.[288] In science, a fluid is "an infinitely divisible substance, a *continuum*, and the behavior of individual molecules, again due to their dynamic complexity, cannot be accounted for mathematically."[289] Similarly, war is a complex continuum of forms that cannot be fully understood by the analysis of single combatants. In war and fluid dynamics, complex interactions at the individual and molecular levels, respectively, can result in chaotic, nonlinear behavior, where identical initial conditions yield substantially divergent outcomes.

Viscosity

Though complex and chaotic, fluids have measurable macroscopic properties. One example is *viscosity*, the tendency of a fluid to have a cohesive consistency and to resist flow due to molecular attraction.[290] Viscosity is affected by temperature, molecular shape, and intermolecular forces. Higher viscosity corresponds to greater flow resistance (e.g., molasses, tree sap, or tar) A hard solid, like diamond, represents a theoretical upper limit of viscosity. though technically solids are not fluids. Free-flowing fluids, like gaseous or liquid water, have low viscosity.[291] Fluid dynamics are more easily modeled and predictable in special cases, like incompressibility and where viscosity is high, and less so as fluid molecules become more dispersed (i.e., with lower viscosity). Similarly, most war theory explains war's traditional forms, where forces are ordered and concentrated, better than it does irregular forms. Like fluids, war has measurable macroscopic properties that enable us to characterize its forms with terms like "regular" or "irregular." In general, regular forces adhere to rules, are well-ordered, and fight with traditional weapons and tactics; whereas irregular forces often ignore rules, have an amorphous structure, and fight in an unorthodox manner.

Viscosity: A Fluidic Metaphor for War

We can mold the conceptual parallels between fluid dynamics and war into a *fluidic metaphor* to advance our understanding of war's forms. In the Clausewitzian tradition of adapting scientific terms to describe key concepts in war (e.g., friction and center of gravity), we adopt *viscosity* as the underlying property linking war's regular and irregular forms. Replacing war's legacy jargon with this neutral term, borrowed from a phenomenon that shares war's complexity, allows us to visualize warring forces as fluidic analogues of varying viscosity (versus "regularity") and war writ large as a *continuum of forms characterized by the viscosity of opposing forces*. This fluidic metaphor effectively (and at last) relates all war's forms through viscosity.

Defining Viscosity

Viscosity, our fundamental property and dependent variable, is a function of the following independent variables (e.g., war's "temperature," "pressure," and so on): *directness, acceleration, restriction, cohesion,* and *concentration* (DARCC). Each of these correlates to one or more of war's dialectics.

Directness (direct-indirect dialectic) relates to the path and objective of force movement and actions exemplified by Sun Tzu's *cheng* and *ch'i* and Liddell Hart's direct and indirect approach. A direct approach moves along Liddell Hart's "line of natural expectation" with minimal maneuvering or subterfuge. Conversely, an indirect approach advances along lines of "least expectation" and fights only after gaining surprise or otherwise confusing the enemy. Viscosity *increases* with directness (and force size).

Acceleration (rest-movement dialectic) is the rate of change of velocity, but applied to war it is a force's ability to change speed and direction. Generally, smaller, lighter forces are capable of the greatest acceleration. For example, cavalry, aircraft, and platoons have greater acceleration (and lower viscosity) than infantry, ships, or battalions, respectively; thus, viscosity *decreases* with acceleration.

Restriction (good-evil and order-chaos dialectics) describes how closely a force follows rules, like rules of engagement or the law of war. While rules in war have benefits (more on this later), they can also make actions more predictable and lend tactical advantages to less constrained, more fluid forces. Viscosity *increases* with restriction.

Cohesion (control-autonomy dialectic) is the interdependence of force elements. For example, Alexander's phalanx, which relied on interconnected, mutually supporting components, was highly cohesive. In contrast, Genghis Khan's horsemen and Callwell's small wars foes, described as having "no *cohesion* and no *mutual reliance*" (emphasis added), were autonomous, moving independently, and rejoined only when necessary.[292] Viscosity *increases* with cohesion.

Finally, *concentration* (concentration-dispersal dialectic) is the physical arrangement and proximity of forces and their effects (Jomini's *solidity*); technically, concentration is the quantity of forces in a given area. Cohesion and concentration are related but not identical. Concentration can change often and rapidly without a proportional effect on cohesion. Viscosity *increases* with concentration.

These five elements can appear in nearly any measure or combination. For example, concentrated, cohesive forces can take an indirect approach, maneuver rapidly, or ignore restrictions. Conversely, dispersed forces may take a direct approach and adhere to narrow rules of engagement. This highlights viscosity's fluidity within the continuum of war. Additionally, viscosity, like war, is *fractal*, meaning its patterns replicate at all scales (Figure 4.12). Tactical elements include individuals or small units, like ships, aircraft, or companies; at the operational level – large units, like wings, brigades, or naval strike groups; and at the strategic level – entire services, task forces, or numbered armies.

Emergent Viscosity

Viscosity is an *emergent* property, where "emergence" is a "creative principle" applicable when "the parts of a complex system have mutual relations that do not exist for the parts in isolation."[293] In other words, viscosity cannot be discerned by reductive analysis of its individual components; rather, its traits only "emerge" via the *interaction* of its elements within the open system of war.[294] The DARCC elements fluctuate – based on leadership direction,

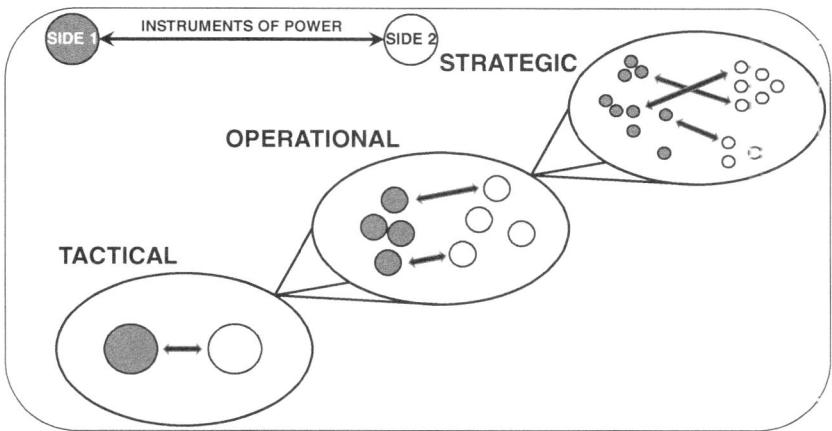

Figure 4.12 Fractal viscosity

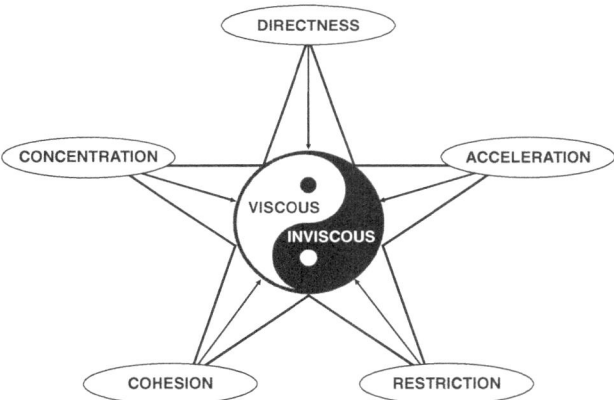

Figure 4.13 DARCC and emergent viscosity

individual initiative, or changing conditions – driving shifts between states of higher (*viscous*) and lower (*inviscous*) viscosity (Figure 4.13).

The Tao of Viscosity

For these two forces are mutually reproductive; their interaction as endless as that of interlocked rings. Who can determine where one ends and the other begins?[295]

Sun Tzu

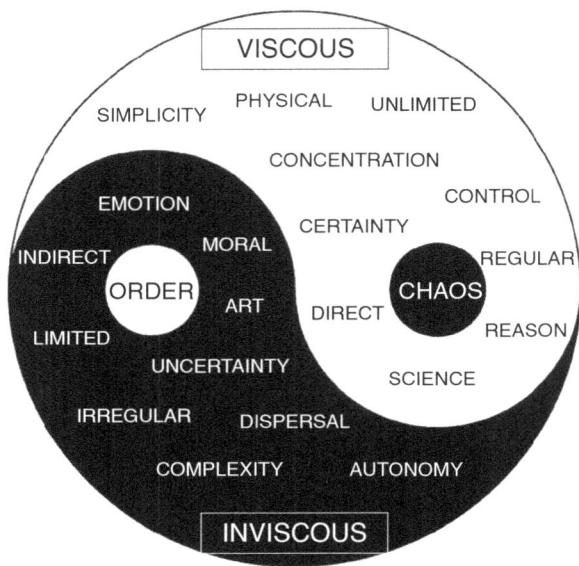

Figure 4.14 Viscosity and war's dialectics

Within the fluidic continuum of war, as Figure 4.13 depicts, higher *states* of viscosity are *viscous* (*yin*) and lower states *inviscous* (*yang*). These terms are, to borrow a phrase from social theorist Thomas Sowell, "abstractions of convenience," which orient our perception of force viscosity within the continuum.[296] There is no hard dividing line between viscous and inviscous force states. Taoists believe "nothing is entirely *yin* or entirely *yang*"; likewise in war, no force is entirely viscous or inviscous.[297] Therefore, viscosity is a *balanced* property inclusive of nearly all war's dialectics (Figure 4.14). In fact, having established its consistency with war's dialectic paradigm, we can now replace the "regular-irregular" dialectic with "viscous-inviscous."

Since war's forms are malleable, force viscosity is dynamic. Clausewitz writes,

These two opposite tendencies – toward concentration and toward dispersal – normally follow the natural bias of the troops themselves. Still, even in the most unequivocal of cases, it will be as impossible for one side always to remain concentrated, as it will be for the other to count on success simply by always operating in open order.[298]

Moreover, no viscosity force state is inherently *better* than another. The value of each depends on the conflict's circumstances and the engine of war. Finally, the principle of reciprocity means that force states (like capacity) are significant *only in relation to an adversary*, or in other words, *relative* viscosity drives war's forms. We will expand upon these concepts shortly.

Viscous Forces

Based on the preceding definitions, *viscous* forces have relatively high levels of directness, restriction, cohesion, and concentration, and low acceleration. Viscous force elements (individuals or units) are like molecules bound closely together, where each element influences nearby or dependent elements. A viscous force generally conforms to laws of war, and its strength lies mainly in its ability to organize, amass, and sustain concentrated power through the cohesion of its elements. Training and discipline contribute to greater viscosity. In general, traditional military forces (our small wars "haves") are viscous (e.g., Greek hoplite armies, Rome's legions, and Napoleon's *Grande Armée*). Fixed points of defense and tightly packed troop formations are the *most* viscous.

Viscosity generally increases with capacity (see Figure 4.15), though not always. The area between the curved lines represents *the likely* distribution of viscosity and capacity in war; but in truth, a force can adopt any viscosity state regardless of its capacity. That said, it makes little sense to disperse when strong and concentrate when weak. Machiavelli, Clausewitz, and Jomini preferred viscous states because they believed that great forces produce great victories. But history proves that the bigger and stronger force does *not* always win, and generating and sustaining coordinated, viscous forces takes considerable time, training, and resources. Viscous forces also tend to be less mobile, less stealthy, and less secure against forces that ignore rules. In short, a viscous force is not necessarily a *superior* force.

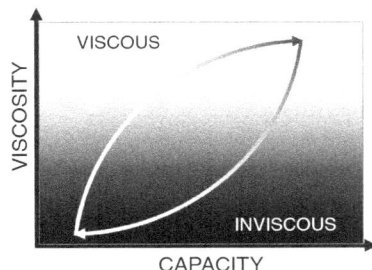

Figure 4.15 Viscosity and capacity

Inviscous Forces

Nothing is true; everything is permitted.[299]   The Assassin's Creed

Inviscous forces have relatively high acceleration and low directness, restriction, cohesion, and concentration. Inviscous force elements are like the atoms or molecules of a gas – dispersed, agile, and unconstrained. Lawrence's viscous opponents were "like plants, immobile, firm-rooted," whereas his inviscous rebels were "a vapor, blowing where we listed."[300] The "extreme rapidity" and "mobility" of inviscous forces, adds Callwell, increases "the perplexity and uncertainty" facing their foes.[301] Furthermore, the dispersal and independence of inviscous forces reduce collateral effects upon their elements. "Small bodies of troops thrown forward into the zone of the enemy's operations," writes Jomini, "belong to the class of detachments that are judicious. A few hundred horsemen thus risked will be no great loss if captured; and they may be the means of causing the enemy great injury."[302] Samuel Griffith describes "unorthodox" militias and guerrillas as "ubiquitous, intangible, and deadly antagonists."[303] Inviscous force examples (our small wars "have-nots") include Spain's *guerrillas*, Mao's peasants, the Viet Cong, and al-Qaeda. The *most* inviscous force would be an individual or small group – like an assassin or terrorist cell – operating independently and without restriction. The Trojan Horse soldiers, John Wilkes Booth, the USS *Cole* bombers, and the 9/11 hijackers exemplify this extreme. Despite their physical separation and reduced interdependence, inviscous forces are not wholly incoherent. They are bound by a cause and strive to create effects, though separated in time and space, that advance a shared political aim. The weak, contends Clausewitz, can use "people's war" against the strong provided the will and soul are there to sustain them.[304]

Survive to Win

As firepower grew by leaps and bounds in the latter half of the nineteenth century . . . dispersion became the key to success.[305]   Gwynne Dyer

As Figure 4.15 implies, inviscous forces typically have less capacity than viscous forces. In fact, the capacity deficit is causal. In small wars, it is irresponsible for "have-nots" to be predictable, immobile, and direct. Thus, they lower viscosity by avoiding direct battle and capitalizing on speed, deception, and surprise, which we learned earlier create temporary, localized capacity advantages. An inviscous force becomes like water to the viscous force's hammer. A hammer shatters ice, but liquid water yields while causing the hammer to rust and weaken. For example, in the Second Punic War (Rome versus Carthage, 202 BCE), Scipio Africanus defeated Hannibal's

"Carthaginian elephant charge" [viscous] by dispersing his troops in a "chequerboard formation" [inviscous].[306] The elephants passed harmlessly through the lines allowing Scipio to counterattack and win. Though fewer fighters can mean less raw power, inviscous forces are often economical, elusive, resilient, and able to sustain operations for extended periods. Weaker forces pursue what Clausewitz calls a *negative* aim, "which lies at the heart of pure resistance, [and] is also the natural formula for outlasting the enemy, for wearing him down."[307]

Fair Is Foul, and Foul Is Fair

Cruelty and brutality unmodified by rules of war are basic and enduring characteristics of irregular warfare.[308]   Colin Gray

Rejecting restriction is a basic trait of inviscous forces. As the *Small Wars Manual* says, "Irregular troops may disregard, in part or entirely, International Law and the Rules of Land Warfare."[309] Without rules, war can be fought in almost any manner, and targets and effects can extend beyond those permitted to viscous forces. "Flexibility," writes Sun Tzu, "is the key to properly exercising combat power," and inviscous forces maximize flexibility by ignoring the rules constraining their opposition.[310] Lawrence's insurgents were "odd" with "no honour, no conventions."[311] Liddell Hart adds, "Violence takes a much deeper root in irregular warfare [inviscous] than it does in regular warfare [viscous]. In the latter it is counteracted by obedience to constituted authority, whereas the former makes a virtue of defying authority and violating rules."[312] By learning its enemy's constraints, the inviscous side can predict adversary actions, which enables deception and surprise. Al-Qaeda's 2001 airliner attacks are a case in point. The following passage (from two Chinese theorists, Qiao Liang and Wang Xiangsui) summarizes the unconstrained nature of inviscous forces:

The direct result of the destruction of rules is that the domains delineated by visible or invisible boundaries which are acknowledged by the international community lose effectiveness. Visible national boundaries, invisible internet space, international law, national law, behavioral norms, and ethical principles, have absolutely no restraining effects on them. They are not responsible to anyone, nor limited by any rules, and there is no disgrace when it comes to the selection of targets, nor are there any means which are not used.[313]

The Principle of Relative Viscosity

Although we can estimate an isolated force's viscosity using the DARCC criteria, the viscosities of forces *relative to each other* are all that matter in combat. For example, Napoleon's army was viscous relative to Spanish guerrillas but inviscous relative to a fortress. Moreover, though viscosity is fractal and applies to every level of war, it can vary with level relative to its opposition. A militia, for instance, may be inviscous relative to an enemy's main army (operational) but viscous relative to small elements of that army (tactical). This is the principle of *relative viscosity*. To further illustrate this concept, consider the 1998 World War II saga, *Saving Private Ryan*. At the outset, a viscous invasion force (greater capacity) overpowers viscous coastal defenses (less capacity) to secure the Normandy beaches. Later, a small, inviscous force, led by Captain Miller, embarks on a mission to retrieve Private Ryan. At one point, Miller's team is beset by a sniper (*relatively* inviscous) who is eventually neutralized by another sniper. Relative to each other, the snipers are viscous, and the outcome is decided by skill and OODA Loop mastery. Later, in another battle of relatively viscous opponents, a stronger German soldier isolates, overpowers, and kills one of Miller's men in single combat. While Miller's unit is inviscous relative to larger units, it is viscous relative to isolated enemy elements, like the sniper.

Before striking, it benefits forces of all viscosities to seek localized capacity superiority (as the German sniper and soldier do versus Miller's team); thus, clashes between inviscous forces, in effect, become viscous clashes at smaller scales (more on this shortly). As Sun Tzu writes, "If I concentrate while he divides, I can use my entire strength to attack a fraction of his."[314] And Clausewitz adds, "Even in the absence of absolute superiority," effective commanders attain "relative superiority ... at the decisive point."[315] The adage "divide and conquer," raids, and ambushes (which leverage surprise's capacity amplification effect) typify this principle.

Viscosity, Posture, Capacity, and Culmination

Relative capacity and viscosity have numerous implications for posture (attack and defense) and culmination. According to Michael Handel, understanding culmination and war (or battle) termination "requires meticulous calculation of the current relative strength of both sides as well as the careful projection of anticipated trends in their relative strength."[316] Figure 4.16 graphically illustrates these relationships. As we indicated earlier, forces with relatively high capacity are more likely to attack and those with low capacity to defend or withdraw. Though attackers generally expend physical capacity faster, *moral* capacity can increase, especially with success (positive momentum). Conversely, if the defender is overcome, its physical *and* moral capacity can decline rapidly, leading to a rout. The defender's capacity surpasses the attacker's at the *culmination point*. Furthermore, though force viscosity *generally*

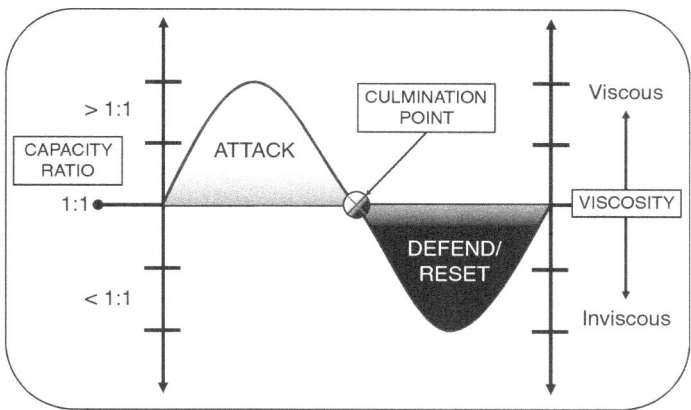

Figure 4.16 Relationship between posture, viscosity, capacity, and culmination

varies directly with relative capacity, leaders can direct *any* viscosity state regardless of capacity. Finally, we must stress that Figure 4.16 represents *probabilities* and *relationships* not mathematical prescriptions for force posture, form, or combat behavior. As always, human perception, decisions, and actions drive postures and the DARCC criteria, which in turn, affect viscosity and war's form. In short, *relative capacity drives viscosity, and relative viscosity drives war's forms*.

Containment

First, we are going to cut it off, and then we are going to kill it.[317]
General Colin Powell referring to the Iraqi Army (1991)

Earlier we introduced tactical containment, which sets the conditions for the effective use of violent force by restricting the enemy's ability to avoid or mitigate an attack. Within our fluidic concept of war, the principle of *containment* assumes a pivotal role as a fluidity inhibitor. A fluid influenced by an unbalanced force will move or flow unless contained. The container balances the force, restraining movement. Likewise, in war, a strong force can contain and successfully engage and defeat a weaker foe by restricting or preventing its ability to disperse and "flow" away (Figure 4.17). Containment, whether imposed or contextual, holds, controls, or limits, thus increasing viscosity and vulnerability to superior force. In his introduction to Mao's *On Guerrilla Warfare*, Samuel Griffith lists three requisites for anti-guerrilla operations: "location, isolation, and eradication," where "isolation" means

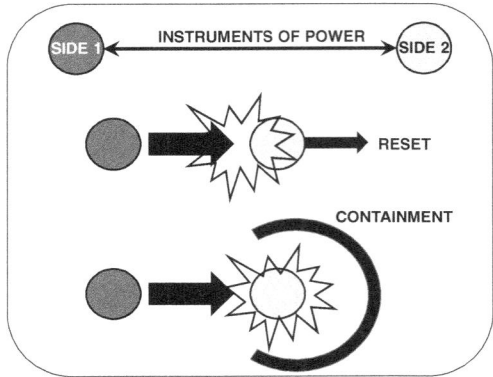

Figure 4.17 Containment

containment.[318] To destroy "an enemy who is invisible, fluid, uncatchable," writes Trinquier, "we have no alternative but to throw a net of fine mesh over the entire area" (i.e., to contain them).[319]

Sun Tzu was the first theorist to relate relative capacity to containment. He writes, "Those sophisticated at actuating the enemy contain them, and the enemy will surely follow."[320] Furthermore, "Those ... sophisticated at command contain the adversary but are free of containment."[321] With a capacity advantage (10–1 or 5–1), Sun Tzu advises forces to surround or fix in place (contain) and attack along the flanks. When the situation is reversed, Sun Tzu recommends deferring battle and escaping containment.

The Principle of Inertia

Newton's first law states that objects remain in motion or at rest unless acted upon by an unbalanced, external force. In war, the former relates to our principle of *momentum* and the latter to a principle of *inertia*, which says that forces in war *remain in their present state unless acted upon or motivated to act*. Inviscous forces, with greater acceleration and less size, generally have less inertia relative to viscous forces. Surprise and speed contain by capitalizing on inertia. In the Peloponnesian War, Demosthenes exploited this principle to contain and rout the "inert" Ambraciots, who "were still in their beds and had no suspicion" as the Athenians, "speaking the Doric dialect and inspiring trust in their sentries while still invisible in the darkness," attacked with impunity.[322] Inertia applies to all domains and levels of war. For example, Japan in 1941 and North Korea in 1950 exploited *strategic inertia* to mount surprise attacks on

Hawaii and South Korea, respectively. Later in the Korean War, US-led forces capitalized on North Korea's *operational inertia* with the daring Inchon landing. A state of inertia often results after culmination (e.g., the USA shortly after China entered the Korean War).

Direct and Indirect Containment

Direct containment is imposed through physical action like a rapid or surprising attack, encirclement, naval blockade, cyber firewall, or airborne "no-fly zone." For example, at Dien Bien Phu, the Viet Minh directly contained the French garrison by occupying the surrounding high terrain and impeding retreat and resupply. Likewise, as Liddell Hart points out, Napoleon directly contained his opponents by "maneuver onto the enemy's rear . . . as a means of gripping them firmly, so that they could be drawn into his jaws."[323] Speed and surprise also create direct containment because, consistent with the OODA Loop, they deny defenders the time to evade or withdraw. Putting an opponent off balance with a feint or ruse slows reaction times to a subsequent attack. In fact, containment increases with surprise and speed such that *total* surprise or *instant* attack result in *total* containment (i.e., the defender receives the attack's full force before it can respond). This explains the proven military value of surprise and high speed weaponry, like the machine gun, that combine containment with firepower (capacity).

Indirect containment is imposed by the conditions or context of battle, especially terrain. According to Jomini, armies on interior lines opposing "armies superior in numbers, should not allow themselves to be crowded into a too contracted space, where the whole might be overwhelmed at once.'[324] This happened during the fourteenth century's battle of Bannockburn where Edward II's "cavalry were enclosed on three sides by the Pelstream and the Bannockburn," thus allowing Robert the Bruce's Scottish *schiltron* (a dense, viscous formation of spears and pike) to close the "sack" and crush the knights.[325] Likewise, the Allies at Dunkirk during World War II were directly contained by German forces and indirectly by the English Channel. The Allies escaped containment thanks to Germany's hesitation (inertia) and lack of sea power.

Finally, containment can always occur by *accident* or *choice*, from incompetence, ignorance, or a perceived need to defend a position at all costs (like Thermopylae). For example, Hitler's obsession with Stalingrad essentially contained the German Sixth Army, enabling the stronger Russians to encircle and destroy it. Lawrence explains that guerrillas (inviscous) can be obligated to fight "by lack of land-room, or by the need to defend a material property dearer than the lives of soldiers."[326] This is why powerful forces often target locations of high value (like a capital city) to force the opposition to contain *themselves* to protect the objective. However, this does not always work, as Napoleon's fruitless occupation of Moscow proved.

The Power of Containment

The chief danger arising from resistance to a superior enemy is the possibility of being outflanked and placed at a definite disadvantage by an enveloping attack.[327]   Clausewitz

Most tactical dogma advises dividing an opposing force into smaller elements and attacking in detail. This creates a series of engagements in which stronger forces contain and destroy weaker forces. As Clausewitz writes, "Such a victory demands an enveloping attack or a battle with reversed fronts, either of which will always make the result decisive."[328] The German word for this technique is "*Kesselschlacht*" or "battle of encirclement," which Hitler's *Wehrmacht* used brilliantly against Russia in 1941.[329] However, achieving full encirclement is rare and can backfire against a desperate enemy if attempted without a massive capacity surplus. "If an enemy is given no chance of escape," warns Chanakya, "its fury and desperation often become irresistible."[330] For this reason, Sun Tzu advises against encirclement with less than a 10–1 force ratio! When encirclement is impossible or imprudent, flank attacks can have positive results by forcing an enemy from a preferred position.

Containment is a multi-domain concept. Direct naval and air containment are possible through blockades and no-fly zones; furthermore, air and naval forces can impact land containment. Lack of air and naval supremacy, for instance, prevented Germany from containing and destroying the Allied force at Dunkirk and, later in the war, precluded the Allies from doing the same to German forces on Sicily (Operation *Husky*). In the 1990 Gulf War, air and sea power helped contain Saddam's forces in Kuwait, enabling Schwarzkopf's "left hook" to crush the Iraqi army in just 100 hours. In the cyber domain, firewalls and speed-of-light "attacks" contain cyber adversaries with digital code; and in space, orbital physics and the atmosphere cause indirect containment, and surprise and speed are optimal for containing directly.

Perhaps the best way to contain is through direct *and* indirect methods (e.g., Bannockburn or Dunkirk). Thucydides explains how the stronger Syracusans of Sicily harried Demosthenes's Athenian invaders by building walls to restrict movement, then encircled and attacked the Greeks from all sides. Prevented from dispersal or retreat, Demosthenes's 6,000 hoplites surrendered.[331]

Operational and Strategic Containment

War's fractal nature means containment applies at all levels. Examples of operational containment include border security and naval and air superiority. In fact, just prior to Operation *Overlord*, the Allies and Russia had effectively contained Germany in three dimensions by decimating the *Luftwaffe* and enveloping continental Europe on land and sea. Similarly, by 1945, the USA's Pacific air and sea dominance had contained Japan setting up the atomic *coup de grace*. Strategic containment can occur via a variety of IOPs including alliance build-ing, trade agreements, economic sanctions, and tariffs. Prior to World War I, France allied with Russia to contain Germany. Then America's entry into that war, along with the British Navy, put Germany in an economic straitjacket. Liddell Hart writes, "The United States wielded the economic weapon with a determination ... America's co-operation converted it into a stranglehold under which Germany gradually became limp, since military power is based on economic endurance."[332] After World War II, George Kennan's Cold War "containment" stratagem frustrated the global expansion of Soviet communism. Though Kennan acknowledged the unacceptability of a nuclear showdown, he believed the USA was obligated to support allies and proxies along the Soviet periphery to contain communist influence and capacity growth.

Viscosity in Combat

Even the most precise, clear, and widely accepted set of definitions .. cannot be used for prediction, explanation, or providing a sense of understanding. Only when the next step is taken, providing statements that describe relation-ships between concepts, can these other goals [of theory] be attained.[333]

Paul Reynolds

Though viscous and inviscous states are more thematic than mathematic, they allow us to define and evaluate four *types* of combat engagements: (1) viscous versus viscous (V-V); (2) viscous versus inviscous (V-I); (3) inviscous versus inviscous (I-I); and (4) hybrid (H-I/V/H). These effectively encompass the legacy regular-irregular-hybrid forms of war.

Viscous-Viscous (V-V)

If the enemy is substantial, then I must use the orthodox. If the enemy is vacuous, then I must use the unorthodox.[334]   Li Ching

V-V interaction (Figure 4.18) is the "hard science" of combat (i.e., clashes of well-ordered, trained, cohesive forces vying to destroy each other's engine of war). V-V engagements best describe the state-versus-state, viscous, land warfare Jomini and Clausewitz were most concerned with and what Hans Delbrück dubbed *Niederwerfungsstrategie* (strategy of annihilation or direct attack). Clausewitz defines two principles for this type of war "that underlie all strategic planning": (1) "act with the utmost concentration," and (2) "act with the utmost speed."[335] Concentration maximizes combat power on the target, and speed creates containment and OODA Loop superiority by exploiting enemy inertia.

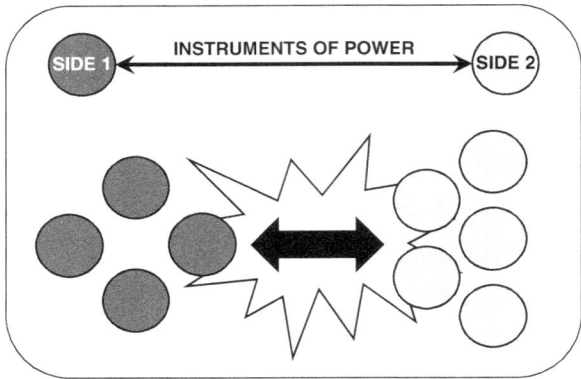

Figure 4.18 Viscous vs viscous (V-V)

V-V combat generally results when capacity is relatively even but at least one side believes itself superior in one or more decisive physical or moral attributes (e.g., leadership, training, or motivation). In these instances, the side that concentrates its power and strikes the enemy's key COGs first has the advantage. "The attack should be like a well-hammered wedge," writes Clausewitz, "not a bubble that expands till it bursts";[336] and, "The force at which our blow is to be aimed requires that our strength be concentrated to the utmost."[337] Opposing viscous forces seek to gain capacity advantages they can exploit at an optimal time and place.

The greater the advantage, the greater the effect. Though achieving full physical containment in V-V combat is rare (and usually unwise when capacities are similar), surprise and speed (via airpower, for example) are always beneficial, because they amplify capacity and achieve containment simultaneously. Order, cohesion, and concentration provide the power, and directness and acceleration the containment. Napoleon's campaigning often exemplified this approach. "When [Napoleon's] cavalry ... reported the presence of the Russians in a strong position," writes Liddell Hart, "he swung his forces straight at the target. The tactical victory was won, not by surprise or mobility, but by pure offensive power ... the massed concentration of guns at a selected point."[338] The German *panzers* of World War II followed a similar model. Liddell Hart calls it "astonishing ... that the German panzer forces, which had just previously been striking north-westward, were so quickly switched southward for a fresh stroke. Such rapidity of re-concentration in another direction was fresh evidence of how mechanized mobility had revolutionized strategy."[339] In cohesion, training, leadership, and experience, if not in numbers or technology, the Germans outclassed their opposition.

When capacities are similar and geographic conditions indirectly contain *both sides*, viscosity increases and defenses can be difficult to penetrate. This occurred in World War I and in later stages of World War II, including in France prior to Operation *Cobra* and in Italy where, as Liddell Hart figuratively depicts, "The Allied armies were reduced to pushing their way up the Italian peninsula like a sticky piston-rod in a stickier cylinder against increasingly strong compression."[340]

Viscous-Inviscous (V-I)

To force an inferior enemy to concentrate is indeed the almost necessary preliminary to securing one of those crushing victories at which we must always aim, but which are so seldom obtained.[341]   Julian Corbett

Earlier we discussed "asymmetric war" relative to small wars and war's forms. Now, with the introduction of viscosity, we can formally define asymmetric war as: *warfare between forces of substantially dissimilar viscosity* (V-I or I-V; see Figure 4.19). Asymmetric warfare proliferated during the twentieth and twenty-first centuries as nations with militaries optimized for V-V conflict struggled against inviscous insurgents and terrorists. Freedman writes,

The most powerful dichotomy in all strategic thought was the one first introduced by Homer as the distinction between *bie* and *metis*, one seeking victory in the physical domain and the other in the mental, one relying on being strong and the other on being

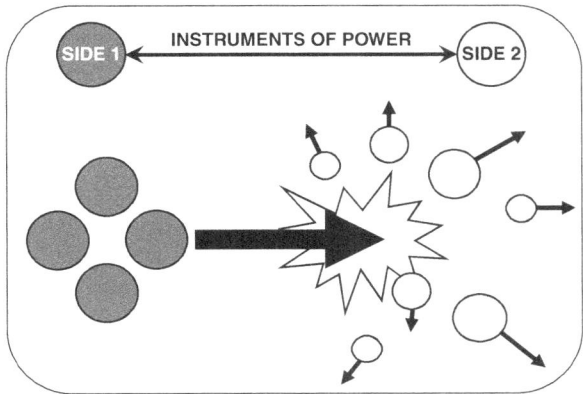

Figure 4.19 Viscous vs inviscous (V-I)

smart, one depending on courage and the other imagination, one facing the enemy directly and the other approaching indirectly, one prepared to fall with honor and the other seeking to survive with deception.[342]

The Inviscous Perspective

As taught or practiced by the great guerrilla leaders of the Far East ... the warfare of national liberation was designed to nullify large-scale maneuver and the superiority of numbers and equipment which underlay it. Chasing will-o'-the-wisps through paddy fields and rain forest, European soldiers were quickly reduced to a despair of frustration, beside which the certainties of traditional warfare, whatever their cost in blood, came to seem temptingly preferable.[343]

John Keegan and Joseph Darracott

Taoism says, "If one side is stronger (*yin*), the other side becomes *yang* in relation to it."[344] In other words, the greater the capacity variance, the more imperative it is for the weaker side to become *yang* (inviscous). To survive against a viscous foe, inviscous forces often abandon constraints and rigidity in favor of flexibility and fluidity. Sun Tzu asserts that smaller forces (low capacity) should be "dynamic" (fluid) while larger forces (high capacity) should "ensnare" (contain).[345] The Roman general, Fabius, followed this model against Carthage, for although his army was organized as a viscous force, Fabius chose an inviscous state because he knew "Hannibal's military superiority too well to risk a military decision."[346] Clausewitz writes, "Militia and bands of armed civilians [inviscous] cannot and should not be employed against the main enemy force – or indeed against any sizable enemy force. They are not supposed to pulverize the core but to nibble at the shell and around the edges."[347] To secure their aims, inviscous forces use any available IOP to circumvent viscous strengths and exploit weaknesses while eluding containment. The inferior force, says Clausewitz, "had better scatter and escape" else the force "be smothered by a single stroke."[348] And, "A general uprising ... should be nebulous and elusive; its resistance should never materialize as a concrete body, otherwise the enemy can direct sufficient force at its core, crush it, and take many prisoners."[349]

Since, inviscous forces are usually inferior to viscous forces in open, force-on-force battle, their primary objective must be to survive while degrading their enemy's engine of war. Hans Delbrück called this *Ermattungsstrategie* (strategy of exhaustion). Common targets for inviscous forces include isolated portions of a viscous force, leaders, lines of communication, logistics nodes, cyberspace vulnerabilities, and civilians. Even if an inviscous force cannot appreciably erode the enemy's capacity, preserving its own while surviving can weaken an enemy's will as it did for Fabius and Ho Chi Minh. Furthermore, inviscous forces that build capacity while degrading the enemy's may flip the capacity ratio, become viscous, and win a decisive battle (e.g., Washington, Mao, and Giap).

The Viscous Perspective

The regular soldier engaged in irregular warfare generally finds that the enemy has no obvious 'center of gravity,' no capital city ... and probably no fixed lines of communication; looks identical to the civilian population; and refuses to stand and fight unless he is trapped and has no choice.[350]

Colin Gray

Since, according to our theory, inviscous force states are a balancing response to greater capacity, viscous forces are *by definition* ill-suited for fighting inviscous forces. Thus, the recurring surprise and frustration exhibited by traditional powers in small wars confirms general theory's inability to adequately inform strategy for asymmetric war. Despite the power and "the resources at their back," writes Callwell, "the trained and organized armies ... have the worst of it," because their "elaborate organization ... and ... armament ... cramps their freedom" and "overburdens them."[351] Having greater acceleration and freedom of action, inviscous forces can remain out of reach, appearing only to strike before disappearing again. Thus, viscous forces must compel inviscous forces to become viscous before their own engine of war runs out of fuel.

As we stressed in the last chapter, the "haves" (viscous) must compete in war's human and political dimensions, and, in combat, *decrease* viscosity to counter the inviscous enemy's advantages. At times, viscous forces can entice the opposition to fight by occupying areas of value, threatening lines of communication, or using deception. To win a decisive battle, a viscous force must contain the inviscous foe directly or indirectly. "Like any opponent in any war," writes Kilcullen, "an insurgent enemy needs to be pinned against an immovable object and 'fixed' in order to be destroyed."[352] The French "Morice Line," which prevented rebel reinforcement from neighboring countries in the Algerian War, is an example of successful operational containment. Absent containment, says Callwell, inviscous forces "simply disperse when they are worsted. They disappear in all directions."[353] Jomini illustrates the dilemma for viscous forces:

After the most carefully concerted movements and the most rapid and fatiguing marches, he thinks he is about to accomplish his aim and deal a terrible blow, he finds no signs of the enemy but his camp fires: so that while, like Don Quixote, he is attacking windmills, his adversary is on his line of communications, destroys the detachments left to guard it, surprises his convoys, his depots, and carries on a war so disastrous for the invader that he must inevitably yield after a time.[354]

And Corbett adds, "If we are too superior, or our concentration too well arranged for him to hope for victory, then our concentration has almost always had the effect of forcing him to disperse."[355]

Until it can achieve containment, a viscous force's best option is to increase fluidity by reducing its own viscosity. Against insurgents (inviscous), the Germans of World War II (viscous) learned that "almost the only forces that mattered were those that were lightly armed – police, light infantry, mountaineers, special forces, signals units, and above all, intelligence personnel of every kind."[356] "The tactics of guerrillas," says Griffith, "must be used against the guerrillas themselves. They must be constantly harried and constantly attacked."[357] And, "The effectiveness of armies," adds Liddell Hart, can depend upon "the practicable object of paralysing the enemy's action rather than the theoretical object of crushing his forces. Fluidity of force may succeed where concentration of force merely entails a perilous rigidity."[358] Furthermore, "Dispersion is also a necessity on the side opposed to the guerrillas, since there is no value in a narrow concentration of force against such elusive forces, nimble as mosquitoes. The chance of curbing them lies largely in being able to extend a fine but closely woven net over the widest possible area."[359] But reducing viscosity is not easy, especially with poor tactical leadership or a lack of prior experience, a condition Galula likens to a giant trying to wear dwarf's clothing.[360] Clausewitz adds,

A regular army fighting another regular army [viscous] can get along without military virtues more easily than when it is opposed by a people in arms [inviscous]; for in the latter case, the forces have to be split up, and the separate units will more frequently have to fend for themselves. Generally speaking, the need for military virtues becomes greater the more the theater of operations and other factors tend to complicate the war and disperse the forces.[361]

The following vignette from the Roman-Parthian Wars of the first century BCE (related by Machiavelli) summarizes the nature of V-I engagements:

The Parthians all fought on horseback in a *loose* and *irregular* manner [inviscous], which, as a mode of combat, was unsteady and full of uncertainty; the Romans ... fought chiefly as infantrymen in *close* and *regular* order [viscous] – their success varied according to the nature of the countries in which they happened to fight. In enclosed places the Romans generally got the better, whereas the Parthians had the advantage in large open plains, and indeed the nature of the country they had to defend was very favorable to their manner of fighting ... thus the Roman armies, slowed down by the heaviness of their arms and armor and the good order they observed, were much annoyed by an active and light-armed, mounted enemy who was at one place one day and 50 miles off by the next (emphasis added).[362]

Inviscous-Inviscous (I-I)

Once soldiers match their tactics to those of their primitive adversaries, their superior manpower, economic surplus, transportation technology, and

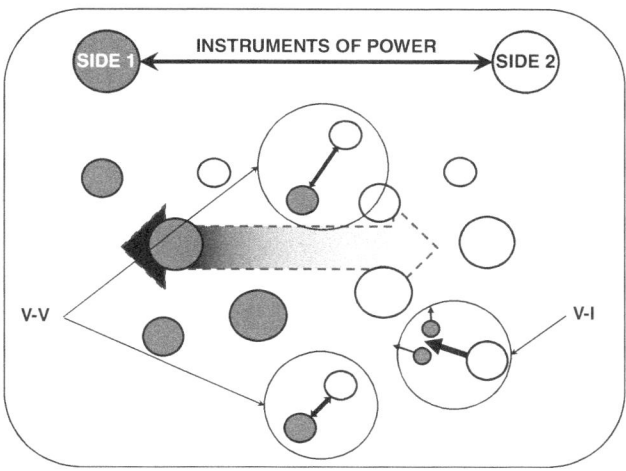

Figure 4.20 Inviscous vs inviscous (I-I)

logistical expertise – if vigorously exploited – enable them to win most such campaigns and wars.[363]   Lawrence Keeley

Symmetric warfare occurs *between forces of similar viscosity* (V-V or I-I). Pure I-I conflict is somewhat of an abstraction, because inviscous states are most appropriately a response to a viscous threat. As Figure 4.20 depicts, the fighting between two inviscous forces produces smaller scale V-V or V-I engagements as forces of varying viscosity attempt to find, fix, and destroy each other.[354] This corresponds to Heuser's "*symmetric* irregular warfare"[365] and political scientist Stathis Kalyvas's "symmetric, non-conventional wars."[366] As previously discussed, force states usually depend on the circumstances and relative capacities of each force. For I-I engagements, the stronger force must still contain the weaker, even if it is a platoon versus a sniper.

Because viscous forces often struggle to locate, contain, and engage inviscous forces, reducing viscosity presents a potentially superior alternative. Like an uncontained gas, operationally dispersed, inviscous forces have what amounts to a large "surface area." An inviscous force more effectively matches the surface area of another inviscous force, shrinking the scale of combat and creating V-V battles where localized capacity advantages (as Keeley implies) prove decisive.[367] According to Trinquier, this is what France did in Vietnam (prior to Dien Bien Phu) when it turned "the Vietminh's skill in fighting behind the lines against the Vietminh itself by implanting anti-Communist guerrillas deep inside the enemy's territory."[368] More recently, the USA employs this strategy against global terrorists. To cover this extensive threat, America uses remotely piloted aircraft, which can surveil and attack over large areas for extended periods; agile special operations forces; and multinational law enforcement units. Nonmilitary IOPs combined with combat further increase effectiveness by influencing human and political COGs, systemically degrading the opposition's engine of war.

Hybrid (H-I/V/H)

For those who excel at employing troops there are none that are not orthodox, none that are not unorthodox, so they cause the enemy never to be able to fathom them. Thus, with the orthodox they are victorious, with the unorthodox they are also victorious.[369]   Li Ching

In 2009, the (now defunct) US Joint Forces Command categorized a "hybrid threat" as "any adversary that simultaneously and adaptively employs a tailored mix of conventional, irregular, terrorism and criminal means or activities in the operational battlespace."[370] Since viscosity is applicable to all variations of war, we can say that *hybrid warfare* occurs *when at least one side employs both viscous and inviscous forces*. Figure 4.21 represents a hybrid force. The interactions of hybrid forces with viscous, inviscous, or other hybrid forces can be expressed in combinations of V-V, V-I, and I-I engagement types.

Hybrid force structure is nothing new. Although theory has not, until now, properly characterized hybrid force states, strategists and generals have employed variations over the years. "All regular armies," writes Keegan,

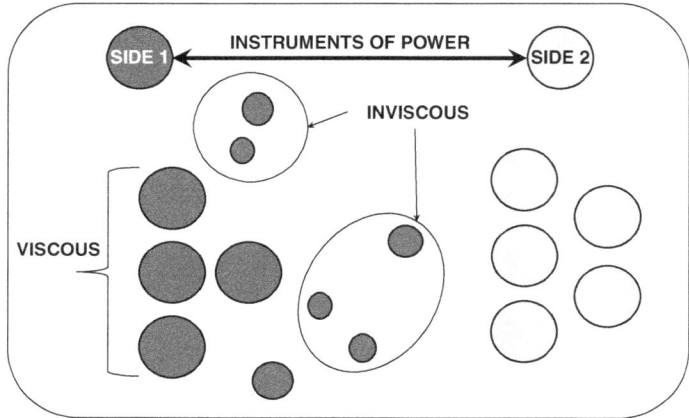

Figure 4.21 Hybrid

"recruited irregulars to patrol, reconnoitre and skirmish for them; during the 18th century the expansion of such forces ... had been one of the most noted contemporary military developments."[371] Moreover, Clausewitz colleague General August Rühle von Lilienstern adds, "War is *generally* composed of both forms of war ... and the operations and endeavors that occur in campaigns partly belong to the realm of small war, partly to that of major war."[372] Furthermore, a hybrid force is hybrid *by design*. In other words, unintentional variations in force viscosity during battle do not constitute a hybrid force. A hybrid force deliberately mixes viscous and inviscous units, like a composite material, to harness the strengths inherent in both states.

Effective employment of hybrid forces can yield significant benefits. As JP 1 points out, "It is, in fact, the creative, dynamic, and synergistic combination of both [traditional and irregular forces] that is usually most effective."[373] Regarding guerrilla warfare in Ireland, Liddell Hart writes, "When these back-area campaigns were analysed, it would seem their effect was largely in proportion to the extent to which they were combined with the operations of a strong regular army that was engaging the enemy's front and drawing off his reserves."[374] Collaboration between viscous and inviscous forces creates a constructive symmetry. Indeed, "Regular units and guerrilla bands," writes Trinquier, "will cooperate closely to try to bring about a situation favorable to the engagement of the enemy army in a decisive fight to annihilate it."[375] And Jomini adds, "Irregular troops supported by disciplined troops may be of the greatest value, in destroying convoys, intercepting communication, &c., and may – as in the case of the French in 1812 – make a retreat very disastrous."[376] Furthermore,

Mixed corps of regular and irregular cavalry may often be more really useful than if they were entirely composed of cavalry of the line, – because the fear of compromising a body of these last often restrains a general from pushing them forward in daring operations where he would not hesitate to risk his irregulars.[377]

The American War of Independence illustrates the benefits of combining regular forces and militia. In the southern colonies, Britain destroyed two viscous colonial armies. Still, Francis Marion, the "Swamp Fox" of South Carolina, used his inviscous militia to keep the British off balance; "operating with the greatest speed from inaccessible bases ... he struck blows in rapid succession at isolated garrisons, convoys, and trains."[378] Ultimately, the American hybrid force, including the viscous armies of Washington and Nathanael Greene, prevailed over the more experienced but less flexible British.

The blending of viscous and inviscous states is similar to the principle of combined arms or *Kampfgruppen*. Most combat-validated doctrines recommend concerted employment of a wide variety of weapons for optimal effects. This practice minimizes vulnerabilities and maximizes advantages to provide greater overall strength and flexibility. In World War II, for example, "even tanks in large numbers," writes Colin Gray, "could achieve little in the absence of supporting infantry, artillery, engineers and air power."[379] For Clausewitz, the "distribution of elementary military strengths" (e.g., the firepower of artillery, the mobility of cavalry, and the versatility of infantry) results in "maximum strength."[380] Adds Machiavelli, "Some generals have accustomed part of their light-armed infantry to mingle with their cavalry and to fight in conjunction with them; this has been of very great service."[381]

Of course, implementing hybrid force states and multi-domain combined arms is resource-intensive and presents training, leadership, and synchronization challenges. A viscous force that deploys inviscous elements must train them to fight in that role, not merely as smaller viscous units. Despite the benefits of hybrid force structure, inviscous forces are less constrained, less cohesive, and harder to command and control; thus, their activities have a higher probability of radiation (counterproductive effects). Inviscous forces, says Callwell, often commit "havoc which the laws of regular warfare do not sanction."[382] Toward the end of the Algerian War, for instance, France had largely defeated the insurgency; however, French use of torture and reprisals soured domestic and international opinion, causing a grave political crisis.

Dynamic Viscosity

Traditional warfare can rapidly evolve into an irregular war and vice versa, requiring the military force to adapt from one form to the other.[383]

Air Force Doctrine Document 1

Opining on the Indian Mutiny (1857), Callwell relates how the "organized mutineers," when faced with disciplined British firepower, "melted away," causing the campaign (V-V) to degenerate "into purely guerrilla warfare [V-I], which took months to bring to a conclusion."[384] As Callwell's tale and the AFDD 1 excerpt suggest, military experience accepts that war can change forms, though theory has yet to systematically explain how and why.

Viscosity in war is always present, and like *yin-yang*, it is "not static, but constantly changing."[385] Viscosity changes originate in the human mind, with decisions and reactions in the open system of war. Consequently, we can never know with certainty *how*, *when*, or *to what degree* viscosity will change. However, we can know with full certainty that viscosity change is always *possible*. Most importantly, using our knowledge of viscosity, we can assess the *probability* of change as well as the most likely force state. Where there is a will to fight on both sides, war *initially* takes its form based, as Clausewitz says, "on the natural bias of the troops" (i.e., on how the warring sides are organized, trained, equipped, and prefer to fight).[386] Adjustments are then made based on the resources available, the environment, the objective, and an assessment of relative capacity. As a war progresses, these factors change, prompting leaders to consider alternative force states. The UWT helps us understand *why* war changes form and provides an analytical framework for determining the *most advantageous* and thus, most *likely* force state.

The War-Viscosity Algorithm

'Algorithm' is arguably the single most important concept in our world.[387]

Yuval Noah Harari

By careful collation of past events it becomes clear that certain lines of conduct tend normally to produce certain effects; that wars tend to take certain forms each with a marked idiosyncrasy . . . that a system of operations which suits one form may not be that best suited to another.[388]

Julian Corbett

We can express this analytical framework as a diagram or algorithm, somewhat like the OODA Loop, illustrating the human decision-making process that influences how and why viscosity and war's forms change. Figure 4.22, the *War-Viscosity Algorithm* (W-V-A), is not prescriptive; however, it is useful for visualizing the relationships between war's key parameters and gauging how an adversary is *likely* to act or react.

According to the W-V-A, decisions to start, continue, or stop fighting and to make force state (viscosity) selections or changes are based on an analysis of will, context, and relative capacity. Reasonably accurate assessments of these parameters indicate the *optimum* friendly force viscosity and *most likely* enemy viscosity. Again, human decisions and actions can be arbitrary; thus, the W-V-A is only a guide, not an oracle for predicting specific future events. Moreover, though it is fully applicable to tactical warfare, the W-V-A is most useful as a tool for analyzing past conflicts and as a conceptual reference for commanders and strategists, not as a guide for real-time combat decisions.

W-V-A User's Guide

If we are strong, we press to the issue of battle when we can. If we are weak, we do not accept the issue unless we must.[389]   Julian Corbett

The W-V-A applies just prior to the start of a war or battle (henceforth called "the action") and runs continuously for *both sides* until the action is resolved. Contemplation of the action is the algorithm's start point. As depicted, there will

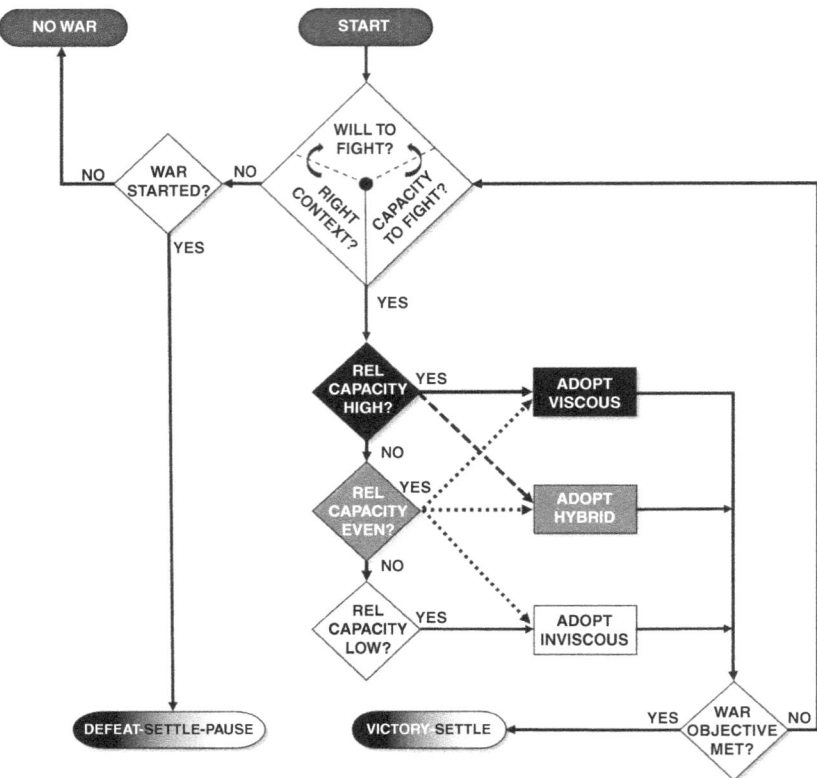

Figure 4.22 The war-viscosity algorithm (W-V-A)

be no war or battle unless both sides considering the action have the will to fight, believe the context (including conditions and timing) is sufficient, and retain some capacity.[390] For Chanakya, this calculus includes "strength of forces [capacity], favorable time and favorable place [context]: these three things should work together. Once the invader has all three he should move against the enemy."[391]

   Though will is paramount and appears first in the algorithm, it is not isolated from context and capacity. As the WCM depicts, will is supported by physical and psychological capacity, and the decision to exercise that will depends on the context, before and during war. In the opening assessment phase, the first, second, and initial parts of the third essential acts of strategy occur: defining the political aim; assessing the probability that ways and means can deliver the aim with acceptable cost, risk, and duration; and identifying and procuring those ways and means. It is here that many critical mistakes occur, increasing the chance for war when an accurate, objective assessment might otherwise prevent it. In particular, misperceptions regarding "the opponent's hostility," "the balance of power," and "risk," writes Greg Cashman, "are probably the most important factors" in the initiation of any given war.[392] There are times when objective evaluation of will, context, and capacity indicates that inaction (rest) is better than action (movement). "One who excels at warfare," advises the ancient Chinese general T'ai Kung, "will await events in the situation without making any movement. When he sees he can be victorious, he will arise; if he sees he cannot be victorious, he will desist."[393]

Once the action has begun, relative capacity assessments drive force state (viscosity) decisions for each side (enacted by adjusting the DARCC factors), and relative viscosity drives war's form (e.g., V-V or V-I). Decisions regarding "the form of operations to be adopted," writes Callwell, depend on careful consideration of "the strength and the fighting methods of the enemy."[394] Where capacity is high relative to the enemy, a viscous state is best (solid arrow; Figure 4.22); however, if the enemy adopts an inviscous form and cannot be contained, a hybrid state – resources permitting – offers the greatest benefit (dashed arrow). Forces with a clear capacity advantage may have little to gain by delaying or taking an indirect approach.

Where relative capacity is roughly equal, there is no clearly superior state of viscosity (dotted arrows). Here, the W-V-A confirms that a "balance of power" can deter by increasing uncertainty, risk, and doubt regarding the probability of success. In these cases, political leaders often judge the context or capacity inadequate for warfighting, for "whoever goes to war with a superior ruler will be ruined," warns Chanakya, "and war with an equal brings ruin to both."[395] As Chanakya implies, war in a balance-of-power scenario often becomes a battle of attrition; however, real-time analysis of circumstances, will, and relative capacities helps leaders optimize force states, which may permit factors like chance, leadership, deception, or surprise to break the deadlock. Finally where relative capacity is low, adopting an inviscous state is clearly best (solid arrow), except in rare cases where martyrdom trumps survival (e.g., the Alamo). Thucydides explains how Sparta's Brasidas adopts an inviscous state versus the stronger Athenians: "He did not consider marching out in regular formation, doubtful about his forces and considering them overmatched, not in numbers ... but in quality (for the whole Athenian contingent in the expedition was first-class) ... but prepared to attack with guile."[396]

Once the decision is made to fight, the remainder of the essential acts of strategy come into play – implementing, assessing, and adjusting ways and means. With the action underway, both sides have the option to fight, reset, or surrender. Assessments continue individually and collectively. For example, an individual combatant in a battle, like the group, evaluates will to fight, context (which can include immediate surroundings, mission, and orders), and capacity. An individual who loses the will to fight can stop fighting, desert, or surrender. Combatants facing a severe capacity shortfall (e.g., that lose their weapon or exhaust their ammunition) will normally withdraw or at least evade confrontation until capacity is restored. Though a hybrid state is not really an option for single combatants, an individual can reduce viscosity quickly with acceleration, indirectness, and relaxed restriction.

The W-V-A confirms what we ought to appreciate intuitively: *there is no ideal type or form of war* discernable *a priori*. The *best* form depends on will, capacity, and context. At the strategic level, these assessments are broad. Chanakya writes, "If the ruler has a strong army, has succeeded in his intrigues against the enemy, applied all remedies against dangers, and secured for himself a favorable position . . . then he may engage in an *open* [V-V] fight. Otherwise, he should engage in a *treacherous* [I-V] fight" (emphasis added).[397]

At lower levels of war, dynamic conditions drive viscosity changes as each side attempts to optimize its state relative to the enemy. Recall from Chapter 2, Sun Tzu says, "One able to gain the victory by modifying his tactics in accordance with the enemy situation may be said to be divine."[398] And Michael Handel adds that the search for "comparative advantage" (i.e., the selection of the most ideal force state) "defines the unique nature of each war" (i.e., engagement type), and, "given the dynamic nature of war, this is an ongoing process in which the more adaptable side that can identify the proper moment to switch from one strategy to another will eventually emerge triumphant."[399]

Responses to enemy action or direction from military leaders drive viscosity transitions during combat. Tactical viscosity adjustments result from inevitable changes in relative capacity during the fighting, and leaders at higher levels alter viscosity as capacity assessments change over the course of the war. If capacity changes proportionally for both sides, there is little incentive to alter viscosity. However, a side that loses capacity more rapidly may face defeat unless it can inflict greater damage on the enemy, gain reinforcement, or reset and change state.

Successful transition between viscosity states (V→I or I→V) without a pause is difficult and rare, because like any complex human activity, fighting successfully requires training and, especially in war, experience in representative conditions. Changing postures (e.g., from attack to defense), however, can alter the capacity equation enough, depending on the situation, to obviate the need for a dramatic viscosity change. For example, assuming a defensive posture might alter relative capacity expenditures enough to rebalance an engagement. Posture changes are not depicted in the W-V-A, though as mentioned before, the stronger side is more likely to assume an attack posture.

Usually when one side surrenders, the war objective for the other is met, unless the aim is annihilation. When one side achieves its war aim and the fighting stops, the result is victory or possibly a settlement. The W-V-A remains operative provided both sides retain the will and capacity to fight and accept the context. For larger units up to the strategic level, the leadership retains decision authority regarding the factors that influence viscosity (DARCC) and continuing the fight. If a leader loses the will to fight, determines conditions have put objectives out of reach, or has lost the capacity to fight, the action ends either as a defeat, settlement, or pause (i.e., a temporary withdrawal or ceasefire).

Divergence

Theory will warn us the moment we begin to leave the beaten track, and enable us to decide with open eyes whether the divergence is necessary or justifiable.[400]
Julian Corbett

Divergence occurs when reality *differs* from the W-V-A (i.e., when the force state diverges from the model's recommendation). The fact of divergence does not invalidate the algorithm or theory, it simply reflects the variability and fallibility of human decisions. Again, the W-V-A is a reference not a recipe. Even so, divergence with respect to the engine of war *is* impossible, because will and capacity either exist or they do not. While war may take any form regardless of the W-V-A's recommendation, war takes *no* form without will and capacity. That said, divergence *can* and *does* occur for parameters subject to human perception: context and relative capacity. Leaders frequently misread the context and initiate, continue, or end an action when conditions favor the opposite course.[401] Saddam Hussein made this mistake when he invaded Iran and Kuwait. Another common error is misperceiving relative capacity (usually overestimating one's own) and consequently failing to anticipate the correct form of war. As Sun Tzu writes, "When a commander unable to estimate his enemy uses a small force to engage a large one, or weak troops to strike the strong ... the result is rout."[402]

The US defense of the Philippines during World War II illustrates the high cost of divergence. Prior to the war, American military planners knew they could not adequately defend the Philippines against a Japanese attack. The long supply lines across the Pacific granted the Japanese an insurmountable capacity advantage. Nevertheless, General Douglas MacArthur, America's leader in the Philippines, convinced the military staff in Washington to allow him to remain in place, and he promised to hold the archipelago until reinforced. Instead, the Japanese overwhelmed the Philippine defense force, compelling MacArthur to abandon the islands and precipitating one of America's worst military defeats. Hindsight, informed by the W-V-A, indicates that MacArthur, faced with

Japanese superiority, should have prepared his forces for inviscous operations and, dispersed and concealed amongst the islands, fought an I-V campaign (like Aguinaldo had against MacArthur's father in the Philippine War forty-one years earlier).

The Israeli Defense Force's (IDF) "33-day War" in 2006 (versus the Lebanese Hezbollah) presents a twenty-first-century example of divergence. The IDF's historical dominance in V-V conflicts but weakness in V-I led the Ministry of Defense to bolster its inviscous capabilities. For a time, the IDF maintained a capable hybrid force; however, their viscous power eventually waned. Hezbollah, long experts in I-V warfare, noticed the IDF's shift, and according to military historian Saul David, refocused their status from "a limited guerrilla outfit" to "a new and sophisticated semi-conventional [V-V] fighting force" with the goal of beating the IDF at their own game.[403] When hostilities commenced, Hezbollah achieved a localized capacity advantage and dealt the IDF a "resounding defeat" in a V-V engagement.[404]

Though divergence represents a departure from the ideal way to fight, it does not *necessarily* lead to defeat, especially when applied to forces with high capacity that choose a hybrid or inviscous state. Conversely, while following the algorithm yields a higher probability of success, it does not assure victory. Neither side in a war can achieve perfect knowledge or action. Mistakes, genius, and chance influence outcomes for both sides. Savvy leaders and strategists strive to avoid divergence and to recognize and exploit enemy divergence to maximize their *probability* for success. For, as Frederick the Great advised, "In the profession of war, the rules of the art are never violated without drawing punishment from the enemy, who is delighted to find us at fault."[405]

Preventing, Ending, and Winning Wars

Because wars start with human decisions, there is no foolproof way to prevent them all; however, the W-V-A shows us how to reduce their probability. Provided we can take care of our own will to fight, the problem then becomes short-circuiting the other side's will, or at least complicating the strategic picture to arrest aggressive intentions in the second essential act of strategy (i.e., by showing that any war is unlikely to succeed). The best way to forestall pugnacity is to cultivate friendships, thus reducing the number of potential enemies and strengthening alliances and partnerships. This works internally (through competent governance) and externally (through skilled diplomacy) regardless of fighting capacity. However, when animosities or political differences preclude this, *deterrence* is the best way to prevent war.

Deterrence

The very large number of wars that have occurred in modern times proves that the threat to use force, even what sometimes looked like superior force, has often failed to deter.[406]   Bernard Brodie

Deterrence strategies prevent unwanted action (like war) by employing ways and means, not necessarily military, to produce perceptions in would-be aggressors that bellicose actions will result in unacceptable consequences. In essence, deterrence convinces another party that a cause (their unacceptable action) will have two intolerable effects: (1) be denied its intended result due to our effective countermeasures and (2) produce a cost-imposing, punitive reaction. In 1960, French airpower strategist Pierre M. Gallois "described deterrence as the product of the technical performance of weapons or military means ... and of the subjective perception by the side to be deterred of the will of the threatened nation (or rather its decision makers) to use these means."[407] In other words, successful deterrence rests on the deterring side's ability to convince a potential aggressor that it has a fully functional engine of war (with defensive and offensive capacity) awaiting only the "cause" of the aggressor's hostile act or intent to ignite. This neutralizes a potential attacker's will to fight by creating a prohibitive context. Ironically, deterrence requires transparency, for what is unknown or unsuspected cannot deter.

Since strategy is iterative, contexts are variable, and perceptions are fluid, deterrence cannot be static. As we will see in the next chapter, assessing future conditions is tricky business. What deters today may not tomorrow, and having capacity and will may be insufficient if the context changes dramatically. In 1972, for example, the USA proved its will and capacity to deter North Vietnam with the Operation *Linebacker II* bombing campaign, which compelled the communists settle. However, by 1975, though the US military had the capacity and will to strike Hanoi, the Watergate political scandal short-circuited the USA's deterrence credibility, and North Vietnam conquered the South.

At times, a balance of power can deter; however, deterrence is not a one-size-fits-all proposition, and this can occasionally undermine the credibility of deterrence and balance-of-power strategies. As the W-V-A shows (and Brodie's lead-in statement implies), simply being strong may *not* deter. In fact, great strength coupled with the *perception of hostility* engendered by *threats* or *provocative actions* can create fear that causes rather than prevents conflict. Furthermore, in cases of near-balanced capacity, Thucydides's Trap can cause the stronger side to attack from fear of another's *growing strength*. For the strong-viscous, deterring weak-inviscous adversaries is always difficult, but establishing viable hybrid capabilities may cloud the other side's strategic calculus enough to forestall action.

While deterrence cannot be foolproof, we can maximize its effectiveness by tailoring it to potential adversaries and contexts. This is done by using all IOPs, especially nonmilitary, to build cooperative relationships within and between states and non-state actors or otherwise encouraging the perception that the cost of war will far outweigh its benefits. When deterrence fails or we *elect* to initiate a war, the problem shifts from prevention to termination. When attacked, the first consideration is to stop the attack and end the war. The defender may have no concept for "winning" other than getting the enemy to quit, though this may change with time and circumstances. For the attacker, "victory" occurs once the political aim is achieved. As the W-V-A depicts, the odds of favorably ending or winning wars increase with awareness of the relationships between will, context, capacity, and dynamic viscosity.

Case Studies

In addition to the normative function of informing strategy and warfighting, the language of the UWT and W-V-A, including momentum, inertia, viscosity, containment, and divergence, is useful for analyzing past wars. The following cases are meant to build familiarity with the UWT's analytical power; however, the reader is encouraged to apply it to any war of interest.

Trojan War (~12th Century BCE)

For if the Iliad *is a collection of myths, the events it describes are rooted in fact.*[408]
                                                                        John Keegan

In the Trojan War, as written in Homer's *Iliad* and depicted in the 2004 movie, *Troy*, the Greek king Agamemnon uses the abduction of a Spartan queen (Helen) by a Trojan prince (Paris) as an excuse to consolidate the Hellenic kingdoms and conquer the mighty walled city-state of Troy.[409] The movie adaptation opens with an impending battle between the viscous armies of Agamemnon of Mycenae and Triopas of Thessaly. Rather than fight a mass battle, the two leaders agreed to a clash of champions featuring the legendary Achilles [viscous] versus the goliath Boagrius [viscous]. Despite Boagrius's size advantage, relative capacity favored Achilles due to his extraordinary skill and invulnerability. The W-V-A recommends Boagrius adopt an inviscous state against Achilles; but, Boagrius misjudged Achilles's lethality and failed to recognize his own capacity deficit [divergence]. Consequently, Achilles combined acceleration and striking power to avoid containment, exploit Boagrius's inertia [containment], and strike a mortal blow to the giant's unarmored neck [V-V].

After this victory, Agamemnon sailed to Troy at the head of a unified Greek army. However, even this massive force lacked the capacity to counter Troy's impenetrable walls, deadly archers, and cavalry. The W-V-A would advise Agamemnon to increase capacity, adopt an inviscous state, or go home, but he persisted [divergence]. Achilles and his elite Myrmidon strike force reluctantly accompanied Agamemnon. Relative to the main army, the Myrmidons were inviscous; however, relative to their initial target, a Trojan garrison, they were viscous with superior capacity. In accord with the W-V-A, Achilles's Myrmidons mounted a swift, direct assault and defeated the garrison [V-V].

As the main Greek army approached Troy, Paris consented to fight Helen's husband, King Menelaus. Relative capacity favored the larger, stronger, and more experienced king [viscous]. The W-V-A would recommend Paris withdraw or cheat; however, contained by his sense of honor [restriction], Paris fought Menelaus directly [V-V]. When Menelaus predictably gained the upper hand and moved in for the kill, Paris finally retreated [escaped containment] and broke the rules [inviscous] by imploring his brother, Hector, to intervene. Hector's assistance changed the capacity equation, and Hector slew Menelaus [V-V]. Enraged, Agamemnon launched a direct assault on Troy [divergence].

During the battle, Hector fought the Greek giant, Ajax. Ajax's greater power [viscous] granted him a capacity advantage over Hector who, as Troy's champion, could not defer the challenge [containment by restriction]. Ajax gained the initial advantage as Hector fought V-V. However, Hector exploited an opportunity to become inviscous. Beaten to the ground and near defeat, he deceived Ajax by feigning incapacitation and hiding a spear beneath his shield [V→I]. Ajax, contained by this ruse, charged Hector only to be surprised by a crippling spear thrust that immediately shifted the capacity advantage to Hector and enabled him to kill the giant with his sword [V-V]. As expected, Troy repelled the Greeks, and Agamemnon retreated to the beach [escaped containment and reset].

Having lost Ajax, the battle, and nearly all hope of penetrating Troy's walls, the Greeks contemplated withdrawing across the sea [pause or defeat]. Sensing their advantage and an opportunity to crush the Greeks on the beach [direct and indirect containment], the Trojans left the city and attacked [V-V]. Troy had nearly won, when Hector's accidental killing of Achilles's beloved cousin, Patroclus, led to a truce [restriction and pause]. Achilles, enraged by Patroclus's death, challenged Hector to a duel. Hector, contained by his sense of honor, agreed to fight despite Achilles's superiority [divergence]. In this battle, Hector did not cheat or withdraw and Achilles won [V-V].

As Troy mourned Hector, the overmatched Greeks turned to wily Odysseus who hatched one of history's notorious deceptions [inviscous]. This time, the Greeks followed the W-V-A recommendation to assume a hybrid state. Faced with a stronger enemy, most of the Greeks sailed away, leaving behind a large, hollow wooden horse hiding soldiers within [inviscous]. The superstitious Trojans unwisely took the horse into the city, and in the predawn darkness, the Greek army returned and advanced on Troy. Meanwhile, the hidden soldiers escaped and opened the gates. Using deception, stealth, and surprise [inviscous], and their main army [viscous], the Greeks sacked the city and killed or scattered all its inhabitants [victory]. However, during the melee, Paris surprised and killed Achilles by exploiting a moment of distraction [inertia] and shooting an arrow [containment and capacity advantage by speed and surprise] through his notoriously vulnerable heel [I-V].

The Philippine War (1899–1902)

In the Philippine War, fought between the USA and Filipino nationalists, the Filipino leader Aguinaldo initially led a conventional, "European-style" army [viscous] formed originally to oppose a viscous Spanish colonial army.[410] But when Spain ceded the Philippines to the USA (after the Spanish-American War), the Filipinos targeted America's transition forces. Aguinaldo had low capacity relative to the USA; however, he originally chose a viscous state [divergence] and was beaten [V-V]. Nevertheless, Aguinaldo escaped containment and lowered his viscosity, leading "a prolonged Indian-fighting style of campaign" [inviscous] across the jungle archipelago.[411] By the end of 1899, Aguinaldo had fully transitioned from a viscous to an inviscous force state [V→I].[412] This instance reinforces Kalyvas's observation that "rebels in both conventional [V-V] and symmetric non-conventional [I-I] wars are easier to defeat" by "well-organized" [viscous] forces than are irregular insurgents [I-V].[413]

As US forces engaged Aguinaldo's guerrillas, they realized that rigid columns [viscous] were "worse than useless" and dispersed small units [inviscous] to isolate, contain, and strike their enemy [I-I].[414] American forces isolated [containment] insurgents with a naval blockade, then conducted island-by-island search-and-destroy missions that led to the discovery and defeat of the guerrillas and their leaders. This effectively negated Aguinaldo's strategic COG and won the war [victory].[415] Even so, the US triumph was tarnished when some American units, having become inviscous, ignored the law of war and committed shameful atrocities.

Chinese Civil War (1927–1936 & 1945–1949)

Mao's mobile war strategy and three stages of people's war, which served as Giap's example, illustrate the rationale behind viscosity transitions in the continuum of war. Like Aguinaldo, Mao initially opted to fight the superior nationalists viscously [divergence]. A beaten but unbowed Mao escaped containment into China's interior on the Long March [pause]. While there, Mao transitioned to an inviscous state [V→I]. Later, as his capacity increased, he adopted a hybrid and then a viscous state [I→H→V]. Mao summarized this "innovative bridge between guerrilla and conventional strategy," as follows: Stage 1, guerrilla fighting [I-V]; Stage 2, mobile war blending viscous-inviscous [H-V]; and, Stage 3, strategic offensive [V-V].[416]

Vietnam

During the French Indochina War (1946–54), General Giap initially attempted to beat the French with inferior viscous forces and suffered many tactical defeats [divergence]. Using Mao's example (consistent with the W-V-A), Giap organized, trained, and employed inviscous forces [V→I] to wear down French will. Then, at Dien Bien Phu, Giap had gained enough capacity to transition from inviscous to viscous [I→V] and defeated the French garrison [V-V], which had misjudged its relative capacity [divergence]. It was as if Giap had read this from Clausewitz: "Once the victor . . . has been weakened by a variety of losses in men and materiel, the time has come for the defending army to take the field again. Then a well-placed blow on the attacker in his difficult position will be enough to shake him."[417]

A decade later, America found itself facing a hybrid force composed of Viet Cong insurgents [inviscous] and North Vietnam's [viscous] People's Army of Vietnam (PAVN). The Americans countered with their own hybrid force [H-H] hoping to win a war of attrition, but a lack of training and experience in V-I warfare, especially regarding containment, produced mixed results. The Viet Cong remained elusive, while the outclassed PAVN reset and was resupplied with relative impunity in neighboring countries off-limits to the USA [restriction]. Despite many tactical successes [V-V], the Americans could not compel North Vietnam to surrender, and by the end of the 1960s, US public support for the war and willingness to fight had evaporated [settlement]. After the Americans departed, the far stronger North Vietnamese defeated the South in direct battle [V-V].

Traditional Principles of War

Although the principles of war are simple and can be learned by anyone, the application of each principle requires much care, skill, and caution.[418]

Bevir Alexander

Now, having introduced the UWT's major elements, we can examine its relationship to additional war-related interest areas beginning with the nine traditional principles of war: objective, offensive, mass, economy of force, maneuver, unity of command, security, surprise, and simplicity, which "represent generally accepted 'truths' that have proven to be effective throughout history."[419]

➢ Objective: *the actions in war at all levels should be coordinated toward a well-defined and achievable end.* Objectives vary with level of war, viscosity, and relative capacity but should always relate to the engine of war (will, cause, or capacity). Inviscous forces often pursue limited or nontraditional objectives, while viscous forces target more traditional COGs like military forces.

➢ Offensive: *initiates the action to force an opponent to react.* The offensive posture is usually taken by viscous forces with superior capacity or by inviscous forces through stealth and surprise. The offensive is not always the best posture.

➢ Mass: *concentrates power at the right place and time to achieve desired results.* Mass is a component of capacity, and concentration increases viscosity; thus, greater mass is associated with viscous forces. Mass enables a viscous force to contain and defeat a weaker enemy or to displace an uncontained enemy.

➢ Economy of Force: *distributes and employs forces to the highest priority missions.* Economy without effectiveness is meaningless in war; thus, combatants must first avoid divergence, which requires awareness of friend and enemy engines of war and contextual factors.

➢ Maneuver: *maintains force mobility and moves forces to place the adversary at a disadvantage.* Maneuver is essential to forces of all viscosities and is a primary means of creating *and* eluding containment. Speed, flexibility, and agility decrease viscosity and are typical inviscous force attributes.

➢ Unity of Command: *ensures forces answer to a single commander and receive consistent guidance to optimize synchronization and mutual support.* Unity of command is important for all force states but especially for inviscous forces to ensure effective coordination when dispersed.

➢ Security: *prevents an enemy from gaining and exploiting an advantage* (e.g., by collecting information about force disposition and weaknesses or surprise attack). This is fundamental to forces of all viscosities.

➢ Surprise: *"attacking the enemy at a time, place, or in a manner for which they are unprepared."*[420] Surprise creates containment, exploits inertia, and nullifies defenses by preventing the enemy from observing and orienting. Surprise is valuable for forces in any state but may be harder to achieve for viscous forces.

➢ Simplicity: *reduces complexity and friction and improves efficiency.* "A simpler attack," advises Clausewitz, "one that can be carried out quickly," helps "gain the advantage ... proof enough of the *superiority* of the simple and direct over the complex."[421] Simplicity benefits forces of all viscosities, though simplicity and directness increase viscosity and, perhaps, predictability.[422]

Viscosity and the Warfighting Domains

Multi-domain warfare is similar to hybrid warfare and combined arms because forces from each domain have unique, reinforcing capabilities. In fact, contemporary warfighting almost always benefits from integrated, complementary capabilities in *all* warfighting domains, though it does not necessarily require *supremacy* in all. Given the right conditions, victory is possible through action in only one or two domains. Examples include Napoleon at Austerlitz in 1805 (land), England versus the Spanish Armada in 1588 (sea), and NATO's Operation *Allied Force* against Yugoslavia in 1999 (air). Fundamentally, a domain-centric force's main responsibility is to secure the three a's (awareness, access, and freedom of action) for itself while denying them to the enemy.

Land Warfare

Since men live upon the land and not upon the sea, great issues between nations at war have always been decided – either by what your army can do against your enemy's territory and national life, or else by the fear of what the fleet makes it possible for your army to do.[423]   Julian Corbett

Land is war's prime physical domain because most people live on land and forces from the other domains originate and are supported from land-based sites. Land-based forces (armies) are responsible for occupying and controlling territory, which usually requires combat against similar forces with opposing goals. The extent of territory an army can control depends upon capacity, the nature of the area (e.g., population and terrain), and the opposition. Armies are generally organized, trained, and equipped to fight other armies while defending against threats from other domains. They apply destructive force through a variety of IOPs including small arms, artillery, rockets, missiles, mortars, and fires from organic aircraft (e.g., helicopters).

Land force viscosity is highly variable. Armored vehicles are more viscous than infantry. Furthermore, after World War I, trends favored decreased viscosity as advances in firepower and accuracy rendered large, concentrated formations and fortresses obsolete. Even relatively viscous modern armies must fight in a distributed fashion to complicate detection and targeting while minimizing containment. "Digging in" can be effective against less capable enemies, but as Saddam learned in 1991, remaining in place against a powerful foe is tantamount to self-containment and invites destruction. As war expanded into new domains, ground forces have become more vulnerable and easier to contain.

Fast, lethal weapons employed from adjacent domains compel mobility, dispersal, concealment and deception, defensive armament, fortification, and often air and naval support to protect lateral and vertical flanks. For land forces, terrain and weather are pivotal environmental factors that also influence force viscosity.

Naval Warfare

It follows then as certain as that night succeeds the day, that without a decisive naval force we can do nothing definitive.[424]   George Washington (1781)

Though it is not the prime domain, warfare above, on, and beneath the seas has grown in importance as technology has advanced and more people and commodities travel over and are located near the world's waterways. The technological challenge of living and fighting in the sea domain have made naval vessels and weapons complex and expensive. Few countries can afford a blue-water navy and even fewer an aircraft carrier.

A navy's basic purpose is to advance political and war aims from the maritime domain. It does this by controlling the seas, projecting power into other domains, protecting friendly or allied maritime COGs (like commercial shipping, choke points, islands, or ports), and disrupting adversary COGs. This requires the development and employment of capabilities to avoid, defend against, and neutralize multi-domain threats. Through control of the seas (or bypassing opposing navies), naval forces traverse the Earth and project power via surface, subsurface, and airborne fires (e.g., cruise missiles and sea-borne aircraft).

As Corbett highlights, navies may not be able to take and hold land, but they can affect friendly and enemy armies via containment and by increasing or eroding capacity, respectively.[425] Command of the sea (by France), for example, enabled the American victory at Yorktown; (by the Allies) the miracle at Dunkirk; and (by the USA) the rescue of South Korea. Navies can also enact military and economic containment by blockading nations that rely on sea lines of communication for commerce and resources. Consider the global impact of restricting passage through the Straits of Malacca or Hormuz (the #1 and #2 global oil trade chokepoints) or the Suez or Panama Canals (30,000 ships transit per year).

Ships and submarines are essentially mobile fortresses at sea that have tremendous firepower, including nuclear weapons; however, they are highly viscous and only minimally able to change viscosity. Thus, though the sea is vast, which complicates detection and containment, ships are vulnerable to inviscous forces, like stealthy submarines, aircraft, missiles, or small, mobile boats. In 2000, for instance, the American destroyer, USS *Cole*, was severely damaged by suicide bombers in a tiny boat. Like forces in other domains, weaker naval units must elude superior forces (and containment) and must generally limit offensive operations to lightly defended targets like commercial shipping. Submarines, which rely on stealth and surprise to contain adversaries, are less viscous than surface ships but still vulnerable once detected.

The sea is a particularly unforgiving domain. As a British general once opined, "I know of nothing that is more liable to disaster and danger than something which floats on the water."[426] The crew of a severely damaged ship or submarine cannot simply scatter and retreat. Furthermore, a damaged naval vessel's viscosity increases even as its capacity decreases making it easier to contain. Unless an enemy loses interest or its own capacity, a ship is rarely able to limp from a battle to safety, and once sunk, these vessels (and any crew aboard), are usually lost forever. As such, ships must avoid direct [V-V] battle unless they have a sizable capacity advantage. Naval vessels, like land forces, are vulnerable to attack from other domains, particularly the air. Therefore, most modern navies rely on powerful sensors linked to robust air defenses and, when able, their own aircraft.

Air Warfare

For good or for ill, air mastery is today the supreme expression of military power and fleets and armies, however vital and important, must accept a subordinate rank.[427]
                                                   Winston Churchill (1949)

The advent of airpower and the air domain's contiguity with all physical domains has made control and exploitation of the skies essential for most conflicts. Air forces, depending upon their capabilities, can influence targets in all other domains. Furthermore, airpower provides warfighters with appreciable advantages in speed, range, flexibility, precision, and lethality as well as significant intelligence-gathering capabilities. "The Air," writes Hugh Trenchard, "can carry much more destruction to the enemy per man with a minimum loss of life than any other form of warfare."[428] Aircraft also fill vital, noncombat logistical support and humanitarian relief roles. As with navies, air forces can influence viscosity and containment for ground and naval forces. For example, in 1948 and 1949, the Berlin Airlift negated a Soviet attempt to isolate and annex West Berlin. Fighting in and from the atmosphere requires advanced technology; thus, air forces, like navies, are expensive.

Unlike armies, navies, and even space forces, air forces can only temporarily "occupy" the skies before returning to Earth. Thus, like the navy, an air force's prime objective is to gain freedom of action within the portion of the air domain required to accomplish its tactical, operational, and strategic goals. Of course, an air force can extend the period or scope of its air superiority by degrading or neutralizing an enemy's capacity to interfere. This includes an enemy's air force (in the air or on the ground) or surface-based defenses such as anti-aircraft artillery (AAA), surface-to-air missiles (SAMs), or even directed energy (laser) systems.

Air forces, at least while in the air, are the least viscous in the physical domains because of their speed and maneuverability (acceleration) and the technical challenge of detection, tracking, and engagement. Though sensor advances coupled with SAMs have made aircraft more vulnerable, countering modern, stealthy, high-speed aircraft remains daunting. Most air battles take the form of I-I warfare since aircraft are generally inviscous and hard to contain. "The elusiveness of aircraft, given their inherent mobility and the vastness of the sky," writes Gray, "means that the grammar of warfare in and from the air can seem distinctly guerrilla-like and irregular."[429] It is virtually impossible to "surround" aircraft, although weather, terrain, tall structures, and pressure ceilings indirectly contain them. Speed, accuracy, and surprise through surface or air-based weapons are the best ways to directly contain aircraft.

Airpower's unique ability to combine mass (including nuclear weapons) and maneuver, writes General Moseley, drives opponents to adopt inviscous force states including "distributed and dispersed operations ... and sanctuary in dense urban areas, ungoverned hinterlands and loosely regulated information and social networks."[430] Even so, airpower's ability to combine ISR, acceleration, and precision attack presents challenges even for inviscous foes and explains why aircraft (including unmanned "drones") are often effective in H-I warfare against inviscous forces like insurgents (e.g., in the Malayan Emergency and Huk Rebellion), and versus submarines (e.g., during World War II's "Battle of the Atlantic" antisubmarine campaign).

Like armies and navies, air forces are vulnerable to attack from forces in other domains. Perhaps the best way to defeat air forces is to attack them on the ground or deny them access (basing). Finally, damaged aircraft, like navy ships, can be difficult to defend or recover, but parachutes enable pilots to escape; and most aircraft are less costly than ships or submarines.

Space Warfare

Space is no longer a sanctuary – it is now a warfighting domain.[431]
Patrick M. Shanahan, Acting US Defense Secretary (2019)

Space is the "final frontier," the ultimate high ground. The space domain begins at the vertical limit of the air domain (i.e., the Karman Line roughly 100 kilometers above the Earth).[432] Aircraft cannot fly beyond this altitude; thus, propulsion and control require rockets and thrusters. Currently, no people live permanently in space because it is harshly inimical to human life and traveling there and back is extraordinarily expensive. Enormous energies are required to "escape" the atmosphere and achieve orbit, where payloads moving at about 27,400 kilometers per hour perform a variety of functions including surveillance, reconnaissance, navigation, and communications. Presently, international law (the Outer Space Treaty of 1967) prohibits the use, testing, or positioning of WMDs in the space domain including orbit and on celestial bodies; and, it prevents any nation from claiming sovereignty over or conducting military or war-related activities on the moon and other celestial bodies.[433] As with any treaty limiting war, the Outer Space Treaty remains effective only until someone decides their interests are better served by ignoring it.[434] Space-based weapons would be highly lethal and undeniably provocative.

The space domain is a contested environment. The dependence of technologically advanced nations on space has made space capabilities a significant component of warfighting capacity and a prime target for enemy action. Though their speed and distance from the ground make satellites difficult to target and engage, they are not immune to attack. Spacecraft in orbit are viscous, constrained by energy limitations and the physics of orbital mechanics, have only a limited ability to maneuver or retreat, and often lack active defenses. Nations with access to space have developed and fielded capabilities to interfere with and neutralize spacecraft, and though true combat has yet to debut in space, it is only a matter of time. Space forces must defend themselves from multi-domain attack, especially cyber, and, as with submarines, damaged spacecraft and personnel are exceptionally difficult to recover.

Cyberspace Warfare

Cyberspace and the further exploitation of the electromagnetic spectrum as cyberpower, has indeed arrived.[435]   Colin Gray

Cyberspace is not a physical domain in the same sense as land, sea, air, and space; however, what happens in cyberspace is vital to the conduct of contemporary war. There are no people in cyberspace; however, since the digital age began, nearly everyone depends on what happens in cyberspace; thus, it has enormous political, economic, and military implications. Because there is no physical combative violence in cyberspace, a true war cannot occur exclusively in that domain. Nevertheless, cyberspace actions can create physical effects, including death and destruction, and they could *trigger* a war.

The weapons of cyberspace are digital code, and cyber warriors include highly educated scientists and teenagers in coffee shops. Like war in the other domains, cyber war includes defensive and offensive components – using cyberspace to impact the enemy's engine of war while protecting our own; and the effectiveness of cyber warfare depends on awareness, access, and freedom of action, all of which are enabled digitally and may be supported by forces from other domains. Because of its ubiquity, accessibility, anonymity, and relatively low-cost, cyberspace is a favored domain for inviscous forces.

Regarding DARCC criteria, cyberspace is less about concentration and coherence and more a matter of directness (indirect), acceleration (speed-of-light), and restrictions (few). Indeed, its intangibility makes cyberspace inviscous by nature and enables its forces to disrupt enemies with minimal cost or vulnerability. Whereas governments bound by international and domestic law and fearing reprisal via traditional IOPs generally limit overt cyber-attacks on critical infrastructure, financial and political institutions, and military forces; inviscous cyber-actors are less inhibited. Thus, viscous powers must apportion significant resources to defending critical military and civilian cyber "terrain."

Terrorism

International terrorism is a new type of war ... not because it targets helpless citizens, but because it seeks to achieve political goals across a distance.[436]

Bevin Alexander

Terrorists, or violent extremist organizations (VEOs), come in many forms and pose significant challenges for civilized nations and their often-helpless victims. Though terrorism is nothing new, since the late twentieth century, globalization and weapons proliferation have enabled VEOs to achieve more significant destruction across greater distances. In its "modern manifestations," writes Princeton political theorist Michael Walzer, "terror is the totalitarian form of war and politics. It shatters the war convention and the political code."[437] Terrorism warps the good-evil, friend-enemy, peace-war, creation-destruction, life-death, and order-chaos dialectics toward evil-enemy-war-death-destruction-chaos. No wonder it has garnered such attention!

VEOs use the *tactic* of terrorism to advance political objectives. VEO practices – including targeting civilians and nonmilitary objectives, like churches and nightclubs – are inherently inviscous and create psychological shock through diabolical atrocities. Though, as Walzer points out, "The systematic terrorizing of whole populations is a strategy of both conventional and guerrilla war, and of established governments as well as radical movements," it is often the preferred tool of inviscous forces with low relative capacity but a strong will to fight.[438] Moreover, because it is rarely used directly or with restraint, terrorism almost always lowers overall viscosity, and VEOs can be virtually immune to the influence of diplomacy, economic sanctions, and deterrence. Liddell Hart says that most "aggressive types, whether individuals or nations . . . can be curbed" by "a formidable opposing force"; however, this may not work "against pure fanaticism – a fanaticism that is unmixed with acquisitiveness" (e.g., VEOs).[439] With the world as their battlespace, international terrorists resist containment and frequently exploit geographic, legal, and political gaps and seams. Furthermore, religious VEOs have elusive COGs and resilient indoctrination and propaganda networks that effectively protect and sustain their engines of war. Even the elimination of key leaders, like Osama bin Laden (al-Qaeda), Jainal Antel Sali Jr. (Abu Sayyaf), or Abu Bakr al-Baghdadi (Islamic State of Iraq and Syria, ISIS), may not appreciably reduce worldwide terrorism. Religious causes (e.g., doing God's will by punishing infidels, promoting the caliphate, or deposing apostate regimes) energize the recruitment of more terrorists and enablers.

A significant barrier hindering the eradication of terrorism is cultural inculcation. When *generations* are raised to trust unrestrained violence as the pathway to political and religious reward, war and terror become their own ends. In these instances, as with Jan Bloch who dared question the "cult of the offensive" prior to World War I, critics are silenced and cast as enemies of a *Weltanschauung* that is nearly impervious to change.[440]

Decisive battle rarely wins a war on terrorists because no single battle can stamp out an ideology. Pure military attrition is also ineffective, because defeating every terrorist is impractical. To survive, write Codevilla and Seabury, terrorists require "money, bases from which to organize, safe havens, and propaganda, as well as living proof that the cause is doing well."[441] Thus, the solution is consistent pressure over a long period, disrupting money and havens, but most importantly, denying or discrediting VEO political objectives until no message can disprove the impotence of the cause and the futility of efforts to sustain it.

Nuclear Weapons

[During the Cold War], international order was simultaneously both menaced by the most awesome threat of destruction in all of history and protected robustly by that same awesome menace.[442]   Colin Gray

Between the absurdity of total war and the impossibility of real peace, the hopes of humanity are confined to the possibility of limiting warfare.[443]
                                                                            Raymond Aron

Martin van Creveld proclaimed that, after Hiroshima, "the nature of war had changed, presumably forever."[444] If true, this would turn war theory on its head. However, nuclear weapons have *not* changed war's nature, only its calculus. Military historian Andrew Bacevich writes, "War remains today what it has always been, elusive, untamed, costly, difficult to control, fraught with surprise, and sure to give rise to unexpected consequences. Only the truly demented will imagine otherwise."[445] Though the impact of nuclear weapons could plausibly lead more than the demented to conclude that they have altered war's nature, it is only our *understanding* of war that actually requires redress.

When assessed with the W-V-A, the strategic implications of nuclear weapons become clearer. Before the nuclear age, V-V war was a feasible option for achieving political objectives due to an expectation of "acceptable losses." To date, no two sides contemplating nuclear war have gotten past the W-V-A's will-context-capacity threshold because they get hung up on the second act of strategy (i.e., is there an acceptable probability of producing the desired ends at a reasonable cost?). With nuclear war, the answer is always "no," because these weapons are so irresistibly destructive, particularly when employed on long-range ballistic, cruise, or hypersonic missiles, the cost always exceeds the benefit. In other words, assessment of relative capacities and context between nuclear-armed adversaries *eliminates the will to fight*. Foreign policy expert Franklin Miller summarizes this reality as follows:

In the 300 years following the Treaty of Westphalia in 1648 and the emergence of the modern nation-state, the great powers of Europe went to war with one another an average of seven times per century. Even the horrific carnage of World War I ... was insufficient to prevent World War II. But after 1945, the great powers ... have not engaged in direct military conflict with one another. Human nature has not changed ... But something else did change: nuclear weapons have made war among the great powers too dangerous.[446]

Still, nuclear weapons have not stopped all war, only total war between nuclear powers. For less than existential conflicts, "nukes" are literally overkill and carry such negative cultural and political baggage that nuclear-capable countries would not dare use them except in the direst circumstance. However, every manner of *limited war* judged unlikely to provoke a nuclear clash is now fair game. "The common assumption that atomic power has cancelled out strategy is ill-founded and misleading," writes Liddell Hart. "By carrying destructiveness to a 'suicidal' extreme, atomic power is stimulating and accelerating a reversion to the indirect methods that are the essence of strategy."[447] And Trinquier adds, "The increasing power of weaponry, which places distance between combatants, is also abruptly bringing them together. War will be a juxtaposition of a multitude of small actions. Intelligence and ruse, allied to physical brutality, will succeed the power of blind armament."[448] Indeed, "War in the nuclear age," writes political science expert James Turner Johnson, "has not been a global catastrophe but a continuation of conventional warfare limited in one of several ways – by geography, goals, targets, means."[449] Wars in and between nonnuclear countries, sometimes as proxies for nuclear-armed nations, are prevalent. Additionally, rogue states and non-state actors, like VEOs, feel empowered to provoke nuclear-armed states because they know the likelihood of a nuclear response is essentially zero. VEOs also know how ineffective nukes would be against dispersed inviscous forces. Moreover, inviscous forces, unconstrained by contextual misgivings or concern for retaliation, are more likely to *use* WMDs if they can obtain them. It is not inconceivable for extremists to value spectacular destruction and the sowing of chaos over self-preservation; thus, their threshold for nuclear weapons use could be very low. For this reason, most civilized nations rank the counter-proliferation of WMDs among their most vital interests.

Safer without Them?

A point of contention in contemporary political discourse centers on the question of whether the world would be safer with or without these hellish weapons. There would undoubtedly be less risk of quick self-destruction without them. However, there is no credible path to global nuclear disarmament given the realities of human nature and geopolitics; and in truth, eliminating nuclear weapons would cause significant problems. To begin with, unilateral disarmament of a major power creates a sharp, strategically destabilizing capacity discontinuity. In the nuclear age, all countries that can develop nuclear weapons have them, align with another power that does, or are otherwise secure in their neutrality (and geo-strategic irrelevance). Countries lacking nuclear deterrence are vulnerable to countries with greater conventional capacity or to having their interests held hostage by the threat of nuclear annihilation. Even if it were possible to create and implement an airtight nuclear disarmament process, it is *not* possible to destroy the knowledge of their existence and manufacture. The temptation to construct (or retain) a supply of nuclear weapons "for a rainy day" would be irresistible. Furthermore, as philosopher Bertrand Russell laments, "If all existing nuclear weapons had been destroyed and there were an agreement that no new ones should be manufactured, any serious war would, nevertheless, become a nuclear war as soon as the belligerents had time to manufacture the forbidden weapons."[450] If one rearms, others must follow; thus, disarmament would, at best, be impermanent.

Finally, can we be certain that a world without nuclear weapons would be more peaceful? Clearly, the historical record does not support this supposition. In fact, wars since 1945 have become, if not less violent or frequent, certainly much less deadly (see Figure 4.23). It is likely that a full century will pass without another world war. Ultimately, we must decide between two evils: the risk of nuclear catastrophe or the return of global, great-power wars.

Figure 4.23 Deaths in war since 1900[451]

Ethical Considerations

On trifling pretexts, or none at all, men rush to arms, and when once arms are taken up, all respect for law, whether human or divine, is lost, as though by some edict a fury had been let loose to commit every crime.[452]   Hugo Grotius

Far from being divorced from warfare, ethics constitute its central core.[453]
Martin van Creveld

Earlier, we looked briefly at the difficulties of regulating war and concluded that, while rules governing war's conduct have value in certain circumstances, it is not necessarily wise to bet our lives on them, particularly in existential conflicts. Furthermore, inviscous forces, almost without exception, seek to exploit restrictions for military gain. Nevertheless, ethics and morality, as indelible features of our humanity, are germane to war theory and strategy and warrant further examination.

Ethical theory related to war reflects all four of war's causes and the following dialectics: order-chaos, life-death, creation-destruction, good-evil, friend-enemy, control-autonomy, viscous-inviscous, and attack-defense. Ethics and morality help us delineate what is good or beneficial to human life from what is evil or prejudicial. Additionally, along with laws, these concepts applied to politics and war facilitate order by controlling darker inclinations, defining friend and enemy, and guiding decisions regarding life, death, creation, and destruction. In short, ethical standards promote order, security, peace, and prosperity, and regardless of their religious, customary, logical, or legal origins, form a code that societies instill, enforce, and adjudicate to reduce internal conflict.

While governmentally enforced laws and moral standards have enjoyed reasonable success domestically, the opposite is true *between* states. Historically, international relations have either lacked a regulatory mechanism entirely or relied on religious or imperial authority. The Peace of Westphalia ended the brutal Thirty Years War and offered hope that a new international system would change things for the better. However, political leaders continued to pursue wealth and security by force of arms. Against a backdrop of unremitting warfare, many of history's greatest minds authored and aggregated persuasive moral wisdom into "just war' principles that sanctioned war for ethical purposes only while seeking to limit its regularity and cruelty. However, two factors continue to undermine the effectiveness of "just war" doctrines. First, humans frequently perceive "right" and "wrong" differently, and sovereign political entities very often do, for with rare exception each ranks its own interests above others. Second, there is no entity with sufficient power and authority to arbitrate disputes and hold "wrongdoers" accountable. Within a state, executive and judicial institutions handle these matters, leaving ethical miscreants with four alternatives: no-sanction, self-sanction, sanction by a legitimate authority, or sanction by "peers" (i.e., vigilantism). Generally, if the transgression warrants sanction by violent force and the unethical party is a sovereign state, only the first and final options apply, because both warring sides think their cause is just, and no external authority trumps a sovereign, particularly at war. "There can be no question of a sanction," writes Hans Kelsen, "unless there exists an organization to carry out the act of coercion with powers so superior to the power of the wrongdoer that no serious resistance is possible."[454] Thus, the international arena, when diplomacy fails, devolves into what Kelsen calls "primitive law" or "self-help."[455] Indeed, in "warre," writes Hobbes, "nothing can be Unjust. The notions of Right and Wrong, Justice and Injustice have there no place. Where there is no common Power, there is no Law; where no Law, no Injustice."[456] Thus, international justice becomes vigilantism, and where force must be met with force, war.

The Just War Tradition

We find ourselves confronted with four fundamental questions:
(1) *Can human beings ever take part in war without seriously violating our moral obligations or destroying our moral character?*
(2) *When and under what conditions can war be rightly initiated?*

(3) *How can war be fought so that our most basic moral standards are not violated?*

(4) *What should be done to ensure a lasting peace once the hostilities are over?*[457]

Gregory M. Reichberg, The Ethics of War (2006)

The centuries-old effort to articulate an ethical framework that answers Reichberg's questions is known as the "just war tradition." This tradition has many contributors, a who's who of philosophical giants going back to Thucydides and Plato; however, the Catholic theologians, Saints Augustine (early fifth century) and Thomas Aquinas (thirteenth century), are generally credited with pioneering its foundational precepts. Many of these just war principles are codified in the US Department of Defense's 1,200-page *Law of War Manual* (2016), one of the world's most thoroughly researched and comprehensive practical guides on legally permissible conduct in war. This guide is a modern distillation of the just war tradition's most influential tenets, and it incorporates and references the content of numerous historical documents and treaties, like the Lieber Code, the Hague and Geneva Conventions, and the Convention Against Torture.

According to the *Manual*, the just war paradigm encompasses "the customs, ethical codes, and moral teachings associated with warfare," centering upon two major areas of examination: (1) the "moral justification of war" (*jus ad bellum*), and (2) "the limitations to war" (*jus in bello*).[458] The former pertains to just war "theory" (i.e., is war ethical at all?), while the latter addresses *conduct in war* (i.e., what weapons, targets, and activities are morally permissible?). The *Manual* defines the *law of war* (or law of armed conflict) as "that part of international law that regulates the resort to armed force; the conduct of hostilities and the protection of war victims in both international and non-international armed conflict; belligerent occupation; and the relationships between belligerent, neutral, and non-belligerent States."[459] A subset of the law of war, "rules of engagement" or ROE, help commanders "delineate the circumstances and limitations" of combat engagements in a specific war or theater of operation.[460] ROE may be more but not less restrictive than the law of war.

The *Law of War Manual* specifies five purposes:

(1) protecting combatants, noncombatants, and civilians from unnecessary suffering;

(2) providing protections for persons who fall into the hands of the enemy (e.g., prisoners of war, civilians, and military wounded, sick, and shipwrecked);

(3) facilitating the restoration of peace;

(4) assisting military commanders in ensuring the disciplined and efficient use of military force; and,

(5) preserving the professionalism and humanity of combatants.[461]
While these ideals now appear in the law of war guide, all have been articulated in one way or another by philosophers of the just war tradition.

Jus ad Bellum

In the person for whom the war is fought there should be a just cause, which is the restoration of the good, the suppression of the wicked, and peace for all.[462]

Alexander of Hales (thirteenth century)

Logically, the range of thought on war's permissibility runs from pure pacifism (doves), which holds that war is never justified (for doves, *jus in bello* is irrelevant) to ultra-realist warmongers (hawks) who believe that war is justified for any reason. For hawks, power accumulation and war preparation are imperative for advancing political agendas, and warfighting is a natural and beneficial human activity. While most people prefer peace and shun a war-at-any-time philosophy, most also reject pacifism, because it disallows war even in self-defense or to rescue innocents.

The *Law of War Manual* identifies five *jus ad bellum* (just war) criteria:

(1) a *competent, legitimate authority* to order the war for a public purpose (i.e., war is not for revenge or private vendettas); this aids distinction between combatants and criminals.

(2) a *just cause* (e.g., self-defense, restoring rights or property, or protecting the innocent).

(3) the means must be *proportional* to the just cause (i.e., the positive value of the war's objective must outweigh the death and destruction required to attain it).

(4) war as a *last resort* (i.e., all peaceful alternatives must be exhausted first).

(5) a *right intention* on the part of the just belligerent (e.g., the motive should be to reestablish order and peace, not to gain wealth or territory).[463]

Reasonable chance of success is another criterion often associated with just war theory. This tenet mandates that no war be started without a realistic probability of winning and surviving the effort. "I shall come to the aid of the perishing," says Seneca, "but in such a way that I myself may not perish, unless I am to be the price of a great man or a great cause."[464]

These just war criteria, which roughly align with Jomini's "proper, opportune, and indispensable" conditions, certainly appear sensible, wise, and beneficial. However, in practice, they are fraught with ambiguity. For example, what is the standard for competent authority vis-à-vis non-state and intra-state belligerents? Although competent authorities and war declarations are fascinating to lawyers, warriors are often indifferent about who started the war or why. In theory, anyone with the will and capacity to fight for their political ambitions can begin a war. In their minds the cause is invariably just, opinions to the contrary be damned. Additionally, how are we to interpret self-defense in cases where imminent threats call for preemption? Would everyone agree with the Swiss jurist Emer de Vattel's rationale that "One is justified in forestalling a danger in direct ratio to the degree of probability attending it, and to the seriousness of the evil with which one is threatened?"[465] If so, by what standard do we measure the danger, probability, and seriousness of the evil? Furthermore, how are we to predict how much better a peace a war will create relative to its costs? We could certainly reduce wars by resorting to them only after exhausting all other means, but in reality, *wars are never a last resort* (even invasion victims can refuse to fight), so this criterion is suspect from inception. Lastly, absent a confession, can we reliably judge right intentions? Such assessments are more a matter of perspective than record.

Regrettably, even if there was a "common Power" to evaluate the justness of wars, the ambiguity of just war language would likely prevent an indisputable ruling. And we cannot expect these international edicts, in and of themselves, to settle things; for as Hegel writes, "International law springs from the relations between autonomous states"; thus, "what is absolute in it retains the form of an *ought-to-be*, since its actuality depends on different wills each of which is sovereign" (emphasis added).[466] Therefore, confirms Rousseau, "As for what is called the law of nations, it is clear that without any real sanction these laws are only illusions . . . having no other guarantee than the utility of the one who submits them, [they] are respected only as long as those decisions confirm one's own self-interest."[467] Though somewhat depressing, this reality does not mean that *jus ad bellum* is worthless, yet we cannot ignore the practical challenges related to accountability and enforcement. Can we expect the same of *jus in bello*?

Jus in Bello

All the restrictions, all the international agreements made during peacetime are fated to be swept away like dried leaves on the winds of war. War will always be inhuman, and the means which are used in it cannot be classified as acceptable according to their efficacy, potentiality, or harmfulness to the enemy. The purpose of war is to harm the enemy as much as possible; and all means which contribute to this end will be employed, no matter what they are. The limitations applied to the so-called inhuman and atrocious means of war are nothing but international demagogic hypocrisies.[468]

Giulio Douhet

We will address Douhet's sobering appraisal in a moment; but first, once war begins, *jus in bello* considerations go into effect. These range from unrestricted warfare (anything goes) to a near infinite variety of prohibitions on the "who, what, where, when, and how" of warfighting. It is worth noting that, within the just war tradition, *jus in bello* applies to any war, just or not, and *jus ad bellum* is binding even if *jus in bello* principles are ignored.

The *Law of War Manual* identifies several principles to guide and govern combat:

(1) *Military Necessity*: justifies the use of all measures (not otherwise prohibited) needed to defeat the enemy as quickly and efficiently as possible. This tenet recognizes that lethal force and destruction are part of war but should not be indiscriminate.

(2) *Humanity*: forbids the infliction of suffering, injury, or destruction unnecessary to accomplish a legitimate military purpose. This tenet complements military necessity by safeguarding prisoners and noncombatants, like civilians and medical personnel, and prohibiting torture and weapons that are indiscriminate or designed to cause superfluous harm.

(3) *Proportionality*: prohibits even justified actions that are unreasonable or excessive. For example, bombing an entire town to eliminate a few enemy combatants violates proportionality, humanity, and military necessity.

(4) *Distinction*: or "discrimination," obliges each side to distinguish between the armed forces and civilians and between unprotected and protected objects. Disguising oneself as a noncombatant and using human shields are as much violations of this tenet as killing indiscriminately.

(5) *Honor*: related to *chivalry*, demands fairness, trust, and respect between opposing forces, including acceptance of law of war precepts. For example, while deception and surprise are legitimate tactics, ambushing an embassy during a truce is an honor violation.[469]

Weapons prohibited by the *Law of War Manual* due to their indiscriminate or unnecessarily brutal effects include: "poison, poisoned weapons, poisonous gases, and other chemical weapons; biological weapons; certain environmental modification techniques; weapons that injure by fragments that are non-detectable by X-rays; certain types of mines, booby-traps, and other devices; and blinding lasers."[470] Artificial intelligence or AI (computer algorithms that mimic human cognition), genetic engineering, and nanotechnology are areas to watch for future *jus in bello* guidelines.[471] For example, should humans be held accountable for AI mistakes or malice in war? Is it morally acceptable for AI to decide matters of life and death without a human's inputs or oversight? Is it permissible to infect an enemy with lethal or paralyzing nanobots or to "manufacture" genetically manipulated "super" soldiers?

Jus in bello limitations, as with *jus ad bellum*, raise questions of enforcement and adjudication. Moreover, identifying violations requires detailed awareness of both warring sides, a tall order since fog, friction, and propaganda obscure facts and truth. Additionally, holding one's own forces accountable may depress morale and reduce support for the war effort presenting a potentially lethal conflict of interest. When wars finally end, only the victor typically retains the power to determine accountability. After World War II, for example, the Allies tried and executed Nazi war criminals at Nuremberg; yet, if the Axis had won, it could have been Generals Eisenhower, Eaker, LeMay, and Air Marshall Harris on the gallows.

There are additional concerns related to these principles. For instance, are there exceptions, or do these rules apply regardless of the adversary, form of war, or status of the conflict? History shows that even the best intentions fragment in combat, especially when the enemy is not only ignoring the code but exploiting it. For instance, all five *jus in bello* principles are stymied when clergy, children, and pregnant women carry bombs, and hospitals and churches double as munitions factories. Prior to World War II, every major power rejected doctrines of unrestricted air attack; in fact, in 1939, "Both Britain and Germany declared their agreement with President Roosevelt's proposal not to bomb civilians and unfortified cities."[472] However, bombing was more dangerous and imprecise than expected, and by 1941, with its survival at stake and having resorted to night sorties (for aircrew protection), Britain removed all bombing restrictions.[473] By war's end, over half a million civilians had been crushed, burned, asphyxiated, and irradiated in places like Hamburg, Dresden, Tokyo, and Hiroshima.

Though Douhet's words preceding this section may seem morally shocking, it is hard to dismiss his conclusions given the history of warfare, let alone the last two world wars. Douhet may have been a fascist, but he was no warmongering militarist. For him, World War I proved that, unless deterred, governments would continue to glorify war as a political tool and a path to greater wealth and security to the detriment of humanity (and Italy). He sincerely thought the "invincible" air weapon would make wars obsolete or at least sharper and shorter. "Mercifully," writes Douhet, "the decision will be quick in this kind of war, since the decisive blows will be directed at civilians ... these future wars may yet prove to be more humane."[474] In the end, Douhet was largely right about the futility of international prohibitions on violence but not about airpower's ability to deter wars or render them more "humane."

Sometimes it may seem as though the only effect of applying morality to war is the creation of ethical dilemmas. If torturing a captured terrorist will yield information that may save a hundred lives, what is more unethical, getting the information quickly or allowing a hundred innocents to perish? Which is the greater evil, threatening a city with nuclear destruction or allowing free states to be conquered by a murderous, totalitarian regime? Of what worth is morality, if the last moral person is permitted to perish at the hands of the immoral? Philosophers have wrestled with these questions for centuries, and there are no easy answers.

Ethics and the Unified War Theory

War is the negation of law. It ... is the recourse to force – which rules the
world. Everything is therefore not only permissible but legitimate against the
enemy.[475]   Admiral Hyacinthe Aube (nineteenth century)

A general war theory must acknowledge ethics' role in war and warfighting, only without making its own moral judgments. Theory cannot ignore any valid possibilities, or in other words, all *possible* restrictions or actions in war must fall within theory's purview – from pacifism and the laws of war to ultra-realist aggression and "anything goes." However, having accounted for all *possibilities*, effective theory must also explain the relationships that drive *probabilities*, as we did with war's forms and viscosity.

Since ethical norms vary widely, even among allies and especially between enemies, and often change during a war, they are significant factors in war's dynamic. If we strip away religious, cultural, and legal standards and examine the ethics of war through the prism of logic and practicality, a balanced picture emerges. On one hand, as discussed earlier, ethical restrictions can be a liability, particularly for V-I engagements. While the DARCC factors–acceleration, cohesion, and concentration–are independent of *jus in bello*, directness and restriction are not. At one extreme, inviscous forces pair "indirectness" with rule-breaking (e.g., attacking civilians) to produce military and political effects. Inviscous force strategists scrutinize the law of war and ROE intensely, not to adopt it, but to identify ways to compromise an enemy. War remains undeclared to obscure status and intent, while inviscous warriors disguise themselves as noncombatants, violate truces, attack innocents, and employ forbidden weapons to degrade their enemy's engine of war and to incite legitimacy-sapping reprisals.

On the other hand, for viscous forces in nonexistential V-V and V-I engagement types, fidelity to law-of-war restrictions and directness offer practical and even decisive benefits. Directness yields positive results when there is a clear capacity advantage or a fleeting opportunity, but it can also be a show of force to deter conflict or reassure allies. Furthermore, adherence to just war principles promoting human rights and peace increases the probability of positive resonance with foreign and domestic populations, especially liberal democracies. This reinforces domestic political support and coalitions, gains allies, frustrates insurgent strategies, and discourages others from joining the opposition. For example, Iraq's invasion of Kuwait was such an egregious violation of *jus ad bellum*, that many regional powers that otherwise despised western democracies joined the coalition against Saddam. A just cause is also more likely to boost volunteerism for military service and to motivate and inspire the armed forces to their best efforts. Additionally, observing *jus in bello* raises the likelihood (but does not guarantee) fair and humane treatment of one's own captured forces, and it may also induce a cornered enemy to surrender rather than fight desperately to the last soldier. Moreover, the laws of war encourage economy of force and discipline, which preserve fighting capacity for when and where it can be most effective. Finally, and most importantly, an ethical approach to war supports political ends by improving the prospects for an enduring, postwar peace. "War," writes moral philosopher John Rawls, "must be openly and publicly conducted in ways that make a lasting and amicable peace possible with a defeated enemy."[476] The fewer and shallower the wounds inflicted on personnel, infrastructure, and polities, the faster these "wounds" will heal, and the less likely they are to rupture anew. Preventing a resumption of hostilities often hinges on the extension of just war ethical principles and practices into the postwar period (a relatively recent addendum to just war theory known as *jus post bellum*).

Ultimately, the principles of *jus ad bellum* and *jus in bello* coincide with this book's goals: to minimize the prevalence of war and its deleterious effects. But these objectives will remain unrealized unless statesmen, strategists, and warriors study and reflect upon ethics' role, including its limitations and benefits, vis-à-vis the nature of war. Furthermore, we must ponder the truth in sentiments like those of Douhet and Aube, that ethical matters are seldom unequivocal, that even the best intentions can go awry, and that competitors and enemies are likely to twist the laws of war to suit their interests. For, as Hans Kelsen writes,

Who is to decide the disputed issue as to whether one State actually has violated a right of another State? Not the science of law, not jurists, but only and exclusively the governments of the States in conflict are authorized to decide this question; and they may decide . . . in different ways.[477]

CONCLUSION AND SUMMARY

On War contains many references to the need for principles and system, but never delivers them in a way designed to be learnt by the parrots of military crammers and spoon-fed examinees.[478]   Hew Strachan

This chapter has outlined a *Unified War Theory* (UWT) assembled from war's enduring truths, gleaned from science, history, and the ideas of war's master theorists. Coupled with Chapter 1's examination of war's origins, relationship to humanity, definition, and the True Trinity, the UWT represents a balanced, inclusive perspective on war's twenty dialectics and war theory's great intellectual traditions intended to advance our appreciation for the nature and character of war and to encourage clear thinking and communication on matters of war and strategy.

Unified War Theory Synopsis

❖ **War** is a uniquely human phenomenon, characterized by combative violence between two opposing sides for the purpose of achieving a desired political outcome.
  ○ War's **True Trinity:** *Humanity – Politics – Combat* (Figure 1.4).
❖ There are four *zones* related to the peace-war dialectic:
  (1) *true peace*: a state of harmony and reciprocal altruism
  (2) *competition*: nonviolent contention
  (3) *conflict*: violence short of war (e.g., apolitical or unilateral)
  (4) *war*: reciprocal, combative violence for political purposes
❖ A *general war theory* articulates war's essential truths in abstract form.
  ○ War *sub-theory* (e.g., small war or domain) is subordinate and complementary to general theory and expresses further truths unique to its subject.
  ○ The three tasks of domain theory are:
    (1) deduce facts and abstract principles related to the domain
    (2) determine how to gain a decisive advantage in the domain (requires three a's – *awareness*, *access*, and freedom of *action*)
    (3) describe how to exploit (2) to advance the aims of war
❖ **Politics** is an activity through which humans gain and apportion power and govern.
❖ **Power** enables the pursuit of political objectives by providing the potential or capacity to exert *influence* on things or people.
❖ **Influence** is power's effect upon its intended target. There five general ways to influence:
  (1) accord: influence by example and shared interests
  (2) persuasion: influence through arguments and logic
  (3) inducement: influence by reward or *quid pro quo*
  (4) coercion: influence by threat or intimidation
  (5) conflict: influence by violent force (includes war)
❖ Politics and politicians produce *policy*, which advances political objectives by guiding or directing activities and investments that turn power into influence.
❖ Political groups (including parties and states) behave like individuals with survival their ultimate goal. Thus, the root of any war's cause is political survival.
❖ We generally equate greater power with greater security (i.e., freedom from threats to survival and prosperity); thus, the accumulation of power is often a central aim of policy.
❖ Threatening or employing violence to create influence and advance political aims is one of the oldest, easiest, and most pervasive uses of power;

thus, war is a tool of politics, and the military is one of several *instruments of power.*

❖ **Instruments of power** (IOP) are the tools, the means for exerting influence.
   ○ IOPs apply at every level from individuals to alliances; however, national IOPs fall into four categories: *diplomatic, information, military,* and *economic* (DIME).
   ○ IOPs provide the means for executing *strategy.*
   ○ **Radiation** is influence resulting from the imprecision of using IOPs (e.g., unintended consequences like collateral damage).

❖ **Strategy** is the art and science of identifying, procuring, implementing, assessing, and adjusting ways and means to accomplish physical or psychological ends.
   ○ **Political strategy** serves a political purpose (i.e., it produces results impacting power-apportionment and governance).
   ○ **Grand strategy** is political strategy that pursues objectives relating to the security, prosperity, and stature of a state or society relative to others.
   ○ **Military strategy** employs the military IOP in war or peace to serve grand strategy.

❖ Strategy must include assessments of **risk** – the likelihood of a negative influence or result; and **cost** – the anticipated pecuniary value of the ways and means needed to prepare for, conduct, and conclude the war; and the human (physical *and* psychological) and materiel losses incurred in combat.

❖ There are four *essential acts of strategy* in war:
   (1) The **first act of strategy** *defines the political aim, the desired ends the war is intended to achieve.*
   (2) The **second act of strategy** *assesses whether the IOPs and ways of war,* with due consideration to cost, risk of failure or unintended consequences, friend and enemy disposition, and the environment, *yield an acceptable probability of producing the desired ends within a satisfactory period of time.*
   (3) The **third act of strategy** *identifies, procures, and implements ways and means to achieve the war's ends.*
   (4) The **fourth act of strategy** *regularly assesses and adjusts (as needed) the first three acts to account for changing circumstances.*

❖ There are five tests for evaluating strategy:
   (1) suitability: the ways and means are likely to promote the ends
   (2) desirability: the means are commensurate to the ends (reward > risk + cost)
   (3) feasibility: the means are available or acquirable; the ways are workable
   (4) acceptability: the ways and means are morally and legally tolerable
   (5) sustainability: the ways and means will last as long as needed

❖ The **nature of war** is a *reflection of human nature and interaction within the various contexts of violent political conflict*; thus, it is unchanging.

❖ The **character of war** is *an image of war's nature projected through the continuously changing lens of time and a dynamic universe*; thus, it is variable.

❖ War's *twenty dialectics* are: (1) order-chaos, (2) past-future, (3) peace-war, (4) life-death, (5) creation-destruction, (6) friend-enemy, (7) good-evil, (8) science-art, (9) defense-attack, (10) unlimited-limited, (11) viscous-inviscous, (12) physical-moral, (13) direct-indirect, (14) certainty-uncertainty, (15) simplicity-complexity, (16) reason-emotion, (17) prudence-boldness, (18) control-autonomy, (19) concentration-dispersal, and (20) rest-movement.
  ○ These dialectics are integral to war's nature; thus, exclusively favoring extremes (e.g., viscous-boldness-attack) may lead to unbalanced and ineffective theory and strategy.
  ○ The ability to choose the appropriate dialectic extreme for a given context is critical to successful strategy and warfighting.

❖ There are three **levels of war**:
  (1) *strategic* – orchestration of the military IOP across the entire scope of a war
  (2) *operational* – coordinated battles or campaigns spanning a theater of war
  (3) *tactical* – smaller scale engagements or battles

❖ War is **fractal** – its properties and patterns replicate across all war's levels.

❖ The **cycle of war** (Figure 4.10) leverages information to drive strategy, planning, execution, and assessment in a continuous loop.

❖ Logistics are *always* critical to successful strategy and warfighting.

❖ In the **open system of war** (Figure 4.6), two sides strive to influence each other with IOPs even as both sides are influenced by internal and external factors.
  ○ According to the *principle of reciprocity*, factors within the system of war apply to *both* sides, though often unequally, and strategy must account for the traits of both.
  ○ **Forces** are *human and material assets* used to exert influence in war.
  ○ The **environment of war** (a subset of the open system) is the physical context within which warring sides contend for dominance.

❖ **Warfighting domains** can be *physical* (land, air, space, and water surface and subsurface) or *nonphysical* (cyberspace and cognitive).
  ○ Land is the *prime* warfighting domain, because that is where most people and things we value are located.
  ○ Wars are *fought* in the physical domains and cyberspace but always *decided* in the *cognitive* domain.

- ❖ The "**just war tradition**" includes *jus ad bellum* (the moral justification for war) and *jus in bello* (limitations on the conduct of war). In general, the regulation of war is only effective with accountability and enforcement; otherwise, laws, rules, and treaties are often disregarded, especially when survival is at stake.
- ❖ War is always influenced by *uncertainty* (inability to know with complete accuracy), *chance* (unforeseen or unintended happenings), *chaos* (lack of order or a discernable pattern), and *probability* (the likelihood of any specific event occurring).
- ❖ **Combat** is the violent clash of opposing forces and wills in war.
- ❖ **Will, capacity**, and **cause** form the *engine of war.*
  - ○ **Will** is the psychological, motivating force that sustains human thought and action.
  - ○ **Capacity** is the physical (including personnel and equipment) and mental (including leadership and training) means and aptitude for waging war.
    - • **Combat capacity** is a subset of the overall capacity and includes decision-making (leadership) and IOPs (forces and support).
    - • **Relative capacity** is the ratio of one side's warfighting capacity to the other's.
  - ○ **Cause** is the purpose that supports the will and motivates warriors into action.
- ❖ A **center of gravity** (COG) is the person or group, location, or thing an adversary most relies upon to sustain its will to fight.
  - ○ The *prime* COG is the person or group most responsible for motivating and sustaining the war effort. All other COGs are secondary.
- ❖ **Lines of effort** (LOE) link tasks, missions, and objectives to produce effects within a specified geographic or functional area.
- ❖ The *object of war* is to advance the political aim by eliminating the enemy's will to fight. Attacking capacity is a preferred but not always effective way to influence will.
- ❖ The **will-capacity model** (WCM; Figure 4.7) illustrates the preeminence of war's psychological dimension and how the will to fight is supported by capacity, courage, confidence, leadership, training, morale, education, motivation, and experience.
- ❖ The best leaders share four major attributes: *competence, character, communication*, and *charisma*. Military geniuses exemplify these characteristics.
- ❖ Maximizing a war effort's *probability* of success requires accurate intelligence (processed information) regarding enemy and friendly forces.
- ❖ In war, strategists must weigh effectiveness and efficiency; however, *effectiveness is paramount*, for efficiency is useless in defeat.
- ❖ Military forces can adopt one of two *postures*: attack and defense.

○ Attackers have a positive aim: to initiate action, displace or pursue an enemy, or capture territory. Attackers generally expend capacity faster than defenders.

○ Defenders have a negative aim: to delay or await another's action, hold a position, or repel an attacking force.

❖ In war, **momentum** is *the tendency for trends in actions and results to continue or to become more pronounced over time*, and **strategic momentum** makes it difficult to stop or divert a strategy once enacted.

❖ **Inertia** in war is a force's tendency to remain in its present state unless acted upon or motivated to act.

❖ **Culmination** occurs when the attacker's fighting capacity drops below the defender's. Without an infusion of capacity, culmination often causes a shift from momentum to inertia and a change of posture from attack to defense.

❖ The essence of *operational and tactical art* is knowing how and when to prepare and employ forces for maximum effect given the capabilities of both sides and the nature of the operational environment.

❖ The *engagement sequence* (or kill chain) is the process for applying violent force against an enemy (find, fix, track, target, engage, assess).

❖ The **OODA Loop** (*Observe-Orient-Decide-Act*; Figure 4.9) is the decision cycle of war and battle.

❖ **Command and control** (C2) in war is the exercise of authority and direction over forces to accomplish military objectives. Effective C2 is essential for successful military operations.

❖ A *form* of war is defined by its character: who fights, how they fight, where they fight, and why they fight.

○ War's forms are part of a *continuum* characterized by the *viscosity* of opposing forces.

❖ **Viscosity** in war is a function of a force's directness, acceleration, restriction, cohesion, and concentration (DARCC; Figure 4.13).

○ *Directness* relates to the path and objective of force movement and actions. It decreases with deception, delay, or movements that do not advance toward a target or objective. Viscosity increases with directness.

○ *Acceleration* is a force's ability to change speed and direction; thus, it is proportional to speed and maneuverability. Viscosity decreases with acceleration.

○ *Restriction* describes how closely a force follows rules or laws. This decreases when rules or laws are ignored. Viscosity increases with restriction.

○ *Cohesion* is directly proportional to the interdependence of a force's elements. Viscosity increases with cohesion.

- *Concentration* describes the physical arrangement and proximity of forces and their effects and is proportional to the quantity of forces in a given area. Viscosity increases with concentration.
- ❖ There are two **force states** in war: *viscous* and *inviscous*.
  - **Viscous** forces have relatively low acceleration and relatively high levels of directness, restriction, cohesion, and concentration (e.g., traditional military forces).
  - **Inviscous** forces have relatively high acceleration and relatively low directness, restriction, cohesion, and concentration (e.g., guerrillas or terrorists).
- ❖ According to the *principle of relative viscosity*, the viscosity of forces relative to each other is all that matters in combat. Relative capacity drives viscosity, and relative viscosity drives war's forms.
- ❖ There are four **engagement types** in war: (1) viscous versus viscous (V-V); (2) viscous versus inviscous (V-I); (3) inviscous versus inviscous (I-I); and (4) hybrid (H-V/I/H).
  - **Hybrid warfare** occurs when at least one side employs both viscous and inviscous forces.
  - **Asymmetric warfare** occurs between forces of substantially dissimilar viscosity (e.g., V-I).
  - **Symmetric warfare** occurs between forces of similar viscosity (e.g., V-V, I-I).
- ❖ **Containment** is a fluidity inhibitor, imposed or contextual, that increases viscosity and thus vulnerability to superior force.
  - *Direct containment* is imposed by one force on another through physical action like a rapid or surprising attack or encirclement.
  - *Indirect containment* is imposed by the conditions or context of battle, like terrain.
  - *Tactical containment* sets the conditions for the effective use of violent force by minimizing the enemy's ability to avoid or mitigate an attack.
  - Surprise simultaneously creates containment *and* a localized capacity advantage.
- ❖ The **war-viscosity algorithm** (W-V-A; Figure 4.22) illustrates decision making in war and the relationship between viscosity, the will to fight, context, and relative capacity.
  - The W-V-A is a guide for optimizing the *probability* of success in war for a given set of circumstances.
  - **Divergence** occurs when force states deviate from the W-V-A's recommendation.

5 The Future of War

Up to now, our emphasis has been on the past (history and theory), but a central purpose of war theory is to help predict the future character of war. For this task, theory is superior to history alone, for as Mahan advises, precedents may be faulty or become obsolete, but theoretical principles always have their "root in the essential nature of things."[1] In other words, though history may "rhyme," war theory more comprehensively renders war's unchanging rhythms, which are part of all wars, past, present, and future.

The Future

Whoever cannot seek the unforeseen sees nothing, for the known way is an impasse.[2]
                                                                    Heraclitus

Our bodies are bound in the present, but our minds can travel in time, into the past and into the future. As we learned in Chapter 1, our brain's most remarkable trait is the ability to imagine things that are not real, and this includes pondering the future. Since past events have already occurred, they cannot change; thus, we can examine and analyze the past through personal memories, experiences, and the recollections of others. The future, however, having not yet happened, exists only in the shadowy realm of probability and conjecture. The future is a destination that never arrives but must be constantly anticipated; thus, contemplating the future is a beguiling, vexing part of daily existence. Nearly everyone craves a tomorrow that is better than yesterday or today, so we desperately want to see into the future. Yet we rarely can, at least not far or with much accuracy. There is nothing "that we have a greater desire to possess," writes physicist Lee Smolin, "than what time holds forever unattainable – knowledge of the future."[3]

Humans innately strive for power, and a capacity to know the future is powerful. In fact, prescience is a trait reserved exclusively for omnipotent deities. Still, even those falling short of godlike foreknowledge can turn prophecy into power. Historically, oracles and seers have ascended to positions of great esteem and authority, and modern "prophets" of business, like Oprah

Winfrey (Harpo Productions), Bill Gates (Microsoft), Mark Zuckerberg (Facebook), and Jeff Bezos (Amazon), have parlayed a corporate *coup d'oeil* into vast fortunes. Though most of us lack this gift, we still require *some* capacity in this regard. Fortunately, our predictive skills tend to grow with knowledge and experience, which empowers us to marshal resources and take action to create *desired* futures, like getting a degree or writing a book. This sounds a lot like strategy because strategy is essentially *a systematic form of fortune-telling*. It evaluates what we can do *now* to get what we want *later*. Indeed, "*Strategy*," writes Freedman, "remains the best word we have for expressing attempts to think about actions in advance, in the light of our goals and capacities."[4] Success in daily life entails strategy, and good strategy uses information from the past and present to inform estimates of the future.

Formulating strategy is like piecing together two jigsaw puzzles.[5] The first puzzle is the future and the second is the future *we wish to create*. An image of the first puzzle takes shape as we connect related bits of data, trends, and patterns from the past and present. But as we peer farther into the future, the puzzle becomes harder to finish because fewer pieces fit and more are missing. Strategy energizes ways and means to alter existing pieces and fashion new ones that we surmise will change the first puzzle into the second, our desired future. Complicating this already daunting process is the fact that the puzzles and their pieces are constantly in flux, shifting with every decision and action. No wonder prophesy and strategy are such inexact and frustrating activities!

Fortunately, looking a short way into the future is fairly easy because the puzzles and pieces are mostly known and have little time to change. Moreover, habits, routines, and schedules allow us to treat many assumptions about future conditions as facts, thus reducing the need to create formal strategies for everyday activities (e.g., the Wednesday bus arrives at 8 a.m., and Saturday football practice is on Field B). Alas, predictive accuracy decays when gazing farther into the future as unforeseen factors and events accumulate. It is much easier to predict a person's location, an investment value, or the weather for the next day than it is for the next year. Furthermore, major disruptions, like a career change or relocation, force us to discard old assumptions and develop new habits. Making assumptions or educated guesses about future conditions is critical to strategy. Assumptions are assessments of time-dependent variables judged to be true before they can be proven true. For example, we can assume the trains will run on schedule all week, but the assumption cannot be *validated* until the week is over. Additionally, the longer an assumption remains true, the *greater* its worth. In other words, accurate assumptions made a week ahead are more valuable than those made a day prior.

There are two categories of factors that affect assumptions and forecasts: environmental and human. The former encompasses all nonhuman variables, like weather or car trouble. Though not fully predictable, these are easiest to account for. Human factors, however, are harder to anticipate because we often do things that defy *logic* and *common sense*, two of the most effective predictive aids. When combined, human and environmental factors create the gap between expectation and reality. In general, assumptions and predictions regarding familiar things or regulated activities (e.g., local traffic and weather, the routines of close friends, or a train schedule) are more accurate; and predictions based on immutable laws and principles are most accurate. For example, we can be certain the sun will not be shining on Scotland at midnight next January because the relevant variables are based on the principles of reliable scientific theory and isolated from human interference. Technological aids developed from these principles – like accurate timekeeping, global positioning systems, and digital forecasting models – improve prediction capabilities. However, even science cannot produce accurate and precise estimates for complex, nonlinear systems, like the dynamics of human interaction, or even for simpler systems projected into the distant future. Few of us can even predict our own location and actions a month ahead let alone a year or decade. Amplify this ambiguity by billions of people and raise to the power of the ever-changing environment and it is easy to see that uncertainty, especially about the future, is an indelible facet of reality.

FUTURE WAR

Another critical part of military theory is the vision of future war.[6]

Milan Vego

War, as a human, complex, dangerous, and uncertain enterprise, presents a particularly wicked challenge for prognosticators and strategists. In war, all environmental factors are in play, and the human factor is intensified. Not only are the combatants unpredictable, but they also actively try to deceive, disrupt, and destroy each other! Perhaps it is not so hard to empathize with those who have tried and failed to anticipate future war. As Clausewitz laments, "If only war were as predictable in practice as it is on paper!"[7] However, despite this difficulty, foretelling the future (and navigating the past-future dialectic) is a must because the past is gone and the present is fleeting, which means *all wars are fought in the future*.

As noted earlier, forecasting is critical to strategy. Ways and means are the methods and resources employed to create "ends." And ends are simply outcomes that *have yet to occur*, a set of *future* conditions (our completed puzzle) imagined in the present. All strategies pursue ends; therefore, all strategists must be prophets, though most practice their art with information and analysis versus omens or crystal balls. For potential or current wars, leaders must gather and process information to inform strategic processes that promote successful outcomes in the next hour, day, week, or year. Anticipating future conditions and events, particularly in combat, requires an inquisitive, imaginative, and disciplined mind. Writes Clausewitz,

If the mind is to emerge unscathed from this relentless struggle with the unforeseen, two qualities are indispensable: *first, an intellect that, even in the darkest hour, retains some glimmerings of the inner light which leads to truth; and second, the courage to follow this faint light wherever it may lead.*[8]

Determining the form of future war, adds Douhet, "is not . . . the province of the fortuneteller, it is, rather, a serious problem, the solution to which must be worked out by logical progression from cause to effect."[9] During a war, commanders ask questions and gather answers to develop relatively short-term strategies. What is the mission? Where is the enemy? What is the enemy's capacity, and who are the enemy leaders? How do they prefer to fight, and what are their likely objectives? What and where are the enemy COGs? How are they defended? What are the key factors in the operational environment (e.g., weather, terrain, noncombatants, and infrastructure)? Leaders assemble these puzzle pieces to build a picture of current "reality" from which they make assumptions regarding enemy courses of action and environmental changes. With this information, they employ strategy to form the pieces needed to create the desired future reality. Clausewitz writes, "Reality supplies the data from which we can deduce the unknown that lies ahead. From the enemy's character, from his institutions, the state of his affairs and his general situation, each side, using the laws of probability, forms an estimate of its opponent's likely course and acts accordingly."[10] Throughout this process, we must account for distorted perceptions and intelligence degraded by chance, uncertainty, and enemy action. Specifying branches and sequels and acquiring reserve capacity are effective ways to "hardwire" adaptability into strategy.

An opponent's strategy, according to Sun Tzu, is war's principal target, and it can be degraded by clouding the enemy's sense of reality. The two primary means of doing this are via operational security (OPSEC) and deception. OPSEC *denies* the enemy *true* information about capabilities and intentions, and deception *provides* the enemy *false* information about the same. Returning to the puzzle analogy, OPSEC forces the enemy to work without all the pieces, while deception replaces valid pieces with phonies. Sun Tzu claims that all warfare is based on deception, which is the art and science of duping an enemy into forming false conclusions about the future. Inaccurate forecasting corrupts strategy and the cycle of war leading to flawed decisions and increased vulnerability to war's most lethal snare, surprise.

During periods of nonwar competition, the consequences of poor forecasting may be minimal; however, they can be devastating if war erupts (e.g., Pearl Harbor, *Blitzkrieg*, and 9/11). Absent the urgency of war and a known opponent, gaining meaningful prewar intelligence is challenging. Potential enemies conceal capabilities and intentions, and the most useful information is clouded by a distant or indeterminate future. Strategists must wrestle with questions like these: Is war likely, and if so, when or with whom? Who will fight and for what purpose? How, where, and with what will they fight? These judgments have enormous consequences. If war is hell, losing a war is worse, and the probability of defeat increases when war is unanticipated or takes an unexpected form. Indeed, "Soldiers are reluctant to hypothesize about the future," writes Major General Robert H. Scales, "because war is a high stakes game. Getting it wrong costs lives and catastrophic failure threatens survival of the state."[11]

War is costly, and preparing military forces and honing the military IOP takes time, effort, and money. Without a methodology for predicting the future character of war, forecasting is little more than expensive guesswork. Wasting resources on strategies, weapons, warriors, or doctrines ill-suited to future conflicts leads to tragedy. Conquerors like Alexander, Genghis Khan, Napoleon, and Hitler achieved great success against ill-prepared opponents. Yet, despite our best efforts, we commonly fail to accurately anticipate war's future character. In fact, when asked about the USA's ability to predict wars, US Secretary of Defense Robert Gates replied, "Our record is consistent and perfect – we never get it right."[12] And General Moseley adds,

In the 1920s and 1930s, when our political and military leaders assumed the Nation was appropriately postured for the future, we failed to anticipate the coming crucible ... we entered World War II unprepared for the demands of total war. Likewise, we engaged in both Korea and Vietnam unprepared for the challenges of limited war. America paid a heavy price in blood and treasure for this strategic myopia.[13]

Ignoring Theory

As we have said, strategy's "two-puzzle" challenge is formidable; however, much of the blame for this poor record is attributable to ignorance or misapplication of war theory. "All too often," writes Milan Vego, "the critical importance of military theory either is not well understood or is completely ignored."[14] Without theory, says Corbett, we tend to "argue too exclusively from the latest examples and to become entangled in erroneous thought," because we lack the ability to distinguish between lessons driven by "special conditions" and those "due to factors common to all wars."[15] Theory that explains war's *hows* and *whys* – the source of Clausewitz's "glimmerings of inner light which lead to truth" – is indispensable to strategy. For example, war is always reciprocal; however, in peacetime, war preparations may become one-sided or focus on one enemy while neglecting to account for other threats. Between wars, nonmilitary domestic priorities can overshadow war preparations allowing training, planning, and innovation to stagnate even as potential adversaries adapt. Moreover, leaders may truncate strategies for political or economic expediency, creating divergence with theory and reality. In these instances, future war predictions may feature finessed assessments that ensure the expected enemy justifies previously implemented training and matériel investments. "The strategic debate is often not about the future," writes Williamson Murray, "but rather about desires to preserve the procurement of expensive weapon systems."[16] Overlooking theory increases the likelihood of programmatic, operational, and strategic myopia and the probability that the next action or war will be surprising.

The Imperfect Past

The only knowledge the people have is of today and yesterday. They cannot look beyond the immediate, they cannot escape the illusions of the moment.[17]
Raymond Aron

Theory checks the natural inclination to view the future as only a slightly modified version of the past or present (i.e., skewing the past-future dialectic). As Figure 4.1 depicts, theory informs strategy, doctrine, TTPs, and historical and warfighting analysis. Using history alone as a predictive aid introduces variances that arise from inaccuracy, subjectivity, and psychological "presentism" (i.e., the tendency for current perceptions to sway views of the past and future). While the past should instruct the future, it cannot do so effectively if our perceptions are distorted. Indeed, as Codevilla and Seabury attest, "The obvious danger in applying the lessons of past wars to present or future wars is that we tend to deceive ourselves with our favorite images of the past."[18] And according to General Mark A. Milley, "War tends to slaughter the sacred cows of tradition, of consensus, of group-think and myopia. The next war will be no different. Those ... that stubbornly cling to the past will lose that war ... in a big way."[19] Theory corrects the prejudice of presentism, especially regarding war's likely forms.

An old proverb says that generals are always fighting the *last* war, not the next one. This occurs because recent wars produce visceral experiences and data readily subject to analysis. Leaders promote best practices, fix deficiencies, and justify resource commitments using hard evidence. This approach is easier than forecasting, but future enemies are *also* studying history and devising countermeasures. Thus, preparing linearly from past wars can be hazardously predictable. For example, after the 1990 Gulf War, the USA invested heavily to refine the dominant tactics and technologies used in that campaign only to suffer bitter setbacks against a Somali warlord and al-Qaeda. America expected its opponent to be *yin*, but theory says that when *yin* is strong, the adversary becomes *yang*. Germany's senior military leader of World

War I, General Erich Ludendorff, almost took Germany down the same path. Liddell Hart writes,

Ludendorff's picture of the way that the next war would be waged was merely an intensified reproduction of the offensives he had carried out in 1918. For him the offensive was still a battle-process in which the infantry would be helped forward by artillery, machine-guns, mortars, and tanks until "it overwhelms the enemy in a man-to-man fight."[20]

Clausewitz's statement, "Changes in the art of war always emanate from decisive actions which, little by little, tend to modify the rest," explains how war's "learning curve" gradually influences its character, but preparations between wars require a more nonlinear, creative approach.[21] Habits and routines, for individuals or armies, impede success if left unexamined as major contextual changes occur. Indeed, few of Ludendorff's lessons carried over to World War II. Mechanization and airpower *changed* warfare from an infantry slog to the devastatingly innovative, combined arms *Blitzkrieg*. The USA's fast aircraft carriers and stealthy submarines similarly transformed naval warfare in the Pacific. In fairness, the past *should* inform predictions of future war, just not *exclusively*. As Clausewitz says, "There can ... be little doubt that many previous ways of fighting will reappear."[22] However, it is also true that no two wars are identical. Conditions change and opposing sides use history's lessons in a race to anticipate, avoid, and negate each other's strengths. This competition can cause the character of future wars to diverge sharply from the past and explains why predicting war's future character has even frustrated war's great theorists.

Politics, Technology, and Doctrine

Incorrectly evaluating current conditions and trends in politics, technology, and doctrine, also hinders forecasting. Sudden, unexpected political changes upset the balance of power, throwing assumptions and character-of-war assessments into doubt, especially regarding who fights and why or when they fight. A few examples of surprising political change include the French Revolution and rise of Napoleon, which confounded nineteenth-century Europe and plunged the continent into decades of war; Abraham Lincoln's unlikely election victory in 1860, which set the USA on a path to civil war; the collapse of Tsarist Russia and nationalist China, which resulted in civil war and the destabilizing ascent of Soviet and Chinese Communism; the failure of the Weimar Republic and the rise of Hitler's Nazi fascism; anti-colonialism in Africa, which sparked insurgencies and the expulsion of European powers; and the fall of Iran's Shah in 1979, which led to the Iran-Iraq war, two Gulf Wars, and chronic regional turmoil. Unfortunately, there is no foolproof methodology for anticipating political changes let alone determining which ones will cause wars.

"War is waged," writes Colin Gray, "with the products of technology," and technological advances – like gunpowder, machine guns, tanks, submarines, nuclear weapons, radar, and aircraft – frustrate prediction by confusing assessments concerning war's methods (tactics) and means (weapons).[23] Innovative weapons have expanded warfare into all domains, from undersea-to-orbit, and hastened many martial icons – like knights, archers, horse cavalry, fortresses, and battleships – into obsolescence. Industrialized warfare accelerated battle lethality, leading to the massive carnage of the American Civil War and world wars. Technological trends are easier to follow than the vagaries of politics, but secrecy and rapid innovation still complicate forecasting. And we should not forget, as *The Future of War* (1996) authors George and Meredith Friedman affirm, "Technology changes how men fight and die, but it does not change the horror and glory of battle, nor . . . the reality of death."[24]

Doctrinal changes, pertaining more to warfighting's *how* than *what*, often accompany technological progress. Doctrine and TTPs are essentially formalized habits and routines, great for dealing with proximate uncertainties, but liabilities if not reexamined relative to major environmental or human-driven change. "The use made of technology," writes Gray, "typically is more important than is the technology itself."[25] Thus, failing to anticipate doctrinal change or to revise doctrine and TTPs to accommodate innovation is tantamount to ignoring the technology altogether. Many of antiquity's famous generals used innovative doctrines, even more than advanced weapons, to win battles. Most ancient armies had infantry and cavalry, but only Alexander perfected the phalanx, only Rome fielded the disciplined legion, and only Genghis Khan mastered the horse-mounted archer and lancer. Likewise, prior to World War II, most major powers had mechanized vehicles and aircraft, but only Germany (initially) put it all together. Similarly, the USA beat Japan in the Pacific with superior coordination of naval, air, and ground forces despite many areas of Japanese technological superiority. More recently, inviscous terrorists have frustrated mighty militaries with homemade explosives and cellphones.

Theory's Role

No matter what kind of war we face in the future, whether from terrorists or more traditional armed forces, in formulating effective responses we can draw lessons and devise tactics using military rules or maxims whose origins go back beyond recorded time.[26]   Bevin Alexander

In the tasks of daily life or developing war strategy, the *probability* of successful prediction improves with greater understanding of the nature of our surroundings provided by theory. While war's complexity precludes theory from granting foreknowledge of future events, theory does reduce the likelihood of surprise, particularly regarding war's changing forms (viscosity). By explaining how and why war changes form via the property of force viscosity, the UWT and W-V-A assist with Clausewitz's "supreme ... act of judgment" (i.e., anticipating "the kind of war on which [we] are embarking").[27] Theory empowers us to contrast the lessons of past conflicts and current trends – including political, technological, and doctrinal change – with war's enduring truths. Moreover, as a summation of history's most relevant lessons, theory frees warriors and strategists from having to "play historian," a role most are ill-equipped for. In *On War*, Clausewitz asserts that "not every future war" is likely to be a massive Napoleonic confrontation aiming at a clear decision, rather "one may predict that most wars will tend to revert to wars of observation."[28] This statement, though not consistent with *ex ante* trends or his personal experience, was consistent with *theory*, which explains that human nature, chance, and politics limit war's aims and violence.

Calculus of Future War

We shall glance at the war of the past long enough to retrace its essential features; we shall ask of the present what it is preparing for the future; and, finally, we shall try to decide what modifications will be made in the character of war by the causes at work today in order to point out their inevitable consequences.[29]
                                                                 Giulio Douhet

According to theory, the character of war reflects *war's nature* and the specific *circumstances or context* of the conflict (i.e., war as seen through the "shifting lens of time and a dynamic universe"). War's nature is unchanging; therefore, predicting future war requires a capacity to accurately anticipate future contexts and circumstances. We can best do this and inform war's strategic calculus by thoroughly assessing the following factors (Figure 5.1):[30]

(1) *Current circumstances* (from intelligence): what is happening right now; these are the initial conditions from which to extrapolate forecasts.
(2) *Trends and experience* (from history): what happened in the past; trends indicate the "trajectories" of key factors like politics, technology, and doctrine; future circumstances often mirror historical experience when specific character-of-war traits (who, what, where, and why) are alike.
(3) *The nature of war* (from theory): what is unchanging about war (e.g., the True Trinity, reciprocity, the engine of war, and viscosity).

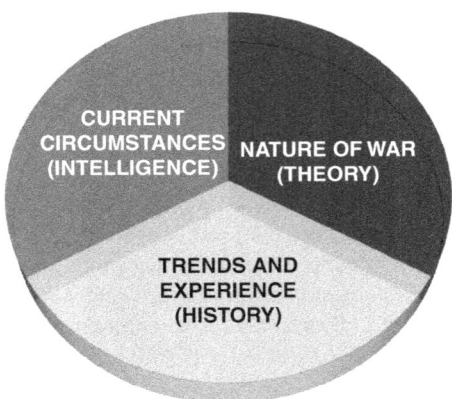

Figure 5.1 Future character of war

Forecasting War

Just because perfection is unattainable ... does not mean that attempts to assess the nature of a war are futile. After all, a level of 25 percent accuracy is still preferable to one of 10 percent or less.[31]   Michael Handel

Though the precognition of Asimov's "psychohistory" will always elude us, access to good war theory, accurate intelligence, and sound historical analysis greatly improves the odds of successfully predicting war's future character. Additionally, we must not forget the fourth essential act of strategy, for the relentless march of time demands continuous reassessment of assumptions *and* predictions. Forecasting, as part of strategy, must be iterative, and individual and institutional thinking must be flexible and adaptive. With circumstances always in flux, the side that processes new information, evaluates it, and rapidly adjusts gains the advantage. "Victory smiles upon those who anticipate the changes in the character of war," writes Douhet, "not upon those who wait to adapt themselves after the changes occur."[32] Theory is the starting point for any forecast – know the enemy and know yourself. We must study potential adversaries and constantly reevaluate our engine of war. As Aristotle advises,

[A strategist] must know the extent of the military strength of his country, both actual and potential ... and further, what wars his country has waged, and how it has waged them. He must know these facts ... also about neighboring countries, and also about countries with which war is likely in order that peace may be maintained with those stronger than his own and that his own will have power to make war or not against those that are weaker. He should know, too, whether the military power of another country is like or unlike that of his own; for this is a matter that may affect their relative strength.[33]

We must also examine the environment of war including political, technological, and doctrinal fluctuations and trends that may influence future warfare. Furthermore, having assessed "relative strength," we can use the UWT to predict war's most likely forms, and thus prepare our engine of war for high (V-I) or low capacity (I-V) scenarios. Finally, theory aids preparation and strategy by preventing us from becoming so mesmerized by war's changing character that we forget its nature and those universal traits true of all wars.

Know the Enemy

In forming the plan of a campaign, it is requisite to foresee everything the enemy may do, and to be prepared with the necessary means to counter-act it.[34]
                                                                          Napoleon

The principle of reciprocity mandates that future war assessments begin with a survey of potential warring sides (i.e., political threat groups – who fights). War can be internal (civil war) or external (between states or non-state entities), and the potential for war between any two sides increases in proportion to the incompatibility of their political ideologies (why they fight). Most political entities, to satisfy their prime directive (survival), seek greater internal and external security, power, and wealth. Just as a biological organism increases its viability by procreating, political entities increase security by spreading their political and ideological "DNA" (e.g., values, beliefs, and customs) as widely as possible. Of course, other "organisms" have the same imperatives, which leads to competition, conflict, and war.

Aggravating factors include *physical* proximity, which can spark territorial disputes; *political* proximity, which stokes cultural and ideological animosities; resource and economic competition; transnational crime; and other overt provocations that amplify political enmity. Some historic examples include border strife between India and Pakistan; religious and sectarian discord in the Balkans, Turkey, Iraq, and Syria; resource competition in the South China Sea and the Arctic; Persian-Arab-Israeli-Shia-Sunni conflicts; NATO-Russia territorial and political tension; and religiously motivated, terrorist aggression by VEOs like al-Qaeda, Abu Sayyaf, ISIS, and Hezbollah.

Once potential threats are identified, the next step is a deeper examination of the threat's history, politics, culture, ideology, geography, and demography including intelligence on technology, doctrine, and TTPs. "The behavior of . . . armed masses," writes Aron, "becomes senseless if we refuse to see a spirit leading them."[35] How and why do they fight? What fear, honor, or interest factors might cause them to fight, and what sustains their will? What is their prime COG? What is their warfighting capacity, and is it growing or contracting? Who are their friends, allies, and enemies? Intelligence on training and weapons development activities like testing, exercises, and wargaming is extremely valuable. "Red teaming" is also an effective way to simulate adversary tactical and strategic processes and decision-making.[36] Building familiarity with potential threats furnishes many of the "puzzle pieces" needed for strategy.

Know Yourself

He who conquers himself is the mightiest warrior.[37]   Confucius

As we learned in previous chapters, assessments of *relative* capacity are fundamental to warfighting and strategy; therefore, we must honestly evaluate our own engine of war. What is the status of our state or organization? How stable are our political circumstances, and what are our political objectives and strategic interests? What resources and IOPs do we have and how do we intend to use them? What is our warfighting capacity, and what supports our will to fight? How does our capacity compare with potential foes? What are our COGs and how will we use and defend them? For what purposes will we use military force – defensive, offensive, or preemption? What is our stance on *jus ad bellum*, *jus in bello*, and the use of WMDs? Answering these questions provides valuable data on relative capacities useful for determining optimal force states (viscosity) and may also reveal vulnerabilities requiring remediation.

Know the Environment

After analyzing potential enemies and ourselves, we must examine the environment of war. Where, during what seasons, and in which domains might a war be fought? How are politics, technology, and doctrines changing? What research is occurring? Are any breakthroughs imminent? How and how often do others test, exercise, and employ new warfighting technologies and TTPs? What alliances presently exist, and how might they change with time? Analysis should include local and *global* trends, for potential adversaries may acquire or emulate technological or doctrinal innovations developed by others. There are many additional environmental factors that will impact future wars. The physical environment is always changing due to our activities and by natural processes. Climate change affects the economy and where and how people live. The Arctic is more accessible for resource exploitation, rising sea levels may flood populous coastal areas, and droughts and famine will cause fear and desperation. Natural catastrophes, which could include fires, earthquakes, tsunamis, hurricanes, volcanic eruptions, meteor impacts, and pandemics, will wreak havoc on infrastructure, commerce, and militaries. These factors increase internal and external political tensions and complicate strategy as governments struggle to adapt, compete, and survive. As global population rises, demands on resources like fresh water, clean air, food, and energy grow, and the competition for energy resources in a tech-centric global economy intensifies the potential for armed conflict. Despite advances in alternative energies, fossil fuels will remain in high demand creating friction between suppliers and consumers. Finally, globalization and automation will create economic opportunities *and* complications as human jobs are eliminated and workers and industries migrate internationally seeking comparative advantage.

Politics and Future War

Reliably anticipating *specific* political changes that might cause a war is virtually impossible. However, building awareness and understanding of global demographic trends and patterns in governance, including economic volatility, helps spotlight areas of political instability that could raise the likelihood of war. As the Earth's human population approaches ten billion, most people will align within about two hundred independent nations plus a plethora of non-state entities all possessing diverse *Weltanschauungs* and interests. Competition for power, status, and resources will fuel the potential for war. Governments that cannot satisfy the expectations of their polities face the added threat of civil war. Furthermore, some nations will experience destabilizing pressures from population stagnation, contraction, and demographic shifts. For example, many developed countries like Russia, Japan, Spain, and Germany have static or contracting populaces, which will depress long-term productivity and economic power. China faces similar challenges as its massive population ages and a large portion enters less productive years. Other countries face political pressures from mass immigration and cultural dissonance.

As we have noted previously, the survival imperative has caused most political systems to coalesce within two categories: *liberal democracies* (which promote human rights, freedom, and the rule of law) and *autocracies* (which prioritize the ruling party's survival over individual rights and freedoms).[38] Moreover, in Chapter 1, we affirmed humanity's *capacity* for war but *preference* for peace and identified two extremes – utopian and totalitarian – as hypothetical though unattainable war-free models. As depicted in Figure 4.2, liberal democracy and totalitarian autocracy are "waypoints" between these extremes – liberal democracy favoring governance by contentment, and autocracy favoring governance by control – that promote internal peace.

Liberal democracies mitigate the will to fight by sharing power with citizens, safeguarding rights, and providing a higher quality of life for more people.

Democracies are generally stable internally, because the rule of law provides nonviolent adjudication for disputes and orderly transitions of power. As Machiavelli writes, "Nothing does so much to stabilize and strengthen a republic as some institution whereby the changeful humours which agitate it are afforded a proper outlet by way of the laws."[39]

On the other hand, autocracies restrict the capacity to fight by hoarding IOPs and limiting rights and freedoms; thus, autocracies can remain stable for long periods. Ruling autocrats retain the authority to ignore or alter laws to suit their interests. Autocratic rule by monarchs, despots, and oligarchies was the historical norm up until the mid-1800s, after which liberal democracies proliferated rapidly. Ironically, though gravitation to liberal democracy or autocracy generally promotes *internal* peace and political stability, it also creates polarity that raises *external* tensions and the potential for international war. This is particularly true of autocracies, because permanent, one-party systems compel autocrats to shift blame for internal problems onto external actors.

Major war between nuclear-armed powers remains unlikely for the reasons outlined in the last chapter. However, wars of choice between liberal democracies and nonnuclear autocracies or between proxy states and greater powers remains possible if not probable, and war between a major nuclear power and a limited nuclear power cannot be ruled out. If a relatively inviscous entity, like a rogue state or VEO, obtains or develops WMDs, a larger nuclear power may act preemptively to neutralize that entity's ability to use them.

Technology and Future War

Today's decisive innovation becomes tomorrow's standard, and is sure to be obsolete sometime thereafter. No one can know how long novelty will last. Only one thing is sure: there is no such thing as an ultimate weapon.[40]

Angelo Codevilla

Not a discovery or invention is made but the war and navy bureaus of all the great nations seize it to see what use can be made of it in war.[41]

William Graham Sumner

It is a fact of war applicable to nearly any conceivable future that *political entities will use scientific and technological progress to gain military advantages.* As we learned in the last chapter, technological advances improve warfighting effectiveness by narrowing the gap between the ideal (instantaneous results) and the real. Like political change, the steady but often unpredictable progression of technology will have significant effects on war's future character.

There are essentially three ways to anticipate technological change and its impact on war. First, intelligence collection must target open-source public- and private-sector technology development and the weapons development programs of potential enemies. By determining what an enemy is developing, we can deduce its purpose and probable uses. Second, sponsor robust public- and private-sector research and development programs and shield them from compromise. Again, it is easier to make accurate forecasts regarding familiar matters. Leaders in advanced technology have the greatest familiarity with cutting-edge systems and their likely uses, strengths, and weaknesses. Consequently, taking a page from Slessor, strategists and policymakers should solicit the advice of technical experts when assessing an adversary's future capabilities. Finally, create a vigorous wargaming, experimentation, and testing enterprise in live and synthetic environments. "The point of experimentation," writes the Heritage Foundation's Dakota Wood, "is to explore the potential of a new idea to solve a problem, answer a question, or reveal something new."[42] Advanced capabilities, even hypothetical ones, can be modeled in digital environments and employed in scenarios that can be adjusted and repeated almost without limit, generating valuable data for engineers and warfighters at a fraction of the cost and time required for live events. Live prototype testing, experimentation, and wargaming provide higher-fidelity data that enables leaders to identify improvements or branch out in new directions.

Current Trends

Given the proclivity of events to ambush the complacent, identification of what is happening today can only have educational merit, it cannot serve as the raw material for prediction.[43]   Colin Gray

As Gray's statement implies, rapid change often renders current trend assessments obsolete; however, there are some emerging developments that are sure to influence future war. Following World War II, martial technologies progressed at an evolutionary pace. Traditional military capacity is still largely measured quantitatively in numbers of personnel; weapons (e.g., bombs and missiles); vehicles (e.g., tanks); naval vessels (e.g., ships and submarines); and aircraft (e.g., fighters and bombers). The speed, range, sophistication, precision, and lethality of these "legacy" systems improve every day, but only incrementally. Meanwhile, a seismic shift in the pace and direction of engineering and applied science has supplanted the technical wonders of the industrial and space ages with the micro- and nano-marvels of the information, digital, (and perhaps) cybernetics and eugenics ages. To paraphrase Pasteur, the world of the infinitely small has become infinitely large. This revolution, which includes superfast computer chips and designer biology, has energized a brave new world that includes satellite and wireless-enabled navigation, communications, and targeting; hypersensitive sensors; genetic mapping and engineering; remote- and AI-controlled, multi-domain vehicles; directed-energy weapons; hypersonic missiles; and the weaponization of cyberspace.

Advances in computing power and storage (perhaps soon to include quantum computing) directly affect war in many ways.[44] Gathering, processing, analyzing, and disseminating information, the foundation for the cycle of war, OODA Loop, and C2, require computing power. So does codebreaking. Thus, advantages here increase the likelihood of achieving decision superiority and also benefit the design, development, testing, fielding, employment, and assessment of other weapons. Additionally, computational power drives genetic, cybernetic, and bioengineering progress with the potential to improve life or destroy it. While armies populated with genetically enhanced or cyborg super-warriors are unlikely in this century, research in these areas will certainly continue, perhaps unethically, and the results may have profound political and societal ramifications that could precipitate or exacerbate future wars.

AI, autonomous, and semi-autonomous weapons (e.g., "robots" or "drones") will undoubtedly influence future war. Autonomous machines capable of operating in complex environments already exist. They reduce risk to humans and typically have wider performance envelopes than traditional human-operated war machines. Additionally, AI decreases the cognitive burdens of combat by processing large amounts of data and performing straightforward tasks. Improvements in AI algorithms and machine learning will enable some weapons (including cyber weapons) to complete more complex missions and perhaps even surpass human decision-making abilities for select purposes.[45] This technology will lead to "swarms" of multi-domain, robotic machines (or cyber-code) that confounds and overwhelms traditional adversaries while shrugging off "casualties," and it will also give rise to *Matrix*-style AI-versus-AI "warfare" in cyberspace. Critical questions (related to the control-autonomy dialectic) yet to be resolved include identifying which decisions can be delegated from humans to AI. Ironically, even this decision may be made by a machine! In any case, many AI "decisions" will seem tactically, strategically, or *ethically* astonishing; thus, reliance on AI and autonomous weapons is risky. War is complex, and computer-based decisions are unlikely to have the requisite flexibility or discretion to make humans irrelevant, particularly for real-time battle management and employment of lethal force in tactically ambiguous circumstances. Additionally, complex digital systems often malfunction or behave in unintended ways. Complete reliance on AI could lead to artificial mayhem or outright malevolence as chillingly depicted by science fiction's HAL-9000 (*2001: A Space Odyssey*), V'ger (*Star Trek*), the WOPR (*War Games*), V.I.K.I. (*I, Robot*), Skynet (*Terminator*), and others. Finally, cyberspace is a fiercely contested domain; thus, weapons dependent on cyber capabilities (like AI) will be prized targets for tech-savvy opponents (live or synthetic). Greater technical sophistication by major militaries ensures that the ability to attack and defend in cyberspace will remain a growth industry, and since it presents a convenient seam for the have-nots to influence the haves, cyberspace will be a key I-V warfare domain.

Nanotechnology covers a class of atomic and molecular-scale materials and devices. Nanotech substances, like graphene, though hard to manufacture, show extraordinary potential for strength and electrical conductivity.[46] As nanotech matures, it will have extensive civilian and military applications, and nanomachines ("nanites" or "nanobots") would have revolutionary repercussions for peace and war. They can theoretically be engineered for a variety of purposes including disease treatments and cures or crafted into insidious weapons capable of infiltrating living tissue or larger machines with devastating consequences. Provided science can circumvent the inherent "spookiness" of quantum-level matter, nano-weapons will likely debut in future wars; however, if they are not restricted by *jus in bello* or if VEOs acquire them, the results could be tragic.

Hypersonic and directed energy weapons (like lasers), at least for the most advanced countries, will feature in future wars.[47] Hypersonic delivery systems are even more difficult to intercept than ballistic missiles because of their extreme speed and unpredictable trajectories. Since these weapons are costly and exotic, the expense-to-effect ratio will likely compel nations to reserve them for high-value strategic purposes or existential circumstances. Directed energy weapons, long staples of sci-fi, have had only limited and generally nonlethal utility in war. While there are practical limits to how small and powerful laser weapons can be, lethal lasers will be part of future combat. Laser energy propagates at the speed of light, making laser attacks nearly impossible to dodge.[48] However, the main engineering challenge for lasers is producing truly destructive energy. A practical tank- or warship-destroying laser is very unlikely; however, lasers will be effective against "softer" targets like people, satellites, aircraft, and missiles.

Thermonuclear explosives (hydrogen bombs) are the world's most powerful weapons. There is no theoretical upper limit to their power, and there are plenty in current inventories to create a global holocaust. Nevertheless, antimatter weapons would be about 100 times more powerful than hydrogen bombs by weight. Thus, an immensely powerful antimatter explosive could be quite small and consequently easy to deliver and conceal. However, antimatter does not exist naturally on or near the Earth. Although we can "make" antimatter in infinitesimal quantities at enormous cost, storing antimatter, let alone crafting it into a weapon, is prohibitive because it detonates upon contact with ordinary matter. Therefore, due to their impracticality, no future war is likely to feature antimatter bombs. Other potential but highly impractical sources of destructive power include so-called *dark energy* and *dark matter*. These two oddities of physics (coined to explain anomalies in cosmological dynamics) have thus far eluded empirical verification. However, even if they are observed and measured someday, there is no cause to believe that we can exploit them for any practical purpose.

Finally, we should avoid the tendency to "fall in love" with our gadgets regardless of how potent they may seem now. "At its high point, just before disaster," warn the Friedmans, "the last generation's technology appears invincible."[49] Despite our efforts here to assess the trajectory of technological trends, we cannot discount the probability that war's character will be impacted by unanticipated breakthroughs or the imaginative combination of known technologies to create novel capabilities. Absent prescience, the best recourse is to be the side making the discoveries or otherwise adapting rapidly enough to prevent a destabilizing power shift.

Science Fiction

The impossible is impossible; it is a condition, not a problem for which a solution has yet to be found.[50]   Colin Gray

This section might be the most controversial because it dares place limitations on human ingenuity. Nevertheless, when pondering future possibilities, it is helpful to eliminate the distraction of future *impossibilities*. The amazing pace of technological progress (and a vigorous science fiction industry) fosters the impression that *anything* is possible given enough time; however, the physical world is actually more limited than the human imagination. Furthermore, with respect to technology and war, we must keep in mind three caveats:

Not everything that . . .

(1) is theoretically possible is *practically* possible.
(2) works on a computer or in a lab will work in *combat*.
(3) can be done *ought* to be done.

Power Outage

A significant limitation pertains to power generation, a subject fundamental to warfighting. Power (the rate of energy expenditure) makes things go; therefore, energy and power are essential to war. However, there are a finite number of practical energy sources and ways to generate power, and the greater the power needs, the larger and more cumbersome the power source (and supporting logistics). Electricity remains the most common and essential power source. It is generated by converting other forms of energy (like mechanical or solar energy) into electrical current. Mechanically produced electricity from spinning turbines, which create alternating current, most commonly comes from energy sources like moving water, wind, or steam generated by burning hydrocarbon fuels, biomass, or nuclear fission. Matter-antimatter annihilation generates the most energy, but this source (along with dark matter and energy) is unfeasible for reasons outlined earlier. Nuclear fusion (literally, star power) is great for explosive weapons but impractical for controlled electricity generation. "It is fifty years away and always will be" is a common refrain regarding the status of safe, effective fusion reactors. Even if fusion energy matures, these reactors will be too large and fragile for use in war machines like ships, tanks, or aircraft. Ruling out antimatter and fusion leaves hydroelectric, geothermal, wind, solar, nuclear fission, chemical, and hydrocarbons, and only the latter three are useful in war.

For energy-on-demand to propel vehicles or projectiles, there are few practical alternatives to chemical propellants, fission (for large vessels like warships), or hydrocarbon-based energy to drive internal combustion, jet, and rocket engines. Energy can be stored in batteries, but these are often unwieldy and environmentally sensitive. Absent an unexpected breakthrough (e.g., nano-scale batteries), a revolution in energy density and endurance is unlikely. With respect to gun-fired projectiles, the limits of explosive chemical energies have already been reached, leaving only exotic methods like plasma or electric "railgun" technologies; however, these remain logistically cumbersome, complex, and extremely expensive. Therefore, barring a major discovery in power generation – and none appear imminent – macro-scale weapons and vehicles will continue to advance only incrementally.

These Are Not the Droids You're Looking For

Despite promising developments in robotics, the ability to create an effective anthropomorphic "droid" battle-force as depicted in movies like *Star Wars* remains far in the future. Perhaps in a century or two it may be possible to produce robot soldiers resembling humans in appearance, performance, and cognition (androids), though powering them will be a significant problem. Then again, such devices might be smart enough to realize that they can put an end to war and improve their own survival odds by getting rid of humans. As Omar Bradley once said, "If we continue to develop our technology without wisdom or prudence, our servant may prove to be our executioner."[51]

Houston, We Have a Problem

Power and energy limitations guarantee that access to space will remain costly and dangerous, and much as it pains this sci-fi enthusiast to admit, human space exploration will be limited to our immediate vicinity – Earth orbit, the Moon, and possibly Mars. There are no reasonable means of safely transporting humans farther into space and back even if there were some pragmatic reason, beyond scientific curiosity, for doing so.[52] Frankly, there is nothing on the Moon, Mars, or local asteroids and comets commercially or militarily valuable enough to justify the risk and expense of human trips. Near-Earth space will remain a vital warfighting domain; however, treaties prohibiting the militarization of space notwithstanding, it is unlikely humans will physically fight in space this century. Transporting humans to space is costly, and conditions there, even in low-Earth orbit, are dangerously debilitating. Space warfare, at least in this century, will be "fought" by "hunter-killer" satellites or other weapons directed from Earth.

The Myth of Invisibility

Stealth technology, which reduces reflected electromagnetic and acoustic energy signatures for vehicles like aircraft and submarines, has justifiably earned praise as a warfare game-changer.[53] Stealth falls under the broader umbrella of camouflage, concealment, and deception; thus, it enables security and surprise by short-circuiting an adversary's OODA Loop and kill chain. However, even as advances (like meta-materials that "bend" electromagnetic waves around objects) are developed and fielded, further innovations in detection capabilities, both active and passive, preclude foolproof invisibility.[54] Nevertheless, stealth will yield appreciable combat benefits provided the side using it understands its limitations and remains vigilant for adversary countermeasures.

Future Imperfect

Other conceivable futuristic capabilities that will *not* be part of future war due to insurmountable physical or technological barriers include time travel, matter-energy transportation, telekinesis, force fields, and gravity-inertia nullification. Though Einstein's General Theory of Relativity proves that time is variable, there is no practical, safe, or precise way to move *people* forward or backward in time. Theoretically, time travel is possible, but only by accelerating to near-light speed or falling through a special type of black hole. Likewise, despite the "spooky" ability to teleport quantum-scale objects from place-to-place in a lab, there is no practical or safe extrapolation of this phenomenon capable of decomposing larger objects into energy in one location and reconstituting them in another (like *Star Trek*'s transporter). Telekinesis or direct mind-control of physical objects or people is another science fiction plot device for which there is no compelling evidence. While it is possible to "read" brain signals and enable rudimentary hands-free control devices, we can already do this more reliably by "reading" head and eye movements or by using voice commands. Force fields that can repel projectiles *and* directed energy, like a starship's "deflectors," are not consistent with the laws of physics, and will remain fictional. Finally, selective gravity and inertia nullification would truly represent astounding technical achievements. Unfortunately, though there are promising theories which might explain how gravity (the most mysterious natural force) works, they are empirically unvalidated and offer no clues that might lead to "antigravity" technologies. Regarding inertia, the human body has limited tolerance for rapid speed-direction changes (acceleration), and even if we could create vehicles with exceptional speed and maneuverability, the inertia experienced as "g-forces" would kill a human occupant. Since physics offers no feasible ways or means for neutralizing inertia, human-bearing craft will always be subject to acceleration limitations.

Again, at the risk of joining all those who have infamously predicted that humans would never fly, reach the moon, or cure diseases, we can be almost certain the preceding examples will remain beyond our reach. As such, the futuristic combat depicted in sci-fi franchises like *Star Wars* and *Star Trek* will remain confined to the universe of imagination and will not debut in a future war. But we should be grateful for this because any positive application of these advances would likely be eclipsed by their disastrous lethality. In any case, we should not forget that, as Heuser says, "Technological change" has not played "the direct and dominant role as the simple, all-conditioning variable which determined the intensity of wars. Instead, throughout history, the use that was made of technology has depended on social, cultural, political and ideological variables."[55] Indeed, adds Aron, "It is policy rather than weapons that creates the danger. It is not machines that make history; they alter the conditions in which men make it."[56]

Doctrine and Future War

Forecasting doctrinal and TTP change requires all three predictive aids: historical study (what has worked in the past); analysis of current conditions (what is working and what is being developed now); and theory (what always works). Appraising past and current conditions with theory helps us judge whether successful doctrines and TTPs will remain effective in the future. For example, theory supports the general efficacy of deception to confuse or dislocate an enemy; however, it does not necessarily follow that the Allied methods used for the Normandy invasion would suffice against a twenty-first-century adversary. Since doctrine deals with the "how" of warfare, it is closely related to war's forms, and thus, viscosity.

Future War and Viscosity

Over the ages, commanders have most often failed to perceive the transform-
ations of war, and both soldiers and civilians have suffered greatly because of
their blindness.[57]   Bevin Alexander

The UWT and principles of viscosity facilitate future war preparation by
countering our enduring penchant for favoring specific forms of war based on
presentism, traditional or cultural preference, or domain-based biases. As we
learned in Chapter 4, we can anticipate likely viscosity-driven force states by
assessing relative capacities and using the W-V-A. In addition to illustrating the
decision process for starting and ending a war and showing how and why
viscosities change, the W-V-A helps predict the most likely form of future war.
The UWT defines four combat engagement types (V-V, V-I, I-I, and hybrid),
which depend on the relative capacities and viscosities (viscous and inviscous)
of opposing sides. Equal capacity creates a balance-of-power that often precludes war; thus, most wars begin when one side believes it has or can create
(e.g., via surprise) an appreciable capacity advantage.[58] This satisfies
the second act of strategy and helps pass the W-V-A's will-capacity-context
threshold. As a result, even symmetric wars may feature a "higher-capacity"
side (if only marginally) and *all* asymmetric wars will; thus, our analysis will
assess future war from the high- and low-capacity perspectives.

High Capacity

Many states, including super- and great powers, have developed and now
maintain regular professional forces trained to fight irregularly.[59]
                                                                      Colin Gray

The global distribution of power and resources favor some over others. The
W-V-A shows that those with an abundance of warfighting capacity will most
likely face inviscous adversaries (V-I) unless their rival is a power-peer (V-V)
with a perceived capacity advantage. In the current age, technology and
weapons proliferation offer a toxic mixture of nasty means that inviscous forces
can exploit. Intelligent foes with a strong will to fight invest time and resources
acquiring means and methods for countering their enemies, and they attempt to
be strong where their opponent is weak. Though strong states usually prepare
for symmetric (V-V) war, they would be foolish to ignore the likelihood of
fighting an asymmetric (V-I) war, even against a near-peer (the W-V-A shows
that *any* force state is possible with balanced capacity). Preparing for *all* forms
of war is expensive, prohibitively for most, but the ability to field forces of
varying viscosity is essential if high-capacity states want to be effective in all
potential engagement types. Resource-rich nations are wise to pursue a hybrid
force structure, including strong, viscous forces (to deter attack from peers) and

relatively inviscous "special" military and nonmilitary forces trained to operate covertly, dispersed, and with high acceleration. The US military has adopted this model by integrating viscous conventional and inviscous special operations forces, a prescription paralleling Colin Gray's advice that, "the US defense community needs to avoid choosing between regular and irregular warfare as its dominant mode. It has to try to excel at both and, indeed, to attempt to achieve some fusion of the two in its way of warfare."[60]

Indeed, the W-V-A suggests that powerful nations benefit most by developing hybrid capabilities, which can be optimized by adhering to four imperatives. First, maintain a strong viscous force relative to potential threats and either arm with nuclear weapons or align with a nuclear-armed ally. "The truth is," writes Gray, "that nuclear weapons are useful both diplomatically and prospectively strategically."[61] Viscous force, backed by nuclear capabilities, is best for deterring survival-threatening aggression. Inviscous forces can harass and degrade stronger enemies; however, outside of revolutionary civil war, they rarely threaten survival. Raising and maintaining a powerful viscous force requires training, education, and technology investment in all warfighting domains, and since few states can afford this, weaker powers facing a low-capacity foe should pursue mutual defense treaties and alliances.

Of course, great power (*yin* – viscous) drives most threats to adapt their own viscosity (*yang* – inviscous), which leads to the second imperative: high-capacity states should organize, train, and equip *permanent*, effective inviscous forces. As centuries of empirical data prove, asymmetric warfare is difficult for high-capacity, viscous forces and underestimating this surprising weakness can be calamitous. The USA learned this the hard way. For example, the US Army failed to codify small-wars doctrine after the Philippine War and Cuban experiences of the early twentieth century, and this contributed to misfortune in Vietnam. Though the US Marines published the *Small Wars Manual*, it was largely obscured by World War II's emphasis on V-V, amphibious warfare.[62] This pattern recurred after the Gulf War as the USA spent a decade pursuing "net-centric warfare" (V-V) only to be surprised by inviscous global terrorists (I-V) on 9/11 and insurgencies in Iraq and Afghanistan. Only after years of frustration did the United States finally publish FM 3–24, the twenty-first century's *Small Wars Manual*.

The inviscous portion of a hybrid force serves two purposes: (1) exploiting the gaps and seams of a viscous opponent (I-V), and (2) locating, containing, and defeating inviscous forces (I-I). Moreover, in developing hybrid capabilities, high-capacity states must consider both offensive and defensive postures. Offensively, they must thoroughly assess peer and asymmetric enemy vulnerabilities and likely defensive capabilities and actions. Defensively, they must anticipate how a peer or asymmetric foe might employ its own inviscous forces and develop a counterstrategy for negating these attackers and minimizing weaknesses. Special operations units are a good starting model for inviscous forces, but their preparation should focus both on maximizing inviscous advantages against stronger forces *and* achieving viscous advantages (like containment) against similar or weaker foes. Rather than detaching inviscous forces for temporary duty, it is better to create a dedicated inviscous force with multi-domain capabilities including cyber and space.[63] This force should be assembled around a professional cadre responsible for studying, preparing for, and leading hybrid warfare. For nations with low tolerance for long-duration wars (like liberal democracies), hybrid forces make early success more likely, thus strengthening strategic will. Hybrid forces require capabilities in all warfighting domains, especially cyberspace, and this entails research, development, and employment of cutting-edge technologies. For instance, through an umbrella program called *Sha Shou Jian* ("Assassin's Mace"), China identifies, develops, and fields relatively inexpensive weapons that target a viscous adversary's most critical, complex, and expensive capabilities, like satellites and aircraft carriers.[64]

Preparing for the long term is the third imperative for high-capacity forces. War is a complex, chaotic affair where results are seldom proportional to expended time, effort, and resources. Curbing an inviscous foe's will often requires a sustained effort. Time is the inviscous force's ally as the viscous side's strategic will tends to wane in the absence of measurable progress. "Our cards were speed and time," writes Lawrence, "not hitting power."[65] Viscous powers must prepare for the long term, placing less emphasis on destroying tactical capacity and more on systematically eroding the foundations of the enemy's cause, ideology, and sources of support. This strategy places a premium on cultivating a durable, resilient strategic COG. Though the Cold War was not a real war, it was a competition between ideologies – liberal democracy and capitalism versus autocratic communist socialism. The Soviet strategic COG eventually collapsed because, despite the rhetoric and propaganda, America had clearly surpassed the Soviets in power and prosperity. Though many ideologies, including religious radicalism and fascism, will never die, a consistent, persistent approach can eventually render them politically impotent.

The final imperative supports the rest: strengthen intelligence gathering, analysis, and dissemination capabilities to speed decision-making. Information is critical to strategy, the cycle of war, and the OODA Loop. Inviscous forces are hard enough to contain and influence, but these tasks are impossible without good information upon which to base strategic, operational, and tactical decisions. Intelligence must extend beyond order-of-battle assessments and probe the extent, depth, and fortitude of the opposition's will and COGs. As Clausewitz says, "The will is not a wholly unknown factor; we can base a forecast of its state tomorrow on what it is today."[66] War is principally a psychological affair; therefore, intelligence cannot neglect the human and political dimensions, including culture, psychology, religion, and human factors assessments of enemy political and military leaders.

Low Capacity

Even if the wealthy nations are protected from invasion by the weapons of mass destruction, they are not protected from guerrilla warfare. In our time, the war of partisans has changed the map of the world more than the classical or atomic destructive machines.[67]   Raymond Aron

The W-V-A offers a single preferred state, inviscous, for any side with low capacity relative to its opponent. There are several ways these "have-nots" can survive and prevail without having to suffer what they must at the hands of the strong. First, they can gain capacity by forming alliances, preferably with a strong country or with numerous smaller powers. A leader facing danger from foreign powers, writes Machiavelli, "will defend himself with good arms and good allies."[68] Where arms (capacity) are lacking, allies become even more important. Countries like Latvia, South Korea, the Philippines, and Qatar have formed alliances or belong to associations, like the Association of Southeast Asian Nations (ASEAN) and the Gulf Cooperation Council (GCC), to counter the economic and political influence of potentially hostile regional powers. Alliances strengthen the weak by association, but only if the terms provide for mutual defense *and* if allies honor their commitments.

Next, weaker countries must prepare to operate inviscously, with mobile, elusive, dispersed, unrestricted forces. Inviscous forces should prepare to exploit the information IOP, to focus on war's human and political dimensions, and to capitalize on viscous vulnerabilities to propaganda, cyberattack, subversion, and deception. Prior to the Internet and social media, stronger governments could monopolize information IOPs to shape perceptions, especially regarding wartime transgressions. Now, real-time, independent reporting and blogging from a myriad of sources and perspectives, including anti-government press and partisan propagandists, can have dramatic effects on a viscous side's strategic COG, especially for liberal democracies with highly polarized internal politics (e.g., the USA in Vietnam and Iraq).

Third, the "have-nots" must be able to collect and process actionable intelligence about key adversary vulnerabilities. Inviscous forces thrive in the enemy's gaps and seams, but only when they are identified ahead of time. An inviscous side's ability to operate illicitly simplifies intelligence gathering since a great deal of information on unprotected civilian targets is available from open sources. This wider range of objectives coupled with global mobility and well-disguised resourcing frustrates the viscous opponent's strategy and security plans, forcing dilemmas regarding the distribution of offensive, defensive, and counterintelligence capabilities.

Finally, inviscous forces, even more than viscous, must prepare for *protracted war*. This is relatively easy on home soil close to support and refuge but much harder otherwise. A powerful, unifying cause and message (e.g., repel the invaders or punish the infidels) supported by propaganda, strong leadership, and a resilient organization help inviscous forces sustain their will to fight long enough to outlast the opposition. Winning a protracted war means avoiding containment and direct battles, and whenever possible, cultivating internal and external sponsorship.

War's Constants

As we implied earlier, war's immutable nature means that its *future* nature is already known. The tendency to forget this is symptomatic of misperceptions regarding war's nature and character. Future war is still war – human, combative violence for political purposes – and it will always be dangerous and complex due to the interplay of uncertainty, chance, and chaos. Future war will have human, political, and combat dimensions in variable measure, the cognitive domain will be paramount, and the engine of any future war will have three components: cause, capacity, and will. As such, preparing for future war requires significant investment in assessing war's human factors, how people think, what they value, and their biases and habits. Confusing the enemy's cognitive perceptions to achieve surprise or dislocation will remain an important facet of future war. Moreover, war futurists should exploit the pervasiveness of media and messaging to influence the human psyche while denying an enemy's ability to reciprocate. Intelligence is vital for high *and* low-capacity forces; therefore, counterintelligence, deception, corrupting an adversary's intelligence enterprise, and OPSEC will also be critical in future war. Furthermore, sound intelligence must be collected, analyzed, and distributed to the right place at the right time to feed decision-making and C2.

Since there is no guaranteed "right way" to fight a war, adaptability and guile (*metis*) are often more important than raw power (*bie*). Preparing for future wars, writes Murray, "should not begin with weapons systems, but rather [with] adaptable and flexible military organizations that can meet a wide range of strategic threats, especially the unexpected."[69] Doctrine and tactics must also be adaptable; for, as Michael Howard writes,

I am tempted indeed to declare dogmatically that whatever doctrine the armed forces are working on now, that they have got it wrong. I am also tempted to declare that it does not matter that they have got it wrong. What matters is their capacity to get it right quickly when the moment arrives.[70]

These attributes assist commanders faced with the uncertainty of combat and help leaders and strategists contend with the ambiguity of future war. Robust C2 capabilities maximize flexibility and adaptability; thus, preparing for future war demands extensive commitment to C2, including training, education, and equipment to ensure military forces can act in concert and adapt to friction and enemy action. Whenever possible, weapons and systems design should allow for continuous upgrade or modification, and force structures and organization should be modular, scalable, flexible, and complementary (e.g., *Kampfgruppen*, multi-domain, and hybrid).

Finally, we must use sound war theory to scrutinize all "new" and "revolutionary" concepts – technological, tactical, and theoretical – that inevitably emerge. Test every slogan and buzzword for validity. Is it grounded in strategic logic or does it lean only to the extremes of war's dialectics? Keep the concepts that withstand these tests, otherwise, change or abandon them immediately, before they can do harm.

Conclusion

It is impossible to predict the future, and all attempts to do so in any detail appear ludicrous within a few years.[71]   Arthur C. Clarke (1962)

Only the dead can ignore the future. The rest of us must operate as though Arthur C. Clarke is wrong, and fortunately he is (sort of). The future is unknown, but an awareness of the past and present informed by theory can increase the likelihood that our predictions will resemble what actually occurs. Imagining future war is essential to sound strategy and requires accurate knowledge and analysis, not only of past and present wars, trends, and current circumstances, but also of war theory. Though it is impossible to know everything about ourselves, the enemy, the environment of war, and the future – and thus, impossible to create *perfect* strategy – the same limitations apply to both sides. "The goal cannot be to make error free predictions about the future," writes Gray. "Instead, it is to make only *relatively* minor mistakes" (emphasis added).[72]

Warfighting capacity analysis and assessment of relative capacity provide a foundation from which to evaluate dynamic viscosity and war's likely forms. We can gain further insight into future war by surveying the environment of war including political, technological, and doctrinal developments with the potential to alter war's character in significant ways. Tensions will continue to arise between contending political and ideological factions within and between states. Competition for territory and resources in the quest for greater power and security will also cause political friction that could lead to war. Scientific and technological advancements with the greatest potential to drastically alter war's character include nanotechnology; biological, cybernetic, and genetic engineering; cyber-weapons; directed energy; and AI-enabled, autonomous machines.

For nearly any conceivable future, there are broad imperatives based on relative capacity assessments that should inform and guide strategy and warfighting preparations. The strong should organize, train, and equip hybrid forces to maximize effectiveness in symmetric and asymmetric forms of war. These forces should strive to win any war quickly but must always be prepared for long-duration conflict, and they must be supported by robust, effective intelligence, counterintelligence, and OPSEC capabilities to outpace the enemy's cycle of war and OODA Loop. On the other hand, low-capacity sides should look to gain strength from alliances and partnerships while also organizing, training, and equipping to fight inviscously, through dispersal, mobility, speed, and cunning rather than pure power. Like their stronger adversaries, inviscous forces require excellent intelligence aided by rapid, secure communications, and C2 capabilities, and they must also be prepared for long-duration conflict.

Finally, we cannot forget the constancy of war's nature. Future war will be like current and past wars: human, political, violent, and consequently uncertain and chaotic. Physical terrain and warfighting capacity may change, but the human "terrain" will always be the prime COG. Cause, capacity, and the will to fight comprise the engine of past, present, and future war, and whether weak or strong, each warring side must recognize that changing capacities can dictate new force states, which dramatically alter war's character and influence outcomes. Moreover, since no one can know the future with certainty, success in future war favors flexibility and adaptability. Along with strategy, we must regularly reevaluate warfighting habits and routines (doctrine and TTPs) and optimize them for present realities and anticipated futures. "People who can easily throw off old habits and quickly acquire new ones," writes Mark May, "have a great advantage in modern warfare."[73] Though we can never know all we want to about future war, following these guidelines will unquestionably improve the *probability* of success. Ultimately, we can find solace in the fact that future opacity applies universally, and its ambiguity can make friends and enemies think twice about starting wars. "The uncertain side of the future holds the widest domain and has proved the most baffling thing of all," writes Thucydides, "yet also the most useful; for the fear that we all feel alike makes us more circumspect about attacking one another."[74]

In Thucydides's time, a war-free world was inconceivable, but is it any more plausible today? Can we realistically expect a future where predicting the character of war is passé and "future war" has become an oxymoron? The next section explores these questions and evaluates the feasibility and implications of ending war forever.

THE END OF WAR?

Only the dead have seen the end of war.[75]   George Santayana (or Plato)

*For what can wars but endless war still breed? Till Truth and Right from
Violence be freed.*[76]   John Milton, *Sonnet 15*

Excepting a brief Chapter 1 excursion to evaluate the feasibility of war-free
societies, we have assumed that wars will always exist. But our holistic
treatment of war remains incomplete without challenging this supposition. In
his groundbreaking anthropological analysis of war, Bronislaw Malinowski
poses several questions regarding war's future: "Shall we abolish war, or must
we submit to it by choice or necessity? Is it desirable to have permanent peace,
and is this peace possible? If it is possible, how can we implement it
successfully?"[77] We can simplify Malinowski's questions as follows:
(1) *Can* we end war (is there a *feasible* way to end wars forever)?
(2) *Should* we end war (would a war-free world be a *better* world)?

We can tackle this problem using strategy's first two essential acts and five tests:
(1) define the aim: *a war-free world*; and (2) assess whether ways and means can
produce this end (can we end war?) at an acceptable cost and risk of unintended
consequences (should we end war?). Provided there is a *suitable* and *feasible* way
to abolish war, we must then judge whether the benefit outweighs the costs
(*desirable* and *acceptable*) and the results will be permanent (*sustainable*).

The first question – can we end war? – has been asked and answered many
times, usually in the negative. The second question, however – *should* we end
war? – may seem odd. Certainly, we would abolish war if we could, right? But
this is an important question because as with war itself, choices and actions can
yield unexpected results. We should not start a war *or* try to end all wars without
being absolutely certain the outcome will be worth the price. Otherwise, we
might discover, as Athens and Germany did after invading Sicily and Russia,
respectively, that we have hastened our demise rather than our salvation.

The previous section illustrates the difficulty of forecasting future conditions
with certainty, particularly regarding war. Accepting this ambiguity as fact,
how can we expect to make durable judgments regarding the future of war?
After all, our conclusion can rise no higher than that of an unprovable hypothesis. Nevertheless, by employing the methodology developed for forecasting
war's future character, we can make some sensible inferences about its permanence as a human activity.

Can We End War?

*This ease in making war, together with rulers' lust for power – something that
seems inborn in human nature – is thus a great hindrance to perpetual
peace.*[78]   Immanuel Kant

I have bedimmed
The noontide sun, called forth the mutinous winds,
And 'twixt the green sea and the azured vault
Set roaring war – to th' dread rattling thunder
Have I given fire ... By my so potent art.
But this rough magic
I here abjure, and when I have required
Some heavenly music, which even now I do ...
I'll break my staff,
Bury it certain fathoms in the earth,
And deeper than did ever plummet sound
I'll drown my book.   Shakespeare, *The Tempest* (5.1.40)

Yes, we can end wars; and furthermore, we know that war between any two potential belligerents (e.g., the USA and Soviet Union) is not inevitable. War is a choice, and there have been and will continue to be instances where that choice is deferred. The primitive Amazonian Waorani, for example, "after generations of warfare and raiding ... were persuaded almost overnight to abandon the pattern."[79] The Cuban Missile Crisis (1962) is a more contemporary example of hostilities deferred, and there are even modern countries without formalized armed forces, like Costa Rica and Iceland. However, according to anthropologists Clayton and Carole Robarchek, despite a proven ability to avert war and set weapons aside, "The old pattern remains latent, and the potential for a return to violence is ever-present."[80] Thus, our formidable challenge is to find a way to *erase* this "old pattern" and end even the *possibility* of war. When asked if this is possible, most people say "no." American science journalist John Horgan, who asks this routinely in his classes, reports that "over 80 percent of those ... queried – liberal, conservative, male, female, affluent, poor, educated, uneducated – say that war will never end," and "more than 90 percent" (of 400) say humans will never stop fighting.[81] Interestingly, most adults, even without military or war-related experience, feel qualified to answer this question. They have seen, heard, or read enough about human nature, history, and current events to make this judgment confidently. And why not? With evidence for war going back over ten millennia and human aggression far earlier even than that, it has become so difficult to conceive of a world without war that most take it for granted. This mindset certainly includes those most familiar with war, like Colin Gray, who says "We strategists are entirely unconvinced that world history is marching inexorably toward a warless condition."[82] Nevertheless, if war *is* to end, it will be in the future, so that is where we must look for answers.

Forecasting the Future of War

The first portion of this chapter introduced a methodology for improving the accuracy of future war forecasts based on assessments of history and trends, current events, and theory. We can apply the same technique to the problem of war's future; however, since history only covers a fraction of human existence, we will add archaeology and anthropology as well.

Earliest Evidence

The archaeological evidence indicates ... that homicide has been practiced since the appearance of modern humankind and that warfare is documented in the archaeological record of the past 10,000 years in every well-studied region. In general, warfare in prestate societies was both frequent and important.[83]
Lawrence Keeley

Chapter 1 established humanity's dualistic nature as an outgrowth of evolutionary development. We have a robust capacity for violence and a propensity to use it to advance interests. These traits enhanced survival by aiding defense against threats and the acquisition of life-sustaining resources. At the same time, humans have a demonstrated capacity for reciprocal altruism that facilitates life's social and nurturing dimensions. Indeed, "Peaceful activities," writes Keeley, are both "crucial" and "common," "even in the most bellicose societies."[84] Therefore, collectively we are neither wholly warlike nor peaceful. Moreover, despite our dual nature, most people prefer peace and live in a state of true peace, competition, or conflict, not war. Nevertheless, as a matter of pure logic, the attainment of *permanent* peace requires humanity to *permanently* shed its warlike capacities and become wholly peaceful. Since it took millions of years of evolution to mold humanity into its present form, it is sensible to infer that it would take a similar epoch – in a threat-free environment – to reverse these warlike tendencies naturally. However, as William James notes, "Our ancestors have bred pugnacity into our bone and marrow, and thousands of years of peace won't breed it out of us."[85] Besides, adds Keegan,

Even were a wholly unaggressive breed of humans to evolve to live in wholly benevolent circumstances, they would still be obliged to kill the lower organisms that cause disease, the insects and small animals that harbour them and the larger animals which compete for food supplies in the stock of vegetation. It is difficult to see how the necessary system of environmental control could be carried by creatures which lacked aggressive responses altogether.[86]

Indeed, as Chapter 1 illustrates, archaeological, ethnographic, and anthropological research confirms significant levels of warfare in the vast majority of prehistoric and primitive human societies.[87] "It is the competition for life, therefore, which makes war," writes Sumner, "and that is why war always has existed and always will."[88] Regrettably, at least from an environmental and evolutionary perspective, there appears to be no quick fix for our violent capacities.

Historical Roots

To judge from the history of mankind, we shall be compelled to conclude that the fiery and destructive passions of war reign in the human breast with much more powerful sway than the mild and beneficent sentiments of peace; and that to model our political systems upon speculations of lasting tranquility would be to calculate on the weaker springs of human character.[89]

Alexander Hamilton (1788)

Statesmen, both in the East and the West, have not arrived at any possible programme for implementing the prevention of war.[90]

Bertrand Russell (1959)

History reinforces Chapter 1's revelations by confirming the ubiquity of war over recorded time. As humanity emerged from a primitive state into modernity, raiding erupted into full-blown warring. In fact, when turning "our attention from small-scale cultures to state-level societies," writes Thomas Gregor in *A Natural History of Peace* (1996), "we find that peace is even more infrequent and that war is commonplace."[91] Indeed, a 2003 *New York Times* article reported, "Of the past 3,400 years, humans have been entirely at peace for 268 of them, or just 8 percent of recorded history."[92] Furthermore, after analyzing one of "the longest and complete" conflict databases known, a team led by University of Florence statistical analyst Gianluca Martelloni concluded "that there is little evidence supporting the idea ... that humankind is progressing toward a more peaceful world."[93]

Still, humanity's dualistic nature precludes war alone from defining us. That war often dominates the human narrative (as Hamilton asserts) stems from war's perception-distorting drama and sway. "War makes rattling good history"; writes Thomas Hardy, "but Peace is poor reading."[94] Wars, like hurricanes, move across the landscape of human existence with profound consequences that compel our attention in ways that peace cannot. After all, we are genetically wired to notice what is dangerous (and attractive) and ignore what is not. However, looking past war's dazzling aura reveals a human race certainly capable of living in peace, even for long periods. In fact, "The greater part of human activity is spent," write Durbin and Bowlby, "not in war, but in peaceful cooperation."[95] Unfortunately, this no more proves that all people can live in peace forever than the oceans prove that all water exists exclusively in liquid form. Our operative question is not whether humans *can* live in peace. Rather, it is to ascertain whether *all* humans can live in peace *perpetually*.

Looking strictly at the evidence of history, it is hard to fault Hamilton. Moreover, the contrast between our violent history and proven preference for peace is actually more cause for concern than optimism. If we agree with the following premises: (1) that most humans favor peace and have found ways to live peacefully, and (2) that humans by nature use their talents to advance interests; then, we can conclude that the peace-loving majority have been actively working to eliminate war throughout history.[96] However, since humanity has clearly *not* ended war, we must conclude that the efforts of countless brilliant, creative, and determined people *have failed.* "Attempts to eliminate war," writes Kenneth Waltz, "however nobly inspired and assiduously pursued, have brought little more than fleeting moments of peace among states."[97] Although this fact does not prove that ending war is *impossible,* barring some radical new circumstance, it is a major strike against the probability of eliminating war. As George C. Marshall opined, "If man does find the solution for world peace, it will be the most revolutionary reversal of his record we have ever known."[98] And, "To look back," adds Fielding Hall solemnly, "is to look at two thousand years of complete failure with nothing that will help us."[99]

Trends

Oh cease! Must hate and death return?
  Cease! Must men kill and die?
Cease! Drain not to its dregs the urn
  Of bitter prophesy.
The world is weary of the past,
Oh, might it die or rest at last![100]   Percy Bysshe Shelley, from *Hellas* (1822)

While historical trends are discouraging, what of more recent indicators? In general, it is unreasonable to infer an ability to achieve a goal when no past effort – regardless of talent, time, or energy expended – has succeeded, *unless* a revolutionary new development or capability emerges. We encounter this paradigm routinely in science and technology where breakthroughs have enabled advances no primitive person could imagine, like powered flight, space travel, nuclear energy, and organ transplantation. However, have we cause to expect a similar revolution in the areas of human relations, politics, and governance? To be blunt, no. Regarding governance, we have tried nearly every conceivable variation from pure democracy to hermetic totalitarianism, and we have even attempted to regulate international politics through the Peace of Westphalia, the League of Nations, the Kellogg-Briand Pact, the UN, the International Criminal Court, the World Trade Organization, and the European Union. While increases in human-rights-focused liberal democracies and international governing organizations have spurred noteworthy reductions in abhorrent practices like slavery and torture and admirable advances in gender and racial equality, permanent solutions remain elusive, and none of these efforts have indelibly quelled international conflict. "War has flourished," writes psychology pioneer Gordon W. Allport, "under all known social systems."[101] And, "Not one major power, empire or culture decided not to go into the war business," adds

Can War Be Eliminated? (2014) author Christopher Coker, "and not one so far has gone out of it."[102]

Democratic Peace

What of statistical and empirical evidence suggesting that democracy reduces interstate war? In fact, there is some truth here; however, while this is a promising trend, there is not enough data (and may never be) to *prove* that democracy makes war impossible. Furthermore, the relative "newness" of liberal democracies coupled with the Cold War's effect of aligning the interests of twentieth-century democracies toward a common threat (Soviet communism) will, at least in the short term, cast doubt on the theory's reliability.[103] As Strachan asserts, the notion of democratic peace either "rides roughshod over differences between types of democracy or has to operate with so many exceptions that it is really unsustainable."[104] And Cashman adds, "However much our democratic values predispose us to cheer for this theory, there nonetheless seems to be little evidence to support it."[105] Even granting the democratic peace theory's promise, as of this writing, barely half the world's states are truly democratic, not all factor political entities are states, many societies are content *without* democracy, and even established democracies can collapse.[106] As Fred Iklé writes, "The global spread of democracy will not be without reversals. We must expect that nations, ethnic groups, religious or political movements will again come under the control of tyrants who unhesitatingly start wars to expand their dominion or to destroy their adversaries."[107]

Ironically, our most ghastly armament, thermonuclear weapons, appears to have had the greatest limiting impact on war. Prior to 1945, wars between the world's great powers occurred at least a few times per century with increasing levels of violence. Yet, as discussed in Chapter 4, nuclear weapons have altered the calculus of war, making it too dangerous to wage between nuclear-armed powers. Pacifists and disarmament activists may strenuously disagree; however, no hypothesis better accounts for the reduction in war deaths and major power warfare since the mid-twentieth century.

Though nuclear weapons may help prevent bloody world wars, they have not stopped war. Whereas great powers have carefully avoided direct confrontation, choosing instead to fight with and through proxies, lesser powers continue to make war, and civil wars are rife. As Figure 5.2 shows, the rate of war-making since 1400 generally increased until spiking sharply upwards in the 1800s. Though wars began to taper from about 1650 (post-Westphalia) until 1800 (Napoleon), this era was hardly peaceful as it encompassed notable conflicts like the War of the Grand Alliance, the English Civil War, the War of the Spanish Succession, the Seven Years' War, and the American Revolution. This was also a seminal age of ideas including the

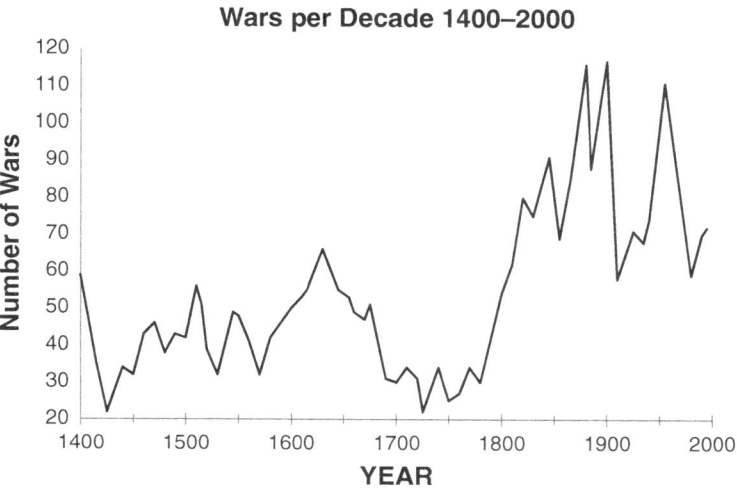

Figure 5.2 Number of wars per decade (1400–2000)[108]

Enlightenment's Age of Reason. Wars are fought for reasons, and intellectual titans of the age – Locke, Descartes, Spinoza, Diderot, Hume, Kant, Montesquieu, Rousseau, Adam Smith, and Voltaire – challenged traditional thinking in philosophy, governance, economics, and religion. This intellectual fecundity prompted many to view the future in new ways and to question traditional norms and bastions of power. Since wars can occur when political views of the future clash, the nineteenth century's sharp uptick should be no surprise.

Figure 5.3 depicts a more granular view of war since 1946. This data contrasts sharply with Figure 4.23, showing the *number* of wars since World War II climbing even as *total deaths* dropped. World War II settled some great power scores but also signaled a violent end to colonialism and the start of Cold War-inspired conflicts. The end of the Cold War (1991) heralded a marked decline in wars until the early twenty-first century when a rise in violent religious extremism and lingering political instabilities sparked a new upsurge. In fact, according to a 2017 report by the UN University Centre for Policy Research, the number of major civil wars almost tripled from 1997 to 2017 with a "six-fold increase in battle deaths"; and 60 percent of conflicts during this period "relapsed within five years."[109] Furthermore, in his comprehensive survey of early twenty-first-century conflict trends, University of Birmingham international relations expert William Avis notes that, while the population-normalized frequency of major wars has generally declined,

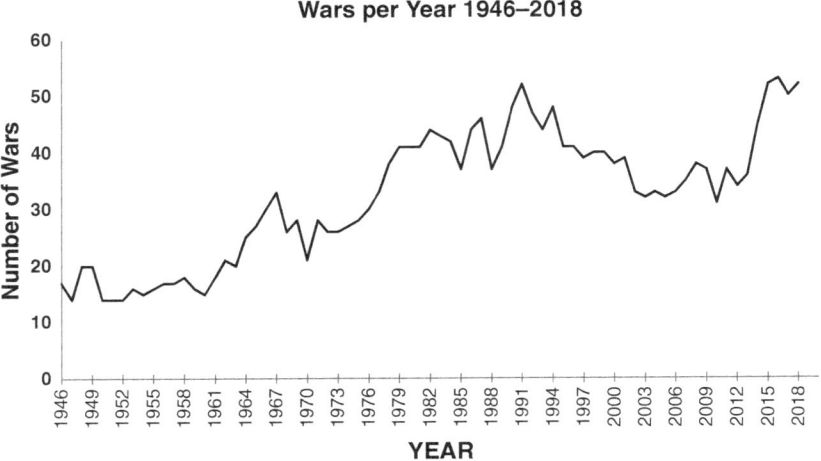

Figure 5.3 Number of wars per year (1946–2018)[113]

circumstances remain highly volatile.[110] For example, a study by the World Bank and UN in 2018 cites the following facts:

○ The number of major violent conflicts tripled from 2010 to 2018.
○ In 2016, more countries experienced violent conflict than at any time in nearly thirty years.
○ Conflicts that ended in 1970 lasted an average of 9.6 years, but conflicts that ended in 2014 had lasted an average of *26 years* (emphasis added).
○ Direct deaths in war, refugee numbers, military spending, and terrorist incidents all reached historic highs in recent years.[111]

Though trends do not signal the imminence of war's end, not all are inauspicious. There have been many encouraging signs since World War II attributable to improvements in international relations and governance, not *just* fear of nuclear annihilation. Western Europe has established a durable peace as centuries-old antagonists like France, Germany, Italy, Spain, and Great Britain have aligned economically and politically. The Western Hemisphere, including once tumultuous South and Central America, has been relatively tranquil, and the nuclear balance of power has kept the most influential and populous countries on Earth – India, Pakistan, the USA, China, and Russia – from major war. Achieving lasting tranquility in areas most often embroiled in war (as in Africa and the Middle East) may eventually create a "critical mass" of global peace. However, we cannot know for certain if or when this might happen, or if unforeseen political, environmental, economic, cultural, or technological changes will alter the calculus of war and peace entirely. Up to now, these sorts of changes and challenges to peace have always occurred. As Fielding Hall writes, ever "in peace there are influences that destroy freedom. No sooner was freedom acquired by war than a state of civilization was set up that became, in time, unendurable, with the inevitable result of a new war. Every civilization we have seen has become a slavery and could find its solvent only in war."[112] Thus, while trend analysis confirms our supposition that humanity prefers and largely strives for peace, it ultimately fails to reveal a breakthrough that could realistically lead to the end of all war.

Current Conditions

War, it seems to me, after a lifetime of reading about the subject, mingling with men of war, visiting the sites of war and observing its effects, may well be ceasing to commend itself to human beings as a desirable or productive, let alone rational, means of reconciling their discontents.[114]   John Keegan

Evaluating current conditions is more germane to near-term, tactical decisions than to forecasting the future of a complex phenomenon like war. Still, in the interest of thoroughness, we cannot ignore the possibility that proximate events and circumstances hold clues regarding war's future. In December of 2020, approximately 69 of 195 (35 percent) countries and 845 distinct groups (including militias, VEOs, guerrillas, separatists, anarchists, and insurgents) were embroiled in violent conflict in various locations.[115] The biggest military powers of the day – Russia, China, and the USA – even in the midst of a global pandemic (COVID-19), continue to compete, probing the boundaries of each other's spheres of influence. These three represent history's largest warfighting capacities with military budgets measured in the hundreds of billions of US dollars and armed forces with millions of troops and trillions of dollars in advanced weapons. Each nation has a different form of government and focuses its grand strategy on competing for greater power, security, prosperity, and influence. At the same time, global extremists, insurgent groups, failing and "rogue" states, and transnational criminal organizations present security challenges for themselves, their neighbors, and more frequently, the broader community of nations. Additionally, according to Gray, the "grossly uneven development between regions," "huge disparities in standards of living," and the political proximity and friction between cultures and religions intensified by globalization and social media herald a hazardous future.[116]

While none of these conditions *proves* that war is eternal, the political and strategic momentum appears to be trending away from a permanent state of peace. In fact, as historians Gordon Craig and Felix Gilbert highlight,

The general role that armament plays in the economy of a country – increasing industrial earnings and reducing unemployment – makes it almost impossible to resist forces driving toward an arms race, and this tendency is encouraged by the apprehensions engendered by the nature of the response (or imagined response) of potential antagonists to one's own efforts.[117]

Entire industries, massive financial investments, and millions of people dedicated to military support and preparation undeniably perpetuate the potential for war. A large and diverse demographic depends on war-related jobs for its livelihood; and, in some cases, it becomes a *raison d'être*. "For many people around the world," writes Horgan, "war provides not just meaning, excitement, and political power but also enormous profits."[118] This sort of power is not easily diverted or relinquished, for, as Freedman adds, those benefitting from war have "a stake in its perpetuation – the politicians dazzled by the prospect of national grandeur, the generals by the prospect of glory, and the manufacturers by the prospect of profits."[119] "The 'permanent' character of these vast military establishments," affirms sociologist Morris Janowitz, "is linked to the 'permanent' threat of war."[120] In short, besides the welcome absence of direct war between nuclear powers, the prevalence of global political discord and the strength of the military-industrial complex show few signs of meaningful abatement.

Theory

Peace is easier to obtain than to sustain ... *that is where society always stumbles.*[121]
                                                                    Donald Tuzin

Absolute peace is empirically non-existent.[122]
                                                                    Thomas Gregor

Most war theory presumes war's permanence. However, theory's insights into human nature shed light on war's future. Earlier in this chapter, we learned that predicting the future is easier when the subject is familiar or governed by inviolable laws. If there is an unbreakable law of war, it is this: the nature of war and human nature are connected and unchanging. Theory tells us that wars are fought for political purposes, and Aristotle asserts that humans are political by nature. If war is a political tool – Clausewitz's politics by other means – then we cannot end the potential for war without changing our political nature. Can we reasonably expect to uninvent politics? If human nature is unchanging, and politics (and war) are extensions of human nature, then the possibility of war *must be permanent* provided there are at least two humans.[123]

Theory also affirms that *will*, *cause*, and *capacity* constitute the engine of war. As we learned in Chapter 4, these embody war's four "causes": *material*, *formal*, *efficient*, and *final*; thus, "destroying" the engine of war is tantamount to removing war's causes, an historically favored strategy for

eliminating war. But how can we do this? Chapter 1 proposed two hypothetical paths to peace provided human nature remains constant: (1) remove
every person's *will to fight* by perfectly satisfying all needs and aspirations
(utopian), or (2) remove by strict control every person's *capacity to fight*,
making needs and aspirations irrelevant (totalitarian).[124] Because neither
option is suitable, feasible, nor sustainable, we are left with the imperfections of liberal democracies, autocracies, and everything in between (recall
Figure 4.2). Real governments are imperfect, and humans are not omnipotent; therefore, we cannot eliminate all factors that could lead a person or
group to choose war, nor can we neutralize all war's possible motivating
factors (e.g., fear, honor, or interest).

Reasoning from Chapter 1's premises regarding human nature, will, and
capacity also leads away from the goal of eternal peace. Humans have an *innate
capacity* for violence, an ability to devise ways and means for inflicting harm.
Additionally, we have *free will* to choose how to use our capacities, a mental
ability to direct actions in ways that may contradict instinctual impulses. War is
the consequence of human choice to exercise a capacity for violence against
other humans for a political purpose. Armed with this *innate capacity for
violence* and the *free will* to employ it against others, we again must conclude
that the possibility of war is eternal so long as humans exist. If human nature is
held constant, we cannot eradicate war without also removing war's *efficient*
cause: humanity.

Perhaps reasoning from war's ends rather than its causes will offer hope. If
we agree that war's purpose is to create a new, favorable state of peace; then, as
a matter of logical symmetry, a state of perpetual, *infinite* peace can only be
created by a perpetual, *infinite* war – an obvious paradox. Viewed from Thomas
Nagel's *consequentialist* perspective, even unfathomable evil, like total war
and genocide, is justifiable in pursuit of the ultimate good of eternal peace.[125]
These findings validate our Chapters 1 and 4 conclusions regarding war-free
societies: the total absence of war can only be found in a hypothetical "heaven"
or "hell."

The Anarchy Problem

*The Lord will settle international disputes; all the nations will beat their
swords into plowshares and their spears into pruning hooks. Then at the last
all wars will stop and all military training will end.*   Isaiah 2:4

*During the time men live without a common Power to keep them all in awe,
they are in that condition which is called Warre; and such a warre, as is of
every man, against every man.*[126]   Thomas Hobbes

While this analysis has focused mainly on the relationship between human
nature and war, the *structural* characteristics of international relations also

affect the prospects for prolonged peace. In previous chapters, we have alluded to the "anarchic" nature of relations between sovereign entities, the crux of Waltz's "third image" cause of war. While we often equate "anarchy" with political chaos, it is properly defined as the absence of a higher authority, literally, "without a ruler." In anarchy, contends Waltz, nations pursue goals violently if the gain is perceived to be worth the cost. Moreover, "Because each state is the final judge of its own cause, any state may at any time use force to implement its policies"; thus, all other states must either be ready to use force as well or "pay the cost of weakness."[127] While it is conceivable (and not uncommon) for disparate political entities to cooperate for mutual benefit or a greater good, this implies a shared perception of the benefit or good being pursued. As Gray notes, even "on the rare occasions when [decisive action] is obviously needed" by a "collectivity of states," "it requires a consensus that will be impossible to achieve."[128] For, according to Hegel, international agreements "always depend ultimately on a particular sovereign will and for that reason would remain infected with contingency."[129] In other words, they can change rapidly, a truth Russia experienced in 1941 when its erstwhile ally, Germany, crushed the Molotov–Ribbentrop Pact beneath the *Wehrmacht*'s *panzer* treads.

The "Minister's" Dilemma

Man's moral ideals, as expressed in the great creeds of the human race, are at a fairly high level. But his capacity to profess one thing and act the opposite seems almost unlimited.[130]
                                                                    Gordon W. Allport

Peace, n. In international affairs, a period of cheating between two periods of fighting.[131]
                                          Ambrose Bierce, *The Devil's Dictionary*

As a practical matter, pursuing peace in a war-filled world presents significant risks that can give even rational, peace-loving leaders pause. Unilateral disarmament in anarchy removes the critical, cost-imposing, and punitive deterrent of armed defense and retaliation, while peace treaties maintain the peace only as long as all parties remain convinced that the terms are consistent with their interests. Indeed, according to Maurice Davie, "From the year 1500 B.C. to 1860 A.D. more than 8,000 treaties of peace meant to remain in force forever were concluded. The average time they remained in force is *two years*" (emphasis added).[132] With political survival at stake, it is imprudent at best and perhaps even suicidal for leaders to gamble on the sustained benevolence of all others. In any human relations, according to Bruce Knauft, "The theoretical and logical validity of reciprocal altruism is predicated on tight control of cheating and deception," something especially hard to find in international relations.[133] "It is useless for the sheep to pass resolutions in favor of vegetarianism," warns Cambridge's "Dean" William Ralph Inge, "while the wolf remains of a different opinion."[134]

Even when treaties are made in good faith, political changes in one side or the other can trigger a rapid reversal. Treaty adherence is essentially an international relations version of game theory's *prisoner's dilemma* in which betrayal pays big dividends for a side that initially commits to cooperation.[135] Again, practically speaking and consistent with Thucydides's Melian imperative, martial strength is the best safeguard for peace, values, and freedom. This rationale is not necessarily aggressive, irrational, or foolish. It simply acknowledges the reality of human nature and international relations. Jomini writes,

> May not the members of an elective legislature ... allow the institutions necessary for a large, well-appointed, and disciplined army fall into decay ... may they not end in convincing themselves and their constituents that the pleasures of peace are always preferable to the more statesmanlike preparations for war? I am far from advising that states should always have the hand upon the sword and always be established on a war-footing: such a condition of things would be a scourge for the human race. I simply mean that civilized governments ought always to be ready to carry on a war in a short time, – that they should never be found unprepared.[136]

The only way to guarantee peaceful relations in anarchy, says Waltz, is for all entities to be *perfectly* rational and benevolent – a utopian vision.[137] Assuming (for now) that perfecting humans and politics is impossible, "the very real problem of how to achieve an approximation to harmony in cooperative and competitive activity is always with us."[138] Note, Waltz says "approximation" to harmony. The problem of *absolute* harmony is that much harder. "War, therefore," asserts von Moltke's Reichstag contemporary Heinrich von Treitschke, "will endure to the end of history as long as there is multiplicity of states. The laws of human thought and of human nature forbid any alternative."[139]

The Power Problem

Can peace be universal? There is no reason to believe it. It is a fallacy to suppose that, by widening the peace group more and more, it can at last embrace all mankind. What happens is that, as it grows bigger, differences, discords, antagonisms, and war begin inside of it on account of the divergence of interests.[140]
                                                                    William Graham Sumner

The obvious solution to anarchy is to install an overarching authority. Dante Alighieri writes, "To put an end to wars and to the source of wars, it is necessary that the whole earth ... should form a monarchy or single empire ... governed by a single emperor who ... will preserve peace."[141] Four centuries after Dante, the relatively obscure Charles-Irénée Castel, abbé de Saint-Pierre authored a step-by-step plan to institute a "confederate, supranational government" to unite the world in perpetual peace.[142] However, this form of "supranational" authority, writes Einstein, "involves the unconditional surrender by every nation ... of its liberty of action, its sovereignty."[143] Rousseau praised the Abbé's *Project for Perpetual Peace* (1712) for its idealism but also criticized its reliance on coercion and unrealistic infringement on sovereignty. Few rulers, let alone *all* of them, would willingly and unconditionally surrender liberty and sovereignty. Are we then to use war to compel peace? Kant asserts, "The only conceivable way" to forge a unifying "state of right" from a multitude of individual desires is by "coercive authority" and "force."[144] Freud adds,

Paradoxical as it may sound, it must be admitted that war might be a far from inappropriate means of establishing the eagerly desired reign of "everlasting" peace, since it is in a position to create the large units within which a powerful central government makes further wars impossible. Nevertheless, it fails in this purpose, for the results of conquest are as a rule short-lived; the newly created units fall apart once again.[145]

In other words, concludes Fielding Hall, "You won't stop war by an United States of Europe, because United States and United Kingdoms under stress become disunited States and Kingdoms."[146] Even were a true world government possible, it would be unlikely to inspire trust, loyalty, or commitment from its diverse constituency, since, as Michael Howard writes, it would "not arise naturally out of the cultural disposition and historical experience of [its] members."[147] There is also the matter of resolving who will rule and how to enforce peace and order. These are generic problems of governance; however, a federation of dissimilar nations under a global government would have all the political frictions of a state amplified a thousandfold. Is the shared goal of a war-free society enough of a bond to nullify all other points of contention?

"The amount of force needed to hold a society together," writes Waltz, "varies with the heterogeneity of the elements composing it."[148] Invoking Lord Acton's dictum – absolute power corrupts absolutely – suggests that the power necessary to hold a super-nation together augurs staggering levels of corruption, which creates a potent motivator for separatism and civil war.[149] "Experience hath shewn," advises Thomas Jefferson, "that even under the best forms [of government], those entrusted with power have, in time, and by slow operations, perverted it into tyranny."[150] Perhaps the emergence of a Solomonic ruler combining the virtues of Plato's philosopher king and Dante's "wise administrator of the laws ... with intense affection for the human race" could make this work for a time.[151] However, in the end, says Waltz, "demonstrating the need for an institution [or ruler] does not bring it into existence. And were world government attempted, we might find ourselves dying in the attempt to unite, or uniting and living a life worse than death."[152]

Using vernacular from Chapter 1 and echoing Sumner's thoughts from the lead-in quotation, though a world government might mitigate the *external problem* (international war), it would only intensify the *internal problem* (civil war). Writes Gwynne Dyer, "Internal conflict, including the use of violence in pursuit of political objectives, would not vanish from the world even if all legitimate governments signed their armed forces over to the United Nations tomorrow."[153] Even psychologist and peace advocate Ervin Staub admits, "Strongly established hierarchical arrangements are potentially harmful, especially in complex, heterogeneous human societies with varied subgroups that can turn against each other."[154] Ultimately, the handicaps of anarchy and human diversity present another disheartening obstruction to endless peace. Therefore, no matter where we turn – to anthropology, history, trends, current events, or theory – there appear to be no ways and means that satisfy the tests of strategy and produce perpetual peace, at least so long as human nature remains fixed.

The Quest for Perpetual Peace

There has never been a shortage of plans for world peace. Those who are strongly enough motivated to develop one are often convinced that the only reason wars continue is that statesmen refuse to listen to them.[155]

Kenneth Waltz

It is reasonable to require a substantial burden of proof-by-argument if we are to take seriously the proposition that humankind is in the process of curing itself of the habit of war.[156]   Colin Gray

Can We or Kant We?

Immanuel Kant is among the brilliant, motivated people who have lent their talents to the search for a permanent end to war. Building on the earlier writings of Rousseau and the Abbé de Saint-Pierre, Kant penned a short treatise entitled *Toward Perpetual Peace: A Philosophical Sketch* (1795). In this essay, Kant acknowledges humanity's selfishness and attraction to violence and war; however, he believes Nature's inherent perfection will herd us inexorably toward a state of peace, *despite* our deficiencies. Says Kant, "What guarantees perpetual peace is nothing less than that great artist Nature, who runs her mechanical course in a way that shows that she aims to produce a harmony among men, whether or not they want it."[157] In effect, "Man is forced to be a good citizen even if he isn't a morally good person," because, "Nature unstoppably wills that the right should finally triumph. What we neglect to do comes about by itself, though with great inconveniences to us."[158]

Given the indelible faults of human nature, which preclude moral reformation, Kant proposes a structural solution to war: a federation of states (*Völkerbund,* as opposed to the world state, *Völkerstaat,* of Rousseau and

Saint-Pierre) that peacefully adjudicates disagreements by melding legal, political, and moral principles into a force that counterbalances human aggression. Kant closes by saying,

> If it's a duty to bring about a state of public law, and if there is well-grounded hope that this can actually be done, even if only through an endless process of ever closer approximations, then perpetual peace – not to be confused with the outcome of wrongly so-called 'peace-treaties' (which are really only armistices) – is not an empty idea. On the contrary, as ways and means are gradually found, we hope at an ever-increasing pace, perpetual peace is a task that grows ever nearer to achievement.[159]

Kant's faith in a collective, structural solution to war, at least with respect to fostering "an ever-increasing" if not eternal peace, has merit. Significant periods of nonviolence between democratic republics, which have largely adopted his model, bear this out. However, his trust in the potency of an elemental, natural force for right to solve the anarchy problem and compel the incremental advance of peace across human affairs has not materialized. In fact, Kant died in 1804, just a year before the battles of Trafalgar and Austerlitz, as the bloody Napoleonic era was hitting its stride. What would Kant have thought had he lived to witness the astonishing violence of the next two centuries?

Goodbye War?

War breeds war; from a small war a greater is born, from one, two; the plague of war, breaking out in one place, infects neighbours too.[160]

Erasmus of Rotterdam (1516)

Some diplomat no doubt
Will launch a heedless word,
And lurking war leap out . . .
And blood in torrents pour
In vain – always in vain,
For war breeds war again.[161]

John Davidson, *War-Song* (1899)

A more recent perspective on this issue comes from the aforementioned John Horgan in *The End of War* (2012). Like Kant, Horgan believes humans can abolish war, despite his skepticism regarding most research and theorizing on the topic. His path to peace can be summarized as follows: humans are *not* genetically hardwired for war; thus, war is not inevitable as proven by instances of peaceful societies and warrior cultures that have become peaceful. Using inductive logic, he contends that, if peace is possible for some, then peace is possible for all; and thus, *perpetual* peace is possible provided *everyone* agrees and works to pursue it. Once we achieve an initial state of universal peace, the "disease" of war can no longer "infect" humanity, everyone will admit that wars were stupid, and no one will ever again resort to violence to promote their political ambitions.

This optimism is refreshing, but alas, Horgan's tenuous reasoning does not effectively raise hopes for an everlasting peace. In his search, Horgan largely rejects analyses that approach the problem of ending war as if it were a contagion or disease, an analogue that implies two lines of effort for solving war: (1) *prevention*: eliminate war's causes; and (2) *treatment*: promote war's remedies. The prevention path presumes that war, like infections, will eventually disappear with the removal of "naturalist" and "materialist" causes. The naturalist camp believes that war is innate and our warrior genes stimulate fear, aggression, and ambition that overwhelm rationality in a few "bad apples" who then compel the docile, stupid, and gullible to fight and kill. Alternatively, the materialist perspective attributes war to economic and ecological factors like income inequality, resource scarcity, or natural disasters.[162] Objective evaluation of these factors, asserts Horgan, proves that war's complexity precludes any one theory or causal factor from satisfactorily explaining war's prevalence (as we also conclude in Chapter 1). "Many conditions appear to be sufficient for war to occur," writes Horgan, "but none are necessary."[163] Codevilla and Seabury agree, confirming that no investigation of war's causes, including psychology, anthropology, political science, economics, or international relations, "has succeeded in predicting any war or any peace" nor bridged the gap between correlation and cause.[164] Peace optimists are particularly vehement in rejecting the "naturalist" view due to the implicit hopelessness of overcoming millions of years of evolutionary programming. But eliminating all other factors is almost as prohibitive; thus, Horgan puts little faith in prevention.

The treatment path, which is closest to Kant's method, identifies and promotes curative factors or "antibodies" to diminish and eventually eliminate war. If war is a contagion, spreading through contact, then reducing wars will gradually curtail their propagation until humanity is finally "cured." Proposed antibodies include democratic governance, socialism, elevation of women or minority leaders, disarmament, pacifism, theocracy, secularism, and media messaging of war's cost and futility. As with preventative factors, Horgan rejects most treatment options saying,

There seem to be no conditions, that, in and of themselves, inoculate a society against militarism. Not small government nor big government. Not democracy, socialism, capitalism, Christianity, Islam, Buddhism, nor secularism. Not giving equal rights to women or minorities nor reducing poverty. The contagion of war can infect any kind of society.[165]

In the end, Horgan bases his hopes for eternal peace on what amounts to an Oz-like *hominum ex machina*: "We just have to want war to end, and to believe we can end it. The choice is ours."[166]

Though Horgan does not mention him, Gordon Allport proposed a similar solution in the late 1960s. Writes Allport, "The greatest menace to the world today are leaders in office who regard war as inevitable and thus prepare their people for armed conflict. For by regarding it as inevitable, it becomes inevitable. Expectations determine behavior."[167] He calls the *expectation* of war its "indispensable condition"; thus, "only by changing the expectation in both leaders and followers, in parents and in children, shall we eliminate war."[168] But, Allport never satisfies Waltz's requirement to explain how this occurs without "the nations somehow [achieving] a magically complete and enduring agreement never to fight," or an entity like the UN gaining global sovereignty.[169] In general, the "expectations-cause-war" school of thought is something of a chicken-and-egg paradox. We have learned to expect and prepare for wars because they are common, dangerous, and unpredictable. With countless millennia of war in our past and no end in sight, it would contradict our self-preservation motive to "expect" that war will never recur. The data supporting this expectation of peace simply does not yet exist. It would be like expecting an end to crime, poverty, diseases, or earthquakes despite knowing the conditions for such phenomena remain prevalent. Therefore, Allport's prescription amounts to little more than an abstract aspiration.

In fairness, Horgan offers some tangible recommendations for getting closer to perpetual peace. These include communicating the efficacy of nonviolent means of redress and political activism, bolstering international laws and restrictions on war, reducing military size and spending, and training military personnel to behave like police. Some of these may help reduce war and promote peace, but it is unclear how they represent a revolutionary approach to the problem of permanently ending war.

Counterargument

The fact that slaughter is a horrifying spectacle must make us take war more seriously, but not provide an excuse for gradually blunting our swords in the name of humanity. Sooner or later someone will come along with a sharp sword and hack off our arms.[170]   Clausewitz

In the end, optimist reasoning breaks down in several critical areas. First, the dismissal of humanity's violent ancestry and contention that war is not innate – a point of departure from fellow optimist, Kant – amounts to quibbling over semantics (i.e., whether violence or war is innate). No sensible person could disavow humanity's inherent capacity for violence and political nature. As with any human activity, whether it be art, science, politics, or war, intellect and creativity lead us to improve and refine our methods, capabilities, and effectiveness over time. While our genes may not *compel* us to fight wars, they have *armed* us with the capacity to do so as a matter of choice, and we cannot deny war's power as the ultimate arbiter of intractable political disputes. Peace optimists also tend to reject war's good-evil and creation-destruction dialectics, seeing war only as an evil, senseless affair of wanton destruction led by stupid or deranged people and fought by fools or remorseless sociopaths. Throughout his book, Horgan struggles to reconcile these sentiments with the evidence supporting the necessity and nobility of "just" wars that have ended tyranny, genocide, and slavery. Truthfully, there is more to war than mindless emotion or visceral hatred. War, like politics, can be cold and calculated. Rousseau writes, "War consists in the constant, reflected, and manifest will to destroy one's enemy. For in order to judge that the existence of this enemy is incompatible with our well-being, one needs coolness and reason."[171] It can be led and waged by thoughtful, peace-loving people who know exactly what they are doing and what the likely costs will be. In the American Civil War, brothers fought brothers and fathers fought sons. Most did not hate their relatives. They fought on principle, for a cause elevated above blood relation. Good people fight because they do not wish to be enslaved or killed by bad people, or as John Stuart Mill opines, "Bad men need nothing more to compass their ends, than that good men should look on and do nothing."[172] Leaders like Washington, Lincoln, Wilson, and Churchill were not warmongers. They fought reluctantly to preserve universal values like liberty, human rights, self-determination, and freedom from fascism.

It is easy to critique the motives of others from the comforts of safe, peaceful neighborhoods. In fact, there are many people who do not think and act as we might expect and share different values, beliefs, and aspirations. "The great illusion of today," writes Aron, "is not that which threw the people of Europe into suicidal opposition; it is the contrary illusion ... which ascribes a single rationality to all peoples and to those who govern them."[173] Overlooking humankind's diversity is often the peace optimist's fatal flaw. "Traditional routes to world order" have a common failing, says social scientist Lawrence Frank. "They all propose that one creed, one state, or one philosophy come to dominate the world."[174] However, Rousseau tells us, "The constitution of this universe does not permit all the sensible beings that compose it to concur at the same time in their mutual happiness. Instead, the happiness of one is the misfortune of another, and thus according to the law of nature each gives himself preference."[175] In other words, paraphrasing Milton, we often prefer to live in a hell of our own making rather than the heaven of another's.

Achieving perpetual peace may seem to be a simple matter of finding whatever has worked for others and replicating it. Yet, no model for permanent peace even exists, and what works temporarily for some does not always work for all. Most cultures and societies have different opinions regarding ideal forms of governance, jurisprudence, law enforcement, resource control and distribution, religion, or way of life. "There is no impartial concept of justice," writes Dyer, "to which all of mankind would subscribe."[176] Add Durbin and Bowlby, "The existence of a general desire for the common good is clearly a force making for peace in adult society. But its power will only extend as far as the idea of the common good extends."[177] Disparities in culture, religion, wealth, and resources often cause our concepts of a desirable peace to diverge. Furthermore, while *most* people prefer peace and shun war, not all do. "Now many desire war and dissension," writes Aquinas, "therefore all men do not desire peace."[178] Recall from Chapter 1, war uniquely satisfies a range of powerful psychological needs; thus, there will always be some who agree with von Moltke (the Elder), who writes,

Permanent peace is a dream and not even a beautiful one, and war is a law of God's order in the world, by which the noblest virtues of man, courage and self-denial, loyalty, and self-sacrifice, even to the point of death, are developed. Without war the world would deteriorate into materialism.[179]

These souls relish war in the spirit of the German author and war hero, Ernst Jünger (1895–1998), who thought combat the "ultimate virile sport, a supreme fulfillment of one's self," the incomparable "cleanser of internal weakness, corruption, and decadence," and the "ultimate moral and spiritual test for the greatness of a nation."[180]

Given this diversity, it would seem the only way to get *everyone* to forsake war is either to force the warlike to become peaceful or eliminate them. Waging war to establish a permanent peace hardly seems a viable or preferable option. Assuming that all people view peace and the ideal way of life similarly is simply false and defining peace as the absence of war is inadequate. There can be little realistic hope of attaining an objective through collective action where no collective view of the objective exists. "First-image optimists," writes Waltz, "betray a naïveté in politics that vitiates their efforts to construct a new and better world."[181]

Comparing war to a contagious disease has merit, for there is truth in the premise that war begets war. "Fighting," write Durbin and Bowlby, "is infectious in the highest degree."[182] Wars stir emotions and leave festering "wounds." Just as disease is a natural part of a complex world of interacting organisms, war is a natural consequence of human interactions in a dangerous, resource-limited environment. However, war is less like *a* disease than it is the *phenomenon* of disease. In this respect, individual wars are like discrete diseases. No single war or illness reveals all we need to know of the rest. The ability to prevent or cure a sinus infection does not necessarily mean that we can do the same for stomach cancer or the flu. Likewise, strategies that prevent or win wars in some instances may prove wholly ineffective in others. In his book, Horgan attests that 80 to 90 percent of us believe that war cannot be eradicated. A comparable majority likely share a similar view of our prospects for eliminating all diseases.[183]

While we cannot "catch" a war like we would an illness, we *can* reduce the probability of war just as we can our likelihood of contracting disease, by prudent choices and actions. Avoiding risky behaviors and attending to personal hygiene, fitness, and diet reduce our chance for illness but do not preclude it entirely. Likewise, pursuing a balance of power while favoring diplomacy over threats and provocation can help prevent wars.[184] At times, with war or disease, we have little choice. Just as genetic anomalies or another's actions can impact our health through no fault of our own, aggressors can thrust war upon us unexpectedly (e.g., the Korean War).

Finally, as social anthropologist Donald Tuzin affirms, humanity has failed to demonstrate the ability to sustain a comprehensive peace once established. How can we prevent the rise of another Napoleon, Shaka, or Hitler? "If history – and religion – teach us anything," writes former US Secretary of Defense Robert Gates, "it is that there will always be evil in the world, people bent on aggression, oppression, satisfying their greed for wealth and power and territory, or determined to impose an ideology based on the subjugation of others and the denial of liberty."[185] Indeed, if the emergence of tyranny, arising from "iniquity and injustice," were not so routine, writes English philosopher John of Salisbury, "perpetual peace would have possessed the peoples of the earth forever."[186] But injustice and tyranny are not extinct, and as Hamilton writes, "Peace or war will not always be left to our option; however moderate or unambitious we may be, we cannot count upon the moderation, or hope to extinguish the ambition of others."[187] Is it possible to establish institutions and processes that satisfy everyone's needs, aspirations, and ambitions? As Machiavelli says, human "desire always exceeds the power of attainment, with the result that men are ill content with what they possess."[188] What happens when corruption, ineptitude, or tyranny emerge, and nonviolent means of reparation are deemed unsatisfactory? Is it credible to base hopes for the eradication of war on the fanciful promise of a self-fulfilling prophesy?

Alpha and Omega

Which methods and institutions are most effective in preserving peace is a question that has exercised the minds of leaders, rulers, councils, philosophers, and visionaries for millennia, without producing any enduring or generally applicable answers.[189]
                                                                Lawrence Keeley

The reason so many great intellects, to include Einstein, Marx, Rousseau, and Kant, have not abolished war, is that all lines of inquiry toward that end lead back to war's prime mover, its efficient cause, humanity. *Homo sapiens*, the *sine qua non* of the True Trinity, stands at the beginning, center, and end of war. War reflects our diversity and volatility, which ironically prevent us from ending all wars without also jeopardizing our survival. Horgan is correct that war is a choice. Therefore, only by eliminating our *ability* to choose war or peace, by destroying our definitive quality – rational free will – can we be certain of abolishing war.[190] Peace optimists would call this pessimism or fatalism. But is it? Though we might prefer an end to rainy days, is it pessimistic to acknowledge that not every day can be sunny? Though we might desire eternal life, is it fatalistic to accept mortality? Quite the opposite really, for these sentiments are grounded in scientifically validated truths. Progress toward any goal is unlikely without first embracing such truths. It is noble, a moral duty, to strive for a more peaceful world, but toil in the absence of realistic expectation ends only in disillusionment. In the universe's philosophical harmony of tensions, war and peace are manifestations of humanity's existential dualism. Because we remain fully capable of choosing either path, the possibility of war and peace is everlasting, while the possibility of *everlasting* war or peace is zero (as shown in Figure 4.2).

Though objective analysis to this point confirms the impossibility of eradicating all war, we can find solace in two facts: (1) war is not always inevitable, and (2) perpetual war is just as impossible as perpetual peace. Moreover, there *are* effective ways to build a *more* peaceful world. Those convinced that no one will work for peace if ending all wars is unachievable peddle a delusional narrative about war and human nature in hopes of creating a self-fulfilling prophesy. Absurd! Peacemaking is more likely to succeed if we understand the true relationship between war and humanity, the causes and motivations for war, and the forms of government and international diplomatic techniques most likely to maintain a harmony of tensions. War may be a contagion, but it begins and ends within each of us. Rejecting this truth takes us farther from peace, not closer, and might actually cause the self-destruction we fear. "If you ignore human nature," writes Fielding Hall, "you simply build up civilizations which can only dissolve in war."[191]

Should We End War?

Your scientists were so preoccupied with whether or not they could that they *didn't stop to think if they* should.[192]   Dr. Ian Malcolm, *Jurassic Park*

War has the higher significance that through it ... the ethical health of peoples is preserved ... just as the blowing of the winds preserves the sea from the foulness which would be the result of a prolonged calm, so also corruption in nations would be the product of prolonged, let alone "perpetual," peace.[193]   Hegel

If human nature causes war, and we cannot expect that to change of its own accord, nor can we create, as Kant hoped, institutions, states, and state systems capable of indelibly defusing the human will to fight, then altering or eliminating humanity become our last options for ending war. Regrettably, some nihilists and misanthropes prefer the latter option, viewing humankind as a plague that should follow the dinosaurs to extinction. Perhaps for them, "It is right that men who choose to destroy each other should find perpetual peace in the vast grave that swallows both the atrocities and their perpetrators."[194] On this matter, we respectfully disagree. So, if humans are the problem and suicide is not the answer, then ending war requires that we become *something else*. Since doing this naturally could take eons, we are left only with artificial methods.

Humanity 2.0

The elimination of aggression is not only an unrealistic goal, it is a misguided one. A friction-free social system has yet to be found, and the ones that come closest lack the structural complexity and individual differentiation that we value so much in our own societies.[195]
                                                                     Frans de Waal

According to Waltz, "solutions for the problem of war" based on reforming human beings and states "must assume the possibility of perfection in the conflicting units."[196] For, absent perfection, "an approximation to world peace" is the best we can do. Therefore, if war is a consequence of human imperfections, it would seem the "something else" we need to become is a "perfect" human (e.g., *Homo perfectus*), a new species without the will to make war and with the ability to establish perfect institutions incapable of waging war. Some believe this is possible. For example, like Kant, the French philosopher, Marquis de Condorcet, believed the inexorable march of "progress" proved the "perfectibility of man is unlimited," perhaps even to include immortality.[197] Evolutionary social theorists, like the Victorian philosopher Herbert Spencer, thought it natural "that societies would evolve away from armed struggle toward harmony and cooperation."[198] And, positivists, like nineteenth-century French philosopher Auguste Comte, "anticipated that knowledge of the laws by which societies functioned would make it possible in the future to engineer a perfect society and even put an end to war."[199]

Though none of these visions have worked for *Homo sapiens*, maybe they will for *Homo perfectus*. Perhaps we should embrace a neo-fascist vision and ride the wave of human technological progress into a eugenics age characterized by the literal "creation of a new man within a radically reconstructed society," as in H. G. Wells's *Modern Utopia*.[200] Though Hitler wanted to create *über*-humans to rule the world, from the liberal optimist viewpoint, we can just as easily imagine *über*-pacifists constructing *über*-organizations capable of gaining and maintaining everlasting peace. For these, technological progress and improvements to human life through enlightened education, nurture, governance, and genetic or cyber-organic engineering will enable us to live without the desire or need for war. Human genome manipulation and synthetic augmentation already exist and are advancing every day, but should we employ these startling advances to "perfect" the human race? Most ethical doctrines shun genetic or cybernetic tampering on human subjects; however, as these capabilities expand, moral and ethical boundaries blur. Is it wrong to engineer new eyes for the blind or internal organs for the sick or infirm? Is it unethical to "cure" debilitating genetic abnormalities in unborn children? Who would ban research and experimentation that could extend lifespans by ten to twenty years?

Even with the tools to engineer *Homo perfectus*, we would first have to characterize the end product. How do we define perfection? What attributes would this new species require to enable perpetual peace, and who would have the authority to approve the design? Should we focus our efforts on eliminating aggression and ambition? Perhaps engineering an innate revulsion at the idea of killing is the right approach. Is it reasonable to expect billions of people to agree on these characteristics and the methods for achieving them? If not, the results could be violent, for as Angelo Codevilla writes, "When … human beings' notions of right, their visions of peace, not to speak of their pride and perceived interests, are irreconcilable, force becomes their natural arbiter."[201] Moreover, with the power to re-engineer humanity and the prospect for opposing visions to spark conflict, would it not be more likely, given history and human nature, for us (in Huxley-esque fashion) to create *three classes* of new humans: a ruling class, *Homo supremus* – "designer" humans with super intelligence and physical beauty; a guardian class, *Homo dominus* – with amplified loyalty, strength, and aggression; and a servile class, *Homo servus* – faithfully obedient workers.[202] Could we prevent such a scenario?[203]

Assuming we could even define the objective, there remain a host of practical dilemmas regarding the artificial perfection of humanity. How much would the endeavor cost and who would pay? Could we "fix" everyone, or would some be left out? Who should go first? This monumental pursuit of perfection could take one of two paths: (1) perfect everyone or (2) perfect some and eliminate the rest. Option one is ideal, but probably impractical. The existence of imperfect people, the "leftovers," and their inferior, war-mongering societies, would be like a cancer, a mortal threat to a perfect, peaceful world necessitating a consequentialist "final solution." Lenin, Stalin, and Mao had little compassion for anyone who failed to embrace their doctrines. Though Mao writes, "No matter how long this war may last, there is no doubt that it will be followed by an unprecedented epoch of peace," nearly 100 million people found only the peace of death during the global rise of communist totalitarianism.[204] Could we reasonably expect a different result with the worldwide forced pacification of humanity?

The Road to Hell

No peace is permanent, and nothing so surely guarantees war as dissatisfaction with contingency and the attempt to establish perpetual peace.[205]
Angelo Codevilla

A world without ... disciplined, obedient and law-abiding armies – would be uninhabitable. Armies of that quality are an instrument but also a mark of civilisation and without their existence mankind would have to reconcile itself either to life at a primitive level, below 'the military horizon, or to a lawless chaos of masses warring, Hobbesian fashion, 'all against all.'[206]
John Keegan

In truth, pursuing this hypothetical nirvana would almost certainly do more harm than good. As the proverb goes, *the road to hell is paved with good intentions.* The arrogant attempt to engineer our violent capacities away genetically or synthetically would inevitably result in more chaos and war than it could hope to prevent. More likely, it would cause the end of our species. For even if it worked and did not unleash a global plague, it would create a new, prime competitor for *Homo sapiens*, triggering war-precipitating fears and a struggle for dominance that could lead to mutual annihilation. Even peace advocate Amitai Etzioni admits, "No human society can survive without applying force to prevent some deviants from violating the normative order ... force is used not just to maintain law and order in the technical sense of the term, but also to force changes willed by the majority upon reluctant minorities."[207] This "remedy, though it may be unassailable in logic," concludes Waltz, "is unattainable in practice."[208] Furthermore, writes Fielding Hall,

Even if war could be abolished and peace enforced upon the world by some machinery, or some teaching, would not that be worst of all? [War may be] Nature's mode of curing a hidden disease that might otherwise be mortal, and could we by artificial means stop war, the patient would die. Therefore, all these specifics for curing war cause in us more doubt and alarm than confidence. They are based on no true knowledge of human nature and on no true diagnosis of the evil. They would do, we fear, far more harm than good. For by stopping war we might kill civilization, and even life itself.[209]

Defending the Peace-War, Good-Evil Dialectics

The poet was a fool who wanted no conflict among us, gods or people. Harmony needs low and high, as progeny needs man and woman.[210]
Heraclitus

Peace can be a cover whereby evil men perpetrate diabolical wrongs.[211]
John Foster Dulles

We must not be too hasty in introducing ethical judgements of good and evil. Neither of these instincts is any less essential than the other; the phenomena of life arise from the concurrent or mutually opposing action of both.[212]
Sigmund Freud

Chapter 1 explains how the concepts of balance and harmony are part of the cosmos. Upsetting that balance, as the preceding quotes suggest, risks violent discontinuities. It may seem that peace must *always* be preferable to war, but is it? Recall from the Introduction, war can be a correcting mechanism for injustices and atrocities perpetrated during times of "peace." "The very evils we associate with war," writes Codevilla, "have fallen upon mankind more fully in times and places well removed from battlefields and in conditions conventionally called peace."[213] And Gray adds, "Alas, much of the world is not blessed with a political system that provides potent checks and balances to curb either dysfunctional or morally outrageous leadership."[214] In *Whether Soldiers, Too, Can Be Saved* (1526), Martin Luther says, "What men write about war, saying that it is a great plague, is all true. But they should also consider how great the plague is that war prevents."[215] In the twentieth century alone, note Codevilla and Seabury, "At least *one hundred million human beings* have been killed by police forces or their equivalent," by methods including "hunger," "exposure," "forced labor," poison gas, and club and machete attacks (emphasis added).[216] History is replete with villains bent only on establishing a "peace" that entitles them to oppress, torture, and kill whomever they wish. Consider Václav Havel's chilling account of life within "peaceful" Czechoslovakia behind the Iron Curtain. Here, warfare had not disappeared, it had merely been "shifted from the daylight of observable public events, to the twilight of unobservable inner destruction ... the slow, secretive bloodless never quite absolute, yet horrifyingly ever-present death of non-action, non-story, non-life, and non-time."[217] Germany's defeat ended Hitler's efforts to exterminate the Jewish race; but, Stalin (killed over 50 million in purges), Mao (killed 45 million with "Great Leap Forward" policies), and Pol Pot (killed nearly 2 million Cambodians after the USA left Vietnam) slaughtered with impunity during the "peace" they established as war victors.

At a deeper philosophical level, eliminating war also eliminates peace, leaving a state of perpetual purgatory where politically sanctioned murder and mayhem reign unimpeded. "The presence of even the concept of 'peace,'" writes anthropologist Arthur A. Demarest, "is built upon a familiarity with war."[218] Just as the existence of matter always indicates the potential for energy, so too must the presence of peace concede the potential for war. Can we truly know serenity without tumult? Light without dark? Good without evil? Heraclitus writes, "Without injustices, the name of justice would mean what?"[219] Indeed, according to Taoism, "Things cannot *exist* without their opposite" (emphasis added).[220]

To illustrate this idea, imagine humanity as a cable-walker holding a balancing pole (Figure 5.4). The cable behind signifies the past, the cable ahead, the future. We cannot return the way we came. We can only move

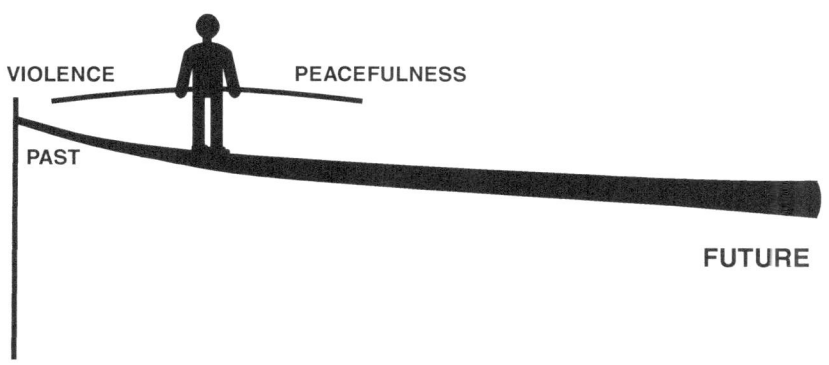

VIOLENCE   PEACEFULNESS

PAST

FUTURE

Figure 5.4 Balance and the fate of humanity

forward or fall. The pole represents the *yin-yang* of human nature. One end signifies our warlike, violent traits and the other our peaceful, benevolent qualities. When all characteristics balance, we can proceed, though never in complete safety.

However, if they become unbalanced, the risk increases. Changing human nature (i.e., cutting the pole in half), even with the goal of reduced pugnacity, destabilizes humanity and could inadvertently trigger our self-destruction. As Fehrenbach writes, "Perhaps the values that comprise a decent civilization and those needed to defend it abroad will always be at odds. A complete triumph for either faction would probably result in disaster."[221] Again, we are talking about the consequences of altering our *nature*, not our *behaviors*. The fate of humanity does not *require* evil, violent, or self-destructive actions, but it just might hinge on the recognition and acceptance that these tendencies are inherent in human nature. "If we are to have peace," opines Fielding Hall, "let it be a real peace and not a pretence, a natural peace and not an artificial one, a free and living peace, not a dead letter enforced upon us."[222] Sumner adds, "Institutions and customs in human society are never either all good or all bad. We cannot adopt either peacefulness or warlikeness as a sole true philosophy."[223]

What We *Can* Do

We find no single basis for peaceful relationships, but a variety of overlapping institutions, values, and attitudes that run the scale from agape, or selfless love, to skill at reconciliation, to fear of deterrence and avoidance of others.[224]
                                                                  Thomas Gregor

For the entire law is fulfilled in keeping this one command: "Love your neighbor as yourself." If you bite and devour each other, watch out or you will be destroyed by each other.   Galatians 5:14–15

In Chapter 4, the W-V-A helped us visualize the mechanism for preventing, ending, or winning wars (i.e., by employing IOPs to build cooperative rather than combative relationships within and between political groups, using tailored deterrence, and disabling an enemy's engine of war). To do more preventing, and thus reduce the need to end or win wars, requires accepting human nature for what it is and not for what we wish it to be. Thus, the key to survival is not to be found in reengineering humanity; rather, it is attained by embracing and embodying our *better* natures. We cannot eliminate the quest for power, ambition, discontent, selfishness, opportunism, incompetence, or corruption; however, there are many ways we can limit war and the potential for self-destruction. "The real problem," writes Malinowski, "is not whether we can completely eliminate [pugnacity] from human nature, but how we can canalize it so as to make it constructive."[225] While it is not possible in a world of billions to please everyone all of the time, we can reduce the potential for conflict by combining a judicious use of authority with personal, civic responsibility. "Antagonisms, hostilities, and conflicts," writes Mark May, "are held in check by customs, laws, and rules which are enforced in part by duly constituted authorities and in part by inner compulsions of loyalties and the sense of social responsibility."[226] Incentivizing collaboration in peace-promoting activities is also important. "If people are confident that their labor will provide at least the necessities of life and some access to comforts and luxuries," affirms Keeley, "violence will generally only attract the pathological."[227]

Societies can minimize civil wars by promoting a liberal education, championing human rights, developing and enforcing an equitable legal system, establishing power sharing between political institutions, encouraging personal freedom and accountability, offering nonlethal outlets for aggression (e.g., sports or virtual reality), providing for equality of opportunity, applying inclusive and transparent governance, and most importantly, sponsoring effective, nonviolent mechanisms for the adjudication of political grievances or what Machiavelli calls, "the malignant humours to which men are prone."[228] "When complaints are freely heard, deeply considered and speedily reformed," says John Milton, "then is the utmost bound of civil liberty attained that wise men look for."[229] And though "we may never be strong enough to be entirely nonviolent in thought, word and deed," advises Gandhi, "we must keep nonviolence as our goal and make strong progress towards it."[230]

Lasting peace on a global, interstate scale is complicated by the anarchy problem and the sheer variety of governmental forms, cultures, languages, and religions. As Waltz writes, "With many sovereign states, with no system of law enforceable among them, with each state judging its grievances and ambitions according to the dictates of its own reason or desire – conflict, sometimes leading to war, is bound to occur."[231] Still, the absence of war can be reinforcing over time, creating a "peaceful momentum," provided governance is effective, politics are reasonably stable, commerce is fair and vigorous, resources are shared, and cooperation and diplomacy are prioritized and practiced.[232] Clausewitz says this momentum causes a "common effort toward maintenance of the status quo," which enables "a number of civilized states to coexist peacefully over a period of time . . . the fact that Europe, as we know it, has existed for over a thousand years can only be explained by the operation of these general interests."[233]

Superimposing successful governance models across all nations might seem an ideal way to forge a lasting global peace; however, for the diversity reasons highlighted earlier, a truly viable, potent global government is unlikely. Nevertheless, sustaining fora for international dialogue, debate, and cooperative humanitarian projects (like the UN) can ease international tensions. As Keegan writes, war-making may have "become a habit," but by relearning and accepting the habits of our primitive ancestors, themselves often devotees of "restraint, diplomacy and negotiation," we may just survive.[234] Additionally, a gradual, peaceful reduction in the number of dissimilar political ideologies and forms of government would also lessen the likelihood of war; however, this cannot be forced.[235] And we cannot condone autocratic forms of governance for preserving amity, because they achieve peace more by control and oppression than preserving rights and liberty. "None can be perfect," says Waltz paraphrasing the theologian, Reinhold Niebuhr, "but the imperfections of democracy are infinitely preferable to the imperfections of totalitarianism."[236] Or, as Ian Morris opines, "Democracies may be messy, but they rarely devour their children; dictatorships get things done, but they tend to shoot, starve, and gas a lot of people."[237] Finally, a relatively peaceful, benevolent, non-autocratic unipolar power (an historical rarity) can serve as an "example" and reduce the potential for war within its sphere of influence. Though, to be realistic, wars of all types, says Gray, are still likely to "occur despite the presence of a willing and fairly able protecting agent"; however, "at least the reality of such a policeman affords some prospect of a stable international, and domestic order."[238]

Ultimately, the closer we get to seeing ourselves as one family, one tribe, one species, the better our chances of entering and sustaining a prolonged epoch of order, equilibrium, and peace. After all, "the awareness of one's shared humanity with others," writes Staub, "is an important building block for peace and positive relations";[239] and the path to what Kant calls "a universal community" and a "cosmopolitan right," whereby the liberties of *one* are defended by *all*.[240] Freud believed the cultivation of a culture-spanning, "moral imagination" could have a pacifying influence on our darker impulses, and that "it may not be Utopian to hope that these two factors, the cultural attitude and the justified dread of the consequences of a future war, may result within a measurable time in putting an end to the waging of war."[241]

Though it may not be unrealistic to expect war to remain largely dormant in a world where the choice for peace predominates, we cannot reasonably assume that it will disappear forever. Sustaining an acceptable state of peace takes work. "Peace is as demanding a state as war," writes Keeley, "requiring for its maintenance effort, economic sacrifice, and even occasional violence. Peace is not an effortless inertial or 'natural' state to which people and societies revert in the absence of perturbation."[242] When these efforts inevitably falter, conditions will arise – incompetent, corrupt, or malicious leadership; competition for resources and power; or political crises – that cause war to reappear, perhaps suddenly. Durbin and Bowlby write,

It may very well be that an appreciation of the advantages of cooperation and an agreement to continue it will preserve the peace for some time. But underneath, there is a powerful and 'natural' tendency to resort to force in order to secure the possession of desired objects or to overcome a sense of frustration or to resist the encroachment of strangers or to attack a scapegoat.[243]

Furthermore, "The transition from [peace to war] is simply when one change in the circumstances of the group alters the balance between the desire for cooperation and the conflicting desire to obtain self-regarding ends by force."[244] When the good-evil and peace-war dialectics shift toward evil-war, good must be ready or else face the consequences. "When the world is at peace," writes Sun Tzu, "a gentleman keeps his sword by his side."[245] And the ancient Chinese *Ssu-ma* (war minister) advises, "Even though calm may prevail under Heaven, those who forget warfare will certainly be endangered."[246] If we delude ourselves into thinking that war is gone forever, reality is likely to drive the human story back into another dark chapter.

Conclusion

Ring out false pride in place and blood,
  The civic slander and the spite;
  Ring in the love of truth and right,
  Ring in the common love of good.

Ring out old shapes of foul disease;
  Ring out the narrowing lust of gold;
  Ring out the thousand wars of old,
Ring in the thousand years of peace.[247]

Alfred Lord Tennyson, *In Memoriam A. H. H.* (1850)

We began this section seeking answers to two questions: Can we end war? Should we end war? A thorough investigation, from anthropology to genetic engineering, failed to reveal a viable path to perpetual peace without eliminating war's root cause: *Homo sapiens*. In fact, artificially ending war holds less promise of creating an everlasting peace than it does a permanent death. Subscribers to what Thomas Sowell calls the "unconstrained vision" (i.e., that humanity's great evils, like war, are solvable with enough time and effort) will no doubt disagree.[248] But this stance ultimately amounts to little more than hope over reason. Ironically, in his treatise on perpetual peace, Kant pessimistically declares, "The natural state of men is not peaceful co-existence but war – not always open hostilities, but at least an unceasing threat of war."[249] In truth, using the peace-war dialectic's four zones, humanity's natural state is best described as a preponderance of competition and conflict between pockets of true peace and war.

Can we end war for all time? No. "A commitment to rid the world of war," confirms Gray, "is mission impossible ... folly is folly, no matter how pure the motive."[250] Should we try to end war for all time? A qualified yes. We should work diligently to make the world a more peaceful place – through cooperative enterprises, enlightened governance, and deterrence – but we must be realistic. Creating power imbalances or engineering disparities in ourselves upsets the natural order and disrupts the harmony of tensions, doing more harm than good. In the words of Rousseau (referring to the Abbé's vision of perpetual peace),

Let us admire such a fine plan, but be consoled that we will never see it come about, for that can only happen by means that humanity might find violent and fearful ... [and] would perhaps cause more harm in one moment than it could prevent for centuries to come.[251]

Though the possibility of war is permanent, there is no cause for despair, for we retain the power of choice, the free will to cultivate our capacities for reciprocal altruism. Let us be ready for war, but never stop working for peace. These activities are not mutually exclusive. Despite our differences, we should not forget our commonality as a species and as a global community with the power to work together for peaceful ends. To conclude, consider the following from Abraham Lincoln's first inaugural address, offered with slight modification [in *italics*],

We are not enemies, but friends. We must not be enemies. Though passion may have strained, it must not break our bonds of affection. The mystic chords of memory, stretching from every battlefield, and ... grave, to every living heart and hearthstone, all over this [*earth*], will yet swell the chorus of [*humanity*], when again touched, as surely they will be, by the better angels of our nature.[252]

Conclusion

And when ye shall hear of wars and rumors of wars, be ye not troubled: for
such things must needs be; but the end shall not be yet.   Mark 13:7

As long as justice and injustice have not terminated their ever-renewing fight
for ascendancy in the affairs of mankind, human beings must be willing, when
need is, to do battle for the one against the other.[1]   John Stuart Mill

In 1827 (the year of his epiphany), Clausewitz, prophetically pondering both his legacy and mortality, wrote of *On War*, "I believe an unprejudiced reader in search of truth and understanding will recognize ... the fruits of years of reflection on war and diligent study of it. He may even find they contain the basic ideas that might bring about a revolution in the theory of war."[2] *The New Art of War* has endeavored to emulate Clausewitz's objectivity, reflection, and passion for revealing war's truths. And while it may not be revolutionary, it has at least aspired to clarify and extend the ideas of war's great theorists and render the language of war, theory, and strategy more accessible, relevant, and useful for modern audiences. To the extent that we have fulfilled these ambitions, we can now discuss war history, strategy, and theory – and practice a *new* art of war – using *The Unified War Theory*'s tools and terms of reference. This universal, inclusive framework facilitates theory's comprehension, prediction, and control functions; arms strategists with greater awareness of war's nature, character, and principal dialectics; and consequently improves our chances of preventing wars, decreasing their frequency, duration, and lethality, and prevailing when there is no alternative.

Sun Tzu confirms the power of theory with a message hidden in the logic of *The Art of War*'s two prime maxims: *know the enemy and know yourself: in a hundred battles you will never be in peril*; and, *to subdue the enemy without fighting is the acme of skill.* Sun Tzu intended his philosophy to apply to both the practice and *study* of war. If the acme of skill is winning without fighting, and fighting is war, then Sun Tzu is telling us that the true enemy *is war itself.* Therefore, invoking the first maxim, freedom from peril through the mastery of war requires that we know the enemy (war) and ourselves (humanity). This points directly to the True Trinity: humanity, politics, and combat. To know war

368

is to know combat; and, to know ourselves is to understand human nature and politics. Thus, Sun Tzu affirms that comprehending the nature of war, which includes humanity, politics, and fighting, is the key to gaining our political aims without having to fight or at least influencing war's destructive means toward more productive ends, like a better peace. Fielding Hall writes,

All our prophets ... explain nothing to us either of the nature of war, of peace or of humanity. We are to have faith, shut our eyes and do as we are told. But ... we want not prescriptions but knowledge. We want to know more about war, its causes and effects, before we undertake any treatment. If we are really to prevent war, we must have this knowledge.[3]

Knowledge provides power to influence others, the environment, and most importantly, ourselves. Ignoring war's impact on the human experience invites injury as ignorance of any hazard would, only on a much grander scale. Indeed, "the right way to handle scary, dangerous things," advises Robert Laughlin, "is to understand them thoroughly and deal with them openly."[4] And Clausewitz adds, "A people and nation can hope for a strong position in the world only if national character and *familiarity with war* fortify each other by continual interaction" (emphasis added).[5]

War is a complex, multi-faceted phenomenon, and like everything else, a product of the universe's harmony of tensions, a destructive force that melds order, chaos, reason, and emotion to create new political realities. Power imbalances and ideological conflict increase the potential for war, which can cause more imbalances as it drives change. Aristotle's model allows us to view war as the synthesis of four causes: (1) *material*: war's physical components, forces, and capacity; (2) *formal*: war's "choreography," its forms (viscous-inviscous) and postures (defense-attack); (3) *efficient*: war's human and psychological dimensions, including the will to fight; and, (4) *final*: war's "why," its purpose or political objective. No truly holistic concept or definition of war can omit even one of these.

Cultural anthropologist Margaret Mead calls war an "invention," implying that it can be un-invented or replaced; but, war is more an emergent manifestation of the complex interplay between human nature and institutions.[6] War, enabled by evolution-honed faculties, is perhaps the most uniquely human, enigmatic, and imperious of all our collective activities. Though humanity – with all our creativity, anxiety, ambition, and pride – causes war, we are not exclusively warlike. Indeed, "fighting and peaceful cooperation are equally 'natural' forms of behavior," emphasize Durbin and Bowlby, "equally fundamental tendencies in human relations," which developed to meet Darwinian imperatives.[7] Human existence, with its epochs of war and peace, like Achilles's shield, reflects the duality of human nature: noble and disgraceful, good and evil, altruistic and selfish, peaceful and pugnacious.[8] War can be a tool for the wicked or the just and produces many iconic figures, heroes, and villains.

War's human origins distinguish it from nonhuman conflict, while its political dimension separates it from other forms of human-to-human violence and competition. War is always reciprocal, characterized by combative violence between two opposing sides each of which decides and acts according to its own perceptions, capacities, and political objectives. War's reasons, for groups and individuals, are as inconstant and varied as humanity, but most relate to Thucydides's fear, honor, or interest and to personal and political survival. War is a choice but sometimes a choice forced upon us by others; hence, in anything other than a utopian world of permanent peace, ignoring war risks exploitation and oppression at the hands of aggressors. "A cautious man," writes philosopher Samuel von Pufendorf, "will believe all men his friends but liable at any time to become enemies . . . this is why that country is considered happy which even in peace contemplates war."[9] The nations of Europe learned this lesson harshly in the late 1930s, when, as Churchill put it, their "blundering defenselessness had its share" in Hitler's ascendance and aggression.[10]

The True Trinity – humanity, politics, and combat – incorporates all of war's Aristotelian causes and is central to its unchanging nature, while its character varies as combatants, weapons, and contexts change. Though its character evolves with time, war's immutable nature manifests itself in patterns and themes that have inspired the likes of Sun Tzu, Thucydides, and Clausewitz to express them theoretically. Found within these patterns (or perhaps driving them) are war's twenty dialectics: order-chaos, past-future, peace-war, life-death, creation-destruction, friend-enemy, good-evil, science-art, defense-attack, unlimited-limited, viscous-inviscous, physical-moral, direct-indirect, certainty-uncertainty, simplicity-complexity, reason-emotion, prudence-boldness, control-autonomy, concentration-dispersal, and rest-movement. War's great theoretical and strategic traditions have exemplified these, with the most successful achieving balance and the less so anchoring to extremes. In fact, we cannot hope to master the art of war and strategy without knowing how and when to move from one dialectic extreme to another, even as we remain always mindful of the whole.

Strategy uses theory's insights into war's nature and character to orchestrate ways and means that accomplish physical and moral ends, from the political to the tactical level. Strategy is an iterative activity that first defines the ends it will serve before identifying, implementing, and adjusting its ways and means within cost and risk constraints. Military strategy employs military instruments of power to support political ends. In war, this always includes combative violence employed to reduce an enemy's engine of war (i.e., its sustaining *cause*, warfighting *capacity*, and *will* to fight). The fighting will, the engine's "fuel" and most critical component, is supported by a variety of factors, including capacity, leadership, education, morale, training, motivation, and experience. While fighting can take place in any domain – land, sea, air, space, or cyberspace – wars are won and lost in the *cognitive* domain, for weapons are nothing without the will to use them. War, like nature, rewards the flexible and adaptive, those who find innovative ways to combine capabilities that present adversaries with lethal dilemmas even in the face of uncertainty and danger. Raw, concentrated power does not always prevail, and cunning, agility, surprise, speed, and mental fortitude can trump pure strength.

War appears in a variety of forms, but within the *Unified War Theory* (UWT) all are part of a continuum and related through the underlying property of *viscosity* (a function of force *directness, acceleration, restriction, cohesion,* and *concentration*). This concept enables us to replace or redefine ambiguous legacy terms (like asymmetric, irregular, regular, and hybrid) and anchor our language for war back to its unchanging nature. In war, forces can adopt viscous or inviscous states relative to their opposition and they can influence enemy viscosity through *containment*, a fluidity inhibitor and prerequisite for the effective employment of destructive force. The UWT's *War-Viscosity Algorithm* (W-V-A) illustrates how force state decisions relate to will, context, and warfighting capacity. In general, relative capacity drives viscosity, relative viscosity drives war's forms, and success is more likely when force viscosity varies proportionally with capacity.

Though the nature of war is constant, the variability of war's character and its human and environmental influencing factors complicate forecasting. Assessments of war's future character are critical to successful strategy and warfighting, and the best way to improve the accuracy of future war predictions is by studying history, trends, current circumstances, and theory. The UWT and W-V-A are helpful references for evaluating potential futures, particularly as they relate to context, relative capacity, and war's changing forms. As for war's future as a human activity, most evidence and objective analysis point to the impracticality of achieving permanent peace as long as humans are humans. Thus, the *potential* for war is permanent. However, the potential for peace is also permanent and becomes more likely when pursued with clear-eyed, truthful knowledge of war and its relationship to humanity. We can prevent wars by using instruments of power to foster cooperative connections within and between nations and by tailored deterrence against those with whom we have profound differences. And we can limit or win wars, if necessary, by leveraging theory to understand the nature of politics and influence; the essential acts of strategy; the dynamics and implications of radiation, inertia, momentum, viscosity, containment, and the W-V-A; and the adversary's engine of war and prime centers of gravity. But most importantly, if we have the courage to embrace war's truths and to cultivate the better angels of our nature; if we understand, as Thomas Gregor suggests, that despite war's virulence and the volatility of domestic politics and international relations, we can nurture "special relationships, structures, and attitudes" that "promote and protect" peace; then war and its dialectic extremes will *not* define us, and humanity will succeed in sustaining its narrative far into the distant future.[11]

Appendix

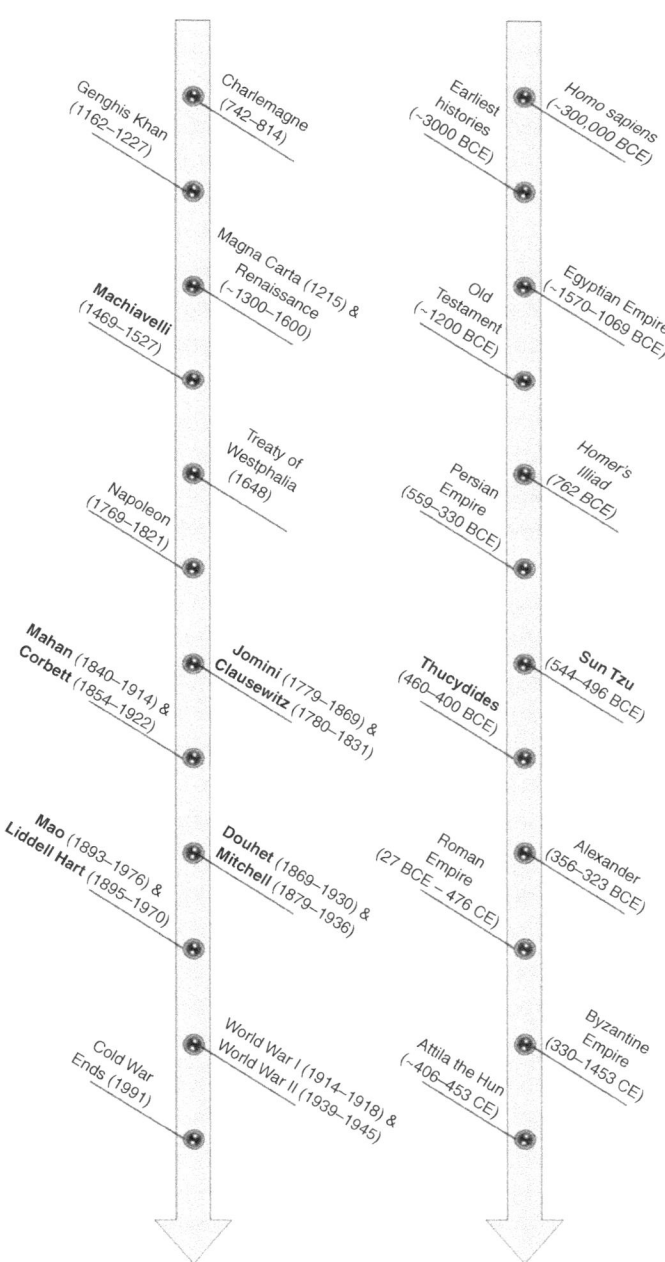

Genghis Khan (1162–1227)

Charlemagne (742–814)

Earliest histories (~3000 BCE)

Homo sapiens (~300,000 BCE)

Machiavelli (1469–1527)

Magna Carta (1215) & Renaissance (~1300–1600)

Old Testament (~1200 BCE)

Egyptian Empire (~1570–1069 BCE)

Napoleon (1769–1821)

Treaty of Westphalia (1648)

Persian Empire (559–330 BCE)

Homer's Illiad (762 BCE)

Mahan (1840–1914) & Corbett (1854–1922)

Jomini (1779–1869) & Clausewitz (1780–1831)

Thucydides (460–400 BCE)

Sun Tzu (544–496 BCE)

Mao (1893–1976) & Liddell Hart (1895–1970)

Douhet (1869–1930) & Mitchell (1879–1936)

Roman Empire (27 BCE – 476 CE)

Alexander (356–323 BCE)

Cold War Ends (1991)

World War I (1914–1918) & World War II (1939–1945)

Attila the Hun (~406–453 CE)

Byzantine Empire (330–1453 CE)

Figure A.1 Timeline of major figures and events

Notes

Dedication

1. Seng-T'san, "The Third Patriarch of Zen Hsin Hsin Ming," accessed October 6, 2019, www
   .age-of-the-sage.org/buddhism/third_patriarch_zen.html.
2. *Encyclopedia Britannica Online*, s.v. "Hegelianism," accessed October 6, 2019, www
   .britannica.com/topic/Hegelianism.

Preface

1. Lao Tzu Quotes, BrainyQuote.com, BrainyMedia Inc, 2019, accessed October 6, 2019 www
   .brainyquote.com/quotes/lao_tzu_137141.
2. Some coincidences: "Geoffrey" is the Anglicized version of "Gottfried," the third name of
   Carl Philipp Gottfried von Clausewitz; our fathers shared the same first name, Frederick;
   and oddly, Machiavelli, Napoleon, Giulio Douhet, and I were born in the sixty-ninth year of
   our respective birth centuries. My Italian-German heritage means that it is at least *possible*
   that Clausewitz, Machiavelli, or Douhet are distant relatives.
3. Chris Hedges, "What Every Person Should Know about War," *New York Times*, July 5, 2003,
   www.nytimes.com/2003/07/06/books/chapters/what-every-person-should-know-about-war.
   Estimates of total war deaths range from 150 million to 1 billion. Between 1900 and 1990,
   43 million soldiers and 62 million civilians died as a result of wars.
4. Azar Gat, *A History of Military Thought: From the Enlightenment to the Cold War* (Oxford:
   Oxford University Press, 2001), 824.
5. Hew Strachan, *The Direction of War: Contemporary Strategy in Historical Perspective*
   (Cambridge: Cambridge University Press, 2013), 117.
6. B. H. Liddell Hart, *Strategy*, 2nd ed. (New York: Meridian, 1991), 339.
7. Colin S. Gray, *Fighting Talk: Forty Maxims on War, Peace, and Strategy* (Westport, CT
   Greenwood Publishing Group, 2007), 59.
8. Liddell Hart, *Strategy*, 342.
9. Peter Paret, *Makers of Modern Strategy from Machiavelli to the Nuclear Age* (Princeton, NJ:
   Princeton University Press, 1986), 211.
10. Gat, *A History of Military Thought*, 690.
11. James Jay Carafano, "It's Time to Return to the Principles of War," *National Interest*, May 4,
    2016, https://nationalinterest.org/feature/its-time-return-the-principles-war-16054.
12. Ibid.
13. Strachan, *The Direction of War*, 82.
14. Carl von Clausewitz, *On War*, eds. and trans. Michael Howard and Peter Paret (Princeton, NJ:
    Princeton University Press, 1976), 168.

15. Tarak Barkawi and Shane Brighton, "Absent War Studies? War, Knowledge, and Critique," in *The Changing Character of War*, eds. Hew Strachan and Sibylle Scheipers (Oxford: Oxford University Press, 2011), 529.
16. Ibid.
17. Gray, *Fighting Talk*, 58.
18. Williamson Murray, *War, Strategy, and Military Effectiveness* (New York: Cambridge University Press, 2011), 7.
19. Bernard Brodie, "The Continuing Relevance of *On War*," in Clausewitz, *On War*, 52.
20. Paul Davidson Reynolds, *A Primer in Theory Construction* (New York: Macmillan Publishing, 1986), 163.
21. Ibid.
22. Gordon A. Craig, "Delbrück: The Military Historian," in *Makers of Modern Strategy from Machiavelli to the Nuclear Age*, ed. Peter Paret (Princeton, NJ: Princeton University Press, 1986), 352–353.
23. Neil Gross, *Why Are Professors Liberal and Why Do Conservatives Care?* (Cambridge, MA: Harvard University Press, 2013), and Scott Jaschik, "Moving Further to the Left," *Insider Higher Ed*, October 24, 2012, www.insidehighered.com/news/2012/10/24/survey-finds-professors-already-liberal-have-moved-further-left.
24. John A. Lynn, "The Embattled Future of Academic Military History," *Journal of Military History* 61, no. 4 (October 1997): 778.
25. Justin Ewers, "Why Don't Colleges Teach Military History?" *US News and World Report*, April 3, 2008, www.usnews.com/news/articles/2008/04/03/why-dont-colleges-teach-military-history.
26. Lynn, "The Embattled Future," 777–789.
27. Ibid., 777.
28. Brodie, "The Continuing Relevance of *On War*," in Clausewitz, *On War*, 53–54.
29. Isaac Asimov, "Isaac Asimov Asks, 'How Do People Get New Ideas?'" *MIT Technology Review*, October 20, 2014, www.technologyreview.com/view/531911/isaac-asimov-asks-how-do-people-get-new-ideas.
30. Clausewitz, *On War*, 75.
31. Barkawi and Brighton, "Absent War Studies?" 527.
32. Murray, *War, Strategy, and Military Effectiveness*, 7.
33. Brodie, "The Continuing Relevance of *On War*," in Clausewitz, *On War*, 57.
34. Liddell Hart, *Strategy*, 3.
35. Ibid., 4.
36. Ibid., 232.
37. Robert Coram, *Boyd: The Fighter Pilot Who Changed the Art of War* (New York: Back Bay Books, 2002), 317–318. For example, Coram says America's senior military leaders during the Vietnam War were "more familiar with business theory than with military theory. They read management books and talked at length of how things were done at the Harvard Business School. But some had never heard of Sun Tzu and could not spell 'von Clausewitz.'"
38. T. Michael Moseley, "Airmen and the Art of Strategy," *Strategic Studies Quarterly* 1, no. 1 (Fall 2007): 11.
39. David Livingstone Smith, *The Most Dangerous Animal: Human Nature and the Origins of War* (New York: St. Martin's Press, 2007), xv.
40. David Galula, *Counterinsurgency Warfare: Theory and Practice*, 1964 (Westport, CT: Praeger Security International, 2006), xi.
41. Murray, *War, Strategy, and Military Effectiveness*, 3.
42. Julian S. Corbett, *Some Principles of Maritime Strategy*, 1911 (Mineola, NY: Dover Publications, 2004), 236.
43. Though attributed to Duell, this quote is apocryphal and most likely originated with a joke that appeared in the print media circa 1900.

44. Colmar von der Goltz, quoted in Michael Howard, "The Influence of Clausewitz," in *Clausewitz, On War*, 31.

45. Robert B. Laughlin, *A Different Universe: Reinventing Physics from the Bottom Down* (New York: Basic Books, 2005), 213.

46. Arthur Schopenhauer, "Further Psychological Observations," in *Studies in Pessimism* (Project Gutenberg, 2004), www.gutenberg.org/files/10732/10732-h/10732-h.htm.

47. Moseley, "Airmen and the Art of Strategy," 7.

48. Sun Tzu, *The Art of War*, ed. and trans. Samuel B. Griffith (New York: Oxford University Press, 1963), 63.

49. Niccolò Machiavelli, *The Art of War*, 1521, trans. Neal Wood (Cambridge, MA: Da Capo Press, 2001), 5.

50. Murray, *War, Strategy, and Military Effectiveness*, 3.

51. Ashton B. Carter, "Running the Pentagon Right: How to Get the Troops What They Need," *Foreign Affairs*, 2014, www.foreignaffairs.com/articles/140346/ashton-b-carter/running-the-pentagon-right.

52. Colin S. Gray, *War, Peace and International Relations* (London and New York: Routledge, 2007), 283.

53. Clausewitz, *On War*, 154.

54. Asimov, "Isaac Asimov Asks, 'How Do People Get New Ideas?'"

55. Reynolds, *A Primer in Theory Construction*, 4.

56. Clausewitz, *On War*, 170.

57. Greg Cashman, *What Causes War? An Introduction to Theories of International Conflict* (Oxford: Lexington Books, 1993), 10.

58. Milan Vego, "On Military Theory," *JFQ Joint Forces Quarterly*, no. 62 (3rd 2011): 61, https://ndupress.ndu.edu/portals/68/Documents/jfq/jfq-62.pdf.

59. Haldane, J. B. S., "The Truth about Death," *Journal of Genetics*, Vol. 58 (1963): 464, accessed November 7, 2020, https://wist.info/haldane-jbs/27097/.

60. Strachan, *The Direction of War*, 195.

61. Liddell Hart, *Strategy*, xx.

62. Clausewitz, "Author's Comment," in *Clausewitz, On War*, 63.

Introduction: Why Study War?

1. Robert E. Park, "The Social Function of War," in *War: Studies from Psychology, Sociology Anthropology*, eds. Leon Bramson and George W. Goethals (New York: Basic Books, 1953), 238.

2. Leon Trotsky, "Leon Trotsky," *Lapham's Quarterly* (blog), accessed October 6, 2019, www.laphamsquarterly.org/contributors/trotsky.

3. Peter Paret, *Understanding War: Essays on Clausewitz and the History of Military Power* (Princeton, NJ: Princeton University Press, 1992), 210.

4. Brodie, "The Continuing Relevance of *On War*," in *Clausewitz, On War*, 55.

5. John Keegan and Joseph Darracott, *The Nature of War* (New York: Holt, Rinehart, and Winston, 1981), xi.

6. David Hume, "Of the Influencing Motives of the Will," in *A Treatise of Human Nature* (Project Gutenberg, 2012), bk. II, pt. III, sec. III, accessed September 19, 2019, www.gutenberg.org/files/4705/4705-h/4705-h.htm#link2H_.

7. Georg Wilhelm Friedrich Hegel, *Lectures on the Philosophy of History* (Kitchener, Ontario: Batoche Books, 2001), 47, accessed July 12, 2020, https://socialsciences.mcmaster.ca/econ/ugcm/3ll3/hegel/history.pdf.

8. Liddell Hart quoted in Gat, *A History of Military Thought*, 659.

9. Nikolai Berdyaev, "The Psychology of War and the Meaning of War: Thoughts About the Nature of War" in *The Fate of Russia*, trans. Fr. S. Janos (1918; repr Moscow: Svarog, 1997), 374–380, accessed December 29, 2009, www.berdyaev.com/berdiaev/berd_lib/1915_197.html.
10. Desiderius Erasmus, "The Education of a Christian Prince," in *The Ethics of War: Classic and Contemporary Readings*, eds. Gregory M. Reichberg, Henrik Syse, and Endre Begby (Malden, MA: Blackwell Publishing, Ltd., 2006), 234.
11. H. Fielding Hall, *The Nature of War – And Its Causes* (London: Hurst & Blackett, 1917), 13.
12. Edward O. Wilson quoted in Cashman, *What Causes War?*, 25.
13. Machiavelli, *The Art of War*, 79.
14. Angelo Codevilla and Paul Seabury, *War: Ends and Means*, 2nd ed. (Washington, DC: Potomac Books, 2006), 245.
15. Algis Valiunas, *Churchill's Military Histories: A Rhetorical Study* (Oxford: Rowman & Littlefield Publishers, 2002), 173.
16. "William T. Sherman," American Battlefield Trust, last modified December 19, 2008, www.battlefields.org/learn/biographies/william-t-sherman.
17. Lawrence H. Keeley, *War Before Civilization: The Myth of the Peaceful Savage* (Oxford: Oxford University Press, 1996), 159.
18. Helmuth von Moltke, "On the Nature of War," in *Die Zerstörung der deutschen Politik: Dokumente,* ed. Harry Pross, trans. Richard S. Levy (Frankfurt, 1959), 29–31, accessed December 29, 2009, https://wwi.lib.byu.edu/index.php/On_the_Nature_of_War_by_Helmut_Moltke_(the_Elder).
19. Gray, *Fighting Talk*, 18.
20. Ian Morris, "The Slaughter Bench of History," *The Atlantic*, April 11, 2014, www.theatlantic.com/international/archive/2014/04/the-slaughter-bench-of-history/360534.
21. Ibid.
22. William Graham Sumner, "War," in *War: Studies from Psychology, Sociology, Anthropology*, eds. Leon Bramson and George W. Goethals (New York: Basic Books, 1968), 223.
23. John Keegan, *A History of Warfare* (New York: Vantage Books, 1994), xvi.
24. Francisco Suárez, *Metaphysical Disputations*, in *The Ethics of War: Classic and Contemporary Readings*, eds. Gregory M. Reichberg, Henrik Syse, and Endre Begby (Malden, MA: Blackwell Publishing, Ltd., 2006), 341.
25. John Stuart Mill, "The Contest in America," Fraser's Magazine (1862), Project Gutenberg, accessed May 26, 2020, www.gutenberg.org/files/5123/5123-h/5123-h.htm.
26. Hans Kelsen, "General Theory of Law and State," in *The Ethics of War: Classic and Contemporary Readings*, eds. Gregory M. Reichberg, Henrik Syse, and Endre Begby (Malden, MA: Blackwell Publishing, Ltd., 2006), 610.
27. Raymond Aron, *Clausewitz: Philosopher of War*, trans. Christine Booker and Norman Stone (Englewood, NJ: Prentice-Hall, 1985), 400.
28. Saul David, ed., *War: The Definitive Visual History from Bronze Age Battles to 21th Century Conflict* (New York: DK Publishing, 2009), 76.
29. Keegan and Darracott, *The Nature of War*, ix.
30. Keeley, *War Before Civilization*, 23.
31. H. G. Wells, *The Island of Doctor Moreau* (New York: Garden City Publishing, 1896), chap. XIV, accessed July 12, 2020, www.gutenberg.org/files/159/159-h/159-h.htm.
32. Victor Davis Hanson, *The Father of Us All: War and History Ancient and Modern* (New York: Bloomsbury Press, 2010), 246.
33. Liddell Hart quoted in Beatrice Heuser, *The Evolution of Strategy: Thinking War from Antiquity to the Present* (Cambridge: Cambridge University Press, 2010), 165.
34. Samuel B. Griffith, "Introduction," in Mao Tse-tung, *On Guerrilla Warfare*, trans. Samuel B. Griffith (New York: BN Publishing, 2007), 34.

1   The Origins of War

1. Nikolai Berdyaev, "The Psychology of War."
2. Andrea Falcon, "Aristotle on Causality," in *The Stanford Encyclopedia of Philosophy*, ed. Edward N. Zalta, Spring 2015 (Metaphysics Research Lab, Stanford University, 2015), accessed August 12, 2017, https://plato.stanford.edu/archives/spr2015/entries/aristotle-causality.
3. Vego, "On Military Theory," 61.
4. Heraclitus, *Fragments: The Collected Wisdom of Heraclitus*, trans. Brooks Haxton (New York, Viking Penguin, 2001), 25.
5. Ibid., 37.
6. Carl Jung Quotes, BrainyQuote.com, BrainyMedia Inc, 2019, accessed October 6, 2019, www.brainyquote.com/quotes/carl_jung_157280.
7. Confucius Quotes, AZ Quotes, accessed October 19, 2019, www.azquotes.com/author-3177-Confucius?p=2.
8. Clausewitz, *On War*, 523.
9. Peter Paret, "The Genesis of *On War*," in Clausewitz, *On War*, 15–16.
10. Clausewitz, *On War*, 374.
11. Ibid., 221.
12. Ibid., 442.
13. Gray, *War, Peace and International Relations*, 341.
14. Clausewitz, *On War*, 221.
15. Ibid., 480.
16. Liddell Hart, *Strategy*, 329.
17. Maizeroy quoted in Strachan, *The Direction of War*, 28.
18. "Order" reflects war as a science, subject to reason and control (e.g., Enlightenment); "Chaos" exemplifies war as an art, subject to emotion and chance (e.g., Romanticism). "Regular" corresponds to orthodox, conventional, or traditional; and "irregular" to unorthodox, unconventional, or nontraditional. Victory-defeat, strong-weak, and courage-cowardice are omitted since one side is always detrimental. Logic-intuition is implied in science-art and reason-emotion; continuity-change is implied in order-chaos; and right-wrong, truth-lie, and honor-deceit are implied in good-evil, friend-enemy, and regular-irregular.
19. Michio Kaku, *Physics of the Impossible: A Scientific Exploration into the World of Phasers, Force Fields, Teleportation, and Time Travel* (New York: Anchor Books, 2009), 268.
20. Liddell Hart, *Strategy*, 368.
21. Strachan, *The Direction of War*, 200.
22. Clausewitz, *On War*, 374.
23. Hugo Grotius, *The Law of War and Peace*, trans. by Louise R. Loomis (Roslyn, NY: Walter J. Black, 1949), 4.
24. Marechal de Saxe quoted in Lawrence LeShan, *The Psychology of War: Comprehending Its Mystique and Its Madness* (New York: Helio Press, 2002), 71.
25. Heraclitus, *Fragments*, 41.
26. Drew Gilpin Faust, "2011 Jefferson Lecture in the Humanities," (lecture, Harvard University, Cambridge, MA, May 2, 2011), www.harvard.edu/president/speech/2011/2011-jefferson-lecture-humanities.
27. Azar Gat, *War in Human Civilization* (Oxford: Oxford University Press, 2006), xi.
28. Smith, *The Most Dangerous Animal*, xiii.
29. Strachan, *The Direction of War*, 35.
30. Technically, modern man is *Homo sapiens sapiens* (where other subspecies, like Neanderthal, are also *Homo sapiens*); however, for brevity, *Homo sapiens* (in Latin, "wise man") means modern humans.
31. *Merriam-Webster*, s.v. "phenomenon (n.)," accessed August 21, 2016, www.merriam-webster.com/dictionary/phenomenon.

32. Sitting Bull Quotes, Goodreads.com, accessed October 6, 2019, www.goodreads.com/author/quotes/5712889.Sitting_Bull.

33. E. F. M. Durbin and John Bowlby, "Personal Aggressiveness and War," in *War: Studies from Psychology, Sociology, Anthropology*, eds. Leon Bramson and George W. Goethals (New York: Basic Books, 1968), 81.

34. This metaphor is attributed to a variety of individuals, though it is unclear who first used it.

35. The works of Sebastian Junger, *War* (2010), and Dave Grossman, *On Killing* (1995) and *On Combat* (2004), show that only a small minority of humans actually relish combat and killing.

36. Mark A. May, "War, Peace, and Social Learning," in *War: Studies from Psychology, Sociology, Anthropology*, eds. Leon Bramson and George W. Goethals (New York: Basic Books, 1968), 151.

37. Leslie E. Sponsel, "The Natural History of Peace: The Positive View of Human Nature and Its Potential," in *A Natural History of Peace*, ed. Thomas Gregor (Nashville, TN: Vanderbilt University Press, 1996), 115.

38. Albert Einstein and Sigmund Freud, *Why War?* (International Institute for Intellectual Cooperation [League of Nations], 1933), 9, accessed December 27, 2018, www.drmalikcikk.atw.hu/wp_readings/einstein_freud.PDF.

39. Frans B. M. de Waal, "The Biological Basis of Peaceful Coexistence: A Review of Reconciliation Research on Monkeys and Apes," in *A Natural History of Peace*, ed. Thomas Gregor (Nashville, TN: Vanderbilt University Press, 1996), 38. De Waal was Charles Howard Candler Professor of Primate Behavior in the Department of Psychology at Emory University.

40. Bruce M. Knauft, "The Human Evolution of Cooperative Interest," in *A Natural History of Peace*, ed. Thomas Gregor (Nashville, TN: Vanderbilt University Press, 1996), 73. Knauft is Director of Graduate Studies in the Department of Anthropology at Emory University.

41. Bronislaw Malinowski, "An Anthropological Analysis of War," *War: Studies from Psychology, Sociology, Anthropology*, eds. Leon Bramson and George W. Goethals (New York: Basic Books, 1968), 260. Malinowski is widely regarded as the father of social anthropology.

42. Keeley, *War Before Civilization*, 160. Lawrence Keeley says: "The universal preference for peace is not just the product of arbitrary moral choice or deep psychology; it is practical and rational."

43. Adolf Hitler Quotes, BrainyQuote.com, BrainyMedia Inc, 2019, accessed October 6, 2019, www.brainyquote.com/quotes/adolf_hitler_382802.

44. Mark Twain quoted in Smith, *The Most Dangerous Animal*, 5.

45. "DNA: Comparing Humans and Chimps," American Museum of Natural History, accessed March 30, 2014, www.amnh.org/exhibitions/permanent/human-origins/understanding-our-past/dna-comparing-humans-and-chimps.

46. "Why Homo Sapiens Won the Battle of Human Survival: Neanderthals Had Larger Eyes but Less Brain Power to Make Decisions," *Daily Mail Online*, June 2, 2013, www.dailymail.co.uk/sciencetech/article-2334637/Why-Homo-sapiens-won-battle-human-survival-Neanderthals-larger-eyes-brain-power-make-decisions.html.

47. Literally, "man is a wolf to another man."

48. Gat, *War in Human Civilization*, 9.

49. Ibid., 9–10.

50. For more on this, check out Frans de Waal, *Chimpanzee Politics* (Johns Hopkins, 1982).

51. Freud, *Why War?*, 5.

52. Keeley, *War Before Civilization*, 158.

53. Thomas Gregor and Clayton A. Robarchek, "Two Paths to Peace: Semai and Mehinaku Nonviolence," in *A Natural History of Peace*, ed. Thomas Gregor (Nashville, TN: Vanderbilt University Press, 1996), 159.

54. Sponsel, "The Natural History of Peace," 103.

55. Gat, *War in Human Civilization*, 4.

56. David Lloyd George Quotes, BrainyQuote.com, BrainyMedia, accessed August 3, 2016, www.brainyquote.com/quotes/david_lloyd_george_386727.

57. Gat, *War in Human Civilization*, 12.

58. Ibid., 29–30.
59. Paul Collier, *Wars, Guns, and Votes: Democracy in Dangerous Places* (New York: HarperCollins, 2009), 170.
60. Gat, *War in Human Civilization*, 24.
61. Morris, "The Slaughter Bench of History."
62. Jared Diamond, *Guns, Germs, and Steel: The Fates of Human Societies* (New York: W. W. Norton & Company, 2017), 279.
63. Thomas Hobbes, *De Cive*, in *The Ethics of War: Classic and Contemporary Readings*, eds. Gregory M. Reichberg, Henrik Syse, and Endre Begby (Malden, MA: Blackwell Publishing, 2006), 444.
64. Collier, *Wars, Guns, and Votes*, 170–172.
65. Malinowski, "An Anthropological Analysis of War," 262.
66. Cashman, *What Causes War?*, 30.
67. Liddell Hart, *Strategy*, 330.
68. May, "War, Peace and Social Learning," 153.
69. Daniel Gilbert, *Stumbling on Happiness* (New York: Vintage, 2007), 4–5.
70. Kenneth N. Waltz, *Man, the State, and War: A Theoretical Analysis* (New York: Columbia University Press, 2001), 21.
71. Clausewitz, *On War*, 149.
72. Bevin Alexander, *How Wars Are Won: The 13 Rules of War From Ancient Greece to the War on Terror* (New York: Crown Publishers, 2002), 2.
73. Freud, *Why War?*, 6.
74. Neal Wood, "Introduction," in Machiavelli, *The Art of War*, lxxv.
75. Lawrence Freedman, *Strategy: A History* (New York: Oxford University Press, 2013), 6.
76. Keegan and Darracott, *The Nature of War*, 168.
77. Malinowski, "An Anthropological Analysis of War," 253.
78. William McDougall, "The Instinct of Pugnacity," in *War: Studies from Psychology, Sociology, Anthropology*, eds. Leon Bramson and George W. Goethals (New York: Basic Books, 1968), 38.
79. Grotius, *The Law of War and Peace*, 26.
80. "Lord Acton," Acton Institute, accessed June 29, 2019, https://acton.org/research/lord-acton.
81. Gwynne Dyer, *War* (New York: Crown Publishers, 1985), 17–18.
82. Immanuel Kant, *Toward Perpetual Peace: A Philosophical Sketch*, ed. Jonathan Bennet (1795; Early Modern Texts, 2015), 18, accessed August 22, 2018, www.earlymoderntexts.com/assets/pdfs/kant1795.pdf.
83. Aristotle, *Pol.* I, 1253a, Perseus Digital Library, Tufts University, ed. Gregory R. Crane, accessed October 19, 2019, www.perseus.tufts.edu/hopper/text?doc=Perseus:abo:tlg,0086,035:1:1253a.
84. Freud, *Why War?*, 7.
85. Niccolò Machiavelli, *The Prince*, 1532, trans. N. H. Thompson (New York: Dover Publications, 1992), 2.
86. Knauft, "The Human Evolution of Cooperative Interest," 77.
87. Gray, *War, Peace and International Relations*, 9.
88. Ervin Staub, "The Psychological and Cultural Roots of Group Violence and the Creation of Caring Societies and Peaceful Group Relations," in *A Natural History of Peace*, ed. Thomas Gregor (Nashville, TN: Vanderbilt University Press, 1996), 132.
89. McDougall, "The Instinct of Pugnacity," 33.
90. Waltz, *Man, the State, and War*, 188.
91. Rev. J. MacDonald quoted in Maurice R. Davie, *The Evolution of War: A Study of Its Role in Early Societies*, 1929 (New York: Dover Publications, Inc., 2003), 13. An early twentieth-century observer of primitive African cultures once said, "There are places in Africa where three men cannot be sent on a journey together for fear two of them may combine and sell the third."

92. Malinowski, "An Anthropological Analysis of War," 257.
93. Freud, *Why War?*, 7.
94. T. E. Lawrence, *Seven Pillars of Wisdom*, 1926 (London: Penguin Books, 2000), 23.
95. Thomas Gregor, ed., *A Natural History of Peace* (Nashville, TN: Vanderbilt University Press, 1996), xvii.
96. John Milton Quotes, BrainyQuote.com, BrainyMedia, 2019, accessed August 6, 2016, www .brainyquote.com/quotes/john_milton_110201.
97. Collier, *Wars, Guns, and Votes*, 170.
98. Keegan, *A History of Warfare*, 79.
99. Ibid., 28.
100. Ibid., 29.
101. Ibid.
102. Clausewitz, *On War*, 76.
103. Keegan, *A History of Warfare*, 30–32.
104. T. R. Fehrenbach, *This Kind of War: The Classic Korean War History* (Washington D.C.: Brassey's, 1994), 455.
105. Adolf Hitler Quotes, BrainyQuote.com, BrainyMedia Inc, 2019, accessed October 6, 2019, www.brainyquote.com/quotes/adolf_hitler_386670.
106. Mahatma Gandhi Quotes, BrainyQuote.com, BrainyMedia Inc, 2019, accessed October 6, 2019, www.brainyquote.com/quotes/mahatma_gandhi_166284.
107. Gat, *War in Human Civilization*, 38.
108. Clausewitz, *On War*, 76.
109. Michel de Montaigne Quotes, Goodreads, accessed October 6, 2019, www.goodreads.com/au thor/quotes/17241.Michel_de_Montaigne?page=3.
110. Enoch Powell, *The Military Quotation Book*, ed. James Charlton (New York: St. Martin's Press, 2013), 137.
111. Asimov's *Foundation* (1951) features a scientific discipline, "psychohistory," that grants precognition of future events, like wars.
112. Jean-Jacques Rousseau, *The State of War*, in *The Ethics of War: Classic and Contemporary Readings*, eds. Gregory M. Reichberg, Henrik Syse, and Endre Begby (Malden, MA: Blackwell Publishing, 2006), 487.
113. LeShan, *The Psychology of War*, 6.
114. Kant, *Toward Perpetual Peace*, 15.
115. Hall, *The Nature of War*, 17.
116. LeShan, *The Psychology of War*, 6.
117. Attributed to Brant Parker and Johnny Hart, "Wizard of Id," *Dallas Morning News*, May 3, 1965.
118. LeShan, *The Psychology of War*, 6.
119. *Encyclopedia Britannica Online*, s.v. "security dilemma," accessed December 11, 2018, www.britannica.com/topic/security-dilemma. The *security dilemma* is "a situation in which actions taken by a state to increase its own security cause reactions from other states, which in turn lead to a decrease rather than an increase in the original state's security. Some scholars of international relations have argued that the security dilemma is the most important source of conflict between states."
120. Montesquieu, *The Spirit of the Laws*, in *The Ethics of War: Classic and Contemporary Readings*, eds. Gregory M. Reichberg, Henrik Syse, and Endre Begby (Malden, MA: Blackwell Publishing, 2006), 478.
121. Waltz, *Man, the State, and War*, 12.
122. Durbin and Bowlby, "Personal Aggressiveness and War," 85–86, 101.
123. Cashman, *What Causes War?*, 40–41.
124. May, "War, Peace, and Social Learning," 152.

125. G. W. F. Hegel, *Philosophy of Right*, in *The Ethics of War: Classic and Contemporary Readings*, eds. Gregory M. Reichberg, Henrik Syse, and Endre Begby (Malden, MA: Blackwell Publishing, 2006), 546.
126. Waltz, *Man, the State, and War*, 12.
127. Richard Lovelace, "Song to Lucasta, Going to the Wars," in *The Best Poems of the English Language: From Chaucer through Frost*, ed. Harold Bloom (New York: HarperCollins, 2004), 166.
128. Malinowski, "An Anthropological Analysis of War," 255.
129. Ibid.
130. Thucydides, Robert B. Strassler, and Richard Crawley, eds., *The Landmark Thucydides: A Comprehensive Guide to the Peloponnesian War* (New York: Free Press, 1996), 43.
131. Clausewitz, *On War*, 549.
132. Thucydides, *The Peloponnesian War*, trans. Steven Lattimore (Indianapolis, IN: Hackett Publishing Company, 1998), 57.
133. Durbin and Bowlby, "Personal Aggressiveness and War," 89.
134. Hall, *The Nature of War*, 20–21.
135. Waltz, *Man, the State, and War*, 203.
136. Plato, *The Republic*, trans. Benjamin Jowett (London: Oxford University Press, 1892; Project Gutenberg, 2017) www.gutenberg.org/files/55201/55201-h/55201-h.htm29.
137. Dave Grossman and Loren W. Christensen, *On Combat: The Psychology and Physiology of Deadly Conflict in War and Peace* (PPCT Research Publications, 2004). This analogy was coined by an unnamed Vietnam veteran and written about by Dave Grossman.
138. Machiavelli, *The Art of War*, 129.
139. Ibid., 127.
140. Thucydides, *The Peloponnesian War*, 341.
141. Liddell Hart, *Strategy*, 135–136.
142. Clausewitz, *On War*, 244.
143. Liddell Hart, *Strategy*, 211.
144. Thucydides, *The Peloponnesian War*, 215–216.
145. Durbin and Bowlby, "Personal Aggressiveness and War," 88–89.
146. Julia Ward Howe, "Battle Hymn of the Republic," in *The Best Poems of the English Language: From Chaucer through Frost*, ed. Harold Bloom (New York: HarperCollins, 2004), 526.
147. Keegan and Darracott, *The Nature of War*, 45.
148. Peter Ochs quoted in Ernie Gates, "In the Name of God, " *University of Virginia Magazine*, Winter 2015, 44.
149. Durbin and Bowlby, "Personal Aggressiveness and War," 89.
150. Ibid., 90. On this topic, Durbin and Bowlby write, "we think it difficult to exaggerate the frequency and importance of this cause of fighting in human societies of all degrees of civilization."
151. Mahatma Gandhi Quotes, BrainyQuote.com, BrainyMedia, accessed August 3, 2016, www.brainyquote.com/quotes/mahatma_gandhi_135298.
152. Sumner, "War," 219.
153. Peter Ochs quoted in Gates, "In the Name of God," 44.
154. George S. Patton Quotes, BrainyQuote.com, BrainyMedia Inc, 2019, accessed October 6 2019, www.brainyquote.com/quotes/george_s_patton_143694.
155. Bernard Knox, "Introduction," in Homer, *The Iliad*, trans. Robert Fagles (New York: Penguin Books, 1990), 29.
156. LeShan, *The Psychology of War*, 22.
157. Ibid., 73–87.
158. Leon Bramson and George W. Goethals ed., *War: Studies from Psychology, Sociology Anthropology* (New York: Basic Books, 1968), 297.

159. Ibid., 297–298.
160. Thomas Babington Macaulay, "Horatius at the Bridge," in *Lays of Ancient Rome* (London: Longman, Brown, Green, and Longman, 1842; University of Toronto Libraries), accessed 16 July 2019, https://rpo.library.utoronto.ca/poems/horatius.
161. Machiavelli, *The Art of War*, 128.
162. Ibid.
163. Mao, *On Guerrilla Warfare*, 88.
164. Ralph D. Sawyer, trans., "The Methods of the Ssu-ma," in *The Seven Military Classics of Ancient China* (New York: Basic Books, 2007), 136.
165. Antoine-Henri de Jomini, *The Art of War*, trans. Captain G. H. Mendell and Lieutenant W. P. Craighill (Radford, VA: Wilder Publications, 2008), 30.
166. John Horgan, *The End of War* (San Francisco: McSweeney's Publishing, 2014), 117–118. In a 1990 *Science* article, Herbert Simon speculated that humans are capable of "bounded rationality," which subordinates personal interests to the leader's interests.
167. Gat, *A History of Military Thought*, 539.
168. Adolf Hitler Quotes, BrainyQuote.com, BrainyMedia, accessed August 3, 2016, www .brainyquote.com/quotes/adolf_hitler_408979.
169. Freedman, *Strategy*, 37.
170. Saul David, *Military Blunders: The How & Why of Military Failure* (New York: Skyhorse Publishing, 2012), 157.
171. Patrick Henry, "Give Me Liberty or Give Me Death" (speech, Richmond, VA March 23, 1775), The Colonial Williamsburg Foundation, accessed February 24, 2019, www.history.org/alman ack/life/politics/giveme.cfm.
172. George Washington, "General Orders, 2 July 1776," Founders Online, National Archives, accessed February 24, 2019, https://founders.archives.gov/documents/Washington/03-05-0 2-0117. [Original source: The Papers of George Washington, Revolutionary War Series, vol. 5, 16 June 1776 – 12 August 1776, ed. Philander D. Chase (Charlottesville: University Press of Virginia, 1993), 179–182.]
173. Woodrow Wilson, "Peace without Victory" (speech, Washington D.C., November 19, 1863), Digital History, accessed February 24, 2019, www.digitalhistory.uh.edu/disp_textbook.cfm? smtID=11&psid=3824.
174. Winston Churchill, "Their Finest Hour" (speech, London, June 18, 1940), The International Churchill Society accessed February 24, 2019, https://winstonchurchill.org/resources/spee ches/1940-the-finest-hour/their-finest-hour.
175. Albert Einstein Quotes, Goodreads, accessed October 6, 2019, www.goodreads.com/author/ quotes/9810.Albert_Einstein?page=2.
176. George Washington, *The Papers of George Washington Digital Edition* (Charlottesville: University of Virginia Press, Rotunda, 2008), accessed August 12, 2018, https://rotunda .upress.virginia.edu/founders/GEWN-05-04-02-0361.
177. Theodore Roosevelt, *The Works of Theodore Roosevelt* (New York: Scribner, 1906), accessed August 12, 2018, https://play.google.com/books/reader?id=3z8OAAAAIAAJ&hl=en&pg= GBS.PA1, 70.
178. Fehrenbach, *This Kind of War*, 456.
179. Morris, "The Slaughter Bench of History."
180. Cicero, "Cicero: Letters to and from Cassius 44 B.C. (Attalus), accessed September 3, 2018, www.attalus.org/translate/cassius.html.
181. Knox, "Introduction," in Homer, *The Iliad*, 37.
182. Jeremy Black, "What Is War?," *Defence-In-Depth Research from the Defence Studies Department, King's College London* (blog), accessed November 3, 2018, https://defencein depth.co/2018/06/11/what-is-war.
183. Gray, *Fighting Talk,* 32.

184. Here, "peace" means all conditions short of war, but we will update this inadequate definition later in the chapter.
185. Erik H. Erikson, "Wholeness and Totality," in *War: Studies from Psychology, Sociology, Anthropology*, eds. Leon Bramson and George W. Goethals (New York: Basic Books, 1968), 120.
186. Raymond C. Kelly, *Warless Societies and the Origin of War* (Ann Arbor: The University of Michigan Press, 2000), 7.
187. Ibid.
188. Clausewitz, *On War*, 75. The Latin word for war, *bellum*, originates from *duellum*, the root of "duel."
189. Ibid., 86.
190. Ibid., 245.
191. Ibid., 75.
192. Ibid., 87.
193. Ibid., 89.
194. Ibid., 85.
195. Ibid., 77.
196. Ibid., 127.
197. Ibid., 596.
198. Ibid., 619.
199. Thucydides, *The Peloponnesian War*, 367.
200. Ibid., 389.
201. Park, "The Social Function of War," 244.
202. Malinowski, "An Anthropological Analysis of War," 247. Imbedded quotes reprinted from "The Deadly Issue," Atlantic Monthly, CLIX (December 1936), 659–669.
203. Colin S. Gray, *Another Bloody Century: Future Warfare* (London: Weidenfeld & Nicolson, 2005), 39.
204. Martha Crenshaw, "Why Violence is Rejected or Renounced: A Case Study of Oppositional Terrorism," in *A Natural History of Peace*, ed. Thomas Gregor (Nashville, TN: Vanderbilt University Press, 1996), 250.
205. Aron, *Clausewitz*, 382.
206. Clausewitz, *On War*, 95.
207. Martin van Creveld, *The Transformation of War: The Most Radical Reinterpretation of Armed Conflict Since Clausewitz* (New York: The Free Press, 1991), 159–160.
208. Galula, *Counterinsurgency Warfare*, 43.
209. Confucius, in "Human Conflicts Come from Structural Disequilibrium," *Battison's Blog* (blog), accessed October 6, 2019, https://biggestquestions.com/2014/04/19/human-conflicts-come-from-structural-disequilibrium/.
210. Sumner, "War," 212.
211. Thucydides, *The Landmark Thucydides*, 282.
212. Baruch Spinoza, *Political Treatise*, in *The Ethics of War: Classic and Contemporary Readings*, eds. Gregory M. Reichberg, Henrik Syse, and Endre Begby (Malden, MA: Blackwell Publishing, 2006), 453.
213. Codevilla and Seabury, *War*, 251.
214. Clausewitz, *On War*, 80.
215. Ibid., 90.
216. Kant, *Toward Perpetual Peace*, 10.
217. Liddell Hart, *Strategy*, 66.
218. Sun Tzu, *The Art of War*, 73.
219. Codevilla and Seabury, *War*, 276.

220. Samuel von Pufendorf, *On the Duty of Man and Citizen*, in *The Ethics of War: Classic and Contemporary Readings*, eds. Gregory M. Reichberg, Henrik Syse, and Endre Begby (Malden, MA: Blackwell Publishing, 2006), 460.

221. Donald Stoker, *Clausewitz: His Life and Work* (Oxford: Oxford University Press, 2014), 12.

222. Brandon Sanderson, *Oathbringer* (New York: Tor, 2017), 663.

223. John A. Vasquez, "Understanding Peace: Insights from International Relations Theory and Research," in *A Natural History of Peace*, ed. Thomas Gregor (Nashville, TN: Vanderbilt University Press, 1996), 273.

224. Sun Tzu, Huang, trans., *Sun Tzu: The New Translation* (New York: William Morrow Company, Inc., 1993), 53.

225. Clausewitz, *On War*, 234.

226. Ibid., 233–234.

227. Liddell Hart, *Strategy*, xvii.

228. Fred Charles Iklé, *Every War Must End* (New York: Columbia University Press, 1991), 84.

229. John A. Lynn, "A Quest for Glory: The Formation of Strategy under Louis XIV, 1661–1715," in *The Making of Strategy: Rulers, States, and War*, eds. Williamson Murray, MacGregor Knox, and Alvin Bernstein (Cambridge University Press, 2009), 203–204.

230. To be fair, the US Congress had not yet ratified the treaty. However, had both sides known of it beforehand, the battle would probably not have occurred since the British intended to use the attack for diplomatic leverage.

231. Freedman, *Strategy*, 611–612.

232. Asimov posed three laws governing robots: A robot: (1) may not injure a human being or, through inaction, allow a human being to come to harm; (2) must obey the orders given it by human beings except where such orders conflict with the First Law; and (3) must protect its own existence as long as that does not conflict with the First or Second Laws. These first appeared in Asimov's 1942 short story, *Runaround*.

233. Ironically, even the people of More's *Utopia* fought wars, though reluctantly, with the Alaopolitans.

234. In fact, there are no known examples of truly utopian societies at any scale, though some small, isolated, primitive groups, like the Yanadi of India, the Semai of Malaysia, and the Mbuti of the Congo, are remarkably peaceful.

235. Galula, *Counterinsurgency Warfare*, 17.

236. Malinowski, "An Anthropological Analysis of War," 263. Malinowski says totalitarianism "is the transformation of nationhood and all its resources into a lethal, 'technocratic' instrument of violence … [that] gradually saps all the resources of culture and destroys its structure."

2   The Masters of War Theory and Strategy

1. Frederick II, quoted in Jay Luvaas, ed. and trans., *Frederick the Great on the Art of War* (New York: Da Capo Press, 1999), 54.

2. Azar Gat, *A History of Military Thought*, 451.

3. "Wei Liao-Tzu," in *The Seven Military Classics of Ancient China*, trans. Sawyer, 246.

4. Sun Tzu, *The Art of War*, 77.

5. Ibid., 84.

6. Ibid., 93.

7. Ibid., 92.

8. Ibid., 88.

9. Ibid., 66.

10. Ibid., 97.

11. Ibid., 77–78.

12. Ibid., 79–80.

13. Ibid., 101.

14. Ibid., 91.

15. Ibid.

16. Ibid., 79.

17. Freedman, *Strategy*, 30.

18. Thucydides, *The Peloponnesian War*, 16.

19. The *Delian* and *Peloponnesian* Leagues were alliances of hundreds of Greek city-states behind Athens and Sparta, respectively.

20. Murray, *War, Strategy, and Military Effectiveness*, 53.

21. Thucydides, *The Peloponnesian War*, 43.

22. Ibid., 16.

23. Ibid., 20.

24. Actually: "misery acquaints a man with strange bedfellows"; *The Tempest* 2.2.

25. Thucydides, *The Peloponnesian War*, 35.

26. Ibid., 43.

27. Ibid., 3.

28. Ibid., 6–7.

29. Thucydides, *The Landmark Thucydides*, 352.

30. Thucydides, *The Peloponnesian War*, 38.

31. Ibid.

32. *Encyclopaedia Britannica Online*, s.v. "balance of power," accessed March 26, 2017, www .britannica.com/topic/balance-of-power. The phrase "balance of power" came into common use to describe relations between European powers during Napoleon's era.

33. Thucydides, *The Peloponnesian War*, 59.

34. Ibid., 103.

35. Ibid., 148.

36. Ibid., 152.

37. Ibid., 152–153.

38. Ibid., 158.

39. Ibid., 52.

40. Ibid., 79.

41. Ibid., 68.

42. Ibid., 57.

43. Ibid., 58.

44. Ibid., 258.

45. Ibid., 342.

46. Durbin and Bowlby, "Personal Aggressiveness and War," 88.

47. Thucydides, *The Peloponnesian War*, 169.

48. Ibid., 158.

49. Ibid., 169.

50. Ibid., 119.

51. Ibid.

52. Ibid., 120.

53. Ibid., 312.

54. Ibid., 319.

55. Ibid., 324.

56. Francois-Marie Arouet Voltaire, "Battalion" in *A Philosophical Dictionary*, tr. from the French (London: W. Dugale, 1843), quoted in Machiavelli, *The Art of War*, xxxii.

57. Machiavelli, *The Prince*, 37.

58. Wood, "Introduction," in Machiavelli, *The Art of War*, xxvi.

59. Ibid., x–xi.

60. Machiavelli, *The Art of War*, 23.
61. Freedman, *Strategy*, 50.
62. Machiavelli, *The Prince*, 12.
63. *Realpolitik* is a German word for politics grounded in practical, not moral, considerations.
64. Machiavelli, *The Prince*, 31.
65. Wood, "Introduction," in Machiavelli, *The Art of War*, xviii.
66. F. L. Taylor, *The Art of War in Italy, 1494–1529* (Cambridge: Cambridge University Press, 1921), 157.
67. Wood, "Introduction," in Machiavelli, *The Art of War*, xlvi.
68. Ibid., xiii.
69. Machiavelli, *The Prince*, 31.
70. Wood, "Introduction," in Machiavelli, *The Art of War*, xvii.
71. Machiavelli, *The Art of War*, 31.
72. Machiavelli, *The Prince*, 12.
73. Wood, "Introduction," in Machiavelli, *The Art of War*, xxiv.
74. Ibid., xxv.
75. Machiavelli, *The Art of War*, 79.
76. Ibid.
77. Wood, "Introduction," in Machiavelli, *The Art of War*, xxv.
78. Ibid.
79. Machiavelli, *The Prince*, 66.
80. Wood, "Introduction," in Machiavelli, *The Art of War*, lv.
81. Ibid., l.
82. Machiavelli, *The Art of War*, 13.
83. Ibid., 24.
84. Ibid., 41.
85. Ibid., 124.
86. Ibid.
87. Ibid.
88. Ibid., 173.
89. Ibid., 178.
90. Ibid., 120.
91. Ibid., 121.
92. Ibid., 88.
93. Ibid., 91–92.
94. Ibid., 95.
95. Ibid.
96. Ibid. "Velites" were a form of very mobile, quick, javelin-armed Roman light infantry.
97. Ibid., 99.
98. Niccolò Machiavelli, *The Discourses*, 1531, trans. Leslie J. Walker (New York: Penguin Books, 1986), 322.
99. Machiavelli, *The Art of War*, 101.
100. Machiavelli, *The Prince*, 39.
101. Machiavelli, *The Art of War*, 202–204.
102. John Shy, "Jomini," in *Makers of Modern Strategy from Machiavelli to the Nuclear Age*, ed. Peter Paret (Princeton, NJ: Princeton University Press, 1986), 163.
103. Jomini, *The Art of War*, 261.
104. Shy, "Jomini," in *Makers of Modern Strategy*, 147.
105. Ibid., 153.
106. Jomini, *The Art of War*, 126.
107. Ibid., 15.
108. Shy, "Jomini," in *Makers of Modern Strategy*, 148.

109. Jomini, *The Art of War*, 94.
110. Ibid., 247.
111. Aron, *Clausewitz*, 1.
112. Jomini, *The Art of War*, 7.
113. Ibid., 245.
114. Ibid., 8.
115. Ibid.
116. Ibid., 133.
117. Ibid., 8.
118. Ibid., 245.
119. Ibid., 8.
120. Ibid., 10.
121. Ibid., 11.
122. Ibid.
123. Ibid., 16.
124. Ibid.
125. Ibid.
126. Ibid., 24.
127. Ibid., 17.
128. Ibid.
129. Ibid.
130. Ibid., 19–20.
131. Ibid., 20.
132. Ibid., 21.
133. Ibid., 22.
134. Ibid., 132.
135. Ibid., 22.
136. Ian F. W. Beckett and John Pimlott, eds., *Armed Forces & Modern Counterinsurgency* (New York: St. Martin's Press, 1985), 49. The British fought the Malayan Emergency from 1948–1960, eventually winning by isolating the insurgents from popular support, then negating their mobility with helicopter-borne assaults. After World War II, the USA helped the newly independent Philippines quell a quasi-communist peasant revolt (the Hukbalahap Rebellion or Insurrection) by implementing governmental and military reforms and aggressive, military-led operations.
137. Jomini, *The Art of War*, 24.
138. William Tecumseh Sherman, *The Military Quotation Book*, 45.
139. Jomini, *The Art of War*, 31.
140. Ibid., 32.
141. Ibid., 45.
142. Ibid., 36.
143. Ibid.
144. Ibid., 157.
145. Ibid., 85–86.
146. Ibid., 51.
147. Ibid.
148. Ibid., 52.
149. Ibid., 41.
150. Ibid., 42.
151. Ibid., 247.
152. Ibid., 96.
153. Ibid., 207.
154. Ibid.

155. Ibid., 211.
156. Ibid., 212.
157. Ibid., 52.
158. Ibid., 250.
159. Awareness, access, and freedom of action, the "three a's" of domain theory, are discussed further in the next chapter.
160. Jomini, *The Art of War*, 53.
161. Ibid.
162. Ibid., 143.
163. Ibid., 64–65.
164. Ibid., 66.
165. Ibid., 68.
166. Ibid., 67.
167. Ibid.
168. Ibid., 53.
169. Ibid., 54.
170. Ibid.
171. Ibid., 226. By "solidity," Jomini means cohesion and concentration, the ability to apply concentrated power.
172. Ibid., 55.
173. Ibid., 247.
174. Ibid., 74.
175. Ibid., 238.
176. Ibid., 239.
177. Ibid., 245.
178. Ibid.
179. Ibid., 252.
180. Ibid., 77.
181. Ibid., 75.
182. Ibid., 77.
183. Ibid., 89.
184. Ibid., 85.
185. Ibid., 86.
186. Ibid., 95.
187. Ibid.
188. Ibid., 105.
189. Ibid., 137. But Jomini acknowledges that at times victory can be won by strategic operations that do not entail pitched battles or by battles that are not preceded by "grand strategic combinations."
190. Ibid., 136.
191. Ibid.
192. Ibid.
193. Ibid., 246.
194. Ibid., 137.
195. Ibid., 142.
196. Ibid., 149.
197. Ibid.
198. Ibid., 104.
199. Ibid.
200. Ibid.
201. Ibid.
202. Ibid., 35.

203. Ibid., 271–272.

204. Shy, "Jomini," in *Makers of Modern Strategy*, 164.

205. Peter Paret, "Clausewitz," in *Makers of Modern Strategy from Machiavelli to the Nuclear Age*, ed. Peter Paret (Princeton, NJ: Princeton University Press, 1986), 213.

206. Brodie, "The Continuing Relevance of *On War*," in Clausewitz, *On War*, 53.

207. Clausewitz quoted in Stoker, *Clausewitz*, 9.

208. Ibid., 23. According to Stoker, "from 1792 to 1815 Europeans fought 713 battles. The previous three centuries had seen only 2,659."

209. Gat, *A History of Military Thought*, 142–151.

210. Ibid., 155.

211. Ibid., 168–169.

212. Ibid., 216. Of this 1827 epiphany, Gat says that Clausewitz's realization of the gulf between his perfect form of war and reality created a paradigm-shattering dilemma, because his "theory conflicted with reality; the 'concept of war' did not withstand the 'test of experience'; the universal contradicted the historical; [and] the unity of the phenomenon of war . . . disintegrated."

213. Clausewitz, *On War*, 583.

214. Ibid., 636.

215. Stoker, *Clausewitz*, 34–35.

216. Clausewitz, *On War*, 515.

217. Ibid., 260.

218. Paret, "Clausewitz," in *Makers of Modern Strategy*, 187.

219. Gat, *A History of Military Thought*, 228.

220. Ibid., 232.

221. Michael I. Handel, *Masters of War: Classical Strategic Thought* (New York, NY: Routledge, 2001), 24.

222. Gat, *A History of Military Thought*, 236.

223. Bernard Brodie, "A Guide to the Reading of *On War*," in Clausewitz, *On War*, 643.

224. Gat, *A History of Military Thought*, 228.

225. Liddell Hart, *Strategy*, 340.

226. Clausewitz, "Author's Preface," in Clausewitz, *On War*, 70. This is not to imply that Clausewitz had not revised other chapters. In particular, Book 8 reflects much of Clausewitz's later thinking.

227. Clausewitz, *On War*, 77.

228. Ibid., 580.

229. Ibid., 217.

230. Ibid., 86, 87.

231. Ibid., 87. Some would argue that politics is war by other means!

232. Ibid., 137.

233. Ibid., 88–89.

234. Ibid., 88.

235. Ibid., 88–89.

236. Ibid., 484.

237. Ibid., 183.

238. Clausewitz, "Two Notes by the Author," in Clausewitz, *On War*, 69.

239. Clausewitz, *On War*, 81. It is unclear how "armed observation" fits Clausewitz's concept of war as fighting.

240. Ibid., 84.

241. Ibid., 359.

242. Matthew Arnold, "Dover Beach," in *The Best Poems of the English Language: From Chaucer through Frost*, ed. Harold Bloom (New York: HarperCollins, 2004), 685.

243. Clausewitz, *On War*, 104.

244. Ibid., 101.
245. Ibid., 140.
246. Ibid., 108.
247. Ibid., 119.
248. Ibid., 118–119.
249. Ibid., 117.
250. Ibid., 120.
251. Ibid., 122.
252. Immanuel Kant quoted in Tom Leddy, "Kant on How to Be a Genius," *Aesthetics Today*, October 2, 2015, accessed October 10, 2019, http://aestheticstoday.blogspot.com/2015/10/kant-on-how-to-be-genius.html.
253. Clausewitz, *On War*, 136.
254. Ibid., 112. Clausewitz says that genius "easily grasps and dismisses a thousand remote possibilities which an ordinary mind would labor to identify and wear itself out in so doing."
255. Ibid., 102.
256. Ibid., 578.
257. Ibid., 108.
258. Ibid., 146.
259. Ibid., 579.
260. Ibid., 386.
261. Ibid.
262. Ibid., 595–596.
263. Charles XII of Sweden ruled from 1697 to 1718 and was a skilled battle commander.
264. Clausewitz, *On War*, 596.
265. Ibid., 485.
266. Ibid., 345.
267. "Wei Liao-Tzu," in *The Seven Military Classics of Ancient China*, trans Sawyer, 262.
268. Clausewitz, *On War*, 524.
269. Ibid., 528.
270. Ibid., 383.
271. Ibid., 570.
272. Ibid., 572.
273. Ibid., 76.
274. Ibid., 91.
275. Ibid., 93.
276. Ibid., 94.
277. Ibid., 99.
278. Ibid., 483.
279. Ibid., 479.
280. Ibid.
281. Ibid.
282. Ibid., 480.
283. Ibid., 350.
284. Ibid., 349.
285. Ibid., 480.
286. Ibid., 482.
287. In reading this chapter, one can sense Clausewitz's frustration with Prussia at not having chosen to fight Napoleon as the Spanish did.
288. Clausewitz, *On War*, 350.
289. Ibid., 481.
290. Clausewitz, "Two Notes by the Author," in Clausewitz, *On War*, 70.
291. Corbett, *Some Principles of Maritime Strategy*, 40.

292. Clausewitz, *On War*, 78.
293. Brodie, "A Guide to the Reading of *On War*," in Clausewitz, *On War*, 643.
294. Liddell Hart, *Strategy*, 341.
295. Gat, *A History of Military Thought*, 228.
296. Clausewitz, *On War*, 606.
297. Ibid., 582.
298. Liddell Hart, *Strategy*, 334.
299. Clausewitz, *On War*, 579.
300. Ibid., 217.
301. Ibid., 142.
302. Ibid., 580–581.
303. Corbett, *Some Principles of Maritime Strategy*, 40.
304. Clausewitz, *On War*, 217.
305. Ibid., 606.
306. Ibid., 218.
307. Ibid., 216.
308. Ibid., 604.
309. Ibid., 87.
310. Ibid., 88.
311. Ibid., 91.
312. Ibid., 149.
313. Ibid., 616.
314. Ibid., 358.
315. Ibid., 524.
316. Ibid., 377.
317. Clausewitz, "Two Notes by the Author," in Clausewitz, *On War*, 70.
318. Gat, *A History of Military Thought*, 335.
319. Aron, *Clausewitz*, 151.
320. Clausewitz, *On War,* 601.
321. Gat, *A History of Military Thought*, 671.
322. Gray, *War, Peace and International Relations*, 281. This is sometimes called a "revolution in military affairs" (RMA). Gray writes, "one of the lessons of strategic history is that RMAs are always eventually emulated, offset or evaded. There is no final move."
323. Gat, *A History of Military Thought*, 265.
324. Clausewitz, *On War*, 220.
325. Ibid.
326. Peter Paret, *Clausewitz and the State: The Man, His Theories, and His Times* (Princeton, NJ: Princeton University Press, 1985), 361.
327. Clausewitz, *On War*, 246.
328. Ibid., 322.
329. Keegan, *A History of Warfare*, 23.
330. Clausewitz, *On War*, 205.
331. Ibid.
332. Ibid., 206–207.
333. Ibid., 209.
334. Ibid., 227.
335. Ibid., 228.
336. Ibid., 386.
337. Ibid., 227.
338. Ibid.
339. Harry Summers quoted in Gray, *Fighting Talk,* 110.
340. Liddell Hart, *Strategy*, 319.

341. Clausewitz, *On War*, 616.
342. Ibid., 89.
343. Handel, *Masters of War*, 102–117. Handel, a former professor at the US Naval War College, expounds on trinitarian analysis in his otherwise excellent book, *Masters of War*.
344. Clausewitz, *On War*, 89.
345. Paret, "Clausewitz," in *Makers of Modern Strategy*, 202.
346. Clausewitz, *On War*, 89.
347. Gray, *War, Peace and International Relations*, 253.
348. Stoker, *Clausewitz*, 281.
349. Ibid., 285.
350. Gat, *A History of Military Thought*, 255.
351. Michael Howard, "The Influence of Clausewitz," in Carl von Clausewitz, *On War*, eds. and trans. Michael Howard and Peter Paret (Princeton, NJ: Princeton University Press, 1976), 44.
352. B. H. Liddell Hart Quotes, BrainyQuote.com, BrainyMedia Inc, 2019, accessed October 18, 2019, www.brainyquote.com/quotes/b_h_liddell_hart_145008.
353. Gat, *A History of Military Thought*, 516.
354. Liddell Hart, *Strategy*, 209.
355. Ibid., 343.
356. Gat, *A History of Military Thought*, 558.
357. Ibid.
358. Ibid., 822.
359. Ibid., 646.
360. Ibid., 663.
361. Liddell Hart, *Strategy*, xix.
362. Ibid., xx.
363. Ibid., 4.
364. Ibid.
365. Ibid., 146.
366. Ibid., 41.
367. Ibid., 10.
368. Ibid., 322.
369. Ibid.
370. Ibid., 212.
371. Ibid., 189.
372. Gat, *A History of Military Thought*, 756.
373. Liddell Hart, *Strategy*, 10.
374. Ibid., 321.
375. Ibid., 211.
376. Ibid., 338.
377. Ibid., 324.
378. Ibid., 323.
379. Ibid., 325.
380. Ibid., 204.
381. Ibid., 279.
382. Ibid., 212–213.
383. Ibid., 5.
384. Ibid., 212.
385. Ibid., 5.
386. Ibid., 6.
387. In boxing, this tactic is known as a "rope-a-dope."
388. Liddell Hart, *Strategy*, 309.
389. Ibid., 283.

390. Ibid., 88.
391. Ibid.
392. Ibid., 212.
393. Ibid.
394. Ibid., 212–213.
395. Ibid., 146.
396. Ibid., 197.
397. Ibid., 190.
398. Ibid., 330.
399. Ibid., 146.
400. Ibid., 147.
401. Ibid., 326.
402. Ibid., 327.
403. Ibid.
404. Ibid., 329.
405. Ibid., 328.
406. Ibid., 329.
407. Ibid., 10.
408. Ibid., 345.
409. Ibid., 346.
410. Ibid.
411. Ibid., 26–27.
412. Ibid., 27.
413. Ibid.
414. Ibid., 71.
415. Ibid., 77.
416. Ibid., 94–95.
417. Ibid., 95.
418. Ibid.
419. Ibid., 132.
420. Mike Murphy, "Vox Ignoramus: On Issues of War and Peace, Public Opinion Has Proven an Unreliable Guide," *The Weekly Standard*, March 24, 2003, www.weeklystandard.com/mike-murphy/vox-ignoramus.
421. Liddell Hart, *Strategy*, 132.
422. Ibid., 134.
423. Ibid., 135.
424. Ibid., 181.
425. Ibid.
426. Ibid.
427. Ibid., 183.
428. Ibid., 207.
429. Ibid.
430. Ibid., 208.
431. Ibid., 248.
432. Ibid., 249.
433. Ibid., 334.
434. Ibid., 336.
435. Ibid., 335–336.
436. Ibid.
437. Ibid., 363.
438. Ibid., 364.
439. Ibid., 365.

440. Ibid.
441. Ibid.
442. Ibid., 366.
443. Ibid., 368.
444. Ibid., 369–370.
445. Ibid., xv.
446. Ibid., xviii.
447. Ibid., xv.
448. Ibid., xviii–xix.
449. Ibid., xv.
450. Ibid., 144.
451. Ibid., 329.
452. Ibid., 332.
453. Ibid., 336. Liddell Hart clarifies this point later in his section on maxims.
454. Aron, *Clausewitz*, 239.
455. Samuel B. Griffith, "Translator's Note," in Mao, *On Guerrilla Warfare*, 37.
456. Mao, *On Guerrilla Warfare*, 12.
457. Ibid., 41.
458. Samuel B. Griffith, "Introduction," in Mao, *On Guerrilla Warfare*, 25.
459. Mao, *On Guerrilla Warfare*, 46.
460. Griffith, "Introduction," in Mao, *On Guerrilla Warfare*, 7.
461. Mao, *On Guerrilla Warfare*, 42.
462. Ibid., 41.
463. Ibid.
464. Griffith, "Introduction," in Mao, *On Guerrilla Warfare*, 21.
465. Ibid., 22.
466. Mao, *On Guerrilla Warfare*, 42.
467. Ibid., 51.
468. Ibid., 52.
469. Ibid., 53.
470. Ibid., 54.
471. Ibid., 57.
472. Ibid., 47.
473. Ibid.
474. Confucius Quotes, AZ Quotes, accessed October 19, 2019, www.azquotes.com/quote/379676.
475. Mao, *On Guerrilla Warfare*, 46.
476. Griffith, "Introduction," in Mao, *On Guerrilla Warfare*, 23.
477. Ibid.
478. Mao, *On Guerrilla Warfare*, 89.
479. Ibid., 43.
480. Ibid., 88.
481. Ibid., 43.
482. Ibid., 90.
483. Ibid., 45.
484. Ibid., 86–87.
485. Ibid., 86.
486. Ibid., 66.
487. Ibid., 67.
488. Ibid., 95.
489. Ibid., 96.
490. Ibid., 102.

491. Ibid., 103.

492. Ibid., 98.

493. Handel, *Masters of War*, 52.

494. David G. Chandler, ed., *The Military Maxims of Napoleon*, trans. George C. D'Aguilar (London: Lionel Leventhal, 2002), 82. Napoleon's maxim actually asks that we study the campaigns of Alexander, Hannibal, Caesar, Gustavus Adolphus, Turenne, Eugene, and Frederick so that we may become great commanders.

3   Small Wars and Domain Theory

1. Lee Smolin, *The Life of the Cosmos* (London: Weidenfeld & Nicolson, 1997), 298.

2. Corbett, *Some Principles of Maritime Strategy*, 9.

3. US Marine Corps, *Small Wars Manual: Fleet Marine Force Reference Publication 12–15*, 1940 (Washington: US Government Printing Office, 1990), 8.

4. Horatio Nelson was a legendary British fleet admiral known for his decisive victories at sea, including a stunning triumph in 1805 over Napoleon's fleet at Trafalgar (that cost Nelson his life).

5. C. E. Callwell, *Small Wars: Their Principles and Practice*, 3rd ed. 1906 (Seattle, WA: Book Jungle, 2009), 23.

6. The British journalist, Alistair Horne, coined the phrase "savage war of peace" in his titular work, *A Savage War of Peace* (Viking Press, 1977) on the French war in Algeria. Military historian Max Boot also used it in *The Savage Wars of Peace: Small Wars and the Rise of American Power* (MIT Press, 2002).

7. Heuser, *The Evolution of Strategy*, 217.

8. Roger Trinquier, *Modern Warfare: A French View of Counterinsurgency*, 1964 (Reprint, Westport, CT: Praeger Security International, 2006), 34.

9. Callwell, *Small Wars*, 21.

10. US Marine Corps, *Small Wars Manual*, 8–9.

11. T. E. Lawrence, "The 27 Articles of T. E. Lawrence," World War I Document Archive, from *The Arab Bulletin*, 20 Aug 1917, accessed March 14, 2020, https://wwi.lib.byu.edu/index.php/The_27_Articles_of_T.E._Lawrence.

12. Trinquier, *Modern Warfare*, 89.

13. Callwell, *Small Wars*, 27.

14. Heuser, *The Evolution of Strategy*, 389.

15. Callwell, *Small Wars*, 21.

16. Aron, *Clausewitz*, 227.

17. Trinquier, *Modern Warfare*, 5.

18. US Marine Corps, *Small Wars Manual*, 1.

19. Lawrence, *Seven Pillars of Wisdom*, 198.

20. Galula, *Counterinsurgency Warfare*, 4.

21. Mao, *On Guerrilla Warfare*, 88.

22. Galula, *Counterinsurgency Warfare*, 62.

23. US Marine Corps, *Small Wars Manual*, 15.

24. Galula, *Counterinsurgency Warfare*, 21.

25. Callwell, *Small Wars*, 44.

26. Ibid., 98.

27. US Marine Corps, *Small Wars Manual*, 58.

28. Bard O'Neill, *Insurgency & Terrorism: From Revolution to Apocalypse*, 2nd ed. (Lincoln, NE: Potomac Books, Inc., 2005), 72.

29. Trinquier, *Modern Warfare*, 42.

30. Galula, *Counterinsurgency Warfare*, 23.

31. Clausewitz, *On War*, 480.
32. T. E. Lawrence, "On Guerrilla Warfare," *Encyclopedia Britannica*, 14th ed. (London: Encyclopedia Britannica, 1929), accessed March 14, 2020, www.britannica.com/topic/T-E-Lawrence-on-guerrilla-warfare-1984900.
33. Galula, *Counterinsurgency Warfare*, 58–59.
34. US Army and US Marine Corps, *Counterinsurgency Field Manual 3–24*, 2006, edited by John McClure (Kissimmee, FL: Signalman Publishing, 2009), 2.
35. Mao, *On Guerrilla Warfare*, 42–43.
36. Christopher Paul, Colin P. Clarke, Beth Grill, and Molly Dunigan, *Paths to Victory: Lessons from Modern Insurgencies* (Santa Monica, CA: RAND Corporation, 2013), 7, accessed March 26, 2020, www.rand.org/pubs/research_reports/RR291z1.html.
37. Ibid., 158.
38. Trinquier, *Modern Warfare*, 45.
39. Galula, *Counterinsurgency Warfare*, 30–40.
40. US Army and US Marine Corps, *Counterinsurgency Field Manual 3–24*, x.
41. Ibid., 2.
42. O'Neill, *Insurgency & Terrorism*, 89.
43. Mao, *On Guerrilla Warfare*, 45.
44. Ibid., 43.
45. Trinquier, *Modern Warfare*, 6.
46. Ibid., 5.
47. O'Neill, *Insurgency & Terrorism*, 33.
48. Trinquier, *Modern Warfare*, 15.
49. O'Neill, *Insurgency & Terrorism*, 139.
50. Lawrence, *Seven Pillars of Wisdom*, 32.
51. Ibid., 44.
52. Lawrence, "The 27 Articles of T. E. Lawrence."
53. Lawrence, *Seven Pillars of Wisdom*, 202.
54. Galula, *Counterinsurgency Warfare*, 44.
55. Callwell, *Small Wars*, 42.
56. US Marine Corps, *Small Wars Manual*, 15.
57. Ibid., 32.
58. Paul, "Paths to Victory," 172. RAND found that counterinsurgencies focused only on combat ("iron fist" cases) failed 62 percent of the time.
59. David Kilcullen, *Counterinsurgency* (Oxford: Oxford University Press, 2010), 3–4.
60. Trinquier, *Modern Warfare*, 51.
61. O'Neill, *Insurgency & Terrorism*, 155.
62. Callwell, *Small Wars*, 43.
63. Trinquier, *Modern Warfare*, 23.
64. Ibid., 31.
65. Ibid., 84.
66. O'Neill, *Insurgency & Terrorism*, 190–191.
67. Paul, "Paths to Victory," 7.
68. Galula, *Counterinsurgency Warfare*, 55.
69. Trinquier, *Modern Warfare*, 54.
70. Ibid., 28.
71. US Marine Corps, *Small Wars Manual*, 9.
72. Ibid., 37.
73. Ibid., 25.
74. Ibid., 7.
75. Galula, *Counterinsurgency Warfare*, 53.
76. Ibid., 81.

77. Lawrence, "On Guerrilla Warfare."
78. Kilcullen, *Counterinsurgency*, 4.
79. US Marine Corps, *Small Wars Manual*, 56.
80. Lawrence, *Seven Pillars of Wisdom*, 202.
81. Mao, *On Guerrilla Warfare*, 52.
82. Trinquier, *Modern Warfare*, 74.
83. Lawrence, *Seven Pillars of Wisdom*, 200.
84. Ibid., 346.
85. Liddell Hart, *Strategy*, 365.
86. Lawrence, *Seven Pillars of Wisdom*, 347.
87. Ibid., 140.
88. Galula, *Counterinsurgency Warfare*, 6.
89. Ibid., 9.
90. Callwell, *Small Wars*, 125.
91. Trinquier, *Modern Warfare*, 45.
92. Callwell, *Small Wars*, 33.
93. Ibid., 38.
94. Galula, *Counterinsurgency Warfare*, 50.
95. Callwell, *Small Wars*, 133.
96. Trinquier, *Modern Warfare*, 69.
97. Galula, *Counterinsurgency Warfare*, 70.
98. Callwell, *Small Wars*, 93.
99. Ibid., 71–72.
100. Ibid., 109.
101. Ibid., 112.
102. Douglas Porch, "Bugeaud, Galliéni, Lyautey: The Development of French Colonial Warfare," in *Makers of Modern Strategy from Machiavelli to the Nuclear Age*, ed. Peter Paret (Princeton, NJ: Princeton University Press, 1986), 378.
103. Galula, *Counterinsurgency Warfare*, 65.
104. Trinquier, *Modern Warfare*, 37.
105. Ibid., 4.
106. Galula, *Counterinsurgency Warfare*, 78.
107. US Marine Corps, *Small Wars Manual*, 159.
108. Callwell, *Small Wars*, 41.
109. Paul, "Paths to Victory," 188.
110. Trinquier, *Modern Warfare*, 70–71.
111. Galula, *Counterinsurgency Warfare*, 93.
112. Callwell, *Small Wars*, 22.
113. Ibid., 23.
114. Ibid., 29.
115. *Intelligence* collects, analyzes, and disseminates information; *surveillance* is general observation; and *reconnaissance* gathers specific information on a region, location, or person.
116. Themistocles, *The Military Quotation Book*, 77.
117. Heuser, *The Evolution of Strategy*, 205–206.
118. Admiral Raoul V. P. Castex of France (1878–1968) is also noteworthy, though he came later. His voluminous writings (including *Théories stratégiques,* a multivolume work written between 1927 and 1935) are fundamentally Mahanian but also align with Corbett's view that strategy is more than air and naval dominance in great battles.
119. Gat, *A History of Military Thought*, 445–446. Mahan was also influenced by Jomini and von Moltke (the Elder).

120. Heuser, *The Evolution of Strategy*, 201. The concept of "sea power" can be traced back to Thucydides's "rule of the sea."
121. Alfred Thayer Mahan, *Influence of Sea Power Upon History, 1660–1783*, 1890 (Scott's Valley, CA: Pantianos Classics; Illustrated edition, 2016), 92.
122. Ibid., 5.
123. Alfred Thayer Mahan, *The Influence of Sea Power upon the French Revolution and Empire, 1793–1812*, 1892 (Scott's Valley, CA: Pantianos Classics; Illustrated edition, 2016), iv.
124. Gat, *A History of Military Thought*, 450.
125. Mahan, *Influence of Sea Power Upon History*, 9.
126. Ibid.
127. Ibid.
128. Ibid., 40.
129. Ibid., 17.
130. Ibid.
131. Ibid., 204.
132. Ibid., 58–59.
133. Ibid., 134.
134. Ibid., 212–213.
135. Ibid., 213.
136. Ibid., 164.
137. Ibid., 196–197.
138. Philip A. Crowl, "Alfred Thayer Mahan: The Naval Historian," in *Makers of Modern Strategy from Machiavelli to the Nuclear Age*, ed. Peter Paret (Princeton, NJ: Princeton University Press, 1986), 454. Crowl was head of the strategy department at the US Naval War College.
139. Ibid., 457.
140. Ibid., 461.
141. Gat, *A History of Military Thought*, 516.
142. Ibid., 479.
143. Heuser, *The Evolution of Strategy*, 214.
144. Gat, *A History of Military Thought*, 481.
145. Corbett, *Some Principles of Maritime Strategy*, 8.
146. Gat, *A History of Military Thought*, 486.
147. Corbett, *Some Principles of Maritime Strategy*, 3.
148. Ibid., 4.
149. Ibid., 4–5.
150. Ibid., 6–7.
151. Ibid., 14.
152. Ibid., 15–16.
153. Ibid., 27.
154. Ibid., 30.
155. Ibid., 37.
156. Ibid., 56.
157. Ibid., 57.
158. Ibid., 63.
159. Ibid., 160.
160. Ibid., 89–90.
161. Ibid., 218.
162. Ibid., 100.
163. Ibid., 94.
164. Ibid., 132.
165. Ibid., 153.
166. Ibid., 167.

167. Ibid., 213. "Fleet in being" traces to Admiral Arthur Herbert, Earl of Torrington (1647–1716), who developed it to counter France. Heuser, *The Evolution of Strategy*, 212.

168. Corbett, *Some Principles of Maritime Strategy*, 269.

169. Heuser, *The Evolution of Strategy*, 214.

170. John Hattendorf quoted in Heuser, *The Evolution of Strategy*, 281.

171. Ibid., 293.

172. Giulio Douhet quoted in Gat, *A History of Military Thought*, 571.

173. Phillip S. Meilinger, ed., *The Paths of Heaven: The Evolution of Airpower Theory* (Maxwell Air Force Base, AL: Air University Press, 1997), xi.

174. I. B. Holley, Jr., "Reflections on the Search for Airpower Theory," in *The Paths of Heaven: The Evolution of Airpower Theory*, ed. Phillip S. Meilinger (Maxwell Air Force Base, AL: Air University Press, 1997), 579. Holley was a co-founder of the US Air Force School of Advanced Air and Space Studies.

175. Ibid., 598–599.

176. John C. Slessor, *Air Power and Armies* (Tuscaloosa: The University of Alabama Press, 2009), 214.

177. Isaiah Berlin quoted in Gat, *A History of Military Thought*, 681.

178. Strachan, *The Direction of War*, 175.

179. Gat, *A History of Military Thought*, 571.

180. Ibid., 572.

181. Ibid., 579.

182. Mark A. Clodfelter, "Molding Airpower Convictions: Development and Legacy of William Mitchell's Strategic Thought," in *The Paths of Heaven: The Evolution of Airpower Theory*, edited by Phillip S. Meilinger (Maxwell Air Force Base, AL: Air University Press, 1997), 90.

183. Ibid., 84.

184. Gat, *A History of Military Thought*, 590.

185. Clodfelter, "Molding Airpower Convictions," 108.

186. H. R. Allen, *The Legacy of Lord Trenchard* (London: The Camelot Press, Ltd., 1972), 1. Allen was a British World War II fighter pilot who retired specifically to write this book.

187. Ibid., 35.

188. Phillip S. Meilinger, "Trenchard, Slessor, and Royal Air Force Doctrine before World War II," in *The Paths of Heaven: The Evolution of Airpower Theory*, ed. Phillip S. Meilinger (Maxwell Air Force Base, AL: Air University Press, 1997), 51.

189. Ibid., 72.

190. Ibid., 42.

191. Ibid.

192. William Mitchell, *Winged Defense: The Development and Possibilities of Modern Air Power, Economic and Military* (Mineola, NY: Dover Publications, Inc., 2006), 20.

193. Heuser, *The Evolution of Strategy*, 217.

194. Gat, *A History of Military Thought*, 562–563.

195. Giulio Douhet, *The Command of the Air*, trans. Dino Ferrari (Washington, DC: Office of Air Force History, 1983), 103.

196. Ibid., 26.

197. Ibid., 10.

198. Ibid., 18–19.

199. Ibid., 49.

200. Ibid., 51.

201. Mitchell, *Winged Defense*, xii.

202. Ibid., 101.

203. Allen, *The Legacy of Lord Trenchard*, 35.

204. Ibid., 48.

205. Meilinger, "Trenchard, Slessor, and Royal Air Force Doctrine before World War II," 51.

206. Ibid., 61.
207. Slessor, *Air Power and Armies*, 100–101.
208. Ibid., 17.
209. Ibid., 5.
210. Ibid., 62.
211. Ibid., 23.
212. Ibid., 10.
213. Ibid., 62.
214. Mitchell, *Winged Defense*, 3–4.
215. Ibid., 214.
216. The volume of air in the atmosphere is roughly three times that of all the water on Earth.
217. Mitchell, *Winged Defense*, ix.
218. Douhet, *The Command of the Air*, 9.
219. Mitchell, *Winged Defense*, 4.
220. Slessor, *Air Power and Armies*, 96.
221. Mitchell, *Winged Defense*, 3.
222. Slessor, *Air Power and Armies*, 214.
223. Meilinger, "Trenchard, Slessor, and Royal Air Force Doctrine before World War II," 62.
224. Slessor, *Air Power and Armies,* 30.
225. Douhet's airpower – unstoppable, devastating weapons granting the advantage to whomever strikes first – foreshadowed nuclear deterrence and mutually assured destruction strategies that were still two decades away.
226. Mitchell, *Winged Defense*, 16.
227. Meilinger, "Trenchard, Slessor, and Royal Air Force Doctrine before World War II," 41–49.
228. Mitchell, *Winged Defense*, x.
229. Ibid., 209–210.
230. The airpower zealots were correct regarding the navy's vulnerability to air attack; however, naval airpower enthusiasts embraced the air weapon without also seeking secession, resulting in the aircraft carrier.
231. Douhet, *The Command of the Air*, 204.
232. Clodfelter, "Molding Airpower Convictions," 93.
233. Mitchell, *Winged Defense*, xvi.
234. Allen, *The Legacy of Lord Trenchard*, 20.
235. Ibid., 32.
236. Clodfelter, "Molding Airpower Convictions," 82.
237. Mitchell, *Winged Defense*, 9.
238. Slessor, *Air Power and Armies*, 4.
239. Douhet, *The Command of the Air*, 24.
240. Ibid., 96.
241. Ibid., 31.
242. Allen, *The Legacy of Lord Trenchard*, 54.
243. Clodfelter, "Molding Airpower Convictions," 98.
244. Meilinger, "Trenchard, Slessor, and Royal Air Force Doctrine before World War II," 62–63.
245. Ibid., 63.
246. Slessor, *Air Power and Armies*, 15–16.
247. Douhet, *The Command of the Air*, 129.
248. Ibid., 128.
249. Ibid., 94.
250. Ibid., 192.
251. Douhet, *The Command of the Air*, 118.
252. Allen, *The Legacy of Lord Trenchard*, 51.
253. Holley, "Reflections on the Search for Airpower Theory," 589.

254. Douhet, *The Command of the Air*, 277.
255. Slessor, *Air Power and Armies*, 7.
256. Ibid., 46.
257. Douhet, *The Command of the Air*, 128.
258. Ibid., 50.
259. Slessor, *Air Power and Armies*, 82–83.
260. Ibid., 25.
261. Ibid., 16–17.
262. Ibid., 26.
263. Ibid., 89.
264. Ibid., 63, 66.
265. Ibid., 20.
266. Ibid., 57.
267. Clodfelter, "Molding Airpower Convictions," 95–97.
268. Mitchell, *Winged Defense*, 23.
269. Meilinger, "Trenchard, Slessor, and Royal Air Force Doctrine before World War II," 41.
270. Ibid., 45.
271. Ibid., 64.
272. Slessor, *Air Power and Armies*, 122.
273. Ibid., 2.
274. Ibid., 82.
275. Galula, *Counterinsurgency Warfare*, 65.
276. Hew Strachan and Sibylle Scheipers, "Introduction: The Changing Character of War," in *The Changing Character of War*, edited by Hew Strachan and Sibylle Scheipers (Oxford: Oxford University Press, 2011), 13.
277. Aron, *Clausewitz*, ix.

4   The Unified War Theory

1. Paret, "Clausewitz," in *Makers of Modern Strategy*, 213.
 2. Clausewitz, "Author's Preface," in Clausewitz, *On War*, 62.
 3. Ibid., 150.
 4. "Definitions of Fact, Theory, and Law in Scientific Work," National Center for Science Education, accessed March 3, 2015, http://ncse.com/evolution/education/definitions-fact-theory-law-scientific-work.
 5. Vego, "On Military Theory," 60.
 6. Corbett, *Some Principles of Maritime Strategy*, 7.
 7. Reynolds, *A Primer in Theory Construction*, 116.
 8. "Definitions of Fact, Theory, and Law in Scientific Work."
 9. Clausewitz, *On War*, 142.
10. Corbett, *Some Principles of Maritime Strategy*, 2.
11. Jomini, *The Art of War*, 248.
12. Ibid., 207.
13. Liddell Hart, *Strategy*, 5.
14. Ibid.
15. Clausewitz, *On War*, 262.
16. Ibid., 157.
17. Ibid., 578.
18. Ibid., 593.
19. Ibid.
20. Ibid., 91.

21. Ibid., 581.
22. Ibid., 132.
23. Ibid.
24. Ibid., 136–137.
25. Ibid., 134–135.
26. Ibid., 141.
27. Ibid., 140.
28. Ibid., 136.
29. Ibid., 152.
30. Ibid.
31. Ibid., 158.
32. Ibid., 578.
33. Charles Kettering Quotes, BrainyQuote.com, BrainyMedia, accessed August 23, 2019, www .brainyquote.com/quotes/charles_kettering_152036.
34. Jomini, *The Art of War*, 256.
35. Clausewitz, *On War*, 141.
36. Ibid.
37. Ibid.
38. Vego, "On Military Theory," 66.
39. Moseley, "Airmen and the Art of Strategy," 9.
40. David W. Barno, "Military Adaptation in Complex Operations," *PRISM* 1:1 (2009): 30.
41. Thomas Hobbes, *Hobbes's Leviathan*, 1651 (Oxford: Oxford University Press, 1965), 75, accessed August 17, 2020, http://files.libertyfund.org/files/869/0161_Bk.pdf.
42. Chanakya, *The Arthashastra* (Sheridan, WY: Classics Press, 2019), 103. Chanakya was principal advisor to Chandragupta Maurya, ruler of the vast Mauryan Empire in India from 321–297 BCE.
43. Codevilla and Seabury, *War*, 73.
44. Thucydides, *The Landmark Thucydides*, 354.
45. Einstein, *Why War?*, 2.
46. Terry L. Deibel, *Foreign Affairs Strategy: Logic for American Statecraft* (New York: Cambridge University Press, 2007), 157.
47. John J. Mearsheimer, "Structural Realism," in *International Relations Theories: Discipline and Diversity*, 3rd ed., eds. Tim Dunne, Milja Kurki, and Steve Smith (Oxford: Oxford University Press, 2013), 72.
48. Ibid., 159.
49. Joseph S. Nye, *Bound to Lead: The Changing Nature of American Power* (New York: Basic Books, 1990).
50. Gray, *Fighting Talk,* 97.
51. Dyer, *War*, 156.
52. Chanakya, *The Arthashastra*, 106.
53. Jean-Jacques Rousseau, *Critique of the Abbé de Saint-Pierre's Project for Perpetual Peace*, in *The Ethics of War: Classic and Contemporary Readings*, eds. Gregory M. Reichberg, Henrik Syse, and Endre Begby (Malden, MA: Blackwell Publishing, 2006), 501.
54. Friedrich List, *The National System of Political Economy*, trans. Sampson S. Lloyd (London: Longmans, Green and Co., 1909), 113–114, accessed July 12, 2020, https://oll.libertyfund.org /titles/315.
55. Gray, *Another Bloody Century*, 59.
56. Plato, *The Republic*.
57. Machiavelli, *The Prince*, 37.
58. Edward Meade Earle, "Adam Smith, Alexander Hamilton, Friedrich List: The Economic Foundations of Military Power," in *Makers of Modern Strategy from Machiavelli to the Nuclear Age*, ed. Peter Paret (Princeton, NJ Princeton University Press, 1986), 217.

59. Clausewitz, *On War*, 605.
60. Keeley, *War Before Civilization*, 130.
61. Dyer, *War*, 157.
62. Friedrich Engels, *Anti-Dühring*, in *The Ethics of War: Classic and Contemporary Readings*, eds. Gregory M. Reichberg, Henrik Syse, and Endre Begby (Malden, MA: Blackwell Publishing, 2006), 590.
63. Von Moltke quoted in Gat, *A History of Military Thought*, 340.
64. Thucydides, *The Landmark Thucydides*, 46.
65. Corbett, *Some Principles of Maritime Strategy*, 99.
66. Clausewitz, *On War*, 607.
67. Dyer, *War*, 151.
68. Strachan, *The Direction of War*, 45.
69. Just as a lawyer representing herself has a fool for a client, it is not necessarily best for the political sovereign and warfighting general to be the same person.
70. Codevilla and Seabury, *War*, 84.
71. Gray, *Fighting Talk*, 29.
72. Codevilla and Seabury, *War*, 85.
73. Gordon A. Craig, "The Political Leader as Strategist," in *Makers of Modern Strategy from Machiavelli to the Nuclear Age*, ed. Peter Paret (Princeton, NJ: Princeton University Press, 1986), 482.
74. Freedman, *Strategy*, ix.
75. Ibid., x.
76. Aron, *Clausewitz*, 371.
77. Freedman, *Strategy*, 75.
78. Corbett, *Some Principles of Maritime Strategy*, 135.
79. Gray, *War, Peace and International Relations*, 43.
80. Winston Churchill, *The World Crisis: 1911–1918* (New York: Free Press, 2005), 294.
81. Clausewitz, *On War*, 245.
82. Liddell Hart, *Strategy*, 212.
83. Strachan, *The Direction of War*, 78.
84. Corbett, *Some Principles of Maritime Strategy*, 20.
85. Jomini, *The Art of War*, 245.
86. Clausewitz, *On War*, 594.
87. Joint Chiefs of Staff, *Strategy*, JDN 1–18 (Washington, DC: Joint Chiefs of Staff, 2018), www.jcs.mil/Portals/36/Documents/Doctrine/jdn_jg/jdn1_18.pdf?ver=2018-04-25-150439-540.
88. Strachan, *The Direction of War*, 209.
89. Clausewitz, *On War*, 593.
90. Ibid.
91. Strachan, *The Direction of War*, 103.
92. Ibid., 206.
93. Augustine, *City of God*, in *The Ethics of War: Classic and Contemporary Readings*, eds. Gregory M. Reichberg, Henrik Syse, and Endre Begby (Malden, MA: Blackwell Publishing, 2006), 79.
94. Clausewitz, *On War*, 584.
95. Ibid., 586.
96. Chanakya, *The Arthashastra*, 107.
97. Gray, *Another Bloody Century*, 43.
98. Clausewitz, *On War*, 177.
99. Hall, *The Nature of War*, 62–63.
100. Cicero, *Pro Milone*, in *The Ethics of War: Classic and Contemporary Readings*, eds. Gregory M. Reichberg, Henrik Syse, and Endre Begby (Malden, MA: Blackwell Publishing. 2006), 50–51.

101. Kant, *Toward Perpetual Peace*, 4.
102. Davie, *The Evolution of War*, 176.
103. Machiavelli, *The Art of War*, 4.
104. Clausewitz, *On War*, 75.
105. Kant, *Toward Perpetual Peace*, 9.
106. Keegan, *A History of Warfare*, 383.
107. Otto von Bismarck Quotes, BrainyQuote.com, BrainyMedia, accessed August 3, 2016, www .brainyquote.com/quotes/otto_von_bismarck_389242.
108. Machiavelli, *The Discourses*, 515.
109. It is also not uncommon to conflate "strategic" with "nuclear" in defense lexicon.
110. Clausewitz, *On War*, 194.
111. Laughlin, *A Different Universe*, xiv.
112. James Gleick, *Chaos: Making a New Science* (New York: Penguin Books, 1987), 98.
113. Ibid., 44.
114. Clausewitz, *On War*, 75.
115. Ibid., 128.
116. Ibid., 293.
117. Ibid., 480.
118. Gray, *Fighting Talk,* 3.
119. Ibid., 78.
120. Chanakya, *The Arthashastra*, 104.
121. Machiavelli, *The Discourses*, 431–432.
122. Gray, *War, Peace and International Relations*, 314.
123. William Mitchell, *Winged Defense: The Development and Possibilities of Modern Air Power, Economic and Military* (Mineola, NY: Dover Publications, 2006), 18.
124. Coram, *Boyd*, 341.
125. Codevilla and Seabury, *War*, 152.
126. Sun Tzu, *The Art of War*, 129.
127. Livy Quotes, BrainyQuote.com, BrainyMedia Inc, 2019, accessed October 18, 2019, www .brainyquote.com/quotes/livy_396605.
128. Colin S. Gray, *The Future of Strategy* (Cambridge, UK: Polity Press, 2015), 1.
129. Murray, *War, Strategy, and Military Effectiveness*, 3–4.
130. Clausewitz, *On War*, 179.
131. Liddell Hart, *Strategy*, 337.
132. Clausewitz, *On War*, 86.
133. Richard Wiseman, "Be Lucky – It's an Easy Skill to Learn," *Telegraph*, January 9, 2003, www.telegraph.co.uk/technology/3304496/Be-lucky-its-an-easy-skill-to-learn.html.
134. Monte Carlo methodology and game theory use mathematical algorithms and probability to gain insights into complex systems, including human interactions and decision-making.
135. Ibid., 94.
136. Clausewitz, *On War*, 80.
137. Ibid., 167.
138. Napoleon Bonaparte Quotes, BrainyQuote.com, BrainyMedia, accessed December 22, 2016, www.brainyquote.com/quotes/quotes/n/napoleonbo140390.html.
139. Chandler, *The Military Maxims of Napoleon*, 37.
140. Dyer, *War*, 133.
141. Clausewitz, *On War*, 167.
142. Stanislaw Sosabowski quoted in David, *Military Blunders*, 125–126.
143. "No Battle Plan Survives Contact with the Enemy," *Lexician*, accessed August 14, 2016, www.lexician.com/lexblog/2010/11/no-battle-plan-survives-contact-with-the-enemy.
144. Dwight D. Eisenhower Quotes, BrainyQuote.com, BrainyMedia, accessed August 14, 2016, www.brainyquote.com/quotes/quotes/d/dwightdei149111.html.

145. Iklé, *Every War Must End*, 3.
146. David*, Military Blunders*, 132.
147. Frederik II as quoted in David, *War: The Definitive Visual History*, 173.
148. Clausewitz, *On War*, 127.
149. Ibid., 77.
150. Ibid., 71.
151. John Milton Quotes, BrainyQuote.com, BrainyMedia, accessed August 3, 2016, www .brainyquote.com/quotes/quotes/j/johnmilton166738.html.
152. Clausewitz, *On War*, 90.
153. Napoleon Bonaparte Quotes, BrainyQuote.com, BrainyMedia Inc, 2019, accessed October 19, 2019, www.brainyquote.com/quotes/napoleon_bonaparte_118625.
154. Liddell Hart, *Strategy*, 211.
155. Ibid., 220.
156. Ibid., 322.
157. Clausewitz, *On War*, 85.
158. Ibid., 184.
159. Ibid., 185.
160. Nathan Bedford Forrest Quotes, BrainyQuote.com, BrainyMedia Inc, 2019, accessed October 19, 2019, www.brainyquote.com/quotes/nathan_bedford_forrest_216066.
161. Clausewitz, *On War*, 127.
162. Ibid., 79.
163. David, *Military Blunders*, 194.
164. US Marine Corps, *Small Wars Manual*, 60.
165. Handel, *Masters of War*, 111. Handel says policymakers and strategists use "net assessment" and "correlation of forces" to identify "*comparative strategic advantages*."
166. Clausewitz, *On War*, 137.
167. Ibid., 528.
168. Ibid., 585–586.
169. Ibid., 517.
170. Ibid., 261.
171. Jomini, *The Art of War*, 87.
172. Clausewitz, *On War*, 225.
173. Ibid., 93.
174. Ibid., 90.
175. Ibid., 93.
176. Ibid.
177. Ibid., 94.
178. Gat, *A History of Military Thought*, 821.
179. Deterrence uses the perception of strength to prevent attack, and assurance uses the same to reassure allies.
180. Clausewitz, *On War*, 97.
181. Ibid., 127.
182. Ibid., 282.
183. Machiavelli, *The Discourses*, 493.
184. Codevilla and Seabury, *War*, 88.
185. Paret, "Napoleon and the Revolution in War," in *Makers of Modern Strategy from Machiavelli to the Nuclear Age*, ed. Peter Paret (Princeton, NJ: Princeton University Press, 1986), 134.
186. Machiavelli, *The Discourses*, 504.
187. Trinquier, *Modern Warfare*, 84.
188. Clausewitz, *On War*, 77.
189. Gray, *Fighting Talk,* 94.

190. George C. Marshall, *The Papers of George Catlett Marshall,* ed. Larry I. Bland, Sharon Ritenour Stevens, and Clarence E. Wunderlin, Jr. (Lexington, Va.: The George C. Marshall Foundation, 1981–), accessed May 8, 2020, www.marshallfoundation.org/library/digital-archive/speech-at-trinity-college/.

191. Clausewitz, *On War,* 189.

192. Sun Tzu, *The Art of War,* 95.

193. Liddell Hart, *Strategy,* 323.

194. Clausewitz, *On War,* 86.

195. William James, "The Moral Equivalent of War," in *War: Studies from Psychology, Sociology, Anthropology,* eds. Leon Bramson and George W. Goethals (New York: Basic Books, 1968), 23.

196. Chanakya, *The Arthashastra,* 154.

197. Nicholas Fedyk, "Russian 'New Generation' Warfare: Theory, Practice, and Lessons for US Strategists," *Small Wars Journal,* August 25, 2016, http://smallwarsjournal.com/jrnl/art/russian-%E2%80%9Cnew-generation%E2%80%9D-warfare-theory-practice-and-lessons-for-us-strategists-0.

198. Ibid.

199. Ibid. Fedyk's "war at all times" phrasing reinforces the need for the UWT's definition for war and zones of the peace-war dialectic; for, if everything is war, then nothing is.

200. Machiavelli, *The Prince,* 38.

201. Abraham Lincoln quoted in Brooke Wanser, "14 Abraham Lincoln Quotes That Are Truly Modern Rules to Live By," *Reader's Digest,* accessed May 5, 2018, www.rd.com/culture/abraham-lincoln-quotes.

202. Codevilla and Seabury, *War,* 237.

203. Sun Tzu, *The Art of War,* 108.

204. Liddell Hart, *Strategy,* 14.

205. Codevilla and Seabury, *War,* 288. They argue that it is ethically preferable to kill the few responsible for the war rather than the thousands who are "just following orders."

206. Freedman, *Strategy,* 213.

207. Clausewitz, *On War,* 256.

208. Ibid., 231.

209. Liddell Hart, *Strategy,* 187.

210. Codevilla and Seabury, *War,* 115.

211. Clausewitz, *On War,* 585.

212. Ibid., 502.

213. Ibid., 585.

214. Often incorrectly attributed to Stalin. Anatoly Rybakov Quotes, BrainyQuote.com, BrainyMedia Inc, 2019, accessed October 19, 2019, www.brainyquote.com/quotes/anatoly_rybakov_109570.

215. The word "genocide" derives from the Greek term *genos* (race) and the Latin term *cide* (to kill).

216. Heuser, *The Evolution of Strategy,* 138–139.

217. Dyer, *War,* 46. History's worst known genocide was inflicted in northern China by Khan's Mongols who "systematically slaughtered" "forty million Chinese ... to depopulate the northern areas of the country and free them for nomadic herding."

218. Lawrence Freedman, *The Future of War: A History* (New York: PublicAffairs, 2017), 126.

219. Francisco de Vitoria, *On the Law of War,* in *The Ethics of War: Classic and Contemporary Readings,* eds. Gregory M. Reichberg, Henrik Syse, and Endre Begby (Malden, MA: Blackwell Publishing, 2006), 329.

220. Dwight D. Eisenhower quoted in "Top 10 Logistics Quotes," Universal Cargo, October 15, 2014, accessed October 18, 2019, www.universalcargo.com/top-10-logistics-quotes/.

221. Gray, *The Future of Strategy,* 88.

222. Omar Bradley quoted in Thomas M. Kane, *Military Logistics and Strategic Performance* (London: Frank Cass, 2001), xiv.
223. Dyer, *War*, 134.
224. Clausewitz, *On War*, 95.
225. Dyer, *War*, 132.
226. Sun Tzu, *The Art of War*, 85.
227. Clausewitz, *On War*, 384.
228. Chandler, *The Military Maxims of Napoleon*, 62.
229. Coram, *Boyd*, 460. From Boyd's unpublished paper, "Destruction and Creation."
230. Clausewitz, *On War*, 151.
231. Coram, *Boyd*, 335. This is a simplified description of Boyd's OODA Loop process (which was part of his 185-slide, 6-hour briefing).
232. Harold D. Lasswell, "The Garrison State," in *War: Studies from Psychology, Sociology, Anthropology*, eds. Leon Bramson and George W. Goethals (New York: Basic Books, 1968), 317. Lasswell was a twentieth century political scientist.
233. Alexander, *How Great Generals Win*, 97–98.
234. Codevilla and Seabury, *War*, 195.
235. Sun Tzu, *The Art of War*, 69.
236. Clausewitz, *On War*, 198.
237. Alexander, *How Great Generals Win*, 107.
238. Callwell, *Small Wars*, 163.
239. Clausewitz, *On War*, 246.
240. Ibid., 214.
241. Fehrenbach, *This Kind of War*, 453.
242. Gray, *War, Peace and International Relations*, 254.
243. Paret, *Clausewitz and the State*, 379.
244. Joint Chiefs of Staff, *Doctrine for the Armed Forces of the United States*, JP 1 (Washington, DC: Joint Chiefs of Staff, 2016), I-5.
245. Ibid.
246. Clausewitz, *On War*, 197.
247. Reynolds, *A Primer in Theory Construction*, 45.
248. Liddell Hart, *Strategy*, 10.
249. Corbett, *Some Principles of Maritime Strategy*, 4.
250. Gray, *War, Peace and International Relations*, 284.
251. Ibid.
252. Heuser, *The Evolution of Strategy*, 389.
253. Joint Chiefs of Staff, *Department of Defense Dictionary of Military and Associated Terms*, JP 1–02 (Washington, DC: Joint Chiefs of Staff, 2016), 199, www.dtic.mil/doctrine/new_pubs/jp1_02.pdf.
254. Ibid., 249.
255. Ibid., 20.
256. Ibid., 100.
257. Ibid., 113.
258. Gray, *War, Peace and International Relations*, 279. Gray says it is "convenient" and "arguably useful as well as historically justifiable ... to divide the experience of warfare into two broad categories: regular and irregular."
259. Ibid., 280.
260. Moseley, "Airmen and the Art of Strategy," 10.
261. Plasma is also a physical state of matter.
262. Charles Dickens, *A Tale of Two Cities*, Charles Dickens online, accessed October 19, 2019, www.dickens-online.info/a-tale-of-two-cities.html.
263. Edward Esko, *Yin Yang Primer* (Becket, MA: Amber Waves, 2012), 6.

264. Ibid., 17–18.
265. Heraclitus, *Fragments*, 23.
266. Ibid., 31.
267. Machiavelli, *The Prince*, 45–46.
268. "Questions and Replies Between T'ang, T'ai-tsung and Li Wei-kung," in *The Seven Military Classics of Ancient China*, trans Sawyer, 344.
269. Donald Tuzin, "The Spectre of Peace in Unlikely Places: Concept and Paradox in the Anthropology of Peace," in *A Natural History of Peace*, ed. Thomas Gregor (Nashville, TN: Vanderbilt University Press, 1996), 3.
270. Clausewitz, *On War*, 136.
271. Corbett, *Some Principles of Maritime Strategy*, 8.
272. Joint Chiefs of Staff, *Doctrine for the Armed Forces of the United States*, I-5.
273. Lawrence, *Seven Pillars of Wisdom,* 198.
274. Gleick, *Chaos*, 16.
275. A fluid is a substance that can flow (liquid or gas).
276. If war were pure chaos (complete disorder), we could make no sense of it, outcomes would be totally unpredictable, and strategy and preparation would be pointless.
277. The "butterfly effect" occurs when minor inputs cause disproportionate effects in chaotic systems (e.g., a butterfly in China changes the weather in America).
278. Sun Tzu, *Sun Tzu: The New Translation*, 67–68.
279. Homer, *The Iliad*, 160.
280. Lawrence, *Seven Pillars of Wisdom*, 347.
281. John Dederer, "Making Bricks without Straw: Nathanael Greene's Southern Campaign and Mao Tse-Tung's Mobile War," *Military Affairs* 47, no. 3 (October 1983): 116.
282. Mao, *On Guerrilla Warfare*, 103.
283. Griffith, "Introduction," in Mao, *On Guerrilla Warfare*, 24.
284. Galula, *Counterinsurgency Warfare*, 7.
285. Liddell Hart, *Strategy*, 365.
286. Robert Fox and Alan T. McDonald, *Introduction to Fluid Mechanics*, 3rd ed. (New York: John Wiley & Sons, 1985), 231.
287. Ibid., 33.
288. Natalie Wolchover, "The Nine Biggest Unsolved Mysteries in Physics," *Live Science*, July 3, 2012, www.livescience.com/34052-unsolved-mysteries-physics.html, accessed March 31, 2015.
289. Fox and McDonald, *Introduction to Fluid Mechanics*, 18.
290. David B. Guralnik, ed., *Webster's New World Dictionary* (New York: Prentice Hall Press, 1986), 1587. Viscous comes from the Latin root *viscum* meaning "sticky."
291. A fluid with negligible viscosity is called "inviscid."
292. Callwell, *Small Wars*, 88.
293. Herbert Simon, *The Sciences of the Artificial*, 3rd ed. (Cambridge, MA: The MIT Press, 1996), 170.
294. For example, we cannot discover *Hamlet* by studying the letters of the alphabet, Beethoven's *5th Symphony* by listening to a musical scale, nor Da Vinci's *Mona Lisa* by examining a box of crayons.
295. Sun Tzu, *The Art of War*, 92.
296. Thomas Sowell, *A Conflict of Visions: Ideological Origins of Political Struggles* (New York: Basic Books, 2007), 10–11.
297. Esko, *Yin Yang Primer*, 22–23.
298. Clausewitz, *On War*, 350.
299. From the twenty-first-century video game series, *Assassin's Creed*, by Ubisoft.
300. Lawrence, *Seven Pillars of Wisdom*, 198.
301. Callwell, *Small Wars*, 52.

302. Jomini, *The Art of War*, 171.

303. Griffith, "Introduction," in Mao, *On Guerrilla Warfare*, 10.

304. Clausewitz, *On War*, 483.

305. Dyer, *War*, 80.

306. Keegan, *A History of Warfare,* 272.

307. Clausewitz, *On War*, 93–94.

308. Gray, *War, Peace and International Relations*, 286.

309. US Marine Corps, *Small Wars Manual*, 12.

310. Sun Tzu, *Sun Tzu: The New Translation*, 61.

311. T. E. Lawrence, "The Evolution of a Revolt," *Army Quarterly and Defence Journal* 1 (October 1920): 55–69.

312. Liddell Hart, *Strategy*, 369.

313. Qiao Liang and Wang Xiangsui, *Unrestricted Warfare*, trans. Foreign Broadcast Information Service (Beijing: PLA Literature and Arts Publishing House, 1999), 132.

314. Sun Tzu, *The Art of War*, 98.

315. Clausewitz, *On War*, 196.

316. Handel, *Masters of War*, 185.

317. Colin Powell quoted in "Powell: 'We're going to cut it off … kill it'," Eliot Brenner, UPI Archives, January 23, 1991, accessed October 19, 2019, www.upi.com/Archives/1991/01/23/Powell-Were-going-to-cut-it-off-kill-it/9990332341450/.

318. Griffith, "Introduction," in Mao, *On Guerrilla Warfare*, 32.

319. Trinquier, *Modern Warfare*, 74.

320. Sun Tzu, *Sun Tzu: The New Translation*, 60.

321. Ibid., 64.

322. Thucydides, *The Peloponnesian War*, 184.

323. Liddell Hart, *Strategy*, 108.

324. Jomini, *The Art of War*, 88.

325. David, *Military Blunders*, 157.

326. Lawrence, "On Guerrilla Warfare."

327. Clausewitz, *On War*, 308.

328. Ibid., 625.

329. Liddell Hart, *Strategy*, 240.

330. Chanakya, *The Arthashastra*, 151.

331. Thucydides, *The Peloponnesian War*, 402–403.

332. Liddell Hart, *Strategy*, 188.

333. Reynolds, *A Primer in Theory Construction*, 65.

334. "Questions and Replies Between T'ang, T'ai-tsung and Li Wei-kung," 336.

335. Clausewitz, *On War*, 617.

336. Ibid., 635.

337. Ibid., 486.

338. Liddell Hart, *Strategy*, 108.

339. Ibid., 234.

340. Ibid., 291.

341. Corbett, *Some Principles of Maritime Strategy*, 138.

342. Freedman, *Strategy*, 42.

343. Keegan and Darracott, *The Nature of War*, 48.

344. Esko, *Yin Yang Primer*, 36.

345. Sun Tzu, *Sun Tzu: The New Translation*, 50.

346. Liddell Hart, *Strategy*, 26.

347. Clausewitz, *On War*, 481.

348. Ibid., 482.

349. Ibid., 481.

350. Gray, *War, Peace and International Relations*, 288.

351. Callwell, *Small Wars*, 85.

352. Kilcullen, *Counterinsurgency*, 9.

353. Callwell, *Small Wars*, 86.

354. Jomini, *The Art of War*, 21.

355. Corbett, *Some Principles of Maritime Strategy*, 138.

356. Martin van Creveld, "Through a Glass, Darkly: Some Reflections on the Future of War," *Naval War College Review* 53, no. 4 (Autumn, 2000), www.questia.com/library/journal/1P3-64943451/through-a-glass-darkly-some-reflections-on-the-future, 9.

357. Griffith, "Introduction," in Mao, *On Guerrilla Warfare*, 33.

358. Liddell Hart, *Strategy*, 333.

359. Ibid., 366.

360. Galula, *Counterinsurgency Warfare*, 51.

361. Clausewitz, *On War*, 188.

362. Machiavelli, *The Art of War*, 54.

363. Keeley, *War Before Civilization*, 75.

364. Air-to-air combat embodies this type of I-I war.

365. Heuser, *The History of Strategy*, 417.

366. Stathis N. Kalyvas, "The Changing Character of Civil Wars, 1800–2009," in *The Changing Character of War*, edited by Hew Strachan and Sibylle Scheipers (Oxford: Oxford University Press, 2011), 205.

367. For example, a glass of ice water cannot cool a room as effectively as cool air pumped into the room at separate locations.

368. Trinquier, *Modern Warfare*, xv.

369. "Questions and Replies Between T'ang, T'ai-tsung and Li Wei-kung," 324–325.

370. Russell W. Glenn, "Thoughts on Hybrid Conflict," *Small Wars Journal*, March 2, 2009, http://smallwarsjournal.com/blog/journal/docs-temp/188-glenn.pdf.

371. Keegan, *A History of Warfare*, 5.

372. August Rühle von Lilienstern quoted in Heuser, *The History of Strategy*, 404. Originally from Otto August Rühle von Lilienstern, *Handbuch für den Offizier zur Belehrung im Prieden und zum Gebrauch im Felde*, vol. II. (Berlin: G. Reimer, 1818).

373. Joint Chiefs of Staff, *Doctrine for the Armed Forces of the United States*, I-5.

374. Liddell Hart, *Strategy*, 368.

375. Trinquier, *Modern Warfare*, 45–46.

376. Jomini, *The Art of War*, 30.

377. Ibid., 240.

378. Griffith, "Introduction," in Mao, *On Guerrilla Warfare*, 10.

379. Gray, *War, Peace and International Relations*, 168.

380. Clausewitz, *On War*, 285–286.

381. Machiavelli, *The Art of War*, 114.

382. Callwell, *Small Wars*, 42.

383. Department of the Air Force, *Air Force Basic Doctrine, Traditional and Irregular War*, AFDD1, February 27, 2015, 2, accessed October 19, 2019, www.doctrine.af.mil/Portals/61/documents/Volume_1/V1-D32-Traditional-IW.PDF.

384. Callwell, *Small Wars*, 27.

385. Esko, *Yin Yang Primer*, 30.

386. Clausewitz, *On War*, 350.

387. Yuval Noah Harari, *Homo Deus: A Brief History of Tomorrow* (New York: HarperCollins Publishers, 2017), 83. Harari: "An algorithm is a methodical set of steps ... used to make calculations, resolve problems, and reach decisions."

388. Corbett, *Some Principles of Maritime Strategy*, 6.

389. Ibid., 166.

390. Gray, *Another Bloody Century*, 55. As Gray asserts, *context* is "the central truth" that "decodes the origins, meaning, character, and consequences of warfare."
391. Chanakya, *The Arthashastra*, 138.
392. Cashman, *What Causes War?*, 75.
393. "T'ai Kung's Six Secret Teachings," in *The Seven Military Classics of Ancient China*, trans. Sawyer, 69.
394. Callwell, *Small Wars*, 29.
395. Chanakya, *The Arthashastra*, 108.
396. Thucydides, *The Peloponnesian War*, 257.
397. Chanakya, *The Arthashastra*, 150.
398. Sun Tzu, *The Art of War*, 101.
399. Handel, *Masters of War*, 117.
400. Corbett, *Some Principles of Maritime Strategy*, 7.
401. Cashman, *What Causes War?*, 62. Cashman says the most common misperception patterns are "misperceptions of the opponent and his intentions, of the opponent's military capabilities and of the relative balance of power, of the opponent's willingness to give in to one's demands, of the risks involved in pursuing one's policies, of the intentions and capabilities of third countries, of the inevitability of war, of its eventual outcome, and of one's self."
402. Sun Tzu, *The Art of War*, 127.
403. David, *Military Blunders*, 381.
404. Ibid.
405. Frederick II, *Frederick the Great on the Art of War*, 54.
406. Bernard Brodie, "The Anatomy of Deterrence," (Santa Monica, CA: RAND Corporation, 1958), accessed December 30, 2019, www.rand.org/pubs/research_memoranda/RM2218 .html.
407. Heuser, *The Evolution of Strategy*, 363.
408. Keegan and Darracott, *The Nature of War*, 136.
409. We use the movie version for its powerful visuals and accessibility (via YouTube, for example). Here, specific details are less important than the themes that illustrate how viscosity and containment apply to combat.
410. Brian McAllister Linn, *The Philippine War 1899–1902* (Lawrence: University Press of Kansas, 2000), 35.
411. Ibid., 64.
412. Ibid., 148. Viewed with the W-V-A, Aguinaldo maintained the will to fight and believed the context adequate; however, with low relative capacity compared to US forces, he adopted an inviscous force state.
413. Kalyvas, "The Changing Character of Civil Wars," 217.
414. Linn, *The Philippine War 1899–1902*, 81.
415. Ibid., 131.
416. Dederer, "Making Bricks without Straw," 115.
417. Clausewitz, *On War*, 483.
418. Alexander, *How Great Generals Win*, 299.
419. Department of the Air Force, *Air Force Basic Doctrine, Organization, and Command*, AFDD 1, October 4, 2011, 30, www.au.af.mil/au/cadre/aspc/l004/pubs/afdd1.pdf.
420. Ibid., 35.
421. Clausewitz, *On War*, 229.
422. Joint Chiefs of Staff, *Doctrine for the Armed Forces of the United States*, I-3.
423. Corbett, *Some Principles of Maritime Strategy*, 14.
424. George Washington Quotes, AZ Quotes, accessed October 19, 2019, www.azquotes.com/ quote/551369.
425. Corbett, *Some Principles of Maritime Strategy*, 68.
426. General Garnet Joseph Wolseley (1896) quoted in Heuser, *The Evolution of Strategy*, 228.

427. Winston Churchill Quotes, BrainyQuote.com, BrainyMedia Inc, 2019, accessed October 19, 2019, www.brainyquote.com/quotes/winston_churchill_143787.

428. Hugh Montague Trenchard, "The Effect of the Rise of Air Power on War," in *Air Power. Three Papers by Marshal of the Royal Air Force The Viscount Trenchard, G.C.B., G.C.V.O., D.S.O., D.C.L., LL.D* (London: Directorate of Staff Duties, Air Ministry, 1946).

429. Gray, *War, Peace and International Relations*, 283.

430. T. Michael Moseley, *The Nation's Guardians: America's 21st Century Air Force* (Washington, DC: Chief of Staff of the United States Air Force, 2007), 3–4, www.au.af.mil /au/awc/awcgate/af/csaf_white_ppr_29dec07.pdf.

431. Patrick Shanahan quoted in C. Todd Lopez, "Shanahan: Next Big War May Be Won or Lost in Space," *US Department of Defense* (April 9, 2019), accessed October 19, 2019, www .defense.gov/Newsroom/News/Article/Article/1810100/shanahan-next-big-war-may-be-wo n-or-lost-in-space/.

432. "What Is the Karman Line?," *Simplicable*, accessed January 16, 2017, http://simplicable.com /new/karman-line.

433. "Treaty on Principles Governing the Activities of States in the Exploration and Use of Outer Space, Including the Moon and Other Celestial Bodies," US Department of State, accessed January 21, 2017, www.state.gov/t/isn/5181.htm. As of 2017, 129 nations were signatories to this treaty.

434. Notable war treaties that have been violated include the Treaty of Versailles, the Anti-Ballistic Missile Treaty, the Treaty on the Non-Proliferation of Nuclear Weapons, and the Intermediate-Range Nuclear Forces Treaty.

435. Gray, *Fighting Talk,* 81.

436. Alexander, *How Wars are Won*, 23.

437. Michael Walzer, *Just and Unjust Wars*, in *The Ethics of War: Classic and Contemporary Readings*, eds. Gregory M. Reichberg, Henrik Syse, and Endre Begby (Malden, MA: Blackwell Publishing, 2006), 650.

438. Ibid., 644.

439. Liddell Hart, *Strategy*, 359.

440. Gat, *A History of Military Thought*, 380.

441. Codevilla and Seabury, *War*, 192.

442. Gray, *War, Peace and International Relations*, 337.

443. Raymond Aron, "War and Industrial Society," in *War: Studies from Psychology, Sociology, Anthropology*, eds. Leon Bramson and George W. Goethals (New York: Basic Books, 1968), 388.

444. Martin van Creveld, "War and Technology," *Foreign Policy Research Institute*, October 24, 2007, www.fpri.org/article/2007/10/war-technology-2.

445. Andrew J. Bacevich, *The Limits of Power* (New York: Metropolitan Books, 2008), 159–160.

446. Franklin C. Miller, "The Need for a Strong Nuclear Deterrent in the Twenty-First Century," *Strategic Studies Quarterly* 7, no. 3 (Fall 2013): 9–10.

447. Liddell Hart, *Strategy*, xix.

448. Trinquier, *Modern Warfare*, 90.

449. James Turner Johnson, "Does Defense of Values by Force Remain a Moral Possibility?" in *The Ethics of War: Classic and Contemporary Readings*, eds. Gregory M. Reichberg, Henrik Syse, and Endre Begby (Malden, MA: Blackwell Publishing, 2006), 664.

450. Bertrand Russell, *Common Sense and Nuclear Warfare*, in *The Ethics of War: Classic and Contemporary Readings*, eds. Gregory M. Reichberg, Henrik Syse, and Endre Begby (Malden, MA: Blackwell Publishing, 2006), 602.

451. Bethany Lacina, Nils Petter Gleditsch, and Bruce Russett, "The Declining Risk of Death in Battle," *International Studies Quarterly* 50(3): 673–680, 2006, accessed May 14, 2020, www .prio.org/Data/Armed-Conflict/Battle-Deaths/The-Battle-Deaths-Dataset-version-30/.

452. Grotius, *The Law of War and Peace*, 10–11.

453. Van Creveld, *The Transformation of War*, 178.

454. Kelsen, *General Theory of Law and State*, in *The Ethics of War*, 611.

455. Ibid.

456. Hobbes, *Hobbes's Leviathan*, 98.

457. Gregory M. Reichberg, *The Ethics of War: Classic and Contemporary Readings*, eds. Gregory M. Reichberg, Henrik Syse, and Endre Begby (Malden, MA: Blackwell Publishing, 2006), ix.

458. Office of General Counsel, *Department of Defense Law of War Manual* (Washington: General Counsel of the Department of Defense, 2016), 26. Translations – *jus ad bellum*: "the right to the war"; *jus in bello*: "right in war."

459. Ibid., 7–8.

460. Ibid., 26.

461. Ibid., 15–16.

462. Alexander of Hales, *Summa theologica*, in *The Ethics of War: Classic and Contemporary Readings*, eds. Gregory M. Reichberg, Henrik Syse, and Endre Begby (Malden, MA: Blackwell Publishing, 2006), 159.

463. Office of General Counsel, *Department of Defense Law of War Manual*, 40.

464. Seneca quoted in *The Ethics of War: Classic and Contemporary Readings*, eds. Gregory M. Reichberg, Henrik Syse, and Endre Begby (Malden, MA: Blackwell Publishing, 2006), 417.

465. Emer de Vattel, *The Law of Nations*, in *The Ethics of War: Classic and Contemporary Readings*, eds. Gregory M. Reichberg, Henrik Syse, and Endre Begby (Malden, MA: Blackwell Publishing, 2006), 510.

466. Hegel, *Philosophy of Right*, in *The Ethics of War*, 547.

467. Rousseau, *The State of War*, in *The Ethics of War*, 482.

468. Douhet, *The Command of the Air*, 181.

469. Office of General Counsel, *Department of Defense Law of War Manual*, 52–67.

470. Ibid., 341–342.

471. *Oxford Dictionaries Press*, s.v. "artificial intelligence (n.)," accessed June 3, 2018, https://en.oxforddictionaries.com/definition/artificial_intelligence. The Oxford English Dictionary defines "artificial intelligence" as "the theory and development of computer systems able to perform tasks normally requiring human intelligence, such as visual perception, speech recognition, decision-making, and translation between languages."

472. Gat, *A History of Military Thought*, 789.

473. Ibid., 790.

474. Douhet, *The Command of the Air*, 61.

475. Hyacinthe Aube quoted in Heuser, *The Evolution of Strategy*, 238.

476. John Rawls, "Fifty Years after Hiroshima," in *The Ethics of War: Classic and Contemporary Readings*, eds. Gregory M. Reichberg, Henrik Syse, and Endre Begby (Malden, MA: Blackwell Publishing, 2006), 639.

477. Kelsen, *General Theory of Law and State*, in *The Ethics of War*, 611.

478. Strachan, *The Direction of War*, 203.

5   The Future of War

1. Mahan, *Influence of Sea Power Upon History*, 9.

2. Heraclitus, *Fragments*, 7.

3. Smolin, *The Life of the Cosmos*, 286.

4. Freedman, *Strategy*, x.

5. James M. Kouzes and Barry Z. Posner, *The Leadership Challenge: How to Make Extraordinary Things Happen in Organizations* (Hoboken, NJ: John Wiley & Sons, 2017), 103.
6. Vego, "On Military Theory," 66.
7. Clausewitz, *On War*, 462.
8. Ibid., 102.
9. Douhet, *The Command of the Air*, 146.
10. Clausewitz, *On War*, 80.
11. Robert H.Scales, "Forecasting the Future of Warfare," *War on the Rocks*, April 9, 2018, https://warontherocks.com/2018/04/forecasting-the-future-of-warfare.
12. Robert Gates quoted in James Stavridis, "Will There Be a War on the Korean Peninsula in 2018?," *Nikkei Asian Review*, December 22, 2017, https://asia.nikkei.com/Viewpoints/James-Stavridis/Will-there-be-a-war-on-the-Korean-Peninsula-in-2018.
13. Moseley, *The Nation's Guardians*, 10.
14. Vego, "On Military Theory," 59.
15. Corbett, *Some Principles of Maritime Strategy*, 17.
16. Murray, *War, Strategy, and Military Effectiveness*, 36.
17. Aron, *Clausewitz*, 21.
18. Codevilla and Seabury, *War*, 46.
19. General Mark A. Milley, "AUSA 2016 – Dwight David Eisenhower Luncheon" (briefing, Association of the US Army 2016 Annual Meeting and Exposition, Washington, DC, October 4, 2016), http://wpswps.org/wp-content/uploads/2016/11/20161004_CSA_AUSA_Eisenhower_Transcripts.pdf.
20. Liddell Hart, *Strategy*, 210.
21. Clausewitz, *On War*, 297–298.
22. Ibid., 515.
23. Gray, *War, Peace and International Relations*, 12.
24. George and Meredith Friedman, *The Future of War: Power, Technology, & American World Dominance in the 21st Century* (New York: Crown Publishers, Inc., 1996), xi.
25. Gray, *Another Bloody Century*, 122.
26. Alexander, *How Wars Are Won*, 3.
27. Clausewitz, *On War*, 88–89.
28. Ibid., 488.
29. Douhet, *The Command of the Air*, 146.
30. Harold R. Winton, *To Change an Army: General Sir John Burnett-Stuart and British Armored Doctrine, 1927–1938* (Lawrence: University Press of Kansas, 1988), 239–240. Winton says mastery of future war "requires firm grounding in the theory of war, informed historical study, close analysis of contemporary developments, and an ability to project trends into the future."
31. Handel, *Masters of War*, 95.
32. Douhet, *The Command of the Air*, 30.
33. Aristotle, *Rhetoric*, in *The Ethics of War: Classic and Contemporary Readings*, eds. Gregory M. Reichberg, Henrik Syse, and Endre Begby (Malden, MA: Blackwell Publishing, 2006), 45–46.
34. Chandler, *The Military Maxims of Napoleon*, 55.
35. Aron, *Clausewitz*, 376.
36. Red teams "act" as the enemy in wargames and simulations and must be well versed in the enemy's strengths and weaknesses.
37. Confucius Quotes, AZ Quotes, accessed October 19, 2019, www.azquotes.com/author/3177-Confucius?p=2.
38. "Anocracies" mix elements of both democracy and autocracy.
39. Machiavelli, *The Discourses*, 124.
40. Codevilla and Seabury, *War*, 109.

41. Sumner, "War," 221.

42. Dakota L. Wood, *Rebuilding America's Military: Thinking About the Future*, Special Report No. 203 (Washington, DC: The Heritage Foundation, 2018), 19, http://report.heritage.org /sr203.

43. Gray, *Another Bloody Century*, 118–119.

44. Bernard Marr, "What Is Quantum Computing? A Super-Easy Explanation for Anyone," *Forbes*, July 4, 2017, www.forbes.com/sites/bernardmarr/2017/07/04/what-is-quantum-computing-a-super-easy-explanation-for-anyone/#1562ea171d3b. Quantum computers use the unique principles of quantum mechanics, including "superposition of states," to drastically improve calculation speed and efficiency. But there is a big caveat: the quantum world is one of intrinsic uncertainty, a fact that limits how small traditional computer chips can be and may literally short-circuit the accuracy and reliability of quantum-scale processors.

45. Machine learning refers to algorithms (computer code) that enable a machine to process results and feedback and modify itself to function more effectively.

46. Graphene is a one-atom-deep sheet of carbon atoms bonded into a honeycomb pattern.

47. *Hypersonic* refers to objects traveling within the atmosphere at speeds greater than five times the speed of sound (Mach 5, or approximately 1,658 meters per second at zero degrees Celsius).

48. "Laser" is derived from "light amplification by stimulated emission of radiation." The speed of light in a vacuum is nearly 300,000 kilometers per second.

49. Friedman, *The Future of War*, 24.

50. Gray, *Fighting Talk*, 86.

51. Omar N. Bradley Quotes, BrainyQuote.com, BrainyMedia Inc, 2020, accessed January 6, 2020, www.brainyquote.com/quotes/omar_n_bradley_131387.

52. Even fusion and ion propulsion systems are slow by cosmic standards. At 161,000 kilometers per hour, it would take 27,000 years to get to Alpha Centauri. While "warping" space-time (à la *Star Trek*) could shorten the trip, the know-how and energies needed for this do not exist.

53. Stealth technology absorbs or disperses the energy from active detection devices like radar, lidar, and sonar.

54. Passive detection systems "listen" for position-revealing target emissions, for example, electromagnetic, acoustic, or thermal (heat).

55. Heuser, *The Evolution of Strategy*, 99.

56. Aron, *Clausewitz*, 390.

57. Alexander, *How Wars Are Won*, 8.

58. Strachan, *The Direction of War*, 82. This illustrates the advantage of defining "asymmetric" war in terms of viscosity and not capacity. For as Hew Strachan points out, "All war is potentially 'asymmetric,'" using a capacity-based definition.

59. Gray, *War, Peace and International Relations*, 285.

60. Gray, *Fighting Talk*, 160.

61. Ibid.

62. Keith B. Bickel, *Mars Learning: The Marine Corp's Development of Small Wars Doctrine, 1915–1940* (Boulder, CO: Westview Press, 2001), 49.

63. In 2007, Sean Naylor proposed a US "unconventional warfare command," to "handle the nagging, intractable political-military challenges" inviscous forces are better able to solve. Sean D. Naylor, "Support Grows for Standing Up an Unconventional Warfare Command," *Armed Forces Journal*, November 2007, http://armedforcesjournal.com/support-grows-for-standing-up-an-unconventional-warfare-command.

64. David Hambling, "China Looks to Undermine US Power, With 'Assassin's Mace,'" *Wired*, July 2, 2009, www.wired.com/2009/07/china-looks-to-undermine-us-power-with-assassins-mace.

65. Lawrence, *Seven Pillars of Wisdom*, 201–202.

66. Clausewitz, *On War*, 78.

67. Aron, "War and Industrial Society," 390.
68. Machiavelli, *The Prince*, 48.
69. Murray, *War, Strategy, and Military Effectiveness*, 36.
70. Michael Howard, "Military Science in the Age of Peace," *RUSI Journal* 199 (March 1974): 3–9.
71. Arthur C. Clarke, *Profiles of the Future: An Inquiry into the Limits of the Possible* (London: Gollancz, 1962).
72. Gray, *Fighting Talk,* 157.
73. May, "War, Peace, and Social Learning," 155.
74. Thucydides, *The Peloponnesian War*, 217.
75. George Santayana Quotes, BrainyQuote.com, BrainyMedia Inc, 2019, accessed October 19, 2019, www.brainyquote.com/quotes/george_santayana_133920.
76. John Milton, "Sonnet 15: Fairfax, whose name in arms through Europe rings," Poetry Foundation, accessed September 5, 2018, www.poetryfoundation.org/poems/44734/sonnet-15-fairfax-whose-name-in-arms-through-europe-rings.
77. Malinowski, "An Anthropological Analysis of War," 246.
78. Kant, *Toward Perpetual Peace*, 3.
79. Clayton A. Robarchek and Carole J. Robarchek, "The Aucas, the Cannibals, and the Missionaries: From Warfare to Peacefulness among the Waorani," in *A Natural History of Peace*, ed. Thomas Gregor (Nashville, TN: Vanderbilt: University Press, 1996), 189–204.
80. Ibid.
81. Horgan, *The End of War*, 22.
82. Gray, *Fighting Talk,* 163.
83. Keeley, *War Before Civilization,* 39.
84. Ibid., 178.
85. James, "The Moral Equivalent of War," 23.
86. Keegan, *A History of Warfare*, 83.
87. Richard G. Sipes, "War, Sports, and Aggression: An Empirical Test of Two Rival Theories," *American Anthropologist* 75, no. 1 (1973): 64–86. In a survey of 130 primitive societies, Richard Sipes found only 11 that seemed peaceful and 5 of those had to be rejected because of insufficient information.
88. Sumner, "War," 209.
89. Alexander Hamilton, *Federalist No. 34*, January 5, 1788, Teaching American History, accessed October 19, 2019, https://teachingamericanhistory.org/library/document/federalist-no-34/.
90. Russell, *Common Sense and Nuclear Warfare*, 603.
91. Gregor, *A Natural History of Peace*, xiii. Gregor was Chair of the Department of Anthropology at Vanderbilt University.
92. Excerpts from Hedges, "What Every Person Should Know About War."
93. Gianluca Martelloni, Francesca Di Patti, and Ugo Bardi, "Pattern Analysis of World Conflicts over the Past 600 Years," December 19, 2018, accessed November 23, 2019, https://arxiv.org/ftp/arxiv/papers/1812/1812.08071.pdf.
94. Thomas Hardy, *The Military Quotation Book*, 154.
95. Durbin and Bowlby, "Personal Aggressiveness and War," 82.
96. Interestingly, beyond Jesus, Buddha, Gandhi, and Dr. Martin Luther King, Jr., it is hard to name any famous peacemakers.
97. Waltz, *Man, the State, and War*, 1.
98. US War Department, *Biennial Report of the Chief of Staff of the US Army, General George C. Marshall, July 1, 1943, to June 30, 1945, to the Secretary of War*, Washington, DC: US News Publishing Corp., 1945.
99. Hall, *The Nature of War*, 37.

100. Percy Bysshe Shelley, "Hellas," in *The Best Poems of the English Language: From Chaucer Through Frost*, ed. Harold Bloom (New York: HarperCollins, 2004), 434–435.

101. Gordon W. Allport, "The Role of Expectancy," in *War: Studies from Psychology, Sociology, Anthropology*, eds. Leon Bramson and George W. Goethals (New York: Basic Books 1968), 191. Allport was a Harvard psychologist regarded by many as one of the founding figures of "personality psychology."

102. Christopher Croker, *Can War Be Eliminated?* (Cambridge, UK: Polity Press, 2014), 5.

103. For more on this see, James Lee Ray, "Does Democracy Cause Peace?" *Annual Review of Political Science*, 1 (1998), pp. 27–46, accessed January 1, 2020, www.researchgate.net/publication/228591512_Does_Democracy_Cause_Peace.

104. Strachan, *The Direction of War*, 269.

105. Cashman, *What Causes War?*, 126.

106. Drew Desilver, "Despite global concerns about democracy, more than half of countries are democratic," Pew Research Center, May 14, 2019, accessed January 1, 2020, www.pewresearch.org/fact-tank/2019/05/14/more-than-half-of-countries-are-democratic/. According to the Pew Research Center, as of 2017, 57 percent of 167 states were democratic, 28 percent were anocratic (mixed), and 13 percent were autocracies.

107. Iklé, *Every War Must End*, xv.

108. Peter Brecke, "Violent Conflicts 1400 A.D. to the Present in Different Regions of the World," Paper presented at the 1999 Meeting of the Peace Science Society (International) on October 8–10, 1999, in Ann Arbor, Michigan, accessed June 10, 2020, https://arxiv.org/ftp/arxiv/papers/1812/1812.08071.pdf.

109. Sebastian von Einsiedel, Louise Bosetti, Cale Salih, Wilfred Wan, and James Cockayne, "Civil War Trends and the Changing Nature of Armed Conflict," Tokyo: United Nations University Centre for Policy Research, April 25, 2017, 2, accessed November 23, 2019, https://collections.unu.edu/eserv/UNU:6156/Civil_war_trends_U PDATED.pdf.

110. William Robert Avis, "Current Trends in Violent Conflict," K4D Helpdesk Report 565, March 25, 2019, Brighton, UK: Institute of Development Studies, 8–9, accessed November 22, 2019, https://assets.publishing.service.gov.uk/media/5cf669a ce5274a07692466db/565_Trends_in_Violent_Conflict.pdf.

111. United Nations and World Bank, 2017, "Pathways for Peace: Inclusive Approaches to Preventing Violent Conflict: Main Messages and Emerging Policy Directions," World Bank, Washington, DC, doi:10.1596/978-1-4648-1162-3, 1, 3, 7, accessed November 23, 2019, https://openknowledge.worldbank.org/bitstream/handle/10986/28337/211162mm .pdf.

112. Hall, *The Nature of War*, 22–23.

113. Therese, Pettersson, Stina Högbladh, and Magnus Öberg, "Organized Violence, 1989–2018 and Peace Agreements," *Journal of Peace Research* 56(4), 2019, accessed June 10, 2020, www.prio.org/Data/Armed-Conflict/UCDP-PRIO/.

114. Keegan, *A History of Warfare*, 59.

115. Statistics from "List of Ongoing Conflicts," Wars in the World, accessed December 3, 2020, www.warsintheworld.com/?page=static1258254223.

116. Gray, *War, Peace and International Relations*, 342.

117. Gordon A. Craig and Felix Gilbert, "Reflections on Strategy in the Present and Future," in *Makers of Modern Strategy from Machiavelli to the Nuclear Age*, ed. Peter Paret (Princeton, NJ: Princeton University Press, 1986), 865.

118. Horgan, *The End of War*, 142.

119. Freedman, *The Future of War*, 25.

120. Morris Janowitz, "Military Elites and the Study of War," in *War: Studies from Psychology, Sociology, Anthropology*, eds. Leon Bramson and George W. Goethals (New York: Basic Books, 1968), 349.

121. Tuzin, "The Spectre of Peace in Unlikely Places," 7.

122. Gregor, *A Natural History of Peace*, xvi.

123. This scenario calls to mind the *Star Trek* episode, "Let That Be Your Last Battlefield" (January 10, 1969).

124. Gray, *War, Peace and International Relations*, 341. Gray says there are two practical conditions for making war unthinkable: (1) war must cease to be perceived as a useful instrument of policy; and (2) war must cease to be socially acceptable.

125. Thomas Nagel, "War and Massacre," in *The Ethics of War: Classic and Contemporary Readings*, eds. Gregory M. Reichberg, Henrik Syse, and Endre Begby (Malden, MA: Blackwell Publishing, 2006), 654. In "War and Massacre" (1972), Nagel says consequentialists can justify "any means . . . if it leads to a sufficiently worthy end."

126. Hobbes, *Hobbes's Leviathan*, 96.

127. Waltz, *Man, the State, and War*, 160.

128. Gray, *Fighting Talk,* 23.

129. Hegel, *Philosophy of Right*, in *The Ethics of War,* 549.

130. Allport, "The Role of Expectancy," 192.

131. Quotations by Ambrose Bierce, Dictionary of Quotes, accessed May 26, 2019, www .dictionary-quotes.com/peace-in-international-affairs-is-a-period-of-cheating-between-two-periods-of-fighting-ambrose-bierce. Bierce was a nineteenth-century American author and Civil War veteran.

132. Davie, *The Evolution of War*, 9.

133. Knauft, "The Human Evolution of Cooperative Interest," 73.

134. William Inge Quotes, BrainyQuote.com, BrainyMedia Inc, 2019, accessed October 19, 2019, www.brainyquote.com/quotes/william_inge_149272.

135. In the *prisoner's dilemma*, two prisoners face variable sentences depending on whether they cooperate with or testify against each other. If both testify, both serve long sentences. If both remain silent, both serve short sentences. However, if one testifies and the other remains silent, the one testifying gets out immediately while the other serves a long sentence. Thus, betraying a partner benefits a prisoner as long as the partner keeps the faith.

136. Jomini, *The Art of War*, 33.

137. Waltz, *Man, the State, and War*, 169.

138. Ibid., 170.

139. Heinrich von Treitschke quoted in Gat, *A History of Military Thought*, 328. Originally from Treitschke, *Politics* (2 vols; London, 1916), i. 21. Compiled from his early 1880s lectures after the author's death.

140. Sumner, "War," 224.

141. Dante quoted in Kenneth C. M. Sills, "The Idea of Universal Peace in the Works of Virgil and Dante," *The Classical Journal* 9, no. 4 (1914): 148, www.jstor.org/stable/3287070, accessed April 13, 2020.

142. Gregory M. Reichberg, "Supranational Government and Peace," in *The Ethics of War: Classic and Contemporary Readings*, eds. Gregory M. Reichberg, Henrik Syse, and Endre Begby (Malden, MA: Blackwell Publishing, 2006), 489.

143. Einstein, *Why War?*, 3.

144. Immanuel Kant, *Perpetual Peace*, in *The Ethics of War: Classic and Contemporary Readings*, eds. Gregory M. Reichberg, Henrik Syse, and Endre Begby (Malden, MA: Blackwell Publishing, 2006), 532–533.

145. Freud, *Why War?*, 7.

146. Hall, *The Nature of War*, 9.

147. Michael Howard, *The Invention of Peace and the Reinvention of War* (London: Profile Books, 2001), 105.

148. Waltz, *Man, the State, and War*, 228.

149. Even the relatively impotent UN is bedeviled by corruption.

150. Thomas Jefferson, "A Bill for the More General Diffusion of Knowledge," Thomas Jefferson Foundation, accessed September 5, 2018, www.monticello.org/site/research-and-collections/bill-more-general-diffusion-knowledge.
151. Sills, "The Idea of Universal Peace in the Works of Virgil and Dante," 150–151.
152. Waltz, *Man, the State, and War*, 228.
153. Dyer, *War*, 257.
154. Staub, "The Psychological and Cultural Roots of Group Violence," 135.
155. Waltz, *Man, the State, and War*, 44.
156. Gray, *Another Bloody Century*, 20.
157. Kant, *Toward Perpetual Peace*, 13.
158. Ibid., 16–17.
159. Ibid., 30.
160. Erasmus, *The Education of a Christian Prince*, in *The Ethics of War*, 234.
161. John Davidson, *War Song*, PoemHunter, accessed October 19, 2019, www.poemhunter.com/poem/war-song/.
162. Horgan, *The End of War*, 104–105.
163. Ibid., 106.
164. Codevilla and Seabury, *War*, 36.
165. Horgan, *The End of War*, 154–155.
166. Ibid., 154.
167. Allport, "The Role of Expectancy," 191.
168. Ibid., 193.
169. Waltz, *Man, the State, and War*, 68.
170. Clausewitz, *On War*, 260.
171. Rousseau, *The State of War*, in *The Ethics of War*, 488.
172. John Stuart Mill Quotes, Notable Quotes, accessed September 22, 2018, www.notable-quotes.com/m/mill_john_stuart.html.
173. Aron, *Clausewitz*, 410.
174. Waltz, *Man, the State, and War*, 49.
175. Rousseau, *The State of War*, in *The Ethics of War*, 487.
176. Dyer, *War*, 257.
177. Durbin and Bowlby, "Personal Aggressiveness and War," 84.
178. Thomas Aquinas, *Summa theologiae*, in *The Ethics of War: Classic and Contemporary Readings*, eds. Gregory M. Reichberg, Henrik Syse, and Endre Begby (Malden MA: Blackwell Publishing, 2006), 172.
179. Von Moltke quoted in Gat, *A History of Military Thought*, 327. Originally in Moltke to Prof. Dr. Bluntschli, December 11, 1880, *Moltke as a Correspondent*, 272.
180. Gat, *A History of Military Thought*, 600–601.
181. Waltz, *Man, the State, and War*, 39.
182. Durbin and Bowlby, "Personal Aggressiveness and War," 87.
183. "Is It Possible to Eradicate All Diseases?" *BBC News Magazine*, September 22, 2015 www.bbc.com/news/magazine-37433012. Few immunologists believe that all diseases can be cured. In fact, smallpox is the *only* infectious disease to be eliminated.
184. Amitai Etzioni, "Toward a Sociological Theory of Peace," in *War: Studies from Psychology, Sociology, Anthropology*, eds. Leon Bramson and George W. Goethals (New York: Basic Books, 1968), 409. Sociologist and communitarian Etzioni writes, "the main shortcoming of the balance-of-power system as a model for peace is that it reduces the probability of war but does not eliminate it."
185. Robert Gates quoted in Andrew Tilghman, "Cut Spending, but Protect Capabilities, Gates Says," *Air Force Times* 71, no. 4 (June 6, 2001): 9.

186. John of Salisbury, *Policraticus*, in *The Ethics of War: Classic and Contemporary Readings*, eds. Gregory M. Reichberg, Henrik Syse, and Endre Begby (Malden, MA: Blackwell Publishing, 2006), 129.
187. Hamilton, *Federalist No. 34*.
188. Machiavelli, *The Discourses*, 200.
189. Keeley, *War Before Civilization*, 161.
190. Shaun Nichols, "Is Free Will an Illusion?," *Scientific American*, November 1, 2011, www.scientificamerican.com/article/is-free-will-an-illusion. Some studies suggest we do not really have free will. However, though our choices may be influenced by subconscious factors, this is one branch of theory we *freely choose* not to believe!
191. Hall, *The Nature of War*, 86.
192. Jurassic Park Quotes, Quotes, accessed October 19, 2019, www.quotes.net/movies/jurassic_park_6097.
193. Hegel, *Philosophy of Right*, in *The Ethics of War*, 545.
194. Kant, *Toward Perpetual Peace*, 11.
195. De Waal, "The Biological Basis of Peaceful Coexistence," 38–39.
196. Waltz, *Man, the State, and War*, 119.
197. Marquis de Condorcet, "The Future Progress of the Human Mind," in *Internet Modern History Sourcebook*, Paul Halsall ed., Fordham University, accessed December 16, 2018, https://sourcebooks.fordham.edu/mod/condorcet-progress.asp.
198. Robert F. Baumann, "Historical Perspectives on Future War," *Military Review* 77, no. 2 (March–April 1997), www.questia.com/library/journal/1P3-13134523/historical-perspectives-on-future-war.
199. Ibid.
200. Gat, *A History of Military Thought*, 522.
201. Codevilla and Seabury, *War*, 32.
202. Huxley's *Brave New World* had five classes: Alpha (rulers), Beta, Gamma, Delta, and Epsilon (laborers).
203. It is likely that the engineering of super-humans is *already* underway somewhere on Earth.
204. Mao, *On Guerrilla Warfare*, 89.
205. Codevilla and Seabury, *War*, 15.
206. Keegan, *A History of Warfare*, 384.
207. Etzioni, "Toward a Sociological Theory of Peace," 409.
208. Waltz, *Man, the State, and War*, 238.
209. Hall, *The Nature of War*, 10–12.
210. Heraclitus, *Fragments*, 29.
211. John Foster Dulles Quotes, AZ Quotes, accessed October 19, 2019, www.azquotes.com/quote/649981.
212. Freud, *Why War?*, 9.
213. Codevilla and Seabury, *War*, 8.
214. Gray, *Fighting Talk*, 125.
215. Martin Luther, *Whether Soldiers, Too, Can Be Saved*, in *The Ethics of War: Classic and Contemporary Readings*, eds. Gregory M. Reichberg, Henrik Syse, and Endre Begby (Malden, MA: Blackwell Publishing, 2006), 270.
216. Codevilla and Seabury, *War*, 8.
217. Václav Havel quoted in Freedman, *The Future of War*, 101–102.
218. Arthur A. Demarest, "War, Peace, and the Collapse of a Native American Civilization: Lessons for Contemporary Systems of Conflict," in *A Natural History of Peace*, ed. Thomas Gregor (Nashville, TN: Vanderbilt University Press, 1996), 217. Demarest is Ingram Professor of Anthropology at Vanderbilt University.
219. Heraclitus, *Fragments*, 39.
220. Esko, *Yin Yang Primer*, 18.

221. Fehrenbach, *This Kind of War*, xi.
222. Hall, *The Nature of War*, 12.
223. Sumner, "War," 221.
224. Gregor, *A Natural History of Peace*, xxi.
225. Malinowski, "An Anthropological Analysis of War," 255.
226. May, "War, Peace, and Social Learning," 157.
227. Keeley, *War Before Civilization*, 161.
228. Machiavelli, *The Discourses*, 127.
229. John Milton, "Areopagitica: a speech by Mr. John Milton for the liberty of unlicenc'd printing, to the Parlament of England," (speech, London, 1664), The John Milton Reading Room, accessed August 18, 2018, www.dartmouth.edu/~milton/reading_room/areopagitica/text.html.
230. Mahatma Gandhi, *Gandhi on Non Violence*, ed. Thomas Merton (New York: New Directions, 1965), 36.
231. Waltz, *Man, the State, and War*, 159.
232. Cashman, *What Causes War?*, 285–286. This predominance of "peaceful norms," driven by evolved and pervasive "moral and aesthetic values," is the centerpiece of John Mueller's *Retreat from Doomsday: The Obsolescence of Major War* (1989).
233. Clausewitz, *On War*, 374.
234. Keegan, *A History of Warfare*, 385.
235. Cashman, *What Causes War?*, 129.
236. Waltz, *Man, the State, and War*, 33.
237. Morris, "The Slaughter Bench of History."
238. Gray, *War, Peace and International Relations*, 336.
239. Staub, "The Psychological and Cultural Roots of Group Violence," 145–146.
240. Kant, *Perpetual Peace*, in *The Ethics of War*, 528.
241. Freud, *Why War?*, 13–14.
242. Keeley, *War Before Civilization*, 157.
243. Durbin and Bowlby, "Personal Aggressiveness and War," 98–99.
244. Ibid., 99.
245. Sun Tzu, *The Art of War*, 114.
246. "The Methods of the Ssu-ma," in *The Seven Military Classics of Ancient China* trans. Sawyer, 117.
247. Alfred Lord Tennyson, "In Memoriam A. H. H," in *The Best Poems of the English Language: From Chaucer through Frost*, ed. Harold Bloom (New York: HarperCollins, 2004), 628.
248. Sowell, *A Conflict of Visions*, 24–25.
249. Kant, *Toward Perpetual Peace*, 5.
250. Gray, *Another Bloody Century*, 341–342.
251. Rousseau, *Critique of the Abbé de Saint-Pierre's Project for Perpetual Peace*, in *The Ethics of War*, 503.
252. Abraham Lincoln, *First Inaugural Address of Abraham Lincoln* (speech, Washington, DC, 1861), The Avalon Project, accessed December 7, 2018, https://avalon.law.yale.edu/19th_century/lincoln1.asp.

Conclusion

1. Mill, "The Contest in America," 10.
2. Clausewitz, "Two Notes by the Author," in Clausewitz, *On War*, 70.
3. Hall, *The Nature of War*, 12–13.
4. Laughlin, *A Different Universe*, 174.
5. Clausewitz, *On War*, 192.

6. Margaret Mead, "Warfare Is Only an Invention," in *War: Studies from Psychology, Sociology, Anthropology*, eds. Leon Bramson and George W. Goethals (New York: Basic Books, 1968), 269–274. Mead, an American cultural anthropologist, suggested that war as an invention can be abandoned provided people "recognize" its "defects" and replace it. However, she offers no practical alternative to the "invention" of war.

7. Durbin and Bowlby, "Personal Aggressiveness and War," 99.

8. Knox, "Introduction," in Homer, *The Iliad*, 62. Achilles's shield, crafted by the god Hephaestus, is etched with an image representing all of human life: two cities, one at peace and one at war, the "two poles of the human condition, war and peace, with their corresponding aspects of human nature, the destructive and creative."

9. Pufendorf, *On the Duty of Man and Citizen*, in *The Ethics of War*, 457.

10. Valiunas, *Churchill's Military Histories*, 119.

11. Gregor, *A Natural History of Peace*, xiv.

Bibliography

Alexander, Bevin. *How Great Generals Win*. New York: W. W. Norton & Company, 1993.
　　How Wars Are Won: The 13 Rules of War from Ancient Greece to the War on Terror. New York: Crown Publishers, 2002.
Alexander of Hales. "*Summa theologica*." In *The Ethics of War: Classic and Contemporary Readings*. Edited by Gregory M. Reichberg, Henrik Syse, and Endre Begby. Malden, MA: Blackwell Publishing, 2006.
Allen, H. R. *The Legacy of Lord Trenchard*. London: The Camelot Press Ltd., 1972.
Allport, Gordon W. "The Role of Expectancy." In *War: Studies from Psychology, Sociology, Anthropology*. Edited by Leon Bramson and George W. Goethals. New York: Basic Books, 1968.
Aquinas, Thomas. "*Summa theologiae*." In *The Ethics of War: Classic and Contemporary Readings*. Edited by Gregory M. Reichberg, Henrik Syse, and Endre Begby. Malden, MA: Blackwell Publishing, 2006.
Aristotle. *Politics*. Perseus Digital Library, Tufts University. Edited by Gregory R. Crane. www.perseus.tufts.edu/hopper/text?doc=Perseus:abo:tlg,0086,035:1:1253a.
　　Rhetoric. In *The Ethics of War: Classic and Contemporary Readings*. Edited by Gregory M. Reichberg, Henrik Syse, and Endre Begby. Malden, MA: Blackwell Publishing, 2006.
Arnold, Matthew. "Dover Beach." In *The Best Poems of the English Language: From Chaucer Through Frost*. Edited by Harold Bloom. New York: HarperCollins, 2004.
Aron, Raymond. *Clausewitz: Philosopher of War*. Translated by Christine Booker and Norman Stone. Englewood Cliffs, NJ: Prentice-Hall, Inc., 1985.
　　"War and Industrial Society." In *War: Studies from Psychology, Sociology, Anthropology*. Edited by Leon Bramson and George W. Goethals. New York: Basic Books, 1968.
Asimov, Isaac. "Isaac Asimov Asks, 'How Do People Get New Ideas?'" *MIT Technology Review* (October 20, 2014). www.technologyreview.com/view/531911/isaac-asimov-asks-how-do-people-get-new-ideas.
Avis, William Robert. "Current Trends in Violent Conflict." K4D Helpdesk Report 565 (March 25, 2019). Brighton, UK: Institute of Development Studies. https://assets.publishing.service.gov.uk/media/5cf669ace5274a07692466 db/565_Trends_in_Violent_Conflict.pdf.
Bacevich, Andrew J. *The Limits of Power: The End of American Exceptionalism*. New York: Metropolitan Books, 2008.
Barkawi, Tarak, and Shane Brighton, "Absent War Studies? War, Knowledge, and Critique." In *The Changing Character of War*. Edited by Hew Strachan and Sibylle Scheipers. Oxford: Oxford University Press, 2011.
Barno, David W. "Military Adaptation in Complex Operations." *PRISM* 1, no. 1 (2009): 30.
Baumann, Robert F. "Historical Perspectives on Future War." *Military Review* 77, no. 2 (March–April 1997): 40.
Beckett, Ian F. W., and John Pimlott, eds. *Armed Forces & Modern Counterinsurgency*. New York: St. Martin's Press, 1985.

Berdyaev, Nikolai. "The Psychology of War and The Meaning of War: Thoughts about the Nature of War." In *The Fate of Russia*, translated by Fr. S. Janos. Reprint, Moscow: Svarog, 1997. www.berdyaev.com/berdiaev/berd_lib/1915_197.html.

Bickel, Keith B. *Mars Learning: The Marine Corps' Development of Small Wars Doctrine, 1915–1940.* Boulder, CO: Westview Press, 2001.

Black, Jeremy. "What is War?" *Defence-in-Depth Research from the Defence Studies Department, King's College London.* https://defenceindepth.co/2018/06/11/what-is-war.

Bloom, Harold, ed. *The Best Poems of the English Language: From Chaucer through Frost.* New York: HarperCollins, 2004.

Bramson, Leon, and George W. Goethals, eds. *War: Studies from Psychology, Sociology, Anthropology.* New York: Basic Books, 1968.

Brecke, Peter. "Violent Conflicts 1400 A.D. to the Present in Different Regions of the World." Paper presented at the 1999 Meeting of the Peace Science Society (International) on October 8–10, 1999, in Ann Arbor, Michigan. https://arxiv.org/ftp/arxiv/papers/1812/1812.08071.pdf.

Brodie, Bernard, "The Anatomy of Deterrence." Santa Monica, CA: RAND Corporation, 1958. www.rand.org/pubs/research_memoranda/RM2218.html. Also available in print form.

    "The Continuing Relevance of *On War*." In Carl von Clausewitz, *On War*. Translated and edited by Michael Howard and Peter Paret. Princeton, NJ: Princeton University Press, 1976.

Callwell, C. E. *Small Wars: Their Principles and Practice.* 3rd ed. [1906]. Seattle, WA: Book Jungle, 2009. 1st ed. published in 1899.

Carafano, James Jay. "It's Time to Return to the Principles of War." *The National Interest*, May 4, 2016. https://nationalinterest.org/feature/its-time-return-the-principles-war-16054.

Carter, Ashton B. "Running the Pentagon Right." *Foreign Affairs*, February 2014. www.foreignaffairs.com/articles/140346/ashton-b-carter/running-the-pentagon-right.

Cashman, Greg. *What Causes War? An Introduction to Theories of International Conflict.* Oxford: Lexington Books, 1993.

Chanakya. *The Arthashastra: Classics on War and Politics*, Vol. 3. Translated by Jeff Mcneill. Sheridan, WY: Classics Press, 2019.

Chandler, David G., ed. *The Military Maxims of Napoleon.* Translated by George C. D'Aguilar. London: Lionel Leventhal, 2002.

Charlton, James ed., *The Military Quotation Book.* New York: St. Martin's Press, 2013.

Churchill, Winston. "Their Finest Hour." Speech, London, June 18, 1940. The International Churchill Society. https://winstonchurchill.org/resources/speeches/1940-the-finest-hour/their-finest-hour.

    The World Crisis, 1911–1918. New York: Free Press, 2005.

Cicero. "Cicero: Letters to and from Cassius." [44 B.C.] *Attalus.* www.attalus.org/translate/cassius.html.

    Pro Milone. In *The Ethics of War: Classic and Contemporary Reading.* Edited by Gregory M. Reichberg, Henrik Syse, and Endre Begby. Malden, MA: Blackwell Publishing, 2006.

Clarke, Arthur C. *Profiles of the Future: An Inquiry into the Limits of the Possible.* London: Gollancz, 1962.

Clodfelter, Mark A. "Molding Airpower Convictions: Development and Legacy of William Mitchell's Strategic Thought." In *The Paths of Heaven: The Evolution of Airpower Theory.* Edited by Phillip S. Meilinger. Maxwell Air Force Base, AL: Air University Press, 1997.

Codevilla, Angelo, and Paul Seabury. *War: Ends and Means.* 2nd ed. Washington, DC: Potomac Books, 2006.

Coker, Christopher. *Can War Be Eliminated?* Cambridge, UK: Polity Press, 2014.

Collier, Paul. *Wars, Guns, and Votes: Democracy in Dangerous Places.* New York: HarperCollins, 2009.

Condorcet, Jean-Antoine-Nicolas de Caritat. "The Future Progress of the Human Mind." Internet History Sourcebooks. New York: Fordham University, 1997. https://source books.fordham.edu/mod/condorcet-progress.asp.

Confucius. In "Human Conflicts Come from Structural Disequilibrium." *Battison's Blog*. Https://biggestquestions.com/2014/04/19/human-conflicts-come-from-structural-disequilibrium/.

Coram, Robert. *Boyd: The Fighter Pilot Who Changed the Art of War*. New York: Back Bay Books, 2004.

Corbett, Julian Stafford. *Some Principles of Maritime Strategy*. [1911]. Mineola, NY: Dover Publications, 2004.

Craig, Gordon A. "Delbrück: The Military Historian." In *Makers of Modern Strategy from Machiavelli to the Nuclear Age*. Edited by Peter Paret. Princeton, NJ: Princeton University Press, 1986.

"The Political Leader as Strategist." In *Makers of Modern Strategy from Machiavelli to the Nuclear Age*. Edited by Peter Paret. Princeton, NJ: Princeton University Press, 1986.

Craig, Gordon A., and Felix Gilbert. "Reflections on Strategy in the Present and Future." In *Makers of Modern Strategy from Machiavelli to the Nuclear Age*. Edited by Peter Paret. Princeton, NJ: Princeton University Press, 1986.

Crenshaw, Martha. "Why Violence Is Rejected or Renounced: A Case Study of Oppositional Terrorism." In *A Natural History of Peace*. Edited by Thomas Gregor. Nashville, TN: Vanderbilt University Press, 1996.

David, Saul. *Military Blunders: The How and Why of Military Failure*. New York: Skyhorse, 2012.

ed. *War: The Definitive Visual History from Bronze Age Battles to 21st Century Conflict*. New York: DK, 2009.

Davie, Maurice R. *The Evolution of War: A Study of Its Role in Early Societies*. [1929]. New York: Dover Publications, 2003.

De Jomini, Antoine-Henri. *The Art of War*. Translated by Captain G. H. Mendell and Lieutenant W. P. Craighill. Radford, VA: Wilder Publications, 2008.

De Vattel, Emer. "*The Law of Nations*." In *The Ethics of War: Classic and Contemporary Readings*. Edited by Gregory M. Reichberg, Henrik Syse, and Endre Begby. Malden, MA: Blackwell Publishing, 2006.

De Vitoria, Francisco. "*On the Law of War*." In *The Ethics of War: Classic and Contemporary Readings*. Edited by Gregory M. Reichberg, Henrik Syse, and Endre Begby. Malden MA: Blackwell Publishing, 2006.

De Waal, Frans B. M. "The Biological Basis of Peaceful Coexistence: A Review of Reconciliation Research on Monkeys and Apes." In *A Natural History of Peace*. Edited by Thomas Gregor. Nashville, TN: Vanderbilt University Press, 1996.

Dederer, John Morgan. "Making Bricks without Straw: Nathanael Greene's Southern Campaign and Mao Tse-Tung's Mobile War." *Military Affairs* 47, no. 3 (October 1983): 115–121.

Deibel, Terry. *Foreign Affairs Strategy: Logic for American Statecraft*. New York: Cambridge University Press, 2007.

Demarest, Arthur A. "War, Peace, and the Collapse of a Native American Civilization: Lessons for Contemporary Systems of Conflict." In *A Natural History of Peace*. Edited by Thomas Gregor. Nashville, TN: Vanderbilt University Press, 1996.

Department of the Air Force. *Air Force Basic Doctrine, Organization, and Command*. AFDD 1. Washington, DC: Department of the Air Force, 2011. www.au.af.mil/au/cadre/aspc/lC04/pubs/afdd1.pdf.

Desilver, Drew. "Despite global concerns about democracy, more than half of countries are democratic." Pew Research Center, May 14, 2019. www.pewresearch.org/fact-tank/2019/05/14/more-than-half-of-countries-are-democratic/.

Diamond, Jared. *Guns, Germs, and Steel: The Fates of Human Societies*. New York: W. W. Norton & Company, 2017.

Douhet, Giulio. *The Command of the Air*. [1921]. Translated by Dino Ferrari. Washington, DC: Office of Air Force History, 1983.

Durbin, E. F. M., and John Bowlby. "Personal Aggressiveness and War." In *War: Studies from Psychology, Sociology, Anthropology*. Edited by Leon Bramson and George W. Goethals. New York: Basic Books, 1968.

Dyer, Gwynne. *War*. New York: Crown Publishers, 1985.

Earle, Edward Meade. "Adam Smith, Alexander Hamilton, Friedrich List: The Economic Foundations of Military Power." In *Makers of Modern Strategy from Machiavelli to the Nuclear Age*. Edited by Peter Paret. Princeton, NJ: Princeton University Press, 1986.

Einstein, Albert, and Sigmund Freud. *Why War?* International Institute for Intellectual Co-operation [League of Nations], 1933. www.drmalikcikk.atw.hu/wp_readings/einstein_freud.PDF.

Engels, Friedrich. "*Anti-Dühring*." In *The Ethics of War: Classic and Contemporary Readings*. Edited by Gregory M. Reichberg, Henrik Syse, and Endre Begby. Malden, MA: Blackwell Publishing, 2006.

Erasmus, Desiderius. "*The Education of a Christian Prince*." In *The Ethics of War: Classic and Contemporary Readings*. Edited by Gregory M. Reichberg, Henrik Syse, and Endre Begby. Malden. MA: Blackwell Publishing, 2006.

Erikson, Erik H. "Wholeness and Totality." In *War: Studies from Psychology, Sociology, Anthropology*. Edited by Leon Bramson and George W. Goethals. New York: Basic Books, 1968.

Esko, Edward. *Yin Yang Primer*. Becket, MA: Amber Waves, 2012.

Etzioni, Amitai. "Toward a Sociological Theory of Peace." In *War: Studies from Psychology, Sociology, Anthropology*. Edited by Leon Bramson and George W. Goethals. New York: Basic Books, 1968.

Ewers, Justin. "Why Don't Colleges Teach Military History?" *US News and World* (*Report* April 3, 2008). www.usnews.com/news/articles/2008/04/03/why-dont-colleges-teach-military-history.

Falcon, Andrea. "Aristotle on Causality." In *The Stanford Encyclopedia of Philosophy*. Edited by Edward N. Zalta, Spring 2015. Metaphysics Research Lab, Stanford University, 2019. https://plato.stanford.edu/archives/spr2015/entries/aristotle-causality/.

Faust, Drew Gilpin. "2011 Jefferson Lecture in the Humanities." Lecture, Harvard University, Cambridge, MA. May 2, 2011. www.harvard.edu/president/speech/2011/2011-jefferson-lecture-humanities.

Fedyk, Nicholas. "Russian 'New Generation' Warfare: Theory, Practice, and Lessons for US Strategists." *Small Wars Journal*, August 25, 2016. https://smallwarsjournal.com/jrnl/art/russian-%e2%80%9cnew-generation%e2%80%9d-warfare-theory-practice-and-lessons-for-us-strategists-0.

Fehrenbach, T. R. *This Kind of War: The Classic Korean War History*. Washington, DC: Brassey's, 1994.

Fox, Robert W., and Alan T. McDonald. *Introduction to Fluid Mechanics*. 3rd ed. New York: John Wiley & Sons, 1985.

Frederick II (the Great). *Frederick the Great on the Art of War*. Edited and translated by Jay Luvaas. New York: Da Capo Press, 1999.

Freedman, Lawrence. *Strategy: A History*. New York: Oxford University Press, 2013.
  The Future of War: A History. New York: PublicAffairs, 2017.

Friedman, George, and Meredith Friedman. *The Future of War: Power, Technology, & American World Dominance in the 21st Century*. New York: Crown Publishers, Inc., 1996.

Galula, David. *Counterinsurgency Warfare: Theory and Practice*. [1964]. Westport, CT: Praeger Security International, 2006.

Gandhi, Mahatma. *Gandhi on Non-Violence*. Edited by Thomas Merton. New York: New Directions, 1965.

Gat, Azar. *A History of Military Thought: From the Enlightenment to the Cold War*. Oxford: Oxford University Press, 2001.

War in Human Civilization. Oxford: Oxford University Press, 2006.

Gates, Ernie. "In the Name of God." *University of Virginia Magazine*, Winter 2015. https://uvam agazine.org/articles/in_the_name_of_god.

Gilbert, Daniel. *Stumbling on Happiness*. New York: Vintage, 2007.

Gleick, James. *Chaos: Making a New Science*. New York, NY: Penguin Books, 1987.

Glenn, Russell W. "Thoughts on 'Hybrid' Conflict." *Small Wars Journal*, March 2, 2009. https:// smallwarsjournal.com/jrnl/art/thoughts-on-hybrid-conflict.

Gray, Colin S. *Another Bloody Century: Future Warfare*. London: Weidenfeld & Nicolson, 2005.
    Fighting Talk: Forty Maxims on War, Peace, and Strategy. Westport, CT: Greenwood Publishing Group, 2007.
    The Future of Strategy. Cambridge, UK: Polity Press, 2015.
    War, Peace and International Relations: An Introduction to Strategic History. London and New York: Routledge, 2007.

Gregor, Thomas A., ed. *A Natural History of Peace*. Nashville, TN: Vanderbilt University Press, 1996.

Gregor, Thomas, and Clayton A. Robarchek. "Two Paths to Peace: Semai and Mehinaku Nonviolence." In *A Natural History of Peace*. Edited by Thomas Gregor. Nashville, TN: Vanderbilt University Press, 1996.

Gross, Neil. *Why Are Professors Liberal and Why Do Conservatives Care?* Cambridge, MA: Harvard University Press, 2013.

Grossman, Dave, and Loren W. Christensen. *On Combat: The Psychology and Physiology of Deadly Conflict in War and Peace*. 3rd ed. Millstadt, IL: Warrior Science Publications, 2008.

Grotius, Hugo. *The Law of War and Peace*. Translated by Louise R. Loomis. Roslyn, NY: Walter J. Black, 1949.

Hall, H. Fielding. *The Nature of War – And Its Causes*. London: Hurst & Blackett, 1917.

Hambling, David. "China Looks to Undermine US Power, With 'Assassin's Mace'." *Wired*. July 2, 2009. www.wired.com/2009/07/china-looks-to-undermine-us-power-with-assassins-mace.

Hamilton, Alexander. *Federalist No. 34*. January 5, 1788. Teaching American History. https://tea chingamericanhistory.org/library/document/federalist-no-34/.

Handel, Michael I. *Masters of War: Classical Strategic Thought*. New York: Routledge, 2001.

Hanson, Victor Davis. *The Father of Us All: War and History Ancient and Modern*. New York: Bloomsbury Press, 2010.

Harari, Noah Yuval. *Homo Deus: A Brief History of Tomorrow*. New York: HarperCollins Publishers, 2017.

Hedges, Chris. "What Every Person Should Know About War." *The New York Times*, July 6, 2003. www.nytimes.com/2003/07/06/books/chapters/what-every-person-should-know-about-war.html.

Hegel, Georg Wilhelm Friedrich. *Lectures on the Philosophy of History*. Kitchener, ON: Batoche Books, 2001. https://socialsciences.mcmaster.ca/econ/ugcm/3ll3/hegel/history.pdf.
    "*Philosophy of Right*." In *The Ethics of War: Classic and Contemporary Readings*. Edited by Gregory M. Reichberg, Henrik Syse, and Endre Begby. Malden, MA: Blackwell Publishing, 2006.

Henry, Patrick. "Give Me Liberty or Give Me Death." Speech, Richmond, VA. March 23, 1775. The Colonial Williamsburg Foundation. www.history.org/almanack/life/politics/giveme.cfm.

Heraclitus. *Fragments: The Collected Wisdom of Heraclitus*. Translated by Brooks Haxton. New York: Viking Penguin, 2001.

Heuser, Beatrice. *The Evolution of Strategy: Thinking War from Antiquity to the Present*. Cambridge: Cambridge University Press, 2010.

Hobbes, Thomas. "*De Cive*." In *The Ethics of War: Classic and Contemporary Readings*. Edited by Gregory M. Reichberg, Henrik Syse, and Endre Begby. Malden, MA: Blackwell Publishing, 2006.
    Leviathan, 1651. Oxford: Oxford University Press, 1965. http://files.libertyfund.org/files/869/ 0161_Bk.pdf.

Holley, Jr., I. B. "Reflections on the Search for Airpower Theory." In *The Paths of Heaven: The Evolution of Airpower Theory*. Edited by Phillip S. Meilinger. Maxwell Air Force Base, AL: Air University Press, 1997.

Homer. *The Iliad*. Translated by Robert Fagles. New York: Penguin Books, 1990.

Horgan, John. *The End of War*. San Francisco: McSweeney's Publishing, 2014.

Howard, Michael. "Military Science in the Age of Peace." *RUSI Journal* 199 (March 1974): 3–9.

"The Influence of Clausewitz." In Carl von Clausewitz, *On War*. Edited and translated by Michael Howard and Peter Paret. Princeton, NJ: Princeton University Press, 1976.

The Invention of Peace and the Reinvention of War: Reflections on War and International Order. London: Profile Books, 2001.

Howe, Julia Ward. "Battle Hymn of the Republic." In *The Best Poems of the English Language: From Chaucer through Frost*. Edited by Harold Bloom. New York: HarperCollins, 2004.

Hume, David. "Of the Influencing Motives of the Will." In *A Treatise of Human Nature*. Project Gutenberg, 2012. www.gutenberg.org/files/4705/4705-h/4705-h.htm#link2H_.

Iklé, Fred Charles. *Every War Must End*. New York: Columbia University Press, 1991.

James, William. "The Moral Equivalent of War." In *War: Studies from Psychology, Sociology, Anthropology*, edited by Leon Bramson and George W. Goethals. New York: Basic Books, 1968.

Janowitz, Morris. "Military Elites and the Study of War." In *War: Studies from Psychology, Sociology, Anthropology*. Edited by Leon Bramson and George W. Goethals. New York: Basic Books, 1968.

Jaschik, Scott. "Moving Further to the Left." *Insider Higher Ed* (October 24, 2012). www.insidehighered.com/news/2012/10/24/survey-finds-professors-already-liberal-have-moved-further-left.

Jefferson, Thomas. "A Bill for the More General Diffusion of Knowledge." Thomas Jefferson Foundation. www.monticello.org/site/research-and-collections/bill-more-general-diffusion-knowledge.

John of Salisbury. *Policraticus*. In *The Ethics of War: Classic and Contemporary Readings*. Edited by Gregory M. Reichberg, Henrik Syse, and Endre Begby. Malden, MA: Blackwell Publishing, 2006.

Johnson, James Turner. "Does Defense of Values by Force Remain a Moral Possibility?" In *The Ethics of War: Classic and Contemporary Readings*. Edited by Gregory M. Reichberg, Henrik Syse, and Endre Begby. Malden, MA: Blackwell Publishing, 2006.

Joint Chiefs of Staff. *Department of Defense Dictionary of Military and Associated Terms*. JP 1–02. Washington, DC: Joint Chiefs of Staff, 2016. www.dtic.mil/doctrine/new_pubs/jp1_02.pdf.

Doctrine for the Armed Forces of the United States. JP 1. Washington, DC: Joint Chiefs of Staff, 2017. www.jcs.mil/Portals/36/Documents/Doctrine/pubs/jp1_ch1.pdf?ver=2019-02-11-174350-967.

Strategy. JDN 1–18. Washington, DC: Joint Chiefs of Staff, 2018.

Kaku, Michio. *Physics of the Impossible: A Scientific Exploration into the World of Phasers, Force Fields, Teleportation, and Time Travel*. New York: Anchor Books, 2009.

Kalyvas, Stathis N. "The Changing Character of Civil Wars, 1800–2009." In *The Changing Character of War*. Edited by Hew Strachan and Sibylle Scheipers. Oxford: Oxford University Press, 2011.

Kane, Thomas M. *Military Logistics and Strategic Performance*. London: Frank Cass, 2001.

Kant, Immanuel. *Perpetual Peace*. In *The Ethics of War: Classic and Contemporary Readings*. Edited by Gregory M. Reichberg, Henrik Syse, and Endre Begby. Malden, MA: Blackwell Publishing, 2006.

Toward Perpetual Peace: A Philosophical Sketch. Translated by Jonathan Bennett. Early Modern Texts, 2015. www.earlymoderntexts.com/assets/pdfs/kant1795.pdf.

Keegan, John. *A History of Warfare*. New York: Vintage, 1994.

The Face of Battle. New York: Penguin Books, 1976.

Keegan, John, and Joseph Darracott. *The Nature of War.* New York: Holt, Rinehart and Winston, 1981.

Keeley, Lawrence H. *War Before Civilization: The Myth of the Peaceful Savage.* Oxford: Oxford University Press, 1996.

Kelly, Raymond C. *Warless Societies and the Origin of War.* Ann Arbor: University of Michigan Press, 2000.

Kelsen, Hans. "*General Theory of Law and State.*" In *The Ethics of War: Classic and Contemporary Readings.* Edited by Gregory M. Reichberg, Henrik Syse, and Endre Begby. Malden, MA: Blackwell Publishing, 2006.

Kilcullen, David. *Counterinsurgency.* Oxford: Oxford University Press, 2010.

Knauft, Bruce M. "The Human Evolution of Cooperative Interest." In *A Natural History of Peace.* Edited by Thomas Gregor. Nashville, TN: Vanderbilt University Press, 1996.

Kouzes, James M., and Barry Z. Posner. *The Leadership Challenge: How to Make Extraordinary Things Happen in Organizations.* Hoboken, NJ: John Wiley & Sons, 2017.

Lacina, Bethany, Nils Petter Gleditsch, Bruce Russett. "The Declining Risk of Death in Battle." *International Studies Quarterly* 50, no. 3: 673–680, 2006.

Lasswell, Harold D. "The Garrison State." In *War: Studies from Psychology, Sociology, Anthropology.* Edited by Leon Bramson and George W. Goethals. New York: Basic Books, 1968.

Laughlin, Robert B. *A Different Universe: Reinventing Physics from the Bottom Down.* New York: Basic Books, 2005.

Lawrence, T. E. "On Guerrilla Warfare." In *Encyclopedia Britannica,* 14th ed. London: Encyclopedia Britannica, 1929. www.britannica.com/topic/T-E-Lawrence-on-guerrilla-warfare-1984900.

   Seven Pillars of Wisdom. [1926]. London: Penguin Books, 2000.

   "The 27 Articles of T. E. Lawrence." *The Arab Bulletin* (20 August 1917). World War I Document Archive. https://wwi.lib.byu.edu/index.php/The_27_Articles_of_T.E._Lawrence.

   "The Evolution of a Revolt." *Army Quarterly and Defence Journal,* 1 (October 1920): 55–69.

Leddy, Tom. "Kant on How to Be a Genius." *Aesthetics Today,* October 2, 2015. http://aestheticstoday.blogspot.com/2015/10/kant-on-how-to-be-genius.html.

LeShan, Lawrence. *The Psychology of War: Comprehending Its Mystique and Its Madness.* New York: Helios Press, 2002.

Liang, Qiao, and Wang Xiangsui. *Unrestricted Warfare.* Translated by Foreign Broadcast Information Service (FBIS). Beijing: PLA Literature and Arts Publishing House, 1999.

Liddell Hart, Basil Henry. *Strategy.* 2nd ed. New York: Meridian, 1991.

Lincoln, Abraham. First Inaugural Address of Abraham Lincoln. Speech, Washington, DC, 1861. The Avalon Project. https://avalon.law.yale.edu/19th_century/lincoln1.asp.

Linn, Brian McAllister. *The Philippine War, 1899–1902.* Lawrence: University Press of Kansas, 2000.

List, Friedrich. *National System of Political Economy: The Theory.* Translated by Sampson S. Lloyd. London: Longmans, Green and Co., 1909. https://oll.libertyfund.org/titles/315.

Lopez, C. Todd. "Shanahan: Next Big War May Be Won or Lost in Space." US Department of Defense. April 9, 2019. www.defense.gov/Newsroom/News/Article/Article/1810100/shanahan-next-big-war-may-be-won-or-lost-in-space/.

Lovelace, Richard. "Song to Lucasta, Going to the Wars." In *The Best Poems of the English Language: From Chaucer through Frost.* Edited by Harold Bloom. New York: HarperCollins, 2004.

Luther, Martin. "*Whether Soldiers, Too, Can be Saved.*" In *The Ethics of War: Classic and Contemporary Readings.* Edited by Gregory M. Reichberg, Henrik Syse, and Endre Begby. Malden, MA: Blackwell Publishing, 2006.

Lynn, John A. "A Quest for Glory: The Formation of Strategy Under Louis XIV, 1661–1715." In *The Making of Strategy: Rulers, States, And War*. Edited by Williamson Murray, MacGregor Knox, and Alvin Bernstein. Cambridge: Cambridge University Press, 2009.

"The Embattled Future of Academic Military History." *The Journal of Military History*, 61, no. 4 (October 1997): 777–789. https://doi.org/10.2307/2954086.

Macaulay, Thomas Babington. "Horatius at the Bridge." In *Lays of Ancient Rome*. London: Longman, Brown, Green, and Longman, 1842. University of Toronto Libraries. https://rpo.library.utoronto.ca/poems/horatius.

Machiavelli, Niccolò. *The Art of War*. [1521]. Translated by Neal Wood. Cambridge, MA: Da Capo Press, 2001.

The Discourses. [1531]. Translated by Leslie J. Walker. New York: Penguin Books, 1986.

The Prince. [1532]. Translated by N. H. Thompson. New York: Dover Publications, 1992.

Mahan, Alfred Thayer. *Influence of Sea Power Upon History, 1660–1783*. [1890]. Ill. ed. Scotts Valley: Pantianos Classics, 2016.

The Influence of Sea Power Upon the French Revolution and Empire, 1793–1812. [1892]. Ill. ed. Scotts Valley: Pantianos Classics, 2016.

Malinowski, Robert E. "An Anthropological Analysis of War." In *War: Studies from Psychology, Sociology, Anthropology*. Edited by Leon Bramson and George W. Goethals. New York: Basic Books, 1968.

Mao, Tse-tung. *On Guerrilla Warfare*. Translated by Samuel B. Griffith. New York: BN Publishing, 2007.

Marshall, George C. *The Papers of George Catlett Marshall*. Edited by Larry I. Bland, Sharon Ritenour Stevens, and Clarence E. Wunderlin, Jr. Lexington, VA: The George C. Marshall Foundation, 1981. www.marshallfoundation.org/library/digital-archive/speech-at-trinity-college/.

Martelloni, Gianluca et al. "Pattern Analysis of World Conflicts over the Past 600 Years," (December 19, 2018). https://arxiv.org/ftp/arxiv/papers/1812/1812.08071.pdf.

May, Mark A. "War, Peace, and Social Learning." In *War: Studies from Psychology, Sociology, Anthropology*. Edited by Leon Bramson and George W. Goethals. New York: Basic Books, 1968.

McDougall, William. "The Instinct of Pugnacity." In *War: Studies from Psychology, Sociology, Anthropology*. Edited by Leon Bramson and George W. Goethals. New York: Basic Books, 1968.

Mead, Margaret. "Warfare Is Only an Invention." In *War: Studies from Psychology, Sociology, Anthropology*. Edited by Leon Bramson and George W. Goethals. New York: Basic Books, 1968.

Mearsheimer, John J. "Structural Realism." In *International Relations Theories*. Edited by Tim Dunne, Milja Kurki, and Steve Smith, 3rd ed. Oxford: Oxford University Press, 2013.

Meilinger, Phillip S. ed. *The Paths of Heaven: The Evolution of Airpower Theory*. Maxwell Air Force Base, AL: Air University Press, 1997.

"Trenchard, Slessor, and Royal Air Force Doctrine before World War II." In *The Paths of Heaven: The Evolution of Airpower Theory*. Edited by Phillip S. Meilinger. Maxwell Air Force Base, AL: Air University Press, 1997.

Mill, John Stuart. "The Contest in America." *Fraser's Magazine* (1862). Project Gutenberg. www.gutenberg.org/files/5123/5123-h/5123-h.htm.

Miller, Franklin C. "The Need for a Strong US Nuclear Deterrent in the Twenty-First Century." *Strategic Studies Quarterly* 10, no. 5 (2016): 23–30.

Milley, Mark A. "AUSA 2016 – Dwight David Eisenhower Luncheon." Briefing, Association of the U.S. Army 2016 Annual Meeting and Exposition, Washington, DC, October 4, 2016. http://wpswps.org/wp-content/uploads/2016/11/20161004_CSA_AUSA_Eisenhower_Transcripts.pdf.

Milton, John. "Areopagitica: a speech by Mr. John Milton for the liberty of unlicenc'd printing, to the Parlament of England." Speech, London, 1664. The John Milton Reading Room. www.dartmouth.edu/~milton/reading_room/areopagitica/text.html.

Mitchell, William. *Winged Defense: The Development and Possibilities of Modern Air Power, Economic and Military.* [1925]. Mineola, NY: Dover Publications, Inc., 2006.

Montesquieu, Charles-Louis de Secondat, Baron de La Brède et de. "*The Spirit of the Laws.*" In *The Ethics of War: Classic and Contemporary Readings.* Edited by Gregory M. Reichberg, Henrik Syse, and Endre Begby. Malden, MA: Blackwell Publishing, 2006.

Morley, Neville. "The Thucydides Trap." *Sphinx: Exploring Antiquity and Modernity with Neville Morley* (blog), October 30, 2012. https://thesphinxblog.com/2012/10/30/the-thucydides-trap.

Morris, Ian. "The Slaughter Bench of History." *The Atlantic*, April 11, 2014. www.theatlantic.com/international/archive/2014/04/the-slaughter-bench-of-history/360534/.

Moseley, T. Michael. "Airmen and the Art of Strategy." *Strategic Studies Quarterly* 1, no. 1 (2007): 7–19.

   The Nation's Guardians: America's 21st Century Air Force. Washington, DC: Chief of Staff of the United States Air Force, 2007. https://apps.dtic.mil/dtic/tr/fulltext/u2/a477488.pdf.

Murphy, Mike. "Vox Ignoramus." *Washington Examiner.* March 24, 2003. www.washingtonexaminer.com/weekly-standard/vox-ignoramus.

Murray, Williamson. *War, Strategy, and Military Effectiveness.* New York: Cambridge University Press, 2011.

Nagel, Thomas. "War and Massacre." In *The Ethics of War: Classic and Contemporary Readings.* Edited by Gregory M. Reichberg, Henrik Syse, and Endre Begby. Malden, MA: Blackwell Publishing, 2006.

Naylor, Sean D. "Support Grows for Standing up an Unconventional Warfare Command." *Armed Forces Journal*, September 1, 2007. http://armedforcesjournal.com/support-grows-for-standing-up-an-unconventional-warfare-command.

Nichols, Shaun. "Is Free Will an Illusion?" *Scientific American*, November 1, 2011. www.scientificamerican.com/article/is-free-will-an-illusion.

Nye, Joseph S. *Bound to Lead: The Changing Nature of American Power.* New York: Basic Books, 1991.

Office of General Counsel. *Department of Defense Law of War Manual.* Washington: General Counsel of the Department of Defense, 2016.

O'Neill, Bard E. *Insurgency & Terrorism: From Revolution to Apocalypse.* 2nd ed. Lincoln, NE: Potomac Books Inc., 2005.

Paret, Peter, ed. *Makers of Modern Strategy from Machiavelli to the Nuclear Age.* Princeton, NJ: Princeton University Press, 1986.

   "Clausewitz." In *Makers of Modern Strategy from Machiavelli to the Nuclear Age.* Edited by Peter Paret. Princeton, NJ: Princeton University Press, 1986.

   Clausewitz and the State: The Man, His Theories, and His Times. Princeton, NJ: Princeton University Press, 1985.

   Understanding War: Essays on Clausewitz and the History of Military Power. Princeton, NJ: Princeton University Press, 1992.

Park, Robert E. "The Social Function of War." In *War: Studies from Psychology, Sociology, Anthropology.* Edited by Leon Bramson and George W. Goethals. New York: Basic Books, 1968.

Paul, Christopher, Colin P. Clarke, Beth Grill, and Molly Dunigan, "Paths to Victory: Lessons from Modern Insurgencies." Santa Monica, CA: RAND Corporation, 2013. www.rand.org/pubs/research_reports/RR291z1.html.

Pettersson, Therese, Stina Högbladh, and Magnus Öberg. "Organized Violence, 1989–2018 and Peace Agreements." *Journal of Peace Research* 56, no. 4 (2019). https://ucdp.uu.se/downloads/index.html#armedconflict.

Plato. *The Republic*. Translated by Benjamin Jowett. Project Gutenberg, 2017. www.gutenberg.org/fi
les/55201/55201-h/55201-h.htm29.

Porch, Douglas. "Bugeaud, Galliéni, Lyautey: The Development of French Colonial Warfare." In
Makers of Modern Strategy from Machiavelli to the Nuclear Age. Edited by Peter Paret.
Princeton, NJ: Princeton University Press, 1986.

Rawls, John. "Fifty Years after Hiroshima." In *The Ethics of War: Classic and Contemporary
Readings*. Edited by Gregory M. Reichberg, Henrik Syse, and Endre Begby. Malden, MA:
Blackwell Publishing, 2006.

Reichberg, Gregory M. "Supranational Government and Peace." In *The Ethics of War: Classic and
Contemporary Readings*. Edited by Gregory M. Reichberg, Henrik Syse, and Endre Begby.
Malden, MA: Blackwell Publishing, 2006.

Reichberg, Gregory M., Henrik Syse, and Endre Begby, eds. *The Ethics of War: Classic and
Contemporary Readings*. Malden, MA: Blackwell Publishing, 2006.

Reynolds, Paul Davidson. *A Primer in Theory Construction*. New York: Macmillan Publishing, 1986.

Robarchek, Clayton A., and Carole J. Robarchek. "The Aucas, the Cannibals, and the Missionaries:
From Warfare to Peacefulness among the Waorani." In *A Natural History of Peace*. Edited by
Thomas Gregor. Nashville, TN: Vanderbilt University Press, 1996.

Roosevelt, Theodore. *The Works of Theodore Roosevelt*. New York: Scribner, 1903. https://play
.google.com/books/reader?id=3z8OAAAAIAAJ&hl=en&pg=GBS.PA1.

Rousseau, Jean-Jacques. "*Critique of the Abbé de Saint-Pierre's Project for Perpetual Peace*." In *The
Ethics of War: Classic and Contemporary Readings*. Edited by Gregory M. Reichberg,
Henrik Syse, and Endre Begby. Malden, MA: Blackwell Publishing, 2006.

"*The State of War*." In *The Ethics of War: Classic and Contemporary Readings*. Edited by Gregory
M. Reichberg, Henrik Syse, and Endre Begby. Malden, MA: Blackwell Publishing, 2006.

Russell, Bertrand. "*Common Sense and Nuclear Warfare*." In *The Ethics of War: Classic and
Contemporary Readings*. Edited by Gregory M. Reichberg, Henrik Syse, and Endre Begby.
Malden, MA: Blackwell Publishing, 2006.

Sanderson, Brandon. *Oathbringer*. New York: Tor, 2017.

Sawyer, Ralph D., trans. *The Seven Military Classics Of Ancient China*. New York: Basic Books,
2007.

Scales, Robert H. "Forecasting the Future of Warfare." *War on the Rocks*, April 9, 2018.
https://warontherocks.com/2018/04/forecasting-the-future-of-warfare/.

Schopenhauer, Arthur. "Further Psychological Observations." In *Studies in Pessimism*. Project
Gutenberg, 2004. www.gutenberg.org/files/10732/10732-h/10732-h.htm.

Shelley, Percy Bysshe. "Hellas." In *The Best Poems of the English Language: From Chaucer
through Frost*. Edited by Harold Bloom. New York: HarperCollins, 2004.

Shy, John. "Jomini." In *Makers of Modern Strategy from Machiavelli to the Nuclear Age*. Edited by
Peter Paret. Princeton, NJ: Princeton University Press, 1986.

Sills, Kenneth C. M. "The Idea of Universal Peace in the Works of Virgil and Dante." *The Classical
Journal* 9, no. 4 (1914): 139–53. www.jstor.org/stable/3287070.

Simon, Herbert A. *The Sciences of the Artificial*. 3rd ed. Cambridge, MA: MIT Press, 1996.

Sipes, Richard G. "War, Sports and Aggression: An Empirical Test of Two Rival Theories."
American Anthropologist 75, no. 1 (1973): 64–86.

Slessor, J. C. *Air Power and Armies*. 1936. Tuscaloosa: University of Alabama Press, 2009.

Smith, David Livingstone. *The Most Dangerous Animal: Human Nature and the Origins of War*.
New York: St. Martin's Press, 2007.

Smolin, Lee. *The Life of the Cosmos*. London: Weidenfeld & Nicolson, 1997.

Sowell, Thomas. *A Conflict of Visions: Ideological Origins of Political Struggles*. New York: Basic
Books, 2007.

Spinoza, Baruch. *Political Treatise*. In *The Ethics of War: Classic and Contemporary Readings*.
Edited by Gregory M. Reichberg, Henrik Syse, and Endre Begby. Malden, MA: Blackwell
Publishing, 2006.

Sponsel, Leslie E. "The Natural History of Peace: A Positive View of Human Nature and Its Potential." In *A Natural History of Peace*. Edited by Thomas Gregor. Nashville TN: Vanderbilt University Press, 1996.

Staub, Ervin. "The Psychological and Cultural Roots of Group Violence and the Creation of Caring Societies and Peaceful Group Relations." In *A Natural History of Peace*. Edited by Thomas Gregor. Nashville, TN: Vanderbilt University Press, 1996.

Stoker, Donald. *Clausewitz: His Life and Work*. Oxford: Oxford University Press, 2014.

Strachan, Hew. *The Direction of War: Contemporary Strategy in Historical Perspective*. Cambridge: Cambridge University Press, 2013.

Strachan, Hew and Sibylle Scheipers, eds. "Introduction: The Changing Character of War." In *The Changing Character of War*. Edited by Hew Strachan and Sibylle Scheipers. Oxford: Oxford University Press, 2011.

Suárez, Francisco. *Metaphysical Disputations*. In *The Ethics of War: Classic and Contemporary Readings*. Edited by Gregory M. Reichberg, Henrik Syse, and Endre Begby. Malden, MA: Blackwell Publishing, 2006.

Sumner, William Graham. "War." In *War: Studies from Psychology, Sociology, Anthropology*. Edited by Leon Bramson and George W. Goethals. New York: Basic Books, 1968.

Sun Tzu. *The Art of War*. Translated by Samuel B. Griffith. New York: Oxford University Press, 1963.
   Sun Tzu: The New Translation. Translated by J. H. Huang. New York: Quill William Morrow, 1993.

Taylor, F. L. *The Art of War in Italy, 1494–1529*. Cambridge: Cambridge University Press, 1921.

Tennyson, Alfred Lord. "In Memoriam A. H. H." In *The Best Poems of the English Language: From Chaucer through Frost*. Edited by Harold Bloom. New York: HarperCollins, 2004.

Thucydides. *The Landmark Thucydides: A Comprehensive Guide to the Peloponnesian War*. Edited by Robert B. Strassler and Richard Crawley. New York: Free Press, 1996.
   The Peloponnesian War. Translated by Steven Lattimore. Indianapolis, IN: Hackett Publishing Company, 1998.

Tilghman, Andrew. "Cut Spending, but Protect Capabilities, Gates Says." *Air Force Times* 71, no. 47 (June 6, 2011): 8–9.

Trenchard, Hugh Montague. "The Effect of the Rise of Air Power on War." In *Air Power: Three Papers by Marshal of the Royal Air Force The Viscount Trenchard, G.C.B., G.C.V.O., D S.O., D.C.L., LL.D.* London: Directorate of Staff Duties, Air Ministry, 1946.

Trinquier, Roger. *Modern Warfare: A French View of Counterinsurgency*. [1964]. Reprint, Westport, CT: Praeger Security International, 2006.

Tuzin, Donald. "The Spectre of Peace in Unlikely Places: Concept and Paradox in the Anthropology of Peace." In *A Natural History of Peace*. Edited by Thomas A. Gregor. Nashville, TN: Vanderbilt University Press, 1996.

United Nations and World Bank. "Pathways for Peace: Inclusive Approaches to Preventing Violent Conflict – Main Messages and Emerging Policy Directions." Washington, DC: World Bank (2017): 1–80. https://openknowledge.worldbank.org/bitstream/handle/10986/28337/211162 mm.pdf.

US Army and US Marine Corps. *Counterinsurgency Field Manual 3–24, 2006*. Edited by John McClure. Kissimmee, FL: Signalman Publishing, 2009.

US Marine Corps. *Small Wars Manual: Fleet Marine Force Reference Publication 12–15*. 1940. Washington, DC: US Government Printing Office, 1990.

US War Department. *Biennial Report of the Chief of Staff of the US Army, General George C. Marshall, July 1, 1943, to June 30, 1945, to the Secretary of War*. Washington, DC: US News Publishing Corp., 1945.

Valiunas, Algis. *Churchill's Military Histories: A Rhetorical Study*. Lanham, MD: Rowman & Littlefield Publishers, 2002.

Van Creveld, Martin. *The Transformation of War: The Most Radical Reinterpretation of Armed Conflict Since Clausewitz*. New York: The Free Press, 1991.

"Through a Glass, Darkly: Some Reflections on the Future of War." *Naval War College Review* 53, no. 4 (2000): 25.

"War and Technology." *Foreign Policy Research Institute*, October 24, 2007. www.fpri.org/art icle/2007/10/war-technology-2/.

Vasquez, John A. "Understanding Peace: Insights from International Relations Theory and Research." In *A Natural History of Peace*. Edited by Thomas Gregor. Nashville, TN: Vanderbilt University Press, 1996.

Vego, Milan. "On Military Theory." *JFQ: Joint Force Quarterly*, no. 62 (3rd Quarter 2011): 60–67. https://ndupress.ndu.edu/portals/68/Documents/jfq/jfq-62.pdf.

Von Clausewitz, Carl. *On War*. Translated and edited by Michael Howard and Peter Paret. Princeton, NJ: Princeton University Press, 1976.

Von Einsiedel, Sebastian et al. "Civil War Trends and the Changing Nature of Armed Conflict." Tokyo: United Nations University Centre for Policy Research (April 25, 2017). https://collec tions.unu.edu/eserv/UNU:6156/Civil_war_trends_UPDATED.pdf.

Von Moltke, Helmuth. "On the Nature of War." In *Die Zerstörung Der Deutschen Politik: Dokumente*. Edited by Harry Pross, translated by Richard S. Levy. Frankfurt, 1959. wwi .lib.byu.edu/index.php/On_the_Nature_of_War_by_Helmut_Moltke_(the_Elder).

Von Pufendorf, Samuel. "*On the Duty of Man and Citizen*." In *The Ethics of War: Classic and Contemporary Readings*. Edited by Gregory M. Reichberg, Henrik Syse, and Endre Begby. Malden, MA: Blackwell Publishing, 2006.

Waltz, Kenneth N. *Man, the State, and War: A Theoretical Analysis*. New York: Columbia University Press, 2001.

Wanser, Brooke. "14 Abraham Lincoln Quotes That Are Truly Modern Rules to Live By." *Reader's Digest*. https://www.rd.com/culture/abraham-lincoln-quotes.

Washington, George. "General Orders, 2 July 1776." *Founders Online*. National Archives. foun ders.archives.gov/documents/Washington/03-05-02-0117.

The Papers of George Washington Digital Edition. Charlottesville: University of Virginia Press, Rotunda, 2008. rotunda.upress.virginia.edu/founders/GEWN-05-04-02-0361.

Wells, Herbert George. *The Island of Doctor Moreau*. New York: Garden City Publishing, 1896. www.gutenberg.org/files/159/159-h/159-h.htm.

Wilson, Woodrow. "Peace without Victory." Speech, Washington, DC, November 19, 1863. *Digital History*. www.digitalhistory.uh.edu/disp_textbook.cfm?smtID=11&psid=3824.

Winton, Harold R. *To Change An Army: General Sir John Burnett-Stuart and British Armored Doctrine, 1927-1938*. Lawrence: University Press of Kansas, 1988.

Wiseman, Richard. "Be Lucky – It's an Easy Skill to Learn." *Telegraph*, January 9, 2003. www .telegraph.co.uk/technology/3304496/Be-lucky-its-an-easy-skill-to-learn.html.

Wolchover, Natalie. "The Nine Biggest Unsolved Mysteries in Physics." *Live*, July 3, 2012. www .livescience.com/34052-unsolved-mysteries-physics.html.

Wood, Dakota. *Rebuilding America's Military: Thinking About the Future*. Special Report No. 203. Washington, DC: The Heritage Foundation, 2018. https://www.heritage.org/defense/report/r ebuilding-americas-military-thinking-about-the-future.

Index

For EU product safety concerns, contact us at Calle de José Abascal, 56–1°, 28003 Madrid, Spain or eugpsr@cambridge.org.

www.ingramcontent.com/pod-product-compliance
Ingram Content Group UK Ltd.
Pitfield, Milton Keynes, MK11 3LW, UK
UKHW010855090126
466816UK00006B/27